Understanding Least Squares Estimation and Geomatics Data Analysis

Understanding Least Squares Estimation and Geomatics Data Analysis

John Olusegun Ogundare, PhD, PEng

Instructor of Geomatics Engineering
Department of Geomatics Engineering Technology
School of Construction and the Environment
British Columbia Institute of Technology (BCIT) – Burnaby, Canada

Registered Office(s)
John Wiley & Sons, Inc., 111 River Street, Hoboken, NJ 07030, USA

Editorial Office
111 River Street, Hoboken, NJ 07030, USA

For details of our global editorial offices, customer services, and more information about Wiley products visit us at www.wiley.com.

Wiley also publishes its books in a variety of electronic formats and by print-on-demand. Some content that appears in standard print versions of this book may not be available in other formats.

Library of Congress Cataloging-in-Publication Data

Names: Ogundare, John Olusegun, author.
Title: Understanding least squares estimation and geomatics data analysis / John Olusegun Ogundare.
Description: 1st edition. | Hoboken, NJ : John Wiley & Sons, 2018. | Includes bibliographical references and index. |
Identifiers: LCCN 2018033050 (print) | LCCN 2018042480 (ebook) | ISBN 9781119501404 (Adobe PDF) | ISBN 9781119501442 (ePub) | ISBN 9781119501398 (hardcover)
Subjects: LCSH: Estimation theory. | Least squares.
Classification: LCC QA276.8 (ebook) | LCC QA276.8 .O34 2018 (print) | DDC 519.5/44–dc23
LC record available at https://lccn.loc.gov/2018033050

Cover Design: Wiley
Cover Illustration: © naddi/iStockphoto

Set in 10/12pt Warnock by SPi Global, Pondicherry, India

Printed in the United States of America

10 9 8 7 6 5 4 3 2 1

Contents

Preface

Paradigm changes are taking place in geomatics with regard to how geomatics professionals function and use equipment and technology. The rise of "automatic surveying systems" and high precision Global Navigation Satellite System (GNSS) networks are changing the focus from how data are captured to how the resultant (usually redundant) data are processed, analyzed, adjusted, and integrated. The modern equipment and technology are continually capturing and storing redundant data of varying precisions and accuracies, and there is an ever-increasing need to process, analyze, adjust, and integrate these data, especially as part of land (or geographic) information systems. The methods of least squares estimation, which are the most rigorous adjustment procedures available today, are the most popular methods of analyzing, adjusting, and integrating geomatics data. Although the concepts and theories of the methods have been developed over several decades, it is not until recently that they are gaining much attention in geomatics professions. This is due, in part, to the recent advancement in computing technology and the various attempts being made in simplifying the theories and concepts involved. This book is to complement the efforts of the geomatics professionals in further simplifying the various aspects of least squares estimation and geomatics data analysis.

My motivation to write this book came from three perspectives: First, my over 15 years of experience in teaching students in the Diploma and Bachelor of Geomatics Engineering Technology (currently, Bachelor of Science in Geomatics) at the British Columbia Institute of Technology (BCIT). Second, my over 10 years as a special examiner and a subject-matter expert for Canadian Board of Examiners for Professional Surveyors (CBEPS) on coordinate systems and map projections, and advanced surveying. Third, as an expert for CBEPS on least squares estimation and data analysis. As a subject-matter expert, I have observed after reviewing syllabus topics, learning outcomes, study guides, and reference and supplementary materials of CBEPS Least Squares Estimation and Data Analysis that there is a definite need for a comprehensive textbook on this subject.

Currently available undergraduate-level books on least squares estimation and data analysis are either inadequate in concepts/theory and content or inadequate in practical and workable examples that are easy to understand. To the best of my knowledge, no specific book in this subject area has synergized concepts/theory and practical and workable examples. Because of this, students and geomatics practitioners are often distracted by having to go through numerous, sometimes irrelevant, materials to extract information to solve a specific least squares estimation problem. Because of this, they end up losing focus and fail to understand the subject and apply it efficiently in practice. My main goal in writing this book is to provide the geomatics community with a comprehensive least squares estimation and data analysis book that is rich in theory/ concepts and examples that are easy to follow. This book is based on *Data Analysis and Least Squares Estimation: The Geomatics Practice*, which I developed and use for teaching students at BCIT for over 15 years. It provides the geomatics undergraduates and professionals with the foundational knowledge that is consistent with the baccalaureate level and also introduces students to some more advanced topics in data analysis.

Compared with other geomatics books in this field, this book is rich in theory/ concepts and provides examples that are simple enough for the students to attempt and manually work through using simple computing devices. The examples are designed to help the students extend their knowledge to solving more practical problems. Moreover, this book assumes that the usually overdetermined geomatics measurements can be formulated generally as three main mathematical models (general, parametric, and conditional), and the number of examples can be limited to the adjustment of these three types of mathematical models.

The book consists of 16 chapters and 6 appendices. Chapter 1 explains survey observables, observations and their stochastic properties, reviews matrix structure and construction, and discusses the needs for geomatics adjustments.

Chapter 2 discusses analysis and error propagation of survey observations, including the application of the heuristic rule for covariance propagation. This chapter explores the concepts and laws of systematic error and random error propagations and applies the laws to some practical problems in geomatics. The use of interactive computing environment for numerical solution of scientific problems, such as Matrix Laboratory (MATLAB) software, is introduced for computing Jacobian matrices for error and systematic error propagations.

In Chapter 3, the important elements of statistical distributions commonly used in geomatics are discussed. The discussion includes an explanation on how statistical problems in geomatics are solved and how statistical decisions are made based on statistical hypothesis tests. The chapter introduces the relevant statistical terms such as statistics, concepts of probability, and statistical distributions.

Chapter 4 discusses the differences among the traditional adjustment methods (transit, compass and Crandall's) and the least squares method of adjustment, including their limitations, advantages, and properties. The concepts of datum definition and the different constraints in least squares adjustment are also introduced in this chapter.

Chapter 5 presents the formulation and linearization of parametric model equations involving typical geomatics observables, the derivation of basic parametric least squares adjustment models, variation functions, and normal equations and solution equations. This chapter also discusses the application of variance–covariance propagation laws in determining the stochastic models of adjusted quantities, such as adjusted parameters, adjusted observations, and observation residuals. The discussion ends with an explanation of how to formulate weight constraint parametric least squares adjustment models, including the solution equations and the associated stochastic models.

In Chapter 6, the concepts of parametric least squares adjustment are applied to various geomatics problems, which include differential levelling, station adjustment, traverse, triangulation, trilateration, resection, and curve fitting. The general formulation of parametric model equations for various geomatics problems, including the determination of stochastic properties of adjusted quantities and the adjustment of weight constraint problems, is also discussed in this chapter.

Chapter 7 discusses the confidence region estimation, which includes the construction of confidence intervals for population means, variances, and ratio of variances, and the construction of standard and confidence error ellipses for absolute and relative cases. Before these, some of the basic statistical terms relating to parameter estimation in geomatics, such as mean squared error, biased and unbiased estimators, mathematical expectation, and point and interval estimators, are defined.

Chapter 8 discusses the problems of network design and pre-analysis. In this chapter, different design variables and how they relate to each other, including their uses and importance, are discussed. The chapter also presents the procedures (with numerical examples) for performing simple pre-analysis of survey observations and for performing network design (or simulation) in one-, two- and three-dimensional cases.

Chapter 9 introduces the concepts of three-dimensional geodetic network adjustment, including the formulation and solution of parametric model equations in conventional terrestrial (CT), geodetic (G), and local astronomic (LA) systems; numerical examples are then provided to illustrate the concepts.

Chapter 10 presents, with examples, the concepts of and the needs for nuisance parameter elimination and the sequential least squares adjustment.

Chapter 11 discusses the steps involved in post-adjustment data analysis and the concepts of reliability. It also includes the procedures for conducting global and local tests in outlier detection and identification and an explanation of the concepts of redundancy numbers, reliability (internal and external), and sensitivity, and their applications to geomatics.

Chapters 12 and 13 discuss the least squares adjustments of conditional models and general models. Included in each of these chapters are the derivation of steps involved in the adjustment, the formulation of model equations for different cases of survey system, the variance–covariance propagation for the adjusted quantities and their functions, and some numerical examples. Also included in Chapter 13 are the steps involved in the adjustment of general models with weight constraints on the parameters.

Chapter 14 discusses the problems of datum and their solution approaches and an approach for performing free network adjustment. It further describes the steps for formulating free network adjustment constraint equations and explains the differences between inner constraint and external constraint network adjustments and how to transform adjusted quantities from one minimal constraint datum to another.

Chapter 15 introduces the dynamic mode filtering and prediction methods, including the steps involved and how simple filtering equations are constructed and solved. The differences between filtering and sequential least squares adjustment are also discussed in this chapter.

Chapter 16 presents an introduction to least squares collocation and the kriging methods, where the theories and steps of least squares collocation and kriging are explained, including their differences and similarities. The book ends with six appendices: Appendices A–C contain sample statistical distribution tables, Appendix D illustrates general partial differentials of typical survey observables, Appendix E presents some important matrix lemmas and identities, and Appendix F lists the commonly used abbreviations in this book.

The topics in this book are designed to meet the needs of the students at the diploma, bachelor, and advanced levels and to support the aspiration of those who work in the geomatics industry and those who are in the process of becoming professional surveyors. Certain aspects of this book are designed to aid the learning and teaching activities: the chapter objectives provide an overview of the material contained in that chapter, and the sample problems with suggested solutions assist readers in understanding the principles discussed.

In general, I expect those who use this book to be familiar with introductory probability and statistics and to have good background in differential calculus, matrix algebra, geometry, and elementary surveying. On this basis, I recommend its use for second- and third-year technological and university undergraduate courses. Some of the topics, such as least squares collocation and kriging methods, will be useful to graduate students and the geomatics practitioners. It is also a valuable tool for readers from a variety of geomatics backgrounds, including practicing surveyors/engineers who are interested in least squares estimation and data analysis, geomatics researchers, software developers for geomatics, and more. Those who are interested in precision surveying will also want to have the book as a reference or complementary material. The professional land surveyors who are gradually discovering the power of least squares method or who are pursuing their continued professional development will likely use the book as a reference material.

John Olusegun Ogundare
Burnaby, BC, Canada

Acknowledgments

This book has benefited from various input in the form of comments, critique and suggestions, from numerous students, educators, and professionals, and the author would like to acknowledge and thank all of them for their help. The author is particularly indebted to the British Columbia Institute of Technology (BCIT) in Canada, for providing the support for the development of the manual on *Data Analysis and Least Squares Estimation: The Geomatics Practice*, on which this book is based. Without this support, this book would not have been possible. The helpful suggestions by the BCIT Geomatics students for continued improvements of the many versions of the manual are also much appreciated.

Special thanks are due to Dr. J.A.R. Blais (Professor Emeritus, Geomatics Engineering Department of the University of Calgary), who provided the author with some valuable comments, suggestions, and reference materials on the last two chapters of this book, on "Introduction to Dynamic Mode Filtering and Prediction" and "Introduction to Least Squares Collocation and The Kriging Methods." The help received from his past technical papers on these topics are also gratefully acknowledged. Others who reviewed material or have assisted in some way in the preparation of this book are Dr. K. Frankich (retired BCIT Geomatics instructor) for allowing the author access to his least squares lecture notes, which he delivered to the BCIT Geomatics students for over several years before his retirement; the faculty members of the Geomatics Department at BCIT, especially Dr. M.A. Rajabi; and Dr. M. Santos of the University of New Brunswick in Canada. The author is grateful to all of them and also to the reviewers, who pointed out problems and identified some areas of improvement for this book.

The Canadian Board of Examiners for Professional Surveyors (CBEPS) is gratefully acknowledged for giving the author the permission to reproduce some of their past exam questions on least squares estimation and data analysis in this book. To those who may have been inadvertently omitted, the author is also grateful.

In spite of the diligent effort of the author, some errors and mistakes are still possible in this edition. The author, therefore, will gratefully accept corrections, comments, and critique to improve future editions.

Finally, the author is grateful to his wife, Eunice, and his children, Joy and Isaac, for their patience, understanding, and support.

About the Author

John Olusegun Ogundare is a practising professional geomatics engineer in British Columbia, Canada; an educator; and author of *Precision Surveying: The Principles and Geomatics Practice*, published by Wiley & Sons, Inc., Hoboken. He received his BSc and MSc degrees in surveying engineering from the University of Lagos, Nigeria, and an MScE and a PhD in high precision and deformation analysis from the University of New Brunswick (UNB) in Canada. He has been in the field of geomatics for over 30 years as a surveyor in various survey engineering establishments in Africa and Canada and as a surveying instructor or teaching assistant in universities and polytechnic institutions also in Africa and Canada.

For over 10 years, John has served as a special examiner for the Canadian Board of Examiners for Professional Surveyors (CBEPS), which includes setting and marking exams on two of their subjects: Coordinate Systems and Map Projections (formerly known as Map Projections and Cartography) and Advanced Surveying. As a subject-matter expert, he has served as a consultant to the Canadian Council of Land Surveyors (CCLS) on those subjects. He has also served as a subject-matter expert in least squares estimation and data analysis for CBEPS, evaluating several Canadian geomatics programs to determine compliance with CBEPS requirements. He sits on the CBEPS Board of Directors and the CBEPS Exemptions and Accreditation Committee. This board with the help of the committee establishes, assesses, and certifies the academic qualifications of individuals who apply to become land surveyors in Canada.

For over 20 years, John has been teaching courses in geomatics technology diploma and degree programs at the British Columbia Institute of Technology (BCIT) in Canada. Some of the courses he teaches or has taught include Advanced Topics in Precision Surveys, Least Squares Adjustments and Data Analysis, Geodetic Positioning, Engineering Surveys, and Coordinate Systems and Mathematical Cartography. He also mentors students pursuing their Bachelor of Science (formerly Bachelor of Technology) in Geomatics in their technical projects and reports. Some of his BCIT-funded works included providing manuals for CBEPS-accredited courses, which he developed and teaches to full-time and web-based students. He has served for over 10 years as a member of the quality committee of the BCIT School of Construction and the Environment and for over 5 years as a member of the School Research committee.

About the Companion Website

This book is accompanied by a companion website:

www.wiley.com/go/ogundare/Understanding-lse-and-gd

The website includes Site for Instructor and Student

The Student Companion Site will have the following:

- Sample multiple-choice questions and answers for all of the chapters (except Chapter 9) – a total of 182 multiple-choice questions and answers.

The Instructor Companion Site will have the following:

- Sample multiple-choice questions and answers for all of the chapters (except Chapter 9) – a total of 182 multiple-choice questions and answers.
- Sample PowerPoint slides for all the chapters of the book – a total of 287 pages.
- "Solutions to the Book Problems" in all of the chapters (except Chapters 9 and 16) of the book – a total of 67 calculation and discussion solutions on a total of 89 pages.

1

Introduction

OBJECTIVES

After studying this chapter, you should be able to:

1) Explain how survey observables (distance, directions, azimuth, angles, elevation differences, etc.) relate to unknown parameters (coordinates).
2) Discuss error properties of measurements, such as random errors, systematic errors, blunders, and mistakes.
3) Explain needs for adjustment and carry out simple station adjustment of survey measurements.
4) Assess precision and accuracy of measurements in terms of residuals, standard errors, etc.
5) Discuss how to determine correlation coefficients from variance–covariance matrices.

Understanding Least Squares Estimation and Geomatics Data Analysis, First Edition.
John Olusegun Ogundare.
© 2019 John Wiley & Sons, Inc. Published 2019 by John Wiley & Sons, Inc.
Companion website: www.wiley.com/go/ogundare/Understanding-lse-and-gda

6) Discus the relationships among variance–covariance matrix, cofactor matrix, and weight matrix.
7) Construct covariance, cofactor, and weight matrices for measurements.

1.1 Observables and Observations

An *observable* is a survey network random variable or a network element that, when measured repeatedly, can have several possible values, e.g. slope or horizontal distance, horizontal direction (or angle), vertical (or zenith) angle, azimuth (or bearing), elevation difference, relative gravity value, and coordinate difference (in GPS surveys). The term observable is used to represent the individual component of a survey network being measured. For example, if the distances AB, BC, CD, and DA, constituting the individual components of a rectangle ABCD, are to be measured, then there are four distance observables AB, BC, CD, and DA. If the observable AB, for example, is measured three times, there will be three observations for one observable AB. In this book, the observables that have been measured will be represented most of the time by a vector ℓ, and those that have not been measured but whose values are to be determined by least squares method (also known as parameters) by the vector x. Matrices and vectors are to be distinguished from constant symbols by considering the context in which they are used.

Observation (or *measurement*) is an operation or numerical outcome of an operation. This should not be confused with the elements (observables) that are to be measured. A numerical value assigned to an observable is an observation or a measurement. The term observation or measurement is considered the same in this course. An observation is referred to as a quantity that varies in its value randomly and for which an estimate is available a priori. Such an estimate must have been derived somehow from direct or previous measurements with some uncertainties clearly associated.

1.2 Significant Digits of Observations

Significant digits or significant figures of a number are the digits of the number that are considered correct together with the first doubtful digit. The concept of significant figures or significant digits is commonly used when rounding is involved, e.g. rounding to certain significant figures. The number of significant figures in a measurement will be based on the least count of the device used to make the measurement. The rules for identifying significant digits in numbers are as follows:

1) All zeros preceding a nonzero digit in a number are not significant, e.g. 0.024 31 and 0.000 491 3 have four significant figures with the leading zeros being

nonsignificant. In the number 0.000 491 3, "4" is the most significant figure and "3" is the least significant figure.

2) All zeros located in a decimal number with no nonzero digit located after them are significant, e.g. 43.2100 has 6 significant figures, 430.00 has 5 significant figures, and 5940 can be considered as having 4 significant figures if the zero is for fixing the position of the implied decimal point or 3 significant figures if not.

3) All zeros located in an integer number with no nonzero digit located after them can have two interpretations, e.g. 5200 may be interpreted as having 4 significant figures if it is accurate to the nearest unit, or it can be interpreted as having 2 significant figures if it is simply shown to the nearest 100 due to uncertainty.

4) All nonzero digits in a number are always significant, e.g. 594 and 12.3 have 3 significant figures.

5) All zeros located within any two nonzero digits are significant, e.g. 201.123 and 594.007 have 6 significant figures, with the zeros considered as significant.

The number of significant figures a number will be reduced to during rounding off process is commonly based on the following rules, assuming n significant figures are required:

i) If the $(n + 1)$th digit of the number is between 0 and 4, disregard all the digits from $n + 1$ (inclusive). For example, the number 65.2443 rounded to $n = 4$ significant figures will be 65.24.

ii) If the $(n + 1)$th digit of the number is 5 and the nth digit is even, disregard the 5 and all the other digits to the right. For example, the number 65.245 rounded to $n = 4$ significant figures will be 65.24.

iii) If the $(n + 1)$th digit of the number is 5 and the nth digit is odd, increase the nth digit by one and disregard all the other digits to the right. For example, the number 65.235 rounded to $n = 4$ significant figures will be 65.24.

iv) If the $(n + 1)$th digit of the number is between 6 and 9, increase the nth digit by one and disregard all the digits from $n + 1$ (inclusive). For example, the number 65.247 rounded to $n = 4$ significant figures will be 65.25.

The usual rules in performing mathematical operations such as addition, subtraction, multiplication, and division with numerical values of different significant figures are as follows:

a) When performing addition (subtraction), the overall result should be rounded off to the number of decimal places that is least in the numbers being added or subtracted. For example, 800 + 136.5 + 9.3 = 945.8 m should be rounded off to 3 significant figures as 946 m based on 800 m having the least number of decimal places; 32.01 + 5.325 + 12 = 49.335 m should be rounded off to 2 significant figures based on 12 m having least number of decimal places.

b) In multiplication or division, the number of significant figures in overall result should be equal to the number of significant figures in the value with the least number of significant figures. This should exclude cases involving exact (not approximate) numerical quantities such as 2, 3, etc. or defined quantities (usually given with no units attached, such as $\pi = 3.1416$); these quantities should not affect significant digits in a calculation. For examples, $2 \times 4.25 \times 25.145$ should be 213 based on the significance figures of 4.25 being three; (30.1 cm + 25.2 cm + 31.3 cm) divided by 3 will give 28.9 cm based on the significance figures of the addition and not on the exact number 3 used for the division.

c) Whenever the operations in (a) and (b) above are carried out together in a calculation, rule (a) should be applied to the addition or subtraction component, and then rule (b) to the multiplication or division component. It is commonly suggested that one extra significant figure be retained in intermediate calculations until the final result is obtained, which is then rounded off to the desired significant figures.

1.3 Concepts of Observation Model

Model (usually mathematical) is a mathematical representation of the observation (outcome). It replaces the observations for the purpose of summarizing and assessing the observations. A model is composed of two parts: *functional model and stochastic model*. Thus, a model representing an observation can be written as

$$\text{Observation} = \text{Functional} + \text{Stochastic} \tag{1.1}$$

Functional model is a set of equations relating *observations* with the *parameters* (unknown) of the model or a set of equations relating observations to each other by satisfying some geometric conditions. The procedure for determining the variances (or covariance matrices) and weights of observations is known as *stochastic model* in least squares adjustment. In adjustment, *parameters* are quantities to be determined or solved for in a given problem. Observations are usually made to assign values to some or all of the components of the functional model. Usually before an observation is linked with a functional model, it must have been corrected for possible errors; errors that have random nature are ignored in order to make the functional model as simple as possible. Typical observations and the corresponding parameters in geomatics are as shown in Table 1.1. In precision surveying in which the main parameters are the 2D (easting, northing) coordinates or 3D (X, Y, Z or latitude, longitude, ellipsoidal) coordinates of network points, *triangulation* or *trilateration* method can be used because of their ability to provide redundancies, provided the appropriate datum has been defined for the network. If triangulation method is used, the

Table 1.1 Typical survey observations and parameters.

Type of survey activity	Typical observations	Typical parameters
Differential leveling	• Elevation differences between sections	Elevations
Trigonometric leveling	• Zenith (or vertical) angles • Slope (or horizontal) distances • Heights of instruments • Heights of targets or staff readings	Elevations
Traverse	• Horizontal directions (or angles) • Horizontal (or slope) distances • Zenith (or vertical) angles • Bearings	Three-dimensional (3D) coordinates (easting, northing, UP) or map projection coordinates (easting, northing)
GPS surveys	• Baseline vectors (coordinate differences of baselines) • Geoid undulations	Earth-fixed-Earth-centered coordinates (X, Y, Z) or map projection coordinates (easting, northing, orthometric or ellipsoidal heights)
Absolute orientation in photogrammetry	• Photo coordinates (x, y) of points • Coordinates of fiducial center (x_0, y_0) of photo • Focal length of camera (f) • Orientation of photo in space (if measured using gyro or inertia system) – Ω_0, Φ_0, K_0 • Translations (X_0, Y_0, Z_0) if measured using GPS	Map projection coordinates (eastings, northings)
Gravimetric leveling	• Relative gravity values	Absolute gravity values of unknown points
Two-dimensional (2D) similarity transformation	• 2D coordinates of known points in one coordinate system • 2D coordinates of the same known points in another coordinate system	• Four unknown constants (2 translations, 1 rotation, and 1 scale) • 2D coordinates of the unknown points
3D similarity transformation	• 3D coordinates of known points in one coordinate system • 3D coordinates of the same known points in another coordinate system	• Seven unknown constants (3 translations, 1 scale, and 3 rotations) • 3D coordinates of the unknown points
Calibration of electronic distance meters (EDM)	• Baseline distances measured by the EDM • Baseline distances published (provided by the government)	• System constant of EDM • Scale factor of EDM

horizontal (and vertical) angles of the network triangles will be the main observations; and if trilateration method is used, the main observations will be horizontal (or slope) distances of the network triangles. It should be noted that when a parameter is given an estimated value with standard deviation associated with it, the parameter can be treated as an observation. Before any adjustment can be performed, a functional model relating observations to parameters must be formulated, or a functional model relating observations to each other must be identified and formulated.

1.4 Concepts of Stochastic Model

Stochastic model is a scientific tool for describing the probability or randomness involved in a quantity. It describes the statistical properties (uncertainties) of survey observations and parameters and explains the expected error distribution of the observations and the parameters. When a quantity is measured repeatedly, the values obtained (observations) will not be the same due to some error sources that may be physical, time dependent, or unknown. The effects of such error sources in observations are of three types: *random errors, systematic errors*, and *blunders*. Blunders (or mistakes) must be detected and removed from observations; they are usually due to carelessness of the observer. Systematic errors are mainly due to the influence of instrument used, procedure of measurement, or atmospheric condition. Examples of systematic errors are prism constant not corrected for, refraction error, collimation error, slope distance used instead of horizontal distance, etc. When systematic errors are present in observations, there is nothing actually wrong with the observations except that the observations might be interpreted wrongly, i.e. not removing the influence of the systematic errors before use. In this case, linking the observations to a functional model will be inconsistent when systematic errors are allowed to remain in the observations. Usually, systematic errors are detected and then corrected for (mathematically or by following certain measurement procedure) in the observations before they are associated with a functional model. Random errors are considered as observational errors that are small and difficult to detect; the errors cannot be removed but can only be minimized through adjustment. This type of error causes repeated measurements to be inconsistent. Least squares method applies when there are only random errors in the measurements with the systematic component already removed.

1.4.1 Random Error Properties of Observations

Variations in repeated observations are considered to be due to random errors. One practical way of estimating statistical properties of a set of observations is to use the concept of statistics. *Statistics* applies the laws of *probability* in obtaining estimates or in making inferences from a given set of observations. Probability uses the laws of chance to describe and predict possible value of a quantity when it is measured repeatedly. If repeated observations of a quantity are not available, the often used approach for estimating or making inferences for an observation is to assume the statistical properties from a general reference to similar observations that were performed under similar circumstances in the past, e.g. statistical distribution. For example, if one observation of an observable is made, the standard deviation of the observation will be considered known from previous measurements made with similar equipment under similar conditions. If a set of observations of an observable expressed as ℓ_i (for $i = 1, 2, \ldots, n$) are made with the true value of the observable known as μ (note that this is never known in reality), the error in each observation can be expressed as

$$\varepsilon_i = \ell_i - \mu \tag{1.2}$$

Since μ is not known, let say the average of the observations is determined as $\bar{\ell}$ and $\mu = \bar{\ell} - \delta$ where δ is the constant error (known as systematic error) between the true value and the average value. Equation (1.2) can be expressed as

$$\varepsilon_i = \ell_i - \bar{\ell} + \delta \tag{1.3}$$

or

$$\varepsilon_i = R_i + \delta \tag{1.4}$$

where $R_i = \ell_i - \bar{\ell}$ is the residual error. Following the usual convention, the *residual correction* to observation will be expressed as $v_i = -R_i$ with R_i as the *residual error*. The *arithmetic mean* ($\bar{\ell}$) is calculated as the average of the observations using the following formula:

$$\bar{\ell} = \frac{\sum\limits_{i=1}^{n} \ell_i}{n} \tag{1.5}$$

or

$$\bar{\ell} = \frac{\ell_1 + \ell_2 + \cdots + \ell_n}{n} \tag{1.6}$$

where n is the number of observations. From Equation (1.4) it can be seen that the true error of an observation can be expressed as the sum of residual error and systematic error. If all the systematic errors in the observations are known to have been removed, then the residual errors can be considered as representing the actual errors of the observations. Usually, the residual errors can be determined from the repeated measurements as illustrated in Example 1.1.

Example 1.1 The repeated measurements of a baseline (in m) are as follows: 20.55, 20.60, 20.50, 20.45, 20.65. Determine the mean measurement and the residual errors of the measurements.

Solution:

$$\text{Mean} = \bar{\ell} = \frac{\sum_{i=1}^{5} \ell_i}{5}$$

$$\bar{\ell} = \frac{20.55 + 20.60 + 20.50 + 20.45 + 20.65}{5} = 20.55\,\text{m}$$

Residual errors:

$$R_i = \ell_i - \bar{\ell} \rightarrow R_1 = \ell_1 - 20.55 \rightarrow 0.00\,\text{m}$$

$$R_2 = \ell_2 - 20.55 \rightarrow 0.05\,\text{m}$$

$$R_3 = \ell_3 - 20.55 \rightarrow -0.05\,\text{m}$$

$$R_4 = \ell_4 - 20.55 \rightarrow -0.10\,\text{m}$$

$$R_5 = \ell_5 - 20.55 \rightarrow 0.10\,\text{m}$$

1.4.2 Standard Deviation of Observations

Errors in observations can be represented by residual errors, or they can be represented by standard deviation (square root of variance) if no systematic errors exist in the measurement. The residuals constitute the random component of the observational errors, and their effects are minimized by adjustment. Since residuals are difficult to assess, they are usually summarized by a quantity known as *standard deviation*. The standard deviation of a random variable is a measure of the spread of its values, i.e. the spread of data about the mean or, specifically, the root mean square (RMS) deviation of values from their arithmetic mean. It is the amount of random errors present in measurements. The usual formula for computing the sample standard deviation (s) of a set of n number of observations ℓ_i (for $i = 1, 2, ..., n$) can be given as

$$s = \sqrt{\frac{\sum_{i=1}^{n} (\ell_i - \bar{\ell})^2}{(n-1)}} \tag{1.7}$$

or

$$s = \sqrt{\frac{\sum\limits_{i=1}^{n} R_i^2}{(n-1)}} \qquad (1.8)$$

where $\bar{\ell}$ is the arithmetic mean expressed by Equation (1.5) or Equation (1.6). The *standard deviation of the mean (or standard error* (SE)) is an estimate of the standard deviation of a statistic (the mean); it is the error in the mean computed from the sample observations. The SE or the standard deviation of the mean ($s_{\bar{\ell}}$) can be given as

$$s_{\bar{\ell}} = \frac{s}{\sqrt{n}} \qquad (1.9)$$

It should be noted that the standard deviation of the mean is really not the amount of random error in the mean, but a measure of the uncertainty of the mean due to random effects when systematic effects have been taken care of through correction; the uncertainty in this case is associated with the correction made. Remember also that a measure of uncertainty does not account for mistakes or blunders in measurements. Another measure of randomness in observations is known as variance. Variance or standard deviation is a measure of the scattering of the values of the random observable about the mean. The *sample variance* is the average of square differences between observations and their mean. It is a measure of precision of a set of sample observations; and it is also determined from residual errors. The square root of the sample variance is the same as the standard deviation. The *population variance* (σ^2) is a measure of the precision of a set of population observations; it should be calculated from true errors; the square root of the population variance is often referred to as the *standard error*. The population variance and SE cannot be determined since the true errors are not known, but can be estimated using sample observations. In this book, however, the standard deviation and SE will be used to mean the same thing.

1.4.3 Mean of Weighted Observations

Mean value of weighted observations is a simple adjusted value of the same observable measured repeatedly. Let an observable ℓ be measured repeatedly n times with the vector of uncorrelated observations as $\ell = (\ell_1, \ell_2, \ell_3, ..., \ell_n)^T$; if the weight matrix of the observations is $P = \text{diag}(p_{11}, p_{22}, p_{33}, ..., p_{nn})$, assuming different means were used to measure the observable so that each measurement has different weight, the weighted mean value ($\bar{\ell}$) of the observations can be given as

$$\bar{\ell} = \frac{\ell_1 p_{11} + \ell_2 p_{22} + \ell_3 p_{33} + \cdots + \ell_n p_{nn}}{p_{11} + p_{22} + p_{33} + \cdots + p_{nn}} \tag{1.10}$$

or

$$\bar{\ell} = \frac{\sum\limits_{i}^{n} \ell_i p_{ii}}{\sum\limits_{i}^{n} p_{ii}} \tag{1.11}$$

By applying the variance–covariance propagation laws on Equation (1.11) and assuming weight is inversely proportional to the observation variance (i.e. $p_{ii} = 1/s_{ii}^2$), the standard deviation ($s_{\bar{\ell}}$) of the mean of weighted observations can be given as

$$s_{\bar{\ell}} = \frac{1}{\sqrt{\sum\limits_{i}^{n} p_{ii}}} \tag{1.12}$$

where weight of observation as inverse of variance of observation is discussed further in Section 1.7.2.

1.4.4   Precision of Observations

Precision is the degree of closeness of an observation to its mean value (represented by residuals or a standard deviation). Precision of observation will depend on the following:

- Precision of measuring instrument.
- Attentiveness of observer, which includes the ability to point, read, level, and center the measuring instrument precisely.
- Stability of environment, such as the effects of the atmospheric refractions on the measurements.
- Overall design of the survey, which includes the geometry of network points, design of targets, adopted procedures or overall methodology in acquiring the observations, etc.

The precision of a measuring instrument usually depends on the least count (smallest directly readable value) of the instrument and its repeatability. It should be noted from the above, however, that the least count of an instrument is not sufficient to describe the precision of observations. Precision and random errors are directly related: Precision increases when random errors decrease, i.e.

standard deviation (or variance) becomes smaller. Precision of measurements can be determined and assessed without detecting and removing systematic errors from observations. In this case, precision can described as the amount of random errors in the observations.

1.4.5 Accuracy of Observations

Theoretically, *accuracy* is the degree of closeness of an estimate or observation to its true value. Since the true value of a quantity is not known, the practical definition of accuracy can be given as a total amount of systematic and random errors present in the measurements. If all of the systematic errors have been removed, then accuracy will be measured by the standard deviation (the amount of random error present) of the measurements. In this case, accuracy will be the same as precision. For example, if a surveyor measured a line AB as 500.725 m and another one measured it as 500.852 m, the more accurate measurement will be determined based on the one who paid more attention to detection and removal of systematic errors and mistakes (e.g. type of instrument, calibration, correction for refraction, etc.). Note that the accuracy of observations cannot be assessed without detecting and removing systematic errors from the observations; in this case, a measurement can be precise but inaccurate. For example, a highly refined instrument may give repeated readings that are very close (precise), but if systematic errors (prism constant, scale factor error, and refraction error) are present, the readings will be precise (since they are very close to each other) but are inaccurate (since systematic errors are present). Similarly, a measurement can be accurate but imprecise. If less refined (but calibrated) instrument is used, it may give a mean value of repeated readings that is closer to the true value, but there will be less agreement among the readings.

Example 1.2 Referring to the measurements in Example 1.1, determine the standard deviations of the measurements and of the mean measurement.

Solution:

$\bar{\ell}$ = 20.55 from Equation (1.8) and the computed residuals in Example 1.1:

$$s = \sqrt{\frac{\sum\limits_{i=1}^{n} R_i^2}{(n-1)}} = \sqrt{\frac{0.025}{4}} = 0.08\,\text{m}$$

From Equation (1.8), the standard deviation of the mean measurement, $s_{\bar{\ell}} = 0.04$ m for $n = 5$.

1.5   Needs for Adjustment

Adjustment is a means of obtaining unique values (mean values) for the unknown parameters to be determined when there are more observations than actually needed (redundant observations); statistical properties (standard deviation, covariance, etc.) may be determined as by-products. The adjustment accounts for any presence of random errors (not systematic errors) in observations and increases precisions of final values computed for the unknown parameters. It should be mentioned that adjustment is only meaningful when observations available exceed the minimum necessary for unique determination of unknown. This means that a least squares adjustment will be required when there are more observations than are actually needed (or when more observables are measured than are actually needed) for the unique determination of the unknown parameters. The extra observations are considered as *redundant data*. Usually, redundant data are inconsistent since each sufficient subset yields different results from another subset (due to statistical properties of data) so that unique solution is not possible with such data. One method of obtaining unique solution is to apply a criterion of least squares to redundant observations. Least squares method is one of the methods of adjustment.

The term *redundancy* (*or number of degrees of freedom*) is used in adjustment to mean the number of independent model equations formed (or independent observations made) minus the number of unknown parameters involved in the equations. Generally, redundancy of models is required before the adjustment of the models can become meaningful; otherwise there will be no need for adjustment. It should also be mentioned that the more redundant measurements there are in an adjustment, the better the precisions of the adjusted measurements and the quantities derived from the measurements.

The traditional station adjustment method can be used to determine the means, residuals, and standard deviations of direction measurements made from one station to a multiple target stations. Station adjustment determines (in the sense of least squares method) the probable values of direction or angular measurements at the station points of the network and assesses the associated measurement precisions. This method is illustrated using Example 1.3.

Example 1.3   Consider Table 1.2 in which four sets of direction measurements are made from a theodolite station P to three target stations Q, R, and S. Each set consists of two measurements (made in face I and face II positions of the theodolite) with releveling and recentering and changing of zero graduation direction of theodolite made between sets. Answer the following.

Table 1.2 Direction measurements from station P to stations Q, R, and S.

Set no.	Direction	Face I	Face II
1	P–Q	0°00'00"	180°00'05"
	P–R	37°30'27"	217°30'21"
	P–S	74°13'42"	254°13'34"
2	P–Q	90°00'00"	270°00'08"
	P–R	127°30'32"	307°30'28"
	P–S	164°13'48"	344°13'42"
3	P–Q	180°00'00"	0°00'15"
	P–R	217°30'26"	37°30'26"
	P–S	254°13'36"	74°13'40"
4	P–Q	270°00'00"	90°00'12"
	P–R	307°30'28"	127°30'30"
	P–S	344°13'45"	164°13'41"

a) Determine the mean (adjusted) direction measurement to each target.
b) Determine the standard deviation of a direction measurement and the standard deviation of the mean direction measurement.

Solution (a):

Station adjustment is performed on the data given in Table 1.2 and the result given in Table 1.3 based on the following steps:

- Find the averages of Face I (column (3)) and Face II (column (4)) measurements in Table 1.3 in each set, and record the corresponding averages in column (5) under "Mean" in Table 1.3.
- Reduce the mean values in each set to the first direction by subtracting the first mean value (for direction P–Q) from each of the other values in that set, and record the corresponding results in column (6) under "Reduced mean."
- Determine the grand mean of the corresponding reduced means in column (6) for each line across the four sets, giving the following:

$$\text{Line P} - \text{Q} : 0°00'00.0''$$

$$\text{Line P} - \text{R} : \frac{37°30'21.5'' + 37°30'26'' + 37°30'18.5'' + 37°30'23''}{4} = 37°30'22.3''$$

$$\text{Line P} - \text{S} : \frac{74°13'35.5'' + 74°13'41'' + 74°13'30.5'' + 74°13'37''}{4} = 74°13'36.0''$$

Table 1.3 Reduction of direction measurements (station adjustment) made from station P.

Set no. (1)	Direction (2)	Face I (3)	Face II (4)	Mean (5)	Reduced mean (6)
1	P–Q	0°00′00″	180°00′05″	0°00′02.5″	0°00′00″
	P–R	37°30′27″	217°30′21″	37°30′24″	37°30′21.5″
	P–S	74°13′42″	254°13′34″	74°13′38″	74°13′35.5″
2	P–Q	90°00′00″	270°00′08″	90°00′04″	0°00′00″
	P–R	127°30′32″	307°30′28″	127°30′30″	37°30′26″
	P–S	164°13′48″	344°13′42″	164°13′45″	74°13′41″
3	P–Q	180°00′00″	0°00′15″	180°00′07.5″	0°00′00″
	P–R	217°30′26″	37°30′26″	217°30′26″	37°30′18.5″
	P–S	254°13′36″	74°13′40″	254°13′38″	74°13′30.5″
4	P–Q	270°00′00″	90°00′12″	270°00′06″	0°00′00″
	P–R	307°30′28″	127°30′30″	307°30′29″	37°30′23″
	P–S	344°13′45″	164°13′41″	344°13′43″	74°13′37″

Solution (b):

The determination of the standard deviation of each direction measurement and the standard deviation of each mean direction measurement is given in Table 1.4 based on the following steps:

- Subtract the grand mean from the corresponding reduced means in column (6) in Table 1.3 to obtain the discrepancies (or misclosures) given in column (3) in Table 1.4, e.g. for line P–R in set 1, the discrepancy is 37°30′21.5″ – 37°30′22.3″ or –0.8″; for line P–R in set 2, it is 37°30′26″ – 37°30′22.3″ or 3.7″.
- Determine the mean discrepancy for each set and subtract that mean from each of the discrepancy in that set, giving the residual (v) for that line as shown in column (4) in Table 1.4, e.g. for set 1, the mean discrepancy is (0.0″ +(–8.0″) + (–0.5″))/3 or –0.4″; subtracting –0.4″ from line P–Q in set 1 gives the residual for that line as 0.4″. The sum of the residuals in each set must add up to zero, e.g. for set 1, (0.4″ +(–0.3″) + (–0.1″)) is 0.0″.
- Square each residual in column (4) and present in column (5); sum the squared residuals in column (5), giving 30.14 s^2.
- Determine the number of degrees of freedom for the station adjustment as follows:
  - Considering the directions P–Q, P–R, and P–S as parameters with the first line fixed as a reference, there will be two unknown direction parameters. In general, for t number of targets to be measured to, there will be $t - 1$ unknown direction parameters if one of the directions is fixed as a reference.

Table 1.4 Determination of standard deviations of direction measurements.

Set no. (1)	Station sighted (2)	Discrepancy (3)	Residual (v_i) (4)	Residual squared (v_i^2) (5)
1	Q	0.0″	0.4″	0.16
	R	−0.8″	−0.3″	0.09
	S	−0.5″	−0.1″	0.01
	Sum	−1.3″	0.0″	
2	Q	0.0″	−2.9″	8.41
	R	3.7″	0.8″	0.64
	S	5.0″	2.1″	4.41
	Sum	8.7″	0.0″	
3	Q	0.0″	3.1″	9.61
	R	−3.8″	−0.7″	0.49
	S	−5.5″	−2.4″	5.76
	Sum	−9.3″	0.0″	
4	Q	0.0″	−0.6″	0.36
	R	0.7″	0.2″	0.04
	S	1.0″	0.4″	0.16
	Sum	1.7″	0.0″	30.14

- Taking each set as a new setup, there will be four unknown orientation parameters for four setups (or four sets) in this problem. In general, for n number of setups, there will be n unknown orientation parameters. The total number of unknown parameters is six (four orientation parameters plus two unknown direction parameters). In general, this will be $n + t - 1$.
- The total number of direction measurements (considering the mean of face I and face II measurements for a line in a set as constituting a single measurement for that line) in this problem is 12. In general, for t targets measured to and n number of sets, the total number of measurements will be nt.
- Taking the number of degrees of freedom as the number of measurements minus the total number of unknown parameters, the degrees of freedom for this problem will be $12 - 6$, giving 6 degrees of freedom. In general, the number of degrees of freedom will be $nt - (n + t - 1)$, which can be expressed as $(t - 1)(n - 1)$.

- Generally, the standard deviation of a direction measurement can be given as

$$s = \sqrt{\frac{\sum_{i=1}^{nt} v_i^2}{(t-1)(n-1)}} \qquad (1.13)$$

In this problem, referring to Table 1.4, $\sum_{i=1}^{12} v_i^2 = 30.14$, and the number of degrees of freedom is $(3-1) \times (4-1)$ or 6, giving the standard deviation of a direction measurement from Equation (1.13) as $s = 2.2''$.

- Since the grand means of direction measurements are based on n number of sets, the standard deviation of each grand mean can be determined from Equation (1.9). In this problem (for $n = 4$ and using Equation (1.13)), the standard deviation of the grand mean direction measurement is $s_{\bar{\ell}} = 1.1''$.

1.6 Introductory Matrices

Matrices are very important in representing and solving a system of model equations in least squares adjustment; they allow the system of equations to be presented in compact forms, making their solutions straightforward. The focus of this section is mainly to review matrix construction and structure to help readers remember how to use them in the adjustment methods to be discussed in this book. Complex matrix operations and manipulations are not discussed since they can be found in many available matrix algebra books. By definition, a matrix A is a rectangular array of numbers contained within a pair of brackets, such as

$$A = \begin{bmatrix} a_{11} & a_{12} & \cdots & a_{1m} \\ a_{21} & a_{22} & \cdots & a_{2m} \\ \vdots & \vdots & \cdots & \vdots \\ a_{n1} & a_{n2} & \cdots & a_{nm} \end{bmatrix} \qquad (1.14)$$

where a_{ij} (for $i = 1, 2, ..., n; j = 1, 2, ..., m$) are the matrix elements; subscript i indicates the row and subscript j the column in which a matrix element is located; and n and m are the numbers of rows and columns in the matrix, respectively, making matrix A an $n \times m$ matrix. The *size* of a matrix is the number of rows and columns of the matrix, written in the form of rows $\times$ columns. The matrix in Equation (1.14) has a size of $n \times m$. As an example, the following system of homogeneous linear equations can be expressed in matrix form:

$$\begin{aligned} 4\ell_1 - 2\ell_2 + \ell_3 = 0 \\ 5\ell_1 + 3\ell_2 = 0 \end{aligned} \qquad (1.15)$$

where the so-called coefficient matrix in the order ℓ_1, ℓ_2, and ℓ_3 can be given as

$$B = \begin{bmatrix} 4 & -2 & 1 \\ 5 & 3 & 0 \end{bmatrix} \tag{1.16}$$

Matrix B is a 2×3 matrix. A *square matrix* will have the same number of rows as that of the columns (e.g. $n = m$), such as the following 3×3 matrix C:

$$C = \begin{bmatrix} c_{11} & c_{12} & c_{13} \\ c_{21} & c_{22} & c_{23} \\ c_{31} & c_{32} & c_{33} \end{bmatrix} \tag{1.17}$$

where the elements c_{11}, c_{22}, and c_{33} are the *diagonal elements* of matrix C with all the other elements as *off-diagonal elements*. A *diagonal matrix* will have zero values as off-diagonal elements with only the diagonal elements as nonzero values as in the following:

$$C = \begin{bmatrix} c_{11} & 0 & 0 \\ 0 & c_{22} & 0 \\ 0 & 0 & c_{33} \end{bmatrix} \tag{1.18}$$

or

$$C = \mathrm{diag}(c_{11}, c_{22}, c_{33}) \tag{1.19}$$

Given an A matrix, the *transpose* (A^T) of the matrix is obtained by changing each row into a corresponding column. This is illustrated as follows:

$$A = \begin{bmatrix} a_{11} & a_{12} \\ a_{21} & a_{22} \\ a_{31} & a_{32} \end{bmatrix} \quad A^T = \begin{bmatrix} a_{11} & a_{21} & a_{31} \\ a_{12} & a_{22} & a_{32} \end{bmatrix} \tag{1.20}$$

A square matrix in which its transpose is the same as itself is a symmetric matrix. For example, the following A matrix is a symmetric matrix since it can be shown that $A = A^T$:

$$A = \begin{bmatrix} 4 & 1 & 3 \\ 1 & 2 & -5 \\ 3 & -5 & 6 \end{bmatrix} \tag{1.21}$$

1.6.1 Sums and Products of Matrices

Two matrices *A* and *B* can be added or subtracted if they have the same size; two matrices of different sizes cannot be added or subtracted. The sum of matrices is demonstrated using the following matrices:

$$C = \begin{bmatrix} a_{11} & a_{12} & a_{13} \\ a_{21} & a_{22} & a_{23} \end{bmatrix} + \begin{bmatrix} b_{11} & b_{12} & b_{13} \\ b_{21} & b_{22} & b_{23} \end{bmatrix} \tag{1.22}$$

giving

$$C = \begin{bmatrix} a_{11} + b_{11} & a_{12} + b_{12} & a_{13} + b_{13} \\ a_{21} + b_{21} & a_{22} + b_{22} & a_{23} + b_{23} \end{bmatrix} \tag{1.23}$$

The multiplication of two matrices $A \times B$ is possible if the number of columns of matrix *A* is the same as the number of rows of *B*. Given the following two matrices *A* and *B*,

$$A = [a_{11} \; a_{12} \; a_{13}] \tag{1.24}$$

$$B = \begin{bmatrix} b_{11} \\ b_{21} \\ b_{31} \end{bmatrix} \tag{1.25}$$

The product of matrices *A* and *B* can be given as follows:

$$A \times B = [a_{11} \; a_{12} \; a_{13}] \begin{bmatrix} b_{11} \\ b_{21} \\ b_{31} \end{bmatrix} \tag{1.26}$$

giving

$$A \times B = [a_{11}b_{11} + a_{12}b_{21} + a_{13}b_{31}] \tag{1.27}$$

Given another two matrices *A* and *B*, their product can be given as follows:

$$A \times B = \begin{bmatrix} a_{11} & a_{12} \\ a_{21} & a_{22} \\ a_{31} & a_{32} \end{bmatrix} \begin{bmatrix} b_{11} & b_{12} \\ b_{21} & b_{22} \end{bmatrix} \tag{1.28}$$

giving

$$A \times B = \begin{bmatrix} a_{11}b_{11} + a_{12}b_{21} & a_{11}b_{12} + a_{12}b_{22} \\ a_{21}b_{11} + a_{22}b_{21} & a_{21}b_{12} + a_{22}b_{22} \\ a_{31}b_{11} + a_{32}b_{21} & a_{31}b_{12} + a_{32}b_{22} \end{bmatrix} \tag{1.29}$$

The product in Equation (1.29) is obtained from Equation (1.28) by following the arrows shown in Equation (1.28) for the multiplication of the elements. The product $B \times A$ will be impossible since the number of columns of B is not the same as the number of rows of A; B is said in this case not to be conformable to A, but A is conformable to B for multiplication.

When an inverse of a matrix is multiplied by the matrix, an identity matrix (I) is obtained. For example, the inverse of a matrix A is A^{-1} such that $AA^{-1} = I$, where the identity matrix (I) is a diagonal matrix with diagonal elements as 1's and non-diagonal elements as zeros. For example, a 3×3 identity matrix (I_3) can be given as

$$I_3 = \begin{bmatrix} 1 & 0 & 0 \\ 0 & 1 & 0 \\ 0 & 0 & 1 \end{bmatrix} \tag{1.30}$$

For example, given a 2×2 A matrix,

$$A = \begin{bmatrix} 3 & 2 \\ 2 & 4 \end{bmatrix} \tag{1.31}$$

The inverse A^{-1} can be given as

$$A^{-1} = \begin{bmatrix} \dfrac{1}{2} & -\dfrac{1}{4} \\ -\dfrac{1}{4} & \dfrac{3}{8} \end{bmatrix} \tag{1.32}$$

Take note from Equation (1.32) that the inverse is not obtained by simply taking the reciprocal value of each element of the matrix. For a square diagonal matrix, however, the inverse is obtained by simply taking the reciprocal of the diagonal elements as can be seen in the case of matrix B in Equation (1.33):

$$B = \begin{bmatrix} 3 & 0 \\ 0 & 4 \end{bmatrix} \tag{1.33}$$

The inverse of matrix B is

$$B^{-1} = \begin{bmatrix} \dfrac{1}{3} & 0 \\ 0 & \dfrac{1}{4} \end{bmatrix} \tag{1.34}$$

Comparing the matrices A and B in Equations (1.31) and (1.33), it can be seen that they both have identical values as diagonal elements, but zero values for the off-diagonal elements in matrix B. By looking at their matrix inverses, it can be seen that the diagonal elements of the inverses are not the same; B^{-1} simply has individual element inverted, but that is not the case in A^{-1}, which has nonzero values as the non-diagonal elements. This is to point out that care should be taken when inverting a matrix with nonzero diagonal elements than when inverting a diagonal matrix.

1.6.2 Vector Representation

Vectors are special forms of matrices; they can come in different kinds, but the basic kind of vector of interest is a row of numbers called *row vector* or a column of numbers called a *column vector*. For example, a row vector with five elements will have five numbers arranged in a matrix of 1×5; a column vector with five elements will have five numbers arranged in a matrix of 5×1. An example of column vector of size $n \times 1$ can be expressed as follows:

$$X = \begin{bmatrix} x_1 \\ x_2 \\ \vdots \\ x_n \end{bmatrix} \quad \text{or} \quad X = \begin{bmatrix} x_1 & x_2 & \cdots & x_n \end{bmatrix}^T \tag{1.35}$$

The following system of equations involving three variables (observables) can be represented in compact matrix and vector forms:

$$\begin{aligned} y_1 &= a_{11}x_1 + a_{12}x_2 + a_{13}x_3 \\ y_2 &= a_{21}x_1 + a_{22}x_2 + a_{23}x_3 \\ y_3 &= a_{31}x_1 + a_{32}x_2 + a_{33}x_3 \end{aligned} \tag{1.36}$$

where the a's are the coefficients and y's are the constant terms. The system of equations can be represented in matrix notation as follows:

$$Y = AX \tag{1.37}$$

where

$$A = \begin{bmatrix} a_{11} & a_{12} & a_{13} \\ a_{21} & a_{22} & a_{23} \\ a_{31} & a_{32} & a_{33} \end{bmatrix} \tag{1.38}$$

$$X = \begin{bmatrix} x_1 \\ x_2 \\ x_3 \end{bmatrix} \tag{1.39}$$

$$Y = \begin{bmatrix} y_1 \\ y_2 \\ y_3 \end{bmatrix} \tag{1.40}$$

with A as a 3×3 matrix and X and Y as 3×1 vectors. If matrix A is invertible, it can be given that

$$X = A^{-1}Y \tag{1.41}$$

Take note in Equation (1.41) that since matrix A comes before vector X in Equation (1.37), it must also come before vector Y when transferred to the other side of the equality sign as shown in Equation (1.41); it will be wrong to give $X = YA^{-1}$. This rule must be obeyed with regard to any matrix manipulation.

1.6.3 Basic Matrix Operations

Given any two matrices A and B, the following basic matrix operations are valid:

a) $\left(A^T\right)^T = A$ (where A^T means a transpose of matrix A) $\tag{1.42}$

b) $(AB)^T = B^T A^T$ $\tag{1.43}$

c) $(A + B)^T = A^T + B^T$ $\tag{1.44}$

Given two square matrices A and B, the following can be established:

a) $\left(A^{-1}\right)^{-1} = A$ $\tag{1.45}$

c) $(AB)^{-1} = B^{-1}A^{-1}$ $\tag{1.46}$

Given a scalar $\varphi = v^T P v$ where v is a vector and P is a matrix, the following partial differentiation of the scalar is valid:

i) $\dfrac{\partial \varphi}{\partial v} = \dfrac{\partial(v^T P v)}{\partial v} + \dfrac{\partial(v^T P v)^T}{\partial v} \rightarrow \dfrac{\partial \varphi}{\partial v} = \dfrac{\partial(v^T P v)}{\partial v} + \dfrac{\partial(v^T P^T v)}{\partial v}$ $\tag{1.47}$

or

$$\frac{\partial \varphi}{\partial v} = v^T P + v^T P^T \rightarrow \frac{\partial \varphi}{\partial v} = v^T \left(P + P^T\right) \tag{1.48}$$

ii) For a symmetric matrix P, then $P = P^T$:

$$\frac{\partial \varphi}{\partial v} = v^T \left(P + P^T\right) \rightarrow \frac{\partial \varphi}{\partial v} = 2v^T P \tag{1.49}$$

Given a scalar $\varphi = v^T P v$ where v is a vector and P is a matrix, the following partial differentiation of the scalar is valid:

i) $\dfrac{\partial \varphi}{\partial v^T} = \dfrac{\partial (v^T P v)}{\partial v^T} + \dfrac{\partial (v^T P v)^T}{\partial v^T} \rightarrow \dfrac{\partial \varphi}{\partial v^T} = \dfrac{\partial (v^T P v)}{\partial v^T} + \dfrac{\partial (v^T P^T v)}{\partial v^T}$ (1.50)

or

$$\dfrac{\partial \varphi}{\partial v^T} = Pv + P^T v \rightarrow \dfrac{\partial \varphi}{\partial v^T} = \left(P + P^T \right) v$$ (1.51)

ii) For a symmetric matrix P, then $P = P^T$:

$$\dfrac{\partial \varphi}{\partial v^T} = \left(P + P^T \right) v \rightarrow \dfrac{\partial \varphi}{\partial v^T} = 2Pv$$ (1.52)

Given a scalar $\alpha = y^T A x$ where y is a vector of size $m \times 1$, x is a vector of size $n \times 1$, and A is a matrix of size $m \times n$ (but is independent of x and y), the following partial differentiation of the scalar is valid:

i) $\dfrac{\partial \alpha}{\partial x} = y^T A$ (1.53)

ii) $\dfrac{\partial \alpha}{\partial y} = x^T A^T$ (1.54)

1.7 Covariance, Cofactor, and Weight Matrices

Standard deviations (discussed in Section 1.6) are derived from variances when dealing with one-dimensional type of observable; in the case of multidimensional types of observables, variance–covariance matrices will replace variances. When observables are stored in vector forms, it is usually convenient to store their variances and covariances in matrix forms. For example, let a vector of observables ℓ consisting of distance, angle, and elevation difference observables be represented symbolically as follows:

$$\ell = \begin{bmatrix} \ell_1 \\ \ell_2 \\ \vdots \\ \ell_n \end{bmatrix}$$ (1.55)

where ℓ_i (for $i = 1, 2, ..., n$) is the sub-vector of measurements of the ith observable in the vector ℓ. Let the vector of the true values of the measurements of the observables be

$$\mu = \begin{bmatrix} \mu_1 \\ \mu_2 \\ \vdots \\ \mu_n \end{bmatrix} \tag{1.56}$$

where μ_i (for $i = 1, 2, ..., n$) is the true value of the ith observable in the vector μ. Using the concepts in Section 1.6, the vector of true errors can be given as

$$\varepsilon = \ell - \mu \tag{1.57}$$

and the variance–covariance matrix Σ (the matrix of the true values of the product of vector of true errors) can be given as follows:

$$\Sigma = E\left(\varepsilon\varepsilon^T\right) \tag{1.58}$$

or

$$\Sigma = \begin{bmatrix} \sigma_{11}^2 & \sigma_{12} & \cdots & \sigma_{1n} \\ \sigma_{21} & \sigma_{22}^2 & \cdots & \sigma_{2n} \\ \cdots & \cdots & \cdots & \cdots \\ \sigma_{n1} & \sigma_{2n} & \cdots & \sigma_{nn}^2 \end{bmatrix} \tag{1.59}$$

where $E(\cdot)$ is a linear operator known as the mathematical expectation or the long-term average value of what is in the bracket (more on this is explained in Section 7.2), the principal diagonal elements of the variance–covariance matrix are the *variances* of the corresponding quantities, and the off-diagonal elements represent the *covariances*. Usually, variance–covariance matrix is simply referred to as *covariance matrix* for the sake of brevity and convenience. As it can be seen in Equation (1.59), the matrix is symmetric, meaning that $\sigma_{ij} = \sigma_{ji}$, e.g. $\sigma_{12} = \sigma_{21}$. Following the approach expressed in Equations (1.3) and (1.4), the true error can be expressed as

$$\varepsilon = \ell - \bar{\ell} + \delta \tag{1.60}$$

or

$$\varepsilon = R + \delta \tag{1.61}$$

where $R = \ell - \bar{\ell}$ is the vector of residual errors (the random component) and δ is a vector of systematic errors (biases or the non-stochastic component).

The true covariance matrix given in Equation (1.59) is usually not known, but the estimate can be made using the residual error vector R instead of the true error vector ε as given in Equation (1.61). With regard to Equations (1.60) and (1.61), let the vector of the mean values of the measurements of the observables in x expressed by Equation (1.55) be given as follows:

$$\bar{\ell} = \begin{bmatrix} \bar{\ell}_1 \\ \bar{\ell}_2 \\ \vdots \\ \bar{\ell}_n \end{bmatrix} \tag{1.62}$$

where $\bar{\ell}_i$ (for $i = 1, 2, ..., n$) is the mean value of the sub-vector of measurements of the ith observable in the vector $\bar{\ell}$. From Equations (1.60) and (1.61), the vector of residual errors can be given as

$$R = \ell - \bar{\ell} \tag{1.63}$$

The estimated variance–covariance matrix (C) of observations can be given as the matrix of the product of vector of residual errors, given as follows:

$$C = E\left(RR^T\right) \tag{1.64}$$

or

$$C = \begin{bmatrix} s_{11}^2 & s_{12} & \cdots & s_{1n} \\ s_{21} & s_{22}^2 & \cdots & s_{2n} \\ \cdots & \cdots & \cdots & \cdots \\ s_{n1} & s_{2n} & \cdots & s_{nn}^2 \end{bmatrix} \tag{1.65}$$

where $E(\cdot)$ is the mathematical expectation of the long-term average of what is in the bracket, the principal diagonal elements of the estimated variance–covariance matrix are the sample *variances* of the corresponding quantities, and the off-diagonal elements represent the sample *covariances*. Usually, the matrix in Equation (1.65) is symmetric with $s_{ij} = s_{ji}$, e.g. $s_{12} = s_{21}$. The variances $(s_{11}^2, s_{22}^2, ..., s_{nn}^2)$ are measures of precisions of the observations $(\ell_1, \ell_2, ..., \ell_n)$, respectively; and $s_{11}, s_{22}, ..., s_{nn}$ (with no squares on them) are known as standard deviations or SE. Covariances usually indicate the nature of the relationship between the corresponding observation sets. For example, covariance s_{12} indicates the nature of relationship between the sub-vector of observations ℓ_1 and the sub-vector of observations ℓ_2. If the covariance s_{12}, for example, is zero, it means there is no relationship or dependency between the two sub-vectors of observations. Covariance of nonzero value means relationship exists between the sub-vectors. In general, a relationship can be said to exist between two sets of measurements if the measurements are made with the same types of instruments or if the measurements are made under similar atmospheric conditions. This relationship between sets of measurements, however, can be better illustrated using *Pearson product-moment correlation coefficient r*. A correlation indicates how much a change in value of one quantity will be reflected in the value of another. For example, the Pearson product-moment correlation coefficient (r_{ij}) between sub-vector of observations x_i and the sub-vector of observations x_j can be expressed as follows:

$$r_{ij} = \frac{s_{ij}}{s_{ii}s_{jj}} \tag{1.66}$$

where r_{ij} varies between -1 and $+1$, inclusive. For example, if r_{ij} is equal to -1 or $+1$, the relationship is considered to be very (negatively or positively) strong and linear; if r_{ij} is equal to zero, it is considered that there is no relationship between the two sets of measurements. It should be remembered, however, that the Pearson correlation coefficient in Equation (1.66) only measures linear relationships. A correlation coefficient of zero does not necessarily mean zero relationship between the two tested variables; it only means that the two variables have *zero linear relationship*, but there could be other types of relationship, such as *curvilinear relationship, moderating relationship*, and *mediating relationship*. Two types of correlation can be identified between measurement sets: *physical correlation* and *mathematical correlation*. Physical correlation is the correlation between pairs of measurement due to common effects of physical processes, such as common atmospheric conditions (changing temperature), using the same instrumentation or making the measurements by the same observers. Parameters are said to be *mathematically correlated* when their calculations have at least one observation in common or if the correlation occurs through error propagation or least squares procedure. In surveying, measurement of a single observable is limited, and measurements are usually considered uncorrelated (i.e. covariance matrix has zero non-diagonal values). Example 1.4 is to show how to determine standard deviations and correlation coefficients from a given covariance matrix; at this stage, readers are advised not to try to figure out how this variance–covariance matrix is formed, but to concentrate on understanding how to extract standard deviations and covariances from covariance matrices.

Example 1.4   Consider a case where two distances of a triangle are measured as 10.2 and 20.4 m and the included horizontal angle is measured as $89°30'00''$. The variance–covariance matrix C of these measurements (in the order given above) is given as follows:

$$C = \begin{bmatrix} 0.01 & 0.008 & 0.000\ 002 \\ 0.008 & 0.04 & 0.000\ 004 \\ 0.000\ 002 & 0.000\ 004 & 0.000\ 02\ \text{rad}^2 \end{bmatrix}$$

Calculate the standard deviations of the measurements and the correlation coefficient between the two measured distances.

Solution:

It should be mentioned here that whenever angular (or directional) measurements and linear (distance) measurements are being used together in an adjustment, the variances of the angular (or directional) measurements must be expressed in radians so as to make their values consistent with those of the linear measurements; the corresponding angular (or directional) measurements must also be expressed in radians. As it can be seen in the above covariance matrix C, the last diagonal element (variance of the angle) is given in radian squared for the angular measurement.

Standard deviations:

$$\text{Distance } 10.2 \text{ m, standard deviation} = \sqrt{0.01} \rightarrow \pm 0.10 \text{ m}$$

$$\text{Distance } 20.4 \text{ m, standard deviation} = \sqrt{0.04} \rightarrow \pm 0.20 \text{ m}$$

$$\text{Angle } 89°30'00'', \text{standard deviation} = \sqrt{0.000\ 02 \text{ rad}^2} \rightarrow$$

$$\pm 0.004\ 47 \text{ rad } \left(\text{or } 0°15'22''\right)$$

Correlation coefficient between the two distances from Equation (1.66):

$$r_{12} = \frac{0.008}{0.10 \times 0.20} = 0.4$$

The computed correlation coefficient (0.4) indicates that the relationship between the two measured distances is weak; the value should be greater than 0.5 or less than −0.5 in order to be considered strong.

1.7.1 Covariance and Cofactor Matrices

In practical application of adjustment procedures, covariance matrix (C) is replaced by the relative covariance matrix called *cofactor matrix of observations* (Q). The relationship between the two matrices can be given as follows:

$$C = kQ \tag{1.67}$$

where k is an arbitrary scalar constant (with no unit) or a factor for scaling the cofactor matrix of observations to make it equivalent to the covariance matrix. For a fully populated covariance matrix

$$C = \begin{bmatrix} s_{11}^2 & s_{12} & \cdots & s_{1n} \\ s_{21} & s_{22}^2 & \cdots & s_{2n} \\ \cdots & \cdots & \cdots & \cdots \\ s_{n1} & s_{2n} & \cdots & s_{nn}^2 \end{bmatrix} \tag{1.68}$$

the cofactor matrix of observations can be given as follows:

$$Q = \frac{1}{k} \begin{bmatrix} s_{11}^2 & s_{12} & \cdots & s_{1n} \\ s_{21} & s_{22}^2 & \cdots & s_{2n} \\ \cdots & \cdots & \cdots & \cdots \\ s_{n1} & s_{2n} & \cdots & s_{nn}^2 \end{bmatrix} \tag{1.69}$$

or

$$Q = \begin{bmatrix} q_{11} & q_{12} & \cdots & q_{1n} \\ q_{21} & q_{22} & \cdots & q_{2n} \\ \cdots & \cdots & \cdots & \cdots \\ q_{n1} & q_{2n} & \cdots & q_{nn} \end{bmatrix} \tag{1.70}$$

where, from Equation (1.70), cofactors are $q_{11} = \dfrac{s_{11}^2}{k}$, $q_{12} = \dfrac{s_{12}}{k}$, ..., $q_{nn} = \dfrac{s_{nn}^2}{k}$.

1.7.2 Weight Matrices

The relationship between weight matrix of observations P and cofactor matrix of observations Q can be given as follows:

$$P = Q^{-1} \tag{1.71}$$

or, from Equations (1.67) and (1.71),

$$P = kC^{-1} \tag{1.72}$$

Measurements in surveying are usually considered to be uncorrelated, i.e. covariances are zero. Assuming the corresponding diagonal covariance matrix is given as follows

$$C = \begin{bmatrix} s_{11}^2 & 0 & \cdots & 0 \\ 0 & s_{22}^2 & \cdots & 0 \\ \cdots & \cdots & \cdots & \cdots \\ 0 & 0 & \cdots & s_{nn}^2 \end{bmatrix} \tag{1.73}$$

the weight matrix of observations can be given from Equation (1.72) as

$$P = k \begin{bmatrix} s_{11}^2 & 0 & \cdots & 0 \\ 0 & s_{22}^2 & \cdots & 0 \\ \cdots & \cdots & \cdots & \cdots \\ 0 & 0 & \cdots & s_{nn}^2 \end{bmatrix}^{-1} \tag{1.74}$$

or after the inversion, the following will be obtained:

$$
P = \begin{bmatrix}
p_{11} & 0 & \cdots & 0 \\
0 & p_{22} & \cdots & 0 \\
\cdots & \cdots & \cdots & \cdots \\
0 & 0 & \cdots & p_{nn}
\end{bmatrix}
\tag{1.75}
$$

where the weights of measurements are given as

$$
p_{ii} = \frac{k}{s_{ii}^2} \quad (\text{for } i = 1,2,\ldots,n)
\tag{1.76}
$$

If the weight (p_{ii}) of each observation is set to unity (one), then the arbitrary scalar constant k will be equal to s_{ii}^2 (usually denoted as σ_0^2 and known as the a priori variance factor of unit weight, since it makes the weight equals to one). Usually, at the beginning of least squares adjustment, the value of this variance factor is taken as one, such that the covariance matrix (C) is the same as the cofactor matrix (Q) in Equation (1.67). It should be mentioned also that high weight means that the corresponding precision is high (small variance) and the observation is good. From Equation (1.74), it can be seen that weight matrix of observations can be constructed from the observation standard deviations. If the measurements are of different types, e.g. distances (in meters) and angles (in degrees, minutes, or seconds), the standard deviations of the angles must be converted to radians before they are used in constructing the covariance matrix.

For $p_{ii} = 1$ in Equation (1.76), the value of $k = \sigma_0^2$ becomes the representative standard deviation of an observation of unit weight. If $\sigma_0^2 = 1$ as a convenient initial value, the weight of each observation can be expressed from Equation (1.76) as

$$
p_{ii} = \frac{1}{s_{ii}^2} \quad (\text{for } i = 1,2,\ldots,n)
\tag{1.77}
$$

and an estimate (s_0^2) of k can be calculated as

$$
s_0^2 = \frac{\sum\limits_{i}^{n} p_{ii}\left(\ell_i - \bar{\ell}\right)^2}{n-1}
\tag{1.78}
$$

If s_{ii}^2 is well estimated in Equation (1.76), i.e. $s_{ii}^2 = k$, then s_0^2 will be close to 1. From Equation (1.76), the scaled standard deviation of a particular observation can be given as

$$
\hat{s}_{ii} = \frac{s_0}{\sqrt{p_{ii}}} \quad (\text{for } i = 1,2,\ldots,n)
\tag{1.79}
$$

and the scaled standard deviation of the mean from Equation (1.12) can be given as

$$\hat{s}_{\bar{x}} = \frac{s_0}{\sqrt{\sum_{i}^{n} p_{ii}}} \tag{1.80}$$

It should also be noted that weights of observations can be based on many factors, such as number of sets of angle measurements, number of repetitions of a measurement, inverses of distance measurements between benchmarks in leveling procedure, etc.

Example 1.5 An angle has been measured independently over five days by different survey crews using the same total station instrument. The mean measurements and the number of sets of measurements for each day are as given in Table 1.5. Answer the following.

a) Calculate the overall mean measurement for the five days.

Solution:

Using Equation (1.12)

$$\bar{\ell} = \frac{\sum_{i}^{n} \ell_i p_{ii}}{\sum_{i}^{n} p_{ii}}$$

$$= 45°00'00'' + \frac{(5 \times 7 + 10 \times 6 + (-2 \times 5) + 8 \times 4 + (-1 \times 3))}{7 + 6 + 5 + 4 + 3}$$

$$= 45°00'00'' + \frac{114}{25} \rightarrow \bar{\ell} = 45°00'04.56''$$

Table 1.5 Field measurements of an angle over five days.

Day	1	2	3	4	5
Mean angle	45°00′05″	45°00′10″	44°59′58″	45°00′08″	44°59′59″
No. of sets	7	6	5	4	3

b) Calculate the representative standard deviation of a single angle observation.

Solution:

Using Equation (1.78)

$$s_0^2 = \frac{\sum_i^n p_{ii}(\ell_i - \bar{\ell})^2}{n-1}$$

$$s_0^2 = \frac{7(0.44)^2 + 6(5.44)^2 + 5(-6.56)^2 + 4(3.44)^2 + 3(-5.56)^2}{5-1}$$

$$s_0 = \sqrt{\frac{534.16}{4}} = 11.56''$$

c) Calculate the standard deviation of the mean observation for the five days.

Solution:

Using Equation (1.80)

$$\hat{s}_{\bar{x}} = \frac{s_0}{\sqrt{\sum_i^n p_{ii}}}$$

$$\hat{s}_{\bar{x}} = \frac{11.56}{\sqrt{7+6+5+4+3}}$$

$$= \frac{11.56}{\sqrt{5}} \text{ or } 2.31''$$

d) Calculate the standard deviation of the mean observation for each day.

Solution:

Using Equation (1.79)

$$\hat{s}_{ii} = \frac{s_0}{\sqrt{p_{ii}}}$$

$$\hat{s}_{11} = \frac{11.56}{\sqrt{7}} \text{ or } 4.37'' \quad \hat{s}_{22} = \frac{11.56}{\sqrt{6}} \text{ or } 4.72''$$

$$\hat{s}_{33} = \frac{11.56}{\sqrt{5}} \text{ or } 5.17'' \quad \hat{s}_{44} = \frac{11.56}{\sqrt{4}} \text{ or } 5.78''$$

$$\hat{s}_{55} = \frac{11.56}{\sqrt{3}} \text{ or } 6.67''$$

e) Assuming the standard deviation for each angle measurement according to the instrument manufacturer's specification is the same as $s_0 = 11.56''$ calculated in (b), answer the following.

i) Calculate the standard deviation of the mean observation for each day.

Solution:

Using Equation (1.9), $\hat{s}_{ii} = \dfrac{s_{ii}}{\sqrt{n}}$ for $s_{ii} = 11.56''$ ($i = 1, 2, \ldots, 5$) with n representing the number of sets of angle measurements for each day, which give the same results as in (d).

ii) Calculate the overall mean measurement for the five days using the standard deviations calculated in e(i) for weighting.

Solution:

Using Equations (1.12) and (1.77)

$$\bar{\ell} = \frac{\sum\limits_{i}^{n} \ell_i p_{ii}}{\sum\limits_{i}^{n} p_{ii}}$$

$$= 45°00'00''$$

$$+ \frac{\left(5 \times \left(\dfrac{1}{4.37}\right)^2 + 10 \times \left(\dfrac{1}{4.72}\right)^2 + \left(-2 \times \left(\dfrac{1}{5.17}\right)^2\right) + 8 \times \left(\dfrac{1}{5.78}\right)^2 + \left(-1 \times \left(\dfrac{1}{6.67}\right)^2\right)\right)}{\left(\dfrac{1}{4.37}\right)^2 + \left(\dfrac{1}{4.72}\right)^2 + \left(\dfrac{1}{5.17}\right)^2 + \left(\dfrac{1}{5.78}\right)^2 + \left(\dfrac{1}{6.67}\right)^2}$$

$$= 45°00'00'' + \frac{0.852\ 845\ 71}{0.187\ 073\ 84} \rightarrow \bar{\ell} = 45°00'04.56''$$

iii) Calculate the standard deviation of the mean observation for the five days.

Solution:

Using Equation (1.12)

$$\hat{s}_{\bar{x}} = \frac{1}{\sqrt{\sum\limits_{i}^{n} p_{ii}}}$$

$$\hat{s}_{\bar{x}} = \frac{1}{\sqrt{0.187\ 073\ 84}} = 2.31''$$

iv) What are main observations made by comparing the results in (a), (c), and (e)?

Solution:

If the standard deviations of the observations are well estimated as stated in (e), the weighted means and their standard deviations will be consistent.

Example 1.6 Three different leveling lines were run between two benchmarks BMA and BMX in order to determine the elevation difference between them as shown in Table 1.6 and Figure 1.1. Answer the following questions.

Table 1.6 Leveling field notes for three lines.

Line	Observed elevation difference (m)	Length (km)
1	−21.270	2.5
2	−21.200	2.0
3	−21.290	2.25

a) Calculate the overall mean elevation difference between the two benchmarks BMA and BMX.

Solution:

Using Equation (1.12)

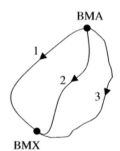

Figure 1.1 Leveling lines between two benchmarks.

$$\bar{\ell} = \frac{\sum\limits_{i}^{n} \ell_i p_{ii}}{\sum\limits_{i}^{n} p_{ii}}$$

$$\bar{\ell} = \frac{\left(\left(-21.270 \times \dfrac{1}{2.5}\right) + \left(-21.200 \times \dfrac{1}{2.0}\right) + \left(-21.290 \times \dfrac{1}{2.25}\right)\right)}{\dfrac{1}{2.5} + \dfrac{1}{2.0} + \dfrac{1}{2.25}}$$

$$= \frac{-28.570\ 222}{1.344\ 44} \rightarrow \bar{\ell} = -21.251\,\text{m}$$

b) Calculate the representative standard deviation of a single elevation difference.

Solution:

Using Equation (1.78)

$$s_0^2 = \frac{\sum_{i}^{n} p_{ii}\left(\ell_i - \bar{\ell}\right)^2}{n-1}$$

$$s_0^2 = \frac{\dfrac{1}{2.5}(-0.019)^2 + \dfrac{1}{2.0}(0.051)^2 + \dfrac{1}{2.25}(-0.039)^2}{3-1}$$

$$s_0 = \sqrt{\frac{0.002\ 120\ 9}{2}} = 0.0326\,\text{m}$$

c) Calculate the standard deviation of the mean elevation difference.

Solution:

Using Equation (1.80)

$$\hat{s}_{\bar{x}} = \frac{s_0}{\sqrt{\sum_{i}^{n} p_{ii}}}$$

$$\hat{s}_{\bar{x}} = \frac{0.032\ 6}{\sqrt{1.344\ 44}} = 0.028\ 1\,\text{m}$$

d) Calculate the standard deviation of each elevation difference.

Solution:

Using Equation (1.79)

$$\hat{s}_{ii} = \frac{s_0}{\sqrt{p_{ii}}}$$

$$\hat{s}_{11} = \frac{0.032\ 6}{\sqrt{0.4}}\text{or}\ 0.051\ 5\,\text{m} \qquad \hat{s}_{22} = \frac{0.032\ 6}{\sqrt{0.500}}\text{or}\ 0.046\ 1\,\text{m}$$

$$\hat{s}_{33} = \frac{0.032\ 6}{\sqrt{0.444\ 44}}\text{or}\ 0.048\ 9\,\text{m}$$

Problems

1.1 The horizontal coordinates of points B (x_B, y_B) and C (x_C, y_C) are to be determined in a traverse survey. In order to determine those coordinates, one bearing (a_1), two angles (a_2, a_3), and one distance (d_1) were measured as shown in the diagram below.

Figure P1.1 (Not to scale.)

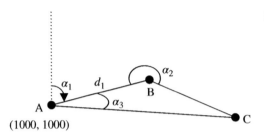

A
(1000, 1000)

The average values of the measurements and their corresponding standard deviations are

$$\hat{a}_1 = 71°29'13'', \quad s_{a_1} = 10''$$
$$\hat{a}_2 = 251°14'50'', \quad s_{a_2} = 5''$$
$$\hat{a}_3 = 35°40'25'', \quad s_{a_3} = 5''$$
$$\hat{d}_1 = 561.025\,\text{m}, \quad s_{d_1} = 0.005\,\text{m}$$

a) Give the covariance matrix C_ℓ of the measurements assuming the angles a_2 and a_3 are correlated by 0.6 and the other measurements are not (provide the matrix elements in simplified numerical values in the order a_1, a_2, a_3, d_1).
b) Explain what will be different in the covariance matrix given in (a) if all the measurements are stochastically independent.
c) If the variance factor of unit weight is 0.86, what is the cofactor matrix (provide the matrix elements in simplified numerical values in the order a_1, a_2, a_3, d_1)?
d) Give the weight matrix (P) of the measurements (providing simplified numerical elements).

1.2 Given the vector of unknown parameters, $\mathbf{x} = [x_B \ y_B \ x_C \ y_C]^T$, and the covariance matrix of the vector of the adjusted parameters ($\hat{x}$) as follows

$$C_{\hat{x}} = \begin{bmatrix} 8.037\ 972\ 7\text{E}-5 & -4.980\ 956\text{E}-5 & 8.037\ 972\ 7\text{E}-5 & -4.980\ 956\text{E}-5 \\ -4.980\ 956\text{E}-5 & 8.071\ 143\text{E}-5 & -4.980\ 956\ 2\text{E}-5 & 8.071\ 143\text{E}-5 \\ 8.037\ 972\ 7\text{E}-5 & -4.980\ 956\ 2\text{E}-5 & 2.904\ 034\ 4\text{E}-4 & 1.178\ 342\ 2\text{E}-4 \\ -4.980\ 956\text{E}-5 & 8.071\ 143\text{E}-5 & 1.178\ 342\ 2\text{E}-4 & 1.406\ 992\ 1\text{E}-3 \end{bmatrix} \text{m}^2$$

compute the correlation coefficient for each of the following pairs of coordinates: (x_B, y_B) and (y_C, y_B).

1.3 An angle has been measured independently five times with the same precision and the observed values are given in the following table.

α_1	α_2	α_3	α_4	α_5
45°00′05″	45°00′10″	44°59′58″	45°00′08″	44°59′59″

a) Calculate the sample standard deviation and the standard deviation of the mean (showing your steps).
b) If one of the measurements given in the table becomes too high because of a constant error, explain how the error will affect the mean and standard deviation estimated in (a).
c) If each of the measurements given in the table becomes too high because of a constant systematic error (of equal magnitude), explain how the error will affect the mean and standard deviation estimated in (a).

1.4 A baseline was measured with a tape and also with total station equipment. Each measuring procedure is repeated five times and summarized as follows: mean distance (tape) = 200.15 m, mean distance (total station) = 200.183 m, and the standard deviations for the tape and the total station measurements are ±0.03 and 0.008, respectively.
a) Which of the measurement sets is more precise? Why?
b) Is it possible to determine which of the two instruments measures more accurately based only on the data given above? Why?
c) The measurements taken with the total station are 20 mm and are found later to be too high because of prism constant error. After correcting these measurements for prism constant, the new computed mean value is 200.163, and the standard deviation of measurements is now ±0.008 m.
   i) Comparing the new results in (c) with the old ones above, explain (using your deduction from the results only) if systematic error affects accuracy.
   ii) Comparing the new results in (c) with the old ones above, explain (using your deduction from the results only) if systematic error affects precision.

1.5 The following field notes are the horizontal direction measurements to targets 1769, 1770, and 1765 when the Nikon NLP 352 total station was set up on a survey station LP1510.

Direction measurements from station LP1510.

Set no.	Station sighted	FL	FR
1	1769	0°00′00″	0°00′00″
	1770	29°43′24″	29°43′27″
	1765	134°34′22″	134°34′27″
2	1769	0°00′00″	0°00′00″
	1770	29°43′25″	29°43′17″
	1765	134°34′21″	134°34′13″
3	1769	0°00′00″	0°00′00″
	1770	29°43′30″	29°43′19″
	1765	134°34′23″	134°34′10″

Determine the standard deviation of a direction measurement and the standard deviation of the mean direction measurement (to one decimal second).

1.6 In order to determine the length of one baseline, it is measured 10 times (providing 10 baseline measurements). Answer the following questions based on this problem.

a) How many observables are in this problem? Explain the reason for your answer.

b) Explain (providing clear reason) if there is a need for adjustment in this problem.

1.7 The three angles and the three lengths of a triangle were measured by a surveyor. Explain with clear reasons if these measurements will need adjustment, e.g. determine if there are redundant measurements and the possible number of redundancies.

1.8 The cofactor matrix of the adjusted coordinates (x, y) of a point is given as follows, and the variance factor of unit weight is 0.8:

$$Q_{\hat{x}} = \begin{bmatrix} 0.05 & 0.004 \\ 0.004 & 0.06 \end{bmatrix}$$

a) Compute the weight matrix of the coordinates and give the weights of the coordinates x and y.

b) Compute the variance–covariance matrix of the coordinates and give the standard deviations of the coordinates x and y.

1.9 Four different leveling lines were run between two benchmarks BM1 and BM2 in order to determine the elevation difference between them as shown in the following table. Answer the following using the inverses of lengths (in km) as weights.

Line	Elevation difference (m)	Length (km)
1	5.405	1.5
2	5.423	3.0
3	5.414	5.0
4	5.392	9.0

a) Calculate the overall mean elevation difference between the two benchmarks.
b) Calculate the representative standard deviation of a single elevation difference.
c) Calculate the standard deviation of the mean elevation difference.
d) Calculate the standard deviation of each elevation difference.

2

Analysis and Error Propagation of Survey Observations

OBJECTIVES

After studying this chapter, you should be able to:

1) Formulate functional models for evaluating observations and unknown parameters.
2) Linearize functional models by Taylor's series.
3) Perform systematic error propagation in measurements.
4) Perform random error (variance–covariance) propagation in measurements.
5) Use heuristic rule to perform covariance propagation between measurements.

2.1 Introduction

In geomatics engineering, observables (distances, angles, azimuth, etc.) are sometimes measured in order to determine more useful quantities (the unknown parameters) such as coordinates of points, calibration parameters, transformation constants, etc. Different types of mathematical models are used

Understanding Least Squares Estimation and Geomatics Data Analysis, First Edition.
John Olusegun Ogundare.
© 2019 John Wiley & Sons, Inc. Published 2019 by John Wiley & Sons, Inc.
Companion website: www.wiley.com/go/ogundare/Understanding-lse-and-gda

in relating measurements of such observables to their unknown parameters; one of such models relates unknown parameters (x) directly as functions of field measurements (ℓ). When a vector of parameters $x = (x_1, x_2, x_3, ..., x_m)^T$ is explicitly expressed as a function of field measurements $\ell = (\ell_1, \ell_2, ..., \ell_n)^T$, the mathematical model so formed is referred to as *special form of general model*. This type of model can be expressed as follows:

$$x = f(\ell) \tag{2.1}$$

where the vector of parameters (x) is the quantity whose error property (the attribute of x) is to be determined. Since field measurements usually contain errors (systematic and random errors), the parameters computed from them will equally have their own share of the errors, propagated into them. The process of evaluating the errors propagated into the computed quantities is known as *error propagation*. Evaluation of random errors is referred to as *random error* (or *variance–covariance*) propagation; the evaluation of systematic errors will be referred to as *systematic error propagation*. The error propagation type discussed in this chapter relates to linear functions of observations. Since linear equations in the forms of Equation (2.1) is not often encountered in practice (except in leveling cases where measurements are related to heights of corresponding points), a way of linearizing nonlinear functions is needed in the process of error propagation.

2.2 Model Equations Formulations

Typical geomatics model equations are formulated in this section. In order to evaluate error properties of parameters, one should be able to first formulate the parameters in terms of the observations, as given in Equation (2.1). This is important since the propagation of errors into measurements will be done mathematically. The general steps for formulating model equations for error propagations can be given as follows:

1) The important elements to be manipulated in Equation (2.1) are the parameters (x) and observations (ℓ). There is need for one to be able to identify from a question what constitutes the parameters (x) and what constitutes the observations (ℓ) and to use suitable symbols to represent each one of them; without this, one will not be able to formulate the necessary equations.

2) Arrange the elements representing the parameters (x) and the observations (ℓ) in equation forms such that each equation has one parameter (x) as a subject. Each equation will involve some or all of the observations (with a possibility of some constant values being present), but no other unknown parameters must be found in the equation so as to satisfy $x = f(\ell)$ model. Each of the identified parameters and the observations should be represented in

the equations as symbols with substitution of numerical values into the equations made later, at the solution stage.

On the basis of the above steps, the following examples are given to illustrate some of the common geomatics model equations that may eventually require an application of error propagation rules.

Example 2.1   Two sides of a right-angled triangular area shown in Figure 2.1 are measured as ℓ_1 and ℓ_2; it is required that the area and the hypotenuse of the triangle be determined. Formulate the equations ($x = f(\ell)$) for computing the area and the hypotenuse.

Figure 2.1 Right-angled triangular area.

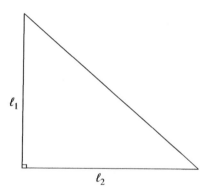

Solution:

The model expected should look like $x = f(\ell)$. The steps provided earlier in this section will be taken as follows:

1) Since the question states that ℓ_1 and ℓ_2 were measured, they represent the observations (ℓ) in symbolic form (not in numerical form) and can be used directly to formulate the equations of area and hypotenuse of the triangle, which are the parameters (x). These parameters can be represented as symbols: A for the area and H for the hypotenuse. The elements of the equations are given in vector forms as follows:

$$x = \begin{bmatrix} A \\ H \end{bmatrix} \quad \ell = \begin{bmatrix} \ell_1 \\ \ell_2 \end{bmatrix}$$

2) Considering the type of model to be formulated ($x = f(\ell)$), it is understood that each parameter is a subject and must form one equation as function

of observations with no other unknown parameters found in that equation. Since there are two parameters, the following two equations are formulated:

$$A = \frac{1}{2}\ell_1 \ell_2 \tag{2.2}$$

$$H = \sqrt{\left(\ell_1^2 + \ell_2^2\right)} \tag{2.3}$$

Example 2.2 In a horizontal traverse survey of a triangular area BAC shown in Figure 2.2, the measurements are the angle (θ) at fixed point A, bearing A–B (α_{AB}), and the distance (S_{AC}). If the coordinates (E_A, N_A) of point A are fixed, give the equations ($x = f(\ell)$) for computing the coordinates (E_C, N_C) of point C.

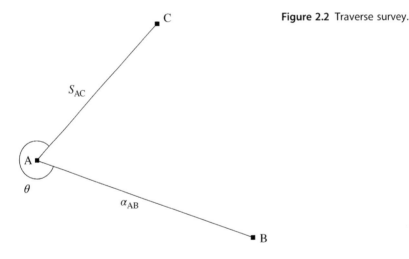

Figure 2.2 Traverse survey.

Solution:

The equations to be formulated should look like $x = f(\ell)$. The solution steps taken in Example 2.1 will be followed in this question:

1) The stated measurements in the question are angle θ, bearing α_{AB}, and distance S_{AC}, which constitute the observations (ℓ). The coordinates (the easting (E_C) and the northing (N_C) of point C are the parameters to be computed. Since the observations are already in symbolic form (not in numerical

form), they can be used directly in formulating the equations as follows. The parameters and observations are given respectively as follows:

$$x = \begin{bmatrix} E_C \\ N_C \end{bmatrix} \qquad \ell = \begin{bmatrix} \theta \\ \alpha_{AB} \\ S_{AC} \end{bmatrix}$$

2) The formulated model equations ($x = f(\ell)$) for the traverse can be given as follows:

$$E_C = E_A + S_{AC} \sin(\alpha_{AB} + \theta) \tag{2.4}$$
$$N_C = N_A + S_{AC} \cos(\alpha_{AB} + \theta) \tag{2.5}$$

It can be seen from the above equations that each parameter forms one equation and only constants (E_A, N_A) and observations are involved on the right-hand sides of the equations, giving a model form of $x = f(\ell)$. Note that the constants are necessary (but not functional) parts of the equations.

Example 2.3 The boundaries of two rectangular lots were determined with measurements of three distances as shown in Figure 2.3. Give the equations for computing the areas (A_1, A_2) of the two lots, expressing them as $x = f(\ell)$.

Figure 2.3 Two rectangular lots of land.

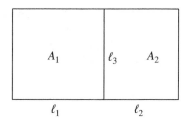

Solution:

The equations to be formed should look like $x = f(\ell)$. The steps in Example 2.1 are similarly applied in this problem as follows. The parameters (x) and observations (ℓ) are as follows:

$$x = \begin{bmatrix} A_1 \\ A_2 \end{bmatrix} \qquad \ell = \begin{bmatrix} \ell_1 \\ \ell_2 \\ \ell_3 \end{bmatrix}$$

Each parameter forms one equation as a function of observations as follows:

$$A_1 = \ell_1 \ell_3 \tag{2.6}$$

$$A_2 = \ell_2 \ell_3 \tag{2.7}$$

Example 2.4   In order to determine the elevation (h_B) of point B, a total station was set up at point A, and the zenith angle z and slope distance s were taken to a target at point B. The height of instrument was measured as h_i, the height of target was measured as h_t, and the elevation of point A is fixed as 30.00 m. Give the model for this problem in the form of $x = f(\ell)$.

Solution:

The equation to be formed must look like $x = f(\ell)$. Applying the same steps as in Example 2.1, the following are obtained:

1) Measured quantities (the observations): zenith angle z, slope distance s, height of instrument h_i, and height of target h_t. The observations can be used directly to determine the elevation (h_B) of point B, which is the only parameter in this problem. The parameter (x) and observations (ℓ) are, respectively, as follows:

$$x = [h_B] \qquad x = [z \ \ s \ \ h_i \ \ h_t]^T$$

2) The model equation ($\hat{x} = f(\hat{\ell})$) can be formulated as a trigonometric leveling problem. The elevation of point A is fixed at 30.0 as a constant value. This constant value is not an observation since no standard deviation is associated with it, and it is also not a parameter since its value is not to be determined. The only parameter in this problem can be expressed as follows:

$$h_B = 30.0 + s \cos z + h_i - h_t \tag{2.8}$$

It can be seen in Equation (2.8) that the numerical value of the constant (fixed elevation of point A) is used directly with no attempt to represent it as symbol, since it is not a parameter and not an observation.

2.3   Taylor Series Expansion of Model Equations

Linearization of nonlinear functional model is usually carried out using *Taylor series*. Taylor series is a representation of a function as a sum of terms determined from the values of its derivatives at approximate values of the parameters involved. Any function satisfying certain criteria can be expressed as a Taylor series. In another words, when a function is nonlinear and its approximate linear form is required, the Taylor series is that approximate linear form. The Taylor series is a useful tool for approximating an original function in linear form when the first two terms (the zero- and first-order terms) in the series are used with the higher-order terms neglected. For example, given a vector of observations $\ell = (\ell_1, \ell_2, ..., \ell_n)^T$ and a vector of parameters $x = (x_1, x_2, x_3, ..., x_m)^T$, with a functional model as in Equation (2.1), the Taylor series expansion of the model can be explained as follows. Let the corrections to be applied to the observations be given as $\Delta \ell = (\Delta \ell_1, \Delta \ell_2, ..., \Delta \ell_n)^T$, and let the corresponding corrections to the parameters be $\Delta x = (\Delta x_1, \Delta x_2, \Delta x_3, ..., \Delta x_m)^T$; Equation (2.1) can be modified to give

$$x + \Delta x = f(\ell + \Delta \ell) \tag{2.9}$$

where $\hat{p} = x + \Delta x$ is the vector of adjusted parameters or adjusted observations. The Taylor series expansion of Equation (2.9) about the observations ℓ can be given as

$$x + \Delta x = f(\ell) + \frac{\partial f}{\partial \ell}(\Delta \ell) \tag{2.10}$$

where $f(\ell)$ is the zero-order term of Taylor series, which is the original function evaluated at the given initial numerical values (ℓ) of the observations, $\Delta \ell$ is a vector of corrections (or errors, following the convention of representing correction as the negative error) to observations, $\frac{\partial f}{\partial \ell}(\Delta \ell)$ is the first-order term of the Taylor series, and $\frac{\partial f}{\partial \ell}$ is evaluated at the initial numerical values of the observations and is otherwise known as the Jacobian (J) matrix that can be expressed as follows:

$$J = \begin{bmatrix} \dfrac{\partial x_1}{\partial \ell_1} & \dfrac{\partial x_1}{\partial \ell_2} & \dfrac{\partial x_1}{\partial \ell_3} & \cdots & \dfrac{\partial x_1}{\partial \ell_n} \\[2ex] \dfrac{\partial x_2}{\partial \ell_1} & \dfrac{\partial x_2}{\partial \ell_2} & \dfrac{\partial x_2}{\partial \ell_3} & \cdots & \dfrac{\partial x_2}{\partial \ell_n} \\[2ex] \dfrac{\partial x_3}{\partial \ell_1} & \dfrac{\partial x_3}{\partial \ell_2} & \dfrac{\partial x_3}{\partial \ell_3} & \cdots & \dfrac{\partial x_3}{\partial \ell_n} \\[2ex] \vdots & \vdots & \vdots & \vdots & \vdots \\[2ex] \dfrac{\partial x_m}{\partial \ell_1} & \dfrac{\partial x_m}{\partial \ell_2} & \dfrac{\partial x_m}{\partial \ell_3} & \cdots & \dfrac{\partial x_m}{\partial \ell_m} \end{bmatrix} \tag{2.11}$$

From Equations (2.1) and (2.10), it can be shown that

$$\Delta x = J(\Delta \ell) \tag{2.12}$$

The type of approximation given in Equation (2.10) is known as first-order approximation by Taylor series. Linearization of functional models using the zero- and first-order terms only in a Taylor series expansion is usually carried out to simplify the numerical solution for parameters and corrected measurements estimated by the least squares procedure. However, the expansion will only be accurate (producing the best values of the parameters) if the second- and higher-order terms of the expansion are insignificant. In this case, the corrections to parameters (Δx) in Equation (2.12) will be accurate if initial values of observations (ℓ) are close to their true values, so that the corrections ($\Delta \ell$) are so small that they can be considered to be at the magnitude of the expected errors. Since initial measurements in surveying are usually close to what their true values should be (after necessary corrections have been made and appropriate precautions have been taken), the corrections can be considered to be at the magnitude of the expected errors. The following examples demonstrate how the linearization method is applied.

Example 2.5 The following field measurements were made from two control points A ($x = 1000.000$ m, $y = 1000.000$ m) and B ($x = 1086.603$, $y = 1050.000$ m) in order to determine the (x, y) coordinates of point C with regard to the survey network in Figure 2.4:

Distance A–C (ℓ_{AC}) : 200.000 m

Distance B–C (ℓ_{BC}) : 124.000 m

Angle C–A–B : 30°02′55″

Angle A–B–C : 126°08′03″

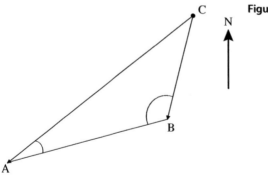

Figure 2.4 Survey network.

The azimuth of line A–B is calculated from the coordinates of the control points A and B as 60°00′00″; the formulas for computing the average coordinates (x, y) of point C are expressed by the following two nonlinear equations, where ℓ_{AC}, ℓ_{AC}, A, and B are the measurements:

$$x = 1043.3015 + \frac{\ell_{AC}}{2} \sin(60° - A) + \frac{\ell_{BC}}{2} \sin(240° + B) \qquad (2.13)$$

$$y = 1025 + \frac{\ell_{AC}}{2} \cos(60° - A) + \frac{\ell_{BC}}{2} \cos(240° + B) \qquad (2.14)$$

Answer the following:

a) Determine the coordinates (x, y) of point A by substituting the field measurements into Equations (2.13) and (2.14).

   Solution:

$$x = 1043.3015 + \frac{200}{2} \sin(60° - 30°02'55'') + \frac{124}{2} \sin(240° + 126°08'03'')$$

$$x = 1043.3015 + 49.9265 + 6.6251 \,(\text{or } \mathbf{1099.8531\,m})$$

$$(2.15)$$

$$y = 1025 + \frac{200}{2} \cos(60° - 30°02'55'') + \frac{124}{2} \cos(240° + 126°08'03'')$$

$$y = 1025 + 86.6449 + 61.6450 \,(\text{or } \mathbf{1173.2899\,m})$$

$$(2.16)$$

b) Linearize the Equations (2.13) and (2.14) using Taylor series expansion about approximate values of the measurements (with unknown corrections to be applied) given as follows:

   Distance $A-C\,(\ell_{AC}^0) : 200.500\,\text{m}$ with correction $\Delta\ell_{AC}$

   Distance $B-C\,(\ell_{BC}^0) : 123.500\,\text{m}$ with correction $\Delta\ell_{BC}$

   Angle $A' : 30°03'00''$ with correction ΔA

   Angle $B' : 126°08'00''$ with correction ΔB

   Solution:

   The linearized model will be in the form of Equation (2.10):

$$\hat{p} = f(\ell) + \frac{\partial f}{\partial \ell}(\Delta\ell)$$

where $\ell = \begin{bmatrix} \ell_{AC}^0 & \ell_{BC}^0 & A' & B' \end{bmatrix}^T$ is a vector of approximate values of the measurements, $\hat{p} = [\hat{x} \ \hat{y}]^T$ is a vector of the adjusted coordinates of point C, $\Delta\ell = [\Delta\ell_{AC} \ \Delta\ell_{BC} \ \Delta A \ \Delta B]$ is a vector of corrections to be applied to the measurements, and $\frac{\partial f}{\partial \ell}$ is the Jacobian matrix, which can be expressed as follows:

$$\frac{\partial f}{\partial \ell} = J = \begin{bmatrix} \dfrac{\partial x}{\partial \ell_{AC}} & \dfrac{\partial x}{\partial \ell_{BC}} & \dfrac{\partial x}{\partial A} & \dfrac{\partial x}{\partial B} \\[2ex] \dfrac{\partial y}{\partial \ell_{AC}} & \dfrac{\partial y}{\partial \ell_{BC}} & \dfrac{\partial y}{\partial A} & \dfrac{\partial y}{\partial B} \end{bmatrix}$$

or

$$J = \begin{bmatrix} \dfrac{\sin(60° - A)}{2} & \dfrac{\sin(240° + B)}{2} & \dfrac{-\ell_{AC}\cos(60° - A)}{2} & \dfrac{\ell_{BC}\cos(240° + B)}{2} \\[3ex] \dfrac{\cos(60° - A)}{2} & \dfrac{\cos(240° + B)}{2} & \dfrac{\ell_{AC}\sin(60° - A)}{2} & \dfrac{-\ell_{BC}\sin(240° + B)}{2} \end{bmatrix}$$

$$(2.17)$$

When the Jacobian matrix J is evaluated at the approximate values of the measurements, the following is obtained:

$$J = \begin{bmatrix} 0.2496 & 0.0534 & -86.8628 & 61.3965 \\ 0.4332 & 0.4971 & 50.0492 & -6.5975 \end{bmatrix}$$

The $f(\ell)$ is a vector of Equations (2.13) and (2.14) evaluated at the approximate values of the measurements and is given as

$$f(\ell) = \begin{bmatrix} 1099.9482 \\ 1173.2593 \end{bmatrix}$$

The linearized equations (Taylor series expansion) can be expressed as follows:

$$\hat{p} = \begin{bmatrix} 1099.9482 \\ 1173.2593 \end{bmatrix} + \begin{bmatrix} 0.2496 & 0.0534 & -86.8628 & 61.3965 \\ 0.4332 & 0.4971 & 50.0492 & -6.5975 \end{bmatrix} \begin{bmatrix} \Delta\ell_{AC} \\ \Delta\ell_{BC} \\ \Delta A \\ \Delta B \end{bmatrix}$$

$$(2.18)$$

(c) Substitute the corrections $\Delta\ell_{AC} = -0.500$ m, $\Delta\ell_{BC} = +0.500$ m, $\Delta A = -5''$, and $\Delta B = +3''$ into the linearized series expansion, and compare the results with those obtained in (a).

Solution:

Expressing the angular values in radians, the corrections can be given as

$$\begin{bmatrix} \Delta\ell_{AC} \\ \Delta\ell_{BC} \\ \Delta A \\ \Delta B \end{bmatrix} = \begin{bmatrix} -0.500 \\ 0.500 \\ -2.424\ 066\ 13\mathrm{E}-5 \\ 1.454\ 439\ 68\mathrm{E}-5 \end{bmatrix}$$

Substituting into Equation (2.18) gives

$$\hat{p} = \begin{bmatrix} 1099.9482 \\ 1173.2593 \end{bmatrix} + \begin{bmatrix} -0.0951 \\ 0.0306 \end{bmatrix} \rightarrow \begin{bmatrix} \hat{x} \\ \hat{y} \end{bmatrix} = \begin{bmatrix} 1099.8531 \\ 1173.2899 \end{bmatrix} \tag{2.19}$$

It can be seen from Equation (2.19) and the result in (a) that the two results are the same, on the basis that the corrections to the measurements are small.

Example 2.6   The following field measurements were made from two control points A ($x = 1000.000$ m, $y = 1000.000$ m) and B ($x = 1086.603$, $y = 1050.000$ m) in order to determine the (x, y) coordinates of point C with regard to the survey network in Figure 2.5:

Distance $A-C(\ell_{AC})$: 200.000 m

Distance $B-C(\ell_{BC})$: 124.000 m

Figure 2.5 A survey network.

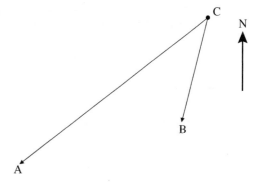

It is required that the following distance equations be used to determine the coordinates (x, y) of point C assuming the initial coordinates for point C are $x' = 1100.000$ m and $y' = 1173.205$ m:

$$d_{AC} = \sqrt{(x-1000)^2 + (y-1000)^2} \tag{2.20}$$

$$d_{BC} = \sqrt{(x-1086.603)^2 + (y-1050)^2} \tag{2.21}$$

Solution:

Equations (2.20) and (2.21) can be presented in Taylor series expansion as follows:

$$\ell = f\left(x^0\right) + \frac{\partial f}{\partial x}(\Delta x) \tag{2.22}$$

where $\ell = [200.000 \quad 124.000]^T$ is a vector of field measurements, $x^0 = [x' \; y']^T$ is a vector of approximate coordinates of point C, $\Delta x = \begin{bmatrix} dx \, dy \end{bmatrix}^T$ is a vector of unknown corrections to be applied to the approximate coordinates, and $\dfrac{\partial f}{\partial x}$ is the Jacobian matrix, which can be expressed as follows:

$$\frac{\partial f}{\partial x} = J = \begin{bmatrix} \dfrac{\partial d_{AC}}{\partial x} & \dfrac{\partial d_{AC}}{\partial y} \\[2mm] \dfrac{\partial d_{BC}}{\partial x} & \dfrac{\partial d_{BC}}{\partial y} \end{bmatrix}$$

or

$$J = \begin{bmatrix} \dfrac{(x-1000)}{\sqrt{(x-1000)^2 + (y-1000)^2}} & \dfrac{(y-1000)}{\sqrt{(x-1000)^2 + (y-1000)^2}} \\[4mm] \dfrac{(x-1086.603)}{\sqrt{(x-1086.603)^2 + (y-1050)^2}} & \dfrac{(y-1050)}{\sqrt{(x-1086.603)^2 + (y-1050)^2}} \end{bmatrix}$$

Substituting the approximate values $x' = 1100.000$ m and $y' = 1173.205$ m into the Jacobian matrix J gives

$$J = \begin{bmatrix} 0.498\ 284 & 0.867\ 013\ 9 \\ 0.107\ 415\ 23 & 0.994\ 214\ 2 \end{bmatrix}$$

Substituting the approximate values into Equations (2.20) and (2.21) gives

$$f(x^0) = \begin{bmatrix} 200.6888 \\ 124.7216 \end{bmatrix}$$

Substituting the above values into Equation (2.22) gives the Taylor series expansion of Equations (2.20) and (2.21) about the approximate values of the coordinates of point C ($x' = 1100.000$ m, $y' = 1173.205$ m) as follows:

$$\begin{bmatrix} 200.000 \\ 124.000 \end{bmatrix} = \begin{bmatrix} 200.6888 \\ 124.7216 \end{bmatrix} + \begin{bmatrix} 0.498\ 284 & 0.867\ 013\ 9 \\ 0.107\ 415\ 23 & 0.994\ 214\ 2 \end{bmatrix} \begin{bmatrix} dx \\ dy \end{bmatrix} \tag{2.23}$$

or

$$200 = 200.6888 + 0.498\ 284dx + 0.867\ 013\ 9dy \tag{2.24}$$

$$124 = 2124.7216 + 0.107\ 415dx + 0.994\ 214dy \tag{2.25}$$

Equation (2.23) can be simplified to give

$$\begin{bmatrix} -0.688\ 8 \\ -0.721\ 6 \end{bmatrix} = \begin{bmatrix} 0.498\ 284 & 0.867\ 013\ 9 \\ 0.107\ 415\ 23 & 0.994\ 214\ 2 \end{bmatrix} \begin{bmatrix} dx \\ dy \end{bmatrix} \tag{2.26}$$

where

$$\begin{bmatrix} dx \\ dy \end{bmatrix} = \begin{bmatrix} 2.471\ 5 & -2.155\ 3 \\ -0.267\ 02 & 1.238\ 68 \end{bmatrix} \begin{bmatrix} -0.688\ 8 \\ -0.721\ 6 \end{bmatrix}$$

or

$$\begin{bmatrix} dx \\ dy \end{bmatrix} = \begin{bmatrix} -0.1471 \\ -0.7099 \end{bmatrix}$$

Adding the corrections to the approximate coordinates of point C gives the adjusted coordinates of point C as

$$\begin{bmatrix} \hat{x} \\ \hat{y} \end{bmatrix} = \begin{bmatrix} x' \\ y' \end{bmatrix} + \begin{bmatrix} dx \\ dy \end{bmatrix} \rightarrow \begin{bmatrix} x \\ y \end{bmatrix} = \begin{bmatrix} 1099.8529 \\ 1173.2901 \end{bmatrix} \text{ m}$$

Back-substitution of these corrected coordinates into Equations (2.20) and (2.21) gives the computed distances as $\hat{d}_{AC} = 200.000\ \text{m}$ and $\hat{d}_{BC} = 124.000\ \text{m}$, which are consistent with the original measurements.

Example 2.7 Figure 2.6 shows a tract of land composed of a semicircle with diameter AF, a rectangle ACDF, and two triangles ABC and FDE. The measured dimensions are $\ell_1 = 50$ m, $\ell_2 = 20$ m, and $\ell_3 = 30$ m. If the total area (A) of the land is expressed as follows, evaluate the Jacobian matrix J for the area:

$$A = \pi \frac{\ell_1^2}{8} + \ell_1 \ell_3 + \ell_2 \ell_3$$

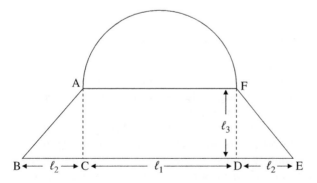

Figure 2.6 Tract of land composed of a semicircle.

Solution:

Given the equation

$$A = \pi \frac{\ell_1^2}{8} + \ell_1 \ell_3 + \ell_2 \ell_3$$

Jacobian matrix is

$$J = \begin{bmatrix} \dfrac{\partial A}{\partial \ell_1} & \dfrac{\partial A}{\partial \ell_2} & \dfrac{\partial A}{\partial \ell_2} \end{bmatrix}$$

$$J = \begin{bmatrix} \left(\dfrac{\pi \ell_1}{4} + \ell_3\right) & \ell_3 & (\ell_1 + \ell_2) \end{bmatrix} \rightarrow J = [69.3 \quad 30 \quad 70]$$

Example 2.8 Given the following direct model for x and y as a function of ℓ_1, ℓ_2, and ℓ_3

$$x = 4\ell_1 - 2\ell_2 + \ell_3$$

$$y = 5\ell_1 + 3\ell_2 - \ell_3$$

express the equations in the form of $x' = J\ell$, giving x' as a vector of x and y, J as the Jacobian matrix, and ℓ as a vector of the observations.

Solution:

Since the two equations are already linear, the following Jacobian matrix is obtained:

$$\begin{bmatrix} x \\ y \end{bmatrix} = \begin{bmatrix} 4 & -2 & 1 \\ 5 & 3 & -1 \end{bmatrix} \begin{bmatrix} \ell_1 \\ \ell_2 \\ \ell_3 \end{bmatrix}$$

2.3.1 Using MATLAB to Determine Jacobian Matrix

MATLAB (**Mat**rix **Lab**oratory), developed by MathWorks, is an interactive computing environment for numerical solution of scientific problems, visualizing results graphically and programming in high-level language. This programming environment will be used in this section to determine the Jacobian of a system of equations. As an example, consider a point P (x_P, y_P), which is to be resected from three control points A, B, and C as shown in Figure 2.7. The coordinates of these control points are given in Table 2.1, and the field measurements are distances d_1, d_2, d_3; bearings PA, PB, PC as α_{PA}, α_{PB}, α_{PC}, respectively; and an angle APB as θ. Formulate the appropriate equations for

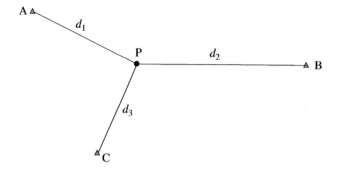

Figure 2.7 Resection of point P.

Table 2.1 Coordinates of control points.

Point	X (m)	Y (m)
A	316.682	505.015
B	500.005	400.040
C	356.804	310.032

these measurements in terms of the coordinates (x_P, y_P) of point P, and give the Jacobian matrix (numerically) of the equations with respect to the coordinates (x_P, y_P), assuming $x_P = 391.165$ m, $y_P = 405.029$ m.

The appropriate equations for the measurements in terms of the coordinates (x_P, y_P) of point P are as follows:

$$\alpha_{PA} = A \tan \left(\frac{x_A - x_P}{y_A - y_P} \right) + 360 \tag{2.27}$$

$$\alpha_{PB} = A \tan \left(\frac{x_B - x_P}{y_B - y_P} \right) + 180 \tag{2.28}$$

$$\alpha_{PC} = A \tan \left(\frac{x_C - x_P}{y_C - y_P} \right) + 180 \tag{2.29}$$

$$\theta = \alpha_{PB} - \alpha_{PA} + 360° = A \tan \left(\frac{x_B - x_P}{y_B - y_P} \right) + 180°$$

$$- \left[A \tan \left(\frac{x_A - x_P}{y_A - y_P} \right) + 360 \right] + 360°$$

or

$$\theta = A \tan\left(\frac{x_B - x_P}{y_B - y_P}\right) - A \tan\left(\frac{316.682 - x_P}{505.015 - y_P}\right) + 180° \tag{2.30}$$

$$d_1 = \sqrt{(x_A - x_P)^2 + (y_A - y_P)^2} \tag{2.31}$$

$$d_2 = \sqrt{(x_B - x_P)^2 + (y_B - y_P)^2} \tag{2.32}$$

$$d_3 = \sqrt{(x_C - x_P)^2 + (y_C - y_P)^2} \tag{2.33}$$

Considering the snippet of MATLAB code given in Table 2.2, **syms** is used in line 1 to declare the variables to be used in equations formulated in lines 2–8, which are consistent with Equations (2.27)–(2.33); in line 9, the **Jacobian** gives

Table 2.2 MATLAB code for determining the Jacobian matrix JM.

```
1:>> syms xA yA xB yB xC yC xP yP

2:>> Apa=atan((xA-xP)/(yA-yP))+2*pi;
3:>> Apb=atan((xB-xP)/(yB-yP))+pi;
4:>> Apc=atan((xC-xP)/(yC-yP))+pi;
5:>> theta=atan((xB-xP)/(yB-yP))-atan((xA-xP)/(yA-yP))
     + pi;
6:>> d1=sqrt((xA-xP)^2+(yA-yP)^2);
7:>> d2=sqrt((xB-xP)^2+(yB-yP)^2);
8:>> d3=sqrt((xC-xP)^2+(yC-yP)^2);
9:>> J=jacobian([Apa;Apb;Apc;theta;d1;d2;d3],[xP yP]);
10:>> xA=316.682;yA=505.015;xB=500.005;yB=400.040;
      xC=356.804;yC=310.032;
11:>> xP=391.165;yP=405.029;
12:>> format long
13:>> JM=eval(J)
JM =
  -0.006432070166759  -0.004791469628055
   0.000420266615083   0.009168534452934
   0.009308769385695  -0.003367039220837
   0.006852336781843   0.013960004080989
   0.597396880060841  -0.801945738621743
  -0.998951094827644   0.045789847593670
   0.340139434154846   0.940375013137217
>>
```

the Jacobian matrix of the given equations with respect to x_P and y_P coordinates of point P; numerical values of the variables are declared in lines 10–11; and line 13 is to evaluate the Jacobian matrix numerically. The Jacobian matrix is given as JM in line 13.

2.4 Propagation of Systematic and Gross Errors

Usually in surveying, measurements are corrected for systematic errors before they are used in any calculation. If this is not done, the systematic errors will be propagated into the calculated quantities in a certain way, which can be derived using the Taylor series linearization approach. Similarly, if gross errors are not detected in observation, the errors will propagate in certain way as in systematic error propagation. By carefully evaluating the linear model in Equation (2.12), it can be seen that the first-order term of the Taylor series constitutes the systematic corrections or systematic (or gross) errors to be applied to the initial parameters (x) estimated from the observations (ℓ). The systematic corrections or systematic errors propagated from the measurements to the initial estimates of the parameters can, therefore, be expressed from Equation (2.12) as

$$\Delta x = J(\Delta \ell) \tag{2.34}$$

where Δx is a vector of systematic corrections or systematic errors to be applied to the initial parameters. To illustrate the systematic error (correction) propagation procedure, refer to Equation (2.1) for a case where one parameter $x = (x_1)$ and three observables $\ell = (\ell_1 \ \ell_2 \ \ell_3)^T$ are measured; the following general mathematical model can be formulated:

$$x_1 = f(\ell_1, \ell_2, \ell_3) \tag{2.35}$$

The systematic error (dx_1) in the parameter x_1 due to the systematic errors $d\ell_1$, $d\ell_2$, and $d\ell_3$ of the measurements ℓ_1, ℓ_2, and ℓ_3, respectively, can be given as

$$dx_1 = \left(\frac{\partial f}{\partial \ell_1}\right) d\ell_1 + \left(\frac{\partial f}{\partial \ell_2}\right) d\ell_2 + \left(\frac{\partial f}{\partial \ell_3}\right) d\ell_3 \tag{2.36}$$

In order to simplify Equation (2.36), matrix notations are commonly employed by using Jacobian approach; in this case, the Equation (2.36) can be expressed as Equation (2.34) as follows:

$$dx_1 = J(\Delta \ell) \tag{2.37}$$

where

$$\Delta \ell = [d\ell_1 \ d\ell_2 \ d\ell_3]^T \tag{2.38}$$

$$J = \left[\frac{\partial f}{\partial \ell_1} \; \frac{\partial f}{\partial \ell_2} \; \frac{\partial f}{\partial \ell_3} \right] \tag{2.39}$$

Equations (2.36) and (2.37) represent the systematic error (correction) propagation laws.

Example 2.9 Consider the tract of land in Figure 2.8 with the measured dimensions as $\ell_1 = 50$ m, $\ell_2 = 20$ m, and $\ell_3 = 30$ m. The tract is to be divided into two parts along the broken line connecting points C and F. The two areas A_1 and A_2 are expressed by the following functions, respectively:

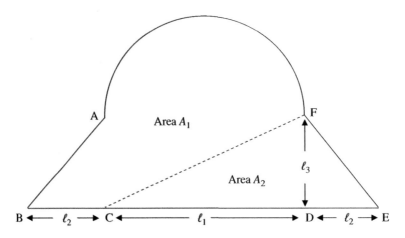

Figure 2.8 Partitioned tract of land.

$$A_1 = \pi \frac{\ell_1^2}{8} + \frac{\ell_1 \ell_3}{2} + \frac{\ell_2 \ell_3}{2}$$

$$A_2 = \frac{\ell_1 \ell_3}{2} + \frac{\ell_2 \ell_3}{2}$$

Answer the following:

a) Evaluate the Jacobian matrix J.

Solution:

Jacobian matrix is formed by considering that there are three measurements (ℓ_1, ℓ_2, ℓ_3) involved in the two equations representing the two areas (A_1, A_2). The three measurements give three columns, while the two equations give two rows so that the Jacobian matrix is 2×3 matrix, given as follows:

$$J = \begin{bmatrix} \dfrac{\partial A_1}{\partial \ell_1} & \dfrac{\partial A_1}{\partial \ell_2} & \dfrac{\partial A_1}{\partial \ell_3} \\[2mm] \dfrac{\partial A_2}{\partial \ell_1} & \dfrac{\partial A_2}{\partial \ell_2} & \dfrac{\partial A_2}{\partial \ell_3} \end{bmatrix} \rightarrow J = \begin{bmatrix} \left(\dfrac{\pi}{4}\ell_1 + \dfrac{1}{2}\ell_3\right) & \dfrac{1}{2}\ell_3 & \left(\dfrac{1}{2}\ell_2 + \dfrac{1}{2}\ell_1\right) \\[2mm] \dfrac{1}{2}\ell_3 & \dfrac{1}{2}\ell_3 & \left(\dfrac{1}{2}\ell_2 + \dfrac{1}{2}\ell_1\right) \end{bmatrix}$$

b) If the corrections to be applied to the dimensions ($\ell_1 = 50$ m, $\ell_2 = 20$ m, and $\ell_3 = 30$ m) are 0.02, –0.04, and 0.03 m, respectively, evaluate the corrections to the areas A_1 and A_2 by applying systematic error propagation laws.

Solution:

Corrections to the areas are to be determined in this problem following the systematic correction propagation law in Equation (2.34):

$$\begin{bmatrix} dA_1 \\ dA_2 \end{bmatrix} = J \begin{bmatrix} \Delta\ell_1 \\ \Delta\ell_2 \\ \Delta\ell_3 \end{bmatrix} \rightarrow \begin{bmatrix} 54.3 & 15 & 35 \\ 15 & 15 & 35 \end{bmatrix} \begin{bmatrix} 0.02 \\ -0.04 \\ 0.03 \end{bmatrix}$$

$$\begin{bmatrix} dA_1 \\ dA_2 \end{bmatrix} = \begin{bmatrix} 1.536 \\ 0.75 \end{bmatrix} m^2$$

Example 2.10 A rectangular tract of land ABCD in Figure 2.9 measures 50.170 m × 61.090 m. If the same 30 m tape (0.030 m too long, i.e. the tape has been stretched but is still displaying the nominal value of 30 m) is used to make the measurements, evaluate the error in the calculated area of the tract:

$$A = \ell_1 \ell_2$$

Figure 2.9 Rectangular tract of land.

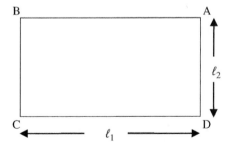

Solution:

$$\text{Jacobian Matrix} : J = \begin{bmatrix} \dfrac{\partial A}{\partial \ell_1} & \dfrac{\partial A}{\partial \ell_2} \end{bmatrix} \rightarrow J = \begin{bmatrix} \ell_2 & \ell_1 \end{bmatrix}$$

Errors are computed on the basis of the 30 m tape being too long by 0.030 m as follows:

$$\Delta\ell_1 = \frac{-0.030}{30}(50.170) = -0.050\,m$$

$$\Delta\ell_2 = \frac{-0.030}{30}(61.090) = -0.061\,m$$

Applying the systematic error propagation rule in Equation (2.34)

$$\text{Error of Area}: dA = J\begin{bmatrix}\Delta\ell_1\\\Delta\ell_2\end{bmatrix} \rightarrow [61.090\ \ 50.170]\begin{bmatrix}-0.050\\-0.061\end{bmatrix}$$

The propagated systematic error in the area is $dA = -6.1\,\text{m}^2$.

2.5 Variance–Covariance Propagation

The equation showing the relationship between variances of measurements and variances and covariances of quantities to be solved for is known as *variance–covariance propagation* or *random error propagation*. In variance–covariance propagation, specific values of the errors are unknown. On this basis, the errors are expressed, for example, as ±5 units so that it is impossible to correct for this type of error or to apply systematic error propagation approach. Instead of working with the actual error values, the probable values of the errors are used in the propagation of variances and covariances. To illustrate the random error propagation procedure, refer to Equation (2.1) for a case where two parameters $x = \begin{pmatrix}x_1\\x_2\end{pmatrix}$ and two observables $\ell = \begin{pmatrix}\ell_1\\\ell_2\end{pmatrix}$ are involved, for the sake of simplicity; the following general mathematical model can be formulated:

$$x = f(\ell) \tag{2.40}$$

For example, if Equation (2.40) is broken into two functions f_1 and f_2 operating on both ℓ_1 and ℓ_2 observables, the following relationships can be obtained:

$$x_1 = f_1(\ell_1, \ell_2) \tag{2.41}$$
$$x_2 = f_2(\ell_1, \ell_2) \tag{2.42}$$

and the variances $\sigma_{x_1}^2$ and $\sigma_{x_2}^2$ of x_1 and x_2, respectively, can be given as

$$\sigma_{x_1}^2 = \left(\frac{\partial f_1}{\partial \ell_1}\right)^2 \sigma_{\ell_1}^2 + \left(\frac{\partial f_1}{\partial \ell_2}\right)^2 \sigma_{\ell_2}^2 + \left(\frac{\partial f_1}{\partial \ell_1}\right)\left(\frac{\partial f_1}{\partial \ell_2}\right)\sigma_{\ell_1\ell_2} \tag{2.43}$$

$$\sigma_{x_2}^2 = \left(\frac{\partial f_2}{\partial \ell_1}\right)^2 \sigma_{\ell_1}^2 + \left(\frac{\partial f_2}{\partial \ell_2}\right)^2 \sigma_{\ell_2}^2 + \left(\frac{\partial f_2}{\partial \ell_1}\right)\left(\frac{\partial f_2}{\partial \ell_2}\right)\sigma_{\ell_1 \ell_2} \tag{2.44}$$

If ℓ_1 and ℓ_2 are measured independently, then the covariance $\sigma_{\ell_1 \ell_2}$ between them will be zero. The covariance $\sigma_{x_1 x_2}$ between the two quantities x_1 and x_2 can be given as

$$\sigma_{x_1 x_2} = \left(\frac{\partial f_1}{\partial \ell_1}\right)\left(\frac{\partial f_2}{\partial \ell_1}\right)\sigma_{\ell_1}^2 + \left(\frac{\partial f_1}{\partial \ell_2}\right)\left(\frac{\partial f_2}{\partial \ell_2}\right)\sigma_{\ell_2}^2$$

$$+ \left[\left(\frac{\partial f_1}{\partial \ell_1}\right)\left(\frac{\partial f_2}{\partial \ell_2}\right) + \left(\frac{\partial f_1}{\partial \ell_2}\right)\left(\frac{\partial f_2}{\partial \ell_1}\right)\right]\sigma_{\ell_1 \ell_2} \tag{2.45}$$

In order to simplify Equations (2.43)–(2.45), matrix notations are commonly employed by using Jacobian approach; in this case, the Equations (2.43)–(2.45) can be given as

$$C_x = JCJ^T \tag{2.46}$$

where

$$C_x = \begin{bmatrix} \sigma_{x_1}^2 & \sigma_{x_1 x_2} \\ \sigma_{x_1 x_2} & \sigma_{x_2}^2 \end{bmatrix} \tag{2.47}$$

$$J = \begin{bmatrix} \dfrac{\partial f_1}{\partial \ell_1} & \dfrac{\partial f_1}{\partial \ell_2} \\ \dfrac{\partial f_2}{\partial \ell_1} & \dfrac{\partial f_2}{\partial \ell_2} \end{bmatrix} \tag{2.48}$$

$$C = \begin{bmatrix} \sigma_{\ell_1}^2 & \sigma_{\ell_1 \ell_2} \\ \sigma_{\ell_1 \ell_2} & \sigma_{\ell_2}^2 \end{bmatrix} \tag{2.49}$$

C_x is the covariance matrix of x_1 and x_2; C is the covariance matrix of ℓ_1 and ℓ_2; and J is the Jacobian matrix of Equations (2.41)–(2.42). Similarly, for a vector of parameters $x = (x_1, x_2, x_3, ..., x_m)^T$ explicitly expressed as a function of field measurements $\ell = (\ell_1, \ell_2, ..., \ell_n)^T$, the following Jacobian matrix can be derived:

$$J = \begin{bmatrix} \dfrac{\partial x_1}{\partial \ell_1} & \dfrac{\partial x_1}{\partial \ell_2} & \dfrac{\partial x_1}{\partial \ell_3} & \cdots & \dfrac{\partial x_1}{\partial \ell_n} \\[2mm] \dfrac{\partial x_2}{\partial \ell_1} & \dfrac{\partial x_2}{\partial \ell_2} & \dfrac{\partial x_2}{\partial \ell_3} & \cdots & \dfrac{\partial x_2}{\partial \ell_n} \\[2mm] \dfrac{\partial x_3}{\partial \ell_1} & \dfrac{\partial x_3}{\partial \ell_2} & \dfrac{\partial x_3}{\partial \ell_3} & \cdots & \dfrac{\partial x_3}{\partial \ell_n} \\[2mm] \vdots & \vdots & \vdots & \vdots & \vdots \\[2mm] \dfrac{\partial x_m}{\partial \ell_1} & \dfrac{\partial x_m}{\partial \ell_2} & \dfrac{\partial x_m}{\partial \ell_3} & \cdots & \dfrac{\partial x_m}{\partial \ell_m} \end{bmatrix} \tag{2.50}$$

Applying the laws of variance–covariance propagation to Equation (2.40) gives

$$\Sigma_x = J\Sigma_\ell J^T \tag{2.51}$$

where Σ_x is the true variance–covariance matrix of the parameters x and Σ_ℓ is the variance–covariance matrix of the observations ℓ. Since the true variance–covariance matrix of the observations is not known, Σ_x cannot be computed. Instead, the estimate of variance–covariance matrix of parameters (C_x) is computed as

$$C_x = JC_\ell J^T \tag{2.52}$$

where C_ℓ is the estimated variance–covariance matrix of the observations. Equations (2.43)–(2.45) and (2.52) constitute what is usually known as the variance–covariance propagation laws. The following examples illustrate how variance–covariance propagation laws are applied.

Example 2.11   Given the following direct model for y_1 and y_2 as a function of ℓ_1, ℓ_2, and ℓ_3

$$
\begin{aligned}
y_1 &= 2\ell_1 - \ell_2 - \ell_3 + 20 \\
y_2 &= \ell_1 + \ell_2 - 3\ell_3 - 50
\end{aligned} \tag{2.53}
$$

where $\ell_1 = \ell_2 = \ell_3 = 1$ and the covariance matrix of ℓ's are as follows:

$$
C_\ell = \begin{bmatrix} 4 & -2 & -1 \\ -2 & 2 & 1 \\ -1 & 1 & 2 \end{bmatrix}
$$

Compute the covariance matrix C_y for y's.

Solution:

Let $y = \begin{bmatrix} y_1 \\ y_2 \end{bmatrix}$ and the covariance of y be C_y

By variance–covariance propagation laws on Equation (2.53),

$$C_y = JC_\ell J^T$$

where $J = \begin{bmatrix} \dfrac{\partial y_1}{\partial \ell_1} & \dfrac{\partial y_1}{\partial \ell_2} & \dfrac{\partial y_1}{\partial \ell_3} \\ \dfrac{\partial y_2}{\partial \ell_1} & \dfrac{\partial y_2}{\partial \ell_2} & \dfrac{\partial y_2}{\partial \ell_3} \end{bmatrix}$ $J = \begin{bmatrix} 2 & -1 & -1 \\ 1 & 1 & -3 \end{bmatrix}$

The snippet of MATLAB code for computing the Jacobian J is given in Table 2.3.

Table 2.3 MATLAB code for the Jacobian JM calculation.

```
>> % syms is to define the variables L1, L2, L3
>> syms L1 L2 L3
>> y1=2*L1-L2-L3+20;
>> y2=L1+L2-3*L3-50;
>> % use jacobian function to determine the Jacobian JM
>> JM=jacobian([y1;y2],[L1 L2 L3])

JM =
[ 2, -1, -1]
[ 1, 1, -3]
```

Variance–covariance propagation of vector y is

$$C_y = JC_\ell J^T = \begin{bmatrix} 2 & -1 & -1 \\ 1 & 1 & -3 \end{bmatrix} \begin{bmatrix} 4 & -2 & -1 \\ -2 & 2 & 1 \\ -1 & 1 & 2 \end{bmatrix} \begin{bmatrix} 2 & 1 \\ -1 & 1 \\ -1 & -3 \end{bmatrix}$$

Covariance matrix of vector y is

$$C_y = \begin{bmatrix} 34 & 19 \\ 19 & 20 \end{bmatrix}$$

where 34 and 20 are the variances of y_1 and y_2, respectively, and 19 is the covariance between y_1 and y_2.

Example 2.12 Given the following direct model for x and y as functions of ℓ_1, ℓ_2, and ℓ_3

$$\begin{bmatrix} x \\ y \end{bmatrix} = \begin{bmatrix} 4 & -2 & 1 \\ 5 & 3 & -1 \end{bmatrix} \begin{bmatrix} \ell_1 \\ \ell_2 \\ \ell_3 \end{bmatrix} \tag{2.54}$$

where the covariance matrix of the ℓ's is

$$C_\ell = \begin{bmatrix} 4 & -2 & -1 \\ -2 & 2 & 1 \\ -1 & 1 & 2 \end{bmatrix}$$

Compute the covariance matrix of x and y.

Solution:

Equation (2.54) can be put in matrix form $\begin{bmatrix} x \\ y \end{bmatrix} = J \begin{bmatrix} \ell_1 \\ \ell_2 \\ \ell_3 \end{bmatrix}$ where the Jacobian

matrix $J = \begin{bmatrix} 4 & -2 & 1 \\ 5 & 3 & -1 \end{bmatrix}$ and

$$C_\ell = \begin{bmatrix} 4 & -2 & -1 \\ -2 & 2 & 1 \\ -1 & 1 & 2 \end{bmatrix} \text{ with the vector } \ell = \begin{bmatrix} \ell_1 \\ \ell_2 \\ \ell_3 \end{bmatrix}$$

The variance–covariance matrix of x and y can be given as

$$C = JC_\ell J^T = \begin{bmatrix} 4 & -2 & 1 \\ 5 & 3 & -1 \end{bmatrix} \begin{bmatrix} 4 & -2 & -1 \\ -2 & 2 & 1 \\ -1 & 1 & 2 \end{bmatrix} \begin{bmatrix} 4 & 5 \\ -2 & 3 \\ 1 & -1 \end{bmatrix}$$

$$C = \begin{bmatrix} 94 & 66 \\ 66 & 64 \end{bmatrix}$$

where 94 and 64 are the variances of x and y, respectively, and 66 is the covariance between x and y.

Example 2.13 The least squares adjusted dimensions of a rectangular piece of land are 75.000 m × 100.000 m with the standard deviations of 0.003 and 0.008 m, respectively. Compute the area of the rectangle and its standard deviation.

Solution:

Let the adjusted dimensions be $\hat{\ell}_1$ and $\hat{\ell}_2$; the area (A) can be given as

$$A = \hat{\ell}_1 \times \hat{\ell}_2 \rightarrow 7500.00\,\text{m}^2$$

By the variance–covariance propagation laws,

$$C_A = JC_{\hat{\ell}}J^T$$

where the Jacobian matrix is

$$J = \frac{\partial A}{\partial \hat{\ell}} = \begin{bmatrix} \hat{\ell}_2 & \hat{\ell}_1 \end{bmatrix} \text{ or } J = \begin{bmatrix} 100.000 & 75.000 \end{bmatrix}$$

The covariance matrix of area A is determined as follows. Use the standard deviations of the adjusted measurements to form the covariance matrix of the adjusted dimensions as follows:

$$C_{\hat{\ell}} = \begin{bmatrix} (0.003)^2 & 0.0 \\ 0.0 & (0.008)^2 \end{bmatrix}$$

Propagated covariance matrix of the area A is $C_A = JC_{\hat{\ell}}J^T = 0.450\,\mathrm{m}^4$:

$$\sigma_A^2 = 0.450\,\mathrm{m}^4 \rightarrow \sigma_A = 0.671\,\mathrm{m}^2$$

Example 2.14 Given n independent measurements of a single quantity as ℓ_1, ℓ_2, ..., ℓ_n and their corresponding standard deviations as σ_1, σ_2, ..., σ_n, derive the expression for the standard deviation of the mean value of the measurements using the law of propagation of variances.

Solution:

Equation for finding the mean can be expressed as

$$\bar{x} = \frac{1}{n}(\ell_1 + \ell_2 + \cdots + \ell_n)$$

Based on the laws of propagation of variance–covariance matrix, the covariance matrix of the mean can be given as

$$C_{\bar{x}} = JC_{\ell}J^T$$

where the Jacobian (J) can be given as

$$J = \begin{bmatrix} \dfrac{\partial \bar{x}}{\partial \ell_1} & \dfrac{\partial \bar{x}}{\partial \ell_2} & \cdots & \dfrac{\partial \bar{x}}{\partial \ell_n} \end{bmatrix} \rightarrow J = \begin{bmatrix} \dfrac{1}{n} & \dfrac{1}{n} & \cdots & \dfrac{1}{n} \end{bmatrix}$$

and the covariance matrix of the observations is a diagonal matrix, which can be given as

$$C_\ell = \begin{bmatrix} \sigma_1^2 & & \cdots & \\ & \sigma_2^2 & \cdots & \\ & & \cdots & \\ & & \cdots & \sigma_n^2 \end{bmatrix}$$

The covariance matrix of the mean can be given as

$$C_{\bar{x}} = \sigma_{\bar{x}}^2 = J \begin{bmatrix} \sigma_1^2 & & \cdots & \\ & \sigma_2^2 & \cdots & \\ & & \cdots & \\ & & \cdots & \sigma_n^2 \end{bmatrix} J^T \quad \text{or} \quad \sigma_{\bar{x}}^2 = \left(\frac{1}{n}\right)^2 (\sigma_1^2 + \sigma_2^2 + \cdots + \sigma_n^2)$$

Assuming each observation has the same variance (i.e. $\sigma_1^2 = \sigma_2^2 = \cdots = \sigma_n^2 = \sigma_\ell^2$), the variance of the mean can be given as follows:

$$\sigma_{\bar{x}}^2 = \left(\frac{1}{n}\right)^2 (n\sigma_\ell^2) \quad \text{or} \quad \sigma_{\bar{x}}^2 = \frac{1}{n}\sigma_\ell^2$$

The standard deviation of the mean is then given as

$$\sigma_{\bar{x}} = \frac{\sigma_\ell}{\sqrt{n}}$$

Example 2.15 Side b and angles A and B in Figure 2.10 were measured independently with their standard deviations and given as follows: $b = 10.00$ m $\pm$ 0.01 m, $A = 30°0'00'' \pm 5''$, and $B = 60°0'00'' \pm 5''$.

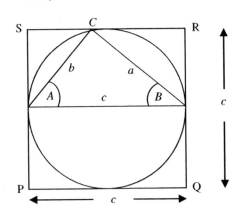

Figure 2.10 A square parcel of land containing a triangular lot in a circular trail.

a) Estimate the areas of triangle ABC and the circle shown inside the square PQRS.

Solution:

First determine lengths a and c using sine law as follows:

$$\frac{\sin B}{b} = \frac{\sin A}{a} \rightarrow a = \frac{b \sin A}{\sin B} \rightarrow a = \frac{10 \sin 30}{\sin 60} = 5.774 \, \text{m}$$

$$\frac{\sin B}{b} = \frac{\sin(A+B)}{c} \rightarrow c = \frac{b \sin(A+B)}{\sin B} \rightarrow c = \frac{10 \sin 90}{\sin 60} = 11.547 \, \text{m}$$

Area of triangle ABC: $A_t = \frac{1}{2}ab \rightarrow A_t = \frac{1}{2} \times 5.774 \times 10 = 28.87 \, \text{m}^2$

Area of circle: $A_c = \pi \times \left(\frac{c}{2}\right)^2 \rightarrow A_c = 3.141\,59 \times \left(\frac{11.547}{2}\right)^2 = 104.72 \, \text{m}^2$

b) Estimate the standard deviations of the areas computed in part (a).

Solution:

$$\text{Area of triangle ABC}: A_t = \frac{1}{2}b^2\frac{\sin A}{\sin B} \tag{2.55}$$

$$\text{Area of circle}: A_c = \pi \times \left(\frac{b\sin(A+B)}{2\sin B}\right)^2 \tag{2.56}$$

$$\sin(A+B) = \sin A \cos B + \cos A \sin B$$

$$A_c = \pi \times \left(\frac{b\sin A\cot B + b\cos A}{2}\right)^2 \tag{2.57}$$

Jacobian (J) of Equations (2.55) and (2.57) is

$$J = \begin{bmatrix} \dfrac{\partial A_t}{\partial b} & \dfrac{\partial A_t}{\partial A} & \dfrac{\partial A_t}{\partial B} \\ \dfrac{\partial A_c}{\partial b} & \dfrac{\partial A_c}{\partial A} & \dfrac{\partial A_c}{\partial B} \end{bmatrix}$$

$$\frac{\partial A_t}{\partial b} = \frac{b\sin A}{\sin B} = 5.7735 \qquad \frac{\partial A_t}{\partial A} = \frac{b^2\cos A}{2\sin B} = 50.0000$$

$$\frac{\partial A_t}{\partial B} = -\frac{b^2\sin A \csc B \cot B}{2} = -16.6667$$

$$\frac{\partial A_c}{\partial b} = \frac{1}{2}\pi b[\sin A\cot B + \cos A]^2 = 20.943\ 95$$

$$\frac{\partial A_c}{\partial A} = \frac{1}{2}\pi b^2[\sin A \cot B + \cos A][\cos A \cot B - \sin A] = 0.0000$$

$$\frac{\partial A_c}{\partial B} = \frac{1}{2}\pi b^2[\sin A \cot B + \cos A]\left[-\sin A\csc^2 B\right] = -120.9200$$

The Jacobian matrix J, also calculated using the MATLAB code in Table 2.4, is given as follows:

$$J = \begin{bmatrix} 5.773\ 50 & 50.000\ 0 & -16.666\ 67 \\ 20.943\ 95 & 0.000\ 00 & -120.920\ 00 \end{bmatrix}; \quad \text{Let}: x = \begin{bmatrix} A_t \\ A_c \end{bmatrix}$$

$$C_\ell = \begin{bmatrix} 1.0E-4 & 0.0 & 0.0 \\ 0.0 & 5.876\ 096\ 6E-10 & 0.0 \\ 0.0 & 0.0 & 5.876\ 096\ 6E-10 \end{bmatrix} m^2$$

Table 2.4 MATLAB code for the calculation of Jacobian matrix JM ($\times 10^2$).

```
>> % Use syms to define the variables b, A, B
>> syms b A B
>> % create the Equations (2.55) and (2.56)
>> At=0.5*b^2*(sin(A)/sin(B));
>> Ac=pi*(b*sin(A+B)/(2*sin(B)))^2;
>> % determine the Jacobian of equations
>> J=jacobian([At;Ac],[b A B])
>> % convert input angles to radians and evaluate the Jacobian
>> A=30*pi/180;
>> b=10.0;
>> B=60*pi/180;
>> JM=eval(J)

JM =
    1.0e+02 *
    0.057735026918963   0.500000000000000   -0.166666666666667
    0.209439510239320   0.000000000000000   -1.209199576156146
>>
```

The variance–covariance propagation (C_x) for areas $x = \begin{bmatrix} A_t \\ A_c \end{bmatrix}$ can be given as follows:

$$C_x = JC_\ell J^T$$

$$C_x = \begin{bmatrix} 5.773\ 50 & 50.000\ 0 & -16.666\ 67 \\ 20.943\ 95 & 0.000\ 00 & -120.920\ 00 \end{bmatrix}$$

$$\times \begin{bmatrix} 10E-4 & 0.0 & 0.0 \\ 0.0 & 5.876\ 096\ 6E-10 & 0.0 \\ 0.0 & 0.0 & 5.876\ 096\ 6E-10 \end{bmatrix}$$

$$\times \begin{bmatrix} 5.773\ 50 & 20.943\ 95 \\ 50.000\ 0 & 0.000\ 00 \\ -16.666\ 7 & -120.920\ 0 \end{bmatrix}$$

$$C_x = \begin{bmatrix} 3.334\ 96E-3 & 1.209\ 32E-2 \\ 1.209\ 32E-2 & 4.387\ 35E-2 \end{bmatrix} m^4$$

Standard deviations are

For area of triangle : $\sigma_{A_t} = \sqrt{3.334\ 96E-3} \rightarrow 0.057\ 7\ m^2$

For area of circle : $\sigma_{A_c} = \sqrt{4.387\ 35E-2} \rightarrow 0.209\ 46\ m^2$

c) Estimate the correlation between the area of the triangle and the area of the circle.

Solution:

From the propagated variance–covariance matrix in (b), the covariance between the area of the triangle and the area of the circle = 1.209 32E-2 m^4, and the correlation coefficient can be given as

$$\text{Correlation coefficient} = \frac{1.209\ 32E-2}{(5.774\ 91E-2)(0.209\ 46)} = 0.999\ 758$$

2.6 Error Propagation Based on Equipment Specifications

The main modern survey equipment consists of total station and leveling instruments. The total station instruments are designed to measure distances and directions (or angles), while leveling instruments are designed to measure height differences. The standard deviation for a distance measurement (S) using a total station equipment is usually specified according to the International Standard Organization (ISO 17123-4) in the form of $a + b$ ppm, and the standard deviation of a direction (angle) measurement (horizontal and vertical) is specified according to ISO 17123-3 as $\pm c$ s. The standard deviation of height measurement is usually specified according to ISO 17123-2 as $\pm d$ mm km^{-1} *double run* or simply $\pm d$ mm.

2.6.1 Propagation for Distance Based on Accuracy Specification

The accuracy specifications for total station equipment are usually stated as follows by the manufacturers:

$$SE_s = \pm (a + bS) \tag{2.58}$$

or

$$SE_s = a + b \tag{2.59}$$

or

$$SE_s = \pm \sqrt{(a^2 + (b \times S)^2)} \tag{2.60}$$

where SE_s is the standard error (or standard deviation) of the distance measurement (S), "a" is the constant part in mm, and "b" is the scale error in ppm. The specifications are usually interpreted in the following two ways:

a) Determine the scale error in mm, and add directly to the constant error to obtain the total error. This approach of interpreting accuracy is generally supported by the instrument manufacturers. This will be valid if one wants to be pessimistic (assuming bigger errors than expected) and be on the safe side of error budgeting. In this case, the constant error and the error due to the scale are considered additive as follows:

$$SE_s = \pm \left(a + b \times 10^{-6} \times S \right) \tag{2.61}$$

b) Determine the scale error in mm, square it, add it to the square of the constant error, and find the square root of the sum. This approach seems to be supported by theory of error propagation by the total station if "a" and "b" are considered as two independent random errors. The random errors are propagated as follows:

$$SE_s = \pm \sqrt{\left(a^2 + \left(b \times 10^{-6} \times S \right)^2 \right)} \tag{2.62}$$

For example, consider a total station having the accuracy specifications for distance measurement (according to ISO17123-4) as 2 mm + 2 ppm (IR Fine mode) and 5 mm + 2 ppm (IR Rapid mode). For IR Rapid mode, for instance, $a = 5$ mm and $b = 2$ ppm. Based on the approach expressed by Equation (2.61), if the distance measured is 1000 m, the standard error for the distance will be

$$SE_s = \pm \left(5 + 2 \times 10^{-6} \times 1000\ 000 \right) \rightarrow \pm 7\,mm$$

Using the example in Equation (2.62) gives the following:

$$SE_s = \pm \sqrt{\left(5^2 + \left(2 \times 10^{-6} \times 1000\ 000 \right)^2 \right)} \rightarrow \pm 5.4\,mm$$

It can be seen from the above examples that Equation (2.61) gives pessimistic estimation; this means using this approach will always give pessimistic estimate that is always larger than the value from Equation (2.62). However, the precise solution is given by Equation (2.62), and its use is suggested rather than using the pessimistic one. It should be stated also that the ISO (17123-4) usually does not decompose distance error into constant and distance-dependent components; rather, the average error of measuring a distance is specified, e.g. the average error of measuring a distance can be stated simply as 5 mm. For educational purpose, the students are advised to use Equation (2.62) for determining the standard deviations of distance measurements. The other random errors that can be considered in propagating the error for a distance measurement are the errors of centering the instrument and the target over the survey markers. For example, if the instrument and target centering errors are σ_i and σ_t (in mm), respectively, the total standard deviation (SE_{Total}) of measuring a distance can be given as

$$SE_{Total} = \pm \sqrt{SE_s^2 + \sigma_i^2 + \sigma_t^2} \tag{2.63}$$

2.6.2 Propagation for Direction (Angle) Based on Accuracy Specification

If a total station instrument has a standard deviation for horizontal (or vertical) angle measurement of 2″ according to ISO 17123-3, the standard deviation is for the average of a set of direction measurement with the standard deviation of a direction measured in one face being $SE_d = 2\sqrt{2}''$ or 2.8″. This error is usually considered as the combined pointing and reading error. The other errors that can be considered with regard to direction (or angle) measurements are due to centering and leveling of the instrument and the corresponding targets. For example, if the instrument centering and target centering errors are σ_i and σ_t (in arcsec), respectively, the total standard deviation (SE_{Total}) of measuring a direction of a line in one face of the instrument's telescope can be given as

$$SE_{Total} = \pm \sqrt{SE_d^2 + \sigma_i^2 + \sigma_t^2} \tag{2.64}$$

For angle measurement, two targets will be involved so that Equation (2.64) will have two σ_t^2 instead of just σ_t^2 if the errors of the two targets are the same or the square of the error of the second target is added if it is different from the first target.

2.6.3 Propagation for Height Difference Based on Accuracy Specification

If the standard deviation of elevation difference over 1 km based on average of double run leveling is ±0.5 mm (according to ISO 17123-3), the standard deviation of the elevation difference over 1 km based on single run leveling will be ±0.5 mm $\sqrt{2}$. The standard deviation over 1 km (double run) can be determined over L km as ±0.5 mm $\sqrt{L}$ with L as the one-way distance expressed in km; the standard deviation over 1 km (single run) can be determined over L km as ±0.5 mm $\sqrt{2L}$ with L as the one-way distance expressed in km.

Example 2.16 In Figure 2.11, two angles B (40°42′50″) and C (80°48′20″) and distance c (21.547 m) were measured with a total station equipment having an angular precision of 2″ and a distance precision of 3 mm ± 2 ppm. If the measured angles B and C are correlated by 0.6, estimate the angle A and length b, their standard deviations, and the correlation coefficient between them.

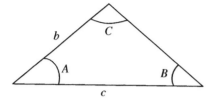

Figure 2.11 A small triangular parcel of land.

Solution:

Equations for the unknown angle A and distance b are

$$A = 180 - B - C$$

$$b = c \left(\frac{\sin B}{\sin C} \right)$$

where $B = 40°42'50''$, $C = 80°48'20''$, and $c = 21.547$ m. The calculated values of A and b are

$$A = 180 - B - C \rightarrow 58°28'50''$$

$$b = c \left(\frac{\sin B}{\sin C} \right) \rightarrow b = 14.2377 \text{ m}$$

With the given correlation coefficient of 0.6 between B and C, the covariance between B and C can be given as

$$\sigma_{BC} = 0.6 \left(\frac{2}{206\ 265} \right)^2 = 5.641\text{E} - 11$$

and the covariance matrix of the observations is formulated in the order of angles B, C, and distance c and given as follows:

$$C_\ell = \begin{bmatrix} (2/206\ 265)^2 & 5.641\text{E} - 11 & 0 \\ 5.641\text{E} - 11 & (2/206\ 265)^2 & 0 \\ 0 & 0 & (0.003)^2 \end{bmatrix}$$

By laws of variance–covariance propagation with respect to $\ell = \begin{bmatrix} B \\ C \\ c \end{bmatrix}$,

$$J = \frac{\partial y}{\partial \ell} = \begin{bmatrix} -1.000 & -1.000 & 0.0 \\ c \cos B & -c \cos C \sin B & \sin B \\ \sin C & \sin^2 C & \sin C \end{bmatrix}$$

For $B = 40°42'50''$, $C = 80°48'20''$, and $c = 21.547$ m, the Jacobian matrix (J) is given as

$$J = \begin{bmatrix} -1.000 & -1.000 & 0.0 \\ 16.544\ 684\ 08 & -2.304\ 579\ 8 & 0.660\ 771\ 92 \end{bmatrix}$$

The snippet of MATLAB code for computing the Jacobian matrix for this problem is given in Table 2.5:

Table 2.5 MATLAB code for the Jacobian matrix calculation given as JM.

```
>> % define the variables B, C, c
>> syms B C c
>> % create the equations for A and b
>> A=180-B-C;
>> b=c*(sin(B)/sin(C));
>> % determine the Jacobian of the equations
>> J=jacobian([A b],[B C c]);
>> B=(40+42/60+50/3600)*pi/180;
>> C=(80+48/60+20/3600)*pi/180;
>> c=21.547;
>> JM=eval(J)

JM =
   -1.000000000000000  -1.000000000000000  0
   16.544684083780819  -2.304579800713474  0.660771915829317
```

$$\text{Given}: C_\ell = \begin{bmatrix} 9.402\text{E}-11 & 5.641\text{E}-11 & 0 \\ 5.641\text{E}-11 & 9.402\text{E}-11 & 0 \\ 0 & 0 & 9.0\text{E}-6 \end{bmatrix}$$

From the laws of variance–covariance propagation, the propagated variance–covariance matrix of $x = [A\ b]^T$ can be given as follows:

$$C_x = JC_\ell J^T = \begin{bmatrix} 3.008\ 6\text{E}-10 & -2.142\ 139\text{E}-9 \\ -2.142\ 139\text{E}-9 & 3.951\ 509\ 2\text{E}-6 \end{bmatrix}$$

The MATLAB code for calculating the covariance matrix using the calculated Jacobian matrix JM in Table 2.5 is shown in Table 2.6.

Table 2.6 MATLAB code for calculating C_x using the Jacobian matrix JM.

```
>> % This is a continuation of code in Table 2.6
>> % Create covariance matrix CL of observations
>> CL = [9.402E-11 5.641E-11 0; 5.641E-11 9.402E-11 0;
   0 0 9.0E-6];
>> % Covariance propagation using Jacobian JM from Table 2.5
>> Cx=JM*CL*JM'
Cx =
   1.0e-05 *
   0.000030086000000  -0.000214213888730
  -0.000214213888730   0.395150918106541
>>
```

Standard deviations of angle A and distance b are extracted from the propagated covariance matrix C_x:

Angle $A = 206\,265^* \sqrt{3.008\,6E-10} = 3.58''$

Distance $b = 0.0020\,\text{m}$

Angle $A = 58°28'50'' \pm 3.6''$

Distance $b = 14.2377\,\text{m} \pm 0.0020\,\text{m}$

Correlation coefficient $= \dfrac{-2.142\,139E-9}{\sqrt{3.008\,6E-10^*}\sqrt{3.951\,509\,18E-6}} = -0.062\,12$

2.7 Heuristic Rule for Covariance Propagation

Heuristic rule is a rule that encourages one to use personal observation of the solution of a problem to solve another problem. The method is referred to as *symbolic multiplication* in Mikhail (1976), who credited Tienstra (1966) as the inventor of the rule. For example, given the following two equations

$$x = \ell_1 + \ell_2 \qquad\qquad (2.65)$$
$$y = \ell_1 + \ell_2 \qquad\qquad (2.66)$$

where ℓ_1 and ℓ_2 are random variables, the covariance s_{xy} of the two variables x and y can be obtained based on the pattern shown by the product of the two variables x and y as follows:

$$\textit{Ordinary product}: \quad xy = (\ell_1 + \ell_2)(\ell_1 + \ell_2) \qquad\qquad (2.67)$$

or

$$\text{Ordinary product}: \quad xy = \ell_1^2 + \ell_2^2 + 2\ell_1\ell_2 \tag{2.68}$$

Using heuristic rule on Equation (2.68), the covariance (s_{xy}) of x and y can be given as

$$\text{Heuristic product}: \quad s_{xy} = s_{\ell_1}^2 + s_{\ell_2}^2 + 2s_{\ell_1\ell_2} \tag{2.69}$$

where $s_{\ell_1}^2$ and $s_{\ell_2}^2$ are the variances of ℓ_1 and ℓ_2, respectively, and $s_{\ell_1\ell_2}$ is the covariance between ℓ_1 and ℓ_2. However, if ℓ_1, ℓ_2, x, and y are random vectors, the same rule can be applied with the variances becoming covariance matrices. In this case, Equation (2.69) will be written, giving the covariance matrix (C_{xy}) of x and y as follows:

$$\text{Heuristic product}: \quad C_{xy} = C_{\ell_1} + C_{\ell_2} + 2C_{\ell_1\ell_2} \tag{2.70}$$

where C_{ℓ_1} and C_{ℓ_2} are the covariance matrices of ℓ_1 and ℓ_2, respectively, and $C_{\ell_1\ell_2}$ is the covariance matrix of ℓ_1 and ℓ_2, since covariance matrix already indicates covariances and variances. Similarly, given the following functions in which A is a constant value and ℓ, X, and Y are random variables

$$X = A\ell \tag{2.71}$$
$$Y = \ell \tag{2.72}$$

The product XY can be given by

$$\text{Ordinary product}: \quad XY = A\ell^2 \tag{2.73}$$

Applying heuristic rule to Equation (2.73), the covariance of X and Y can be given as

$$\text{Heuristic product}: \quad s_{XY} = As_\ell^2 \tag{2.74}$$

where s_ℓ^2 is the variance of ℓ. The same rule can be applied for a case where A is a matrix whose elements are constant values, and X, Y, and ℓ are vectors of random variables, giving the covariance matrix (C_{XY}) of X and Y as follows:

$$\text{Heuristic product}: \quad C_{XY} = AC_\ell \tag{2.75}$$

where C_ℓ is the covariance matrix of ℓ.

Example 2.17 Given the variance–covariance matrix of the measurement vector $\lambda = \begin{bmatrix} \lambda_1 \\ \lambda_2 \end{bmatrix}$

$$C_\lambda = \begin{bmatrix} \dfrac{2}{3} & \dfrac{1}{3} \\ \dfrac{1}{3} & \dfrac{2}{3} \end{bmatrix}$$

and two functions of λ, $x = \lambda_1 + \lambda_2$, and $y = 3\lambda_1$, determine the following:

i) The covariance between x and y (or C_{xy}) using variance–covariance propagation and heuristic rule approaches.

Solution (i):

Covariance between x and y (or C_{xy}) using variance–covariance propagation approach.

Let $x = \begin{bmatrix} x \\ y \end{bmatrix}$ $\lambda = \begin{bmatrix} \lambda_1 \\ \lambda_2 \end{bmatrix}$

Determine the Jacobian (J) of the following equations:

$$x = \lambda_1 + \lambda_2 \tag{2.76}$$

$$y = 3\lambda_1 \tag{2.77}$$

$$J = \begin{bmatrix} 1 & 1 \\ 3 & 0 \end{bmatrix} \quad C_\lambda = \begin{bmatrix} \dfrac{2}{3} & \dfrac{1}{3} \\ \dfrac{1}{3} & \dfrac{2}{3} \end{bmatrix}$$

From the laws of variance–covariance propagation,

$$C = JC_\lambda J^T$$

$$C = \begin{bmatrix} 1 & 1 \\ 3 & 0 \end{bmatrix} \begin{bmatrix} \dfrac{2}{3} & \dfrac{1}{3} \\ \dfrac{1}{3} & \dfrac{2}{3} \end{bmatrix} \begin{bmatrix} 1 & 3 \\ 1 & 0 \end{bmatrix} \rightarrow C = \begin{bmatrix} 2 & 3 \\ 3 & 6 \end{bmatrix}$$

where the covariance between x and y from C matrix above is 3. The covariance between x and y can also be found using the heuristic rule on Equations (2.76) and (2.77) as follows. Give the ordinary product of the two equations as

Ordinary product : $\quad xy = 3\lambda_1^2 + 3\lambda_1\lambda_2$ $\tag{2.78}$

which translates to the covariance (s_{xy}) between x and y in the following heuristic rule:

Heuristic product : $\quad s_{xy} = 3s_{\lambda_1}^2 + 3s_{\lambda_1\lambda_2}$ $\tag{2.79}$

From the given covariance matrix C_λ, $s_{\lambda_1}^2 = \dfrac{2}{3}$, and $s_{\lambda_1\lambda_2} = \dfrac{1}{3}$, substituting in Equation (2.79) gives $s_{xy} = 3$ (the same result as obtained previously).

ii) The covariance between x and λ (or $C_{x\lambda}$) using heuristic rule approach.

Solution (ii):

Covariance between x and λ (or $C_{x\lambda}$) can be determined by applying the heuristic rule in Equation (2.75):

Given:

$$\lambda = \begin{bmatrix} \lambda_1 \\ \lambda_2 \end{bmatrix}$$

$x = \lambda_1 + \lambda_2$, which can be rewritten as $x = A\lambda$, where $A = \begin{bmatrix} 1 & 1 \end{bmatrix}$ is the Jacobian of the equation with respect to vector λ. This can be expressed as

$$x = \begin{bmatrix} 1 & 1 \end{bmatrix} \begin{bmatrix} \lambda_1 \\ \lambda_2 \end{bmatrix} \quad \text{or} \quad x = A\lambda$$

Ordinary product : $\quad x\lambda = A\lambda^2$

Heuristic product : $\quad C_{x\lambda} = AC_\lambda$

Substitution into heuristic product : $\quad C_{x\lambda} = \begin{bmatrix} 1 & 1 \end{bmatrix} \begin{bmatrix} \dfrac{2}{3} & \dfrac{1}{3} \\ \dfrac{1}{3} & \dfrac{2}{3} \end{bmatrix} \rightarrow C_{x\lambda} = \begin{bmatrix} 1 & 1 \end{bmatrix}$

iii) The covariance between y and λ (or $C_{y\lambda}$) using heuristic rule approach.

Solution (iii):

Covariance between y and λ (or $C_{y\lambda}$) can be determined by applying the heuristic rule in Equation (2.75):

Given:

$$\lambda = \begin{bmatrix} \lambda_1 \\ \lambda_2 \end{bmatrix}$$

$y = 3\lambda_1$, which can be rewritten as $x = B\lambda$, where $B = \begin{bmatrix} 3 & 0 \end{bmatrix}$ is the Jacobian of the equation with respect to vector λ. This can be expressed as

$$y = \begin{bmatrix} 3 & 0 \end{bmatrix} \begin{bmatrix} \lambda_1 \\ \lambda_2 \end{bmatrix} \quad \text{or} \quad x = B\lambda$$

Ordinary product : $\quad y\lambda = B\lambda^2$

Heuristic product : $\quad C_{y\lambda} = BC_\lambda$

Substitution into heuristic product : $C_{y\lambda} = \begin{bmatrix} 3 & 0 \end{bmatrix} \begin{bmatrix} \dfrac{2}{3} & \dfrac{1}{3} \\ \dfrac{1}{3} & \dfrac{2}{3} \end{bmatrix} \rightarrow C_{y\lambda} = \begin{bmatrix} 2 & 1 \end{bmatrix}$

Problems

2.1 a) Two angles A and B of a triangle were measured as $75°40'35'' \pm 3''$ and $44°50'20'' \pm 3''$, respectively, with a correlation between the measurements as 0.6. Determine the third angle C and its standard deviation (to 1 decimal second).

b) If it is later detected that each angle measurement in (a) has a systematic error of $-40''$, use systematic error propagation approach to determine the propagated systematic error in the derived angle C, and correct all the angles for the corresponding systematic errors.

2.2 The angle at point O between point P and point Q has been independently measured three times with the same standard deviation using total station equipment, and the standard deviation of a measured angle is $1.5''$.

a) Using Jacobian approach, determine the standard deviation of the mean of the first two angle measurements, standard deviation of the difference between the last measurement and that mean, and the correlation coefficient between the mean and the difference.

b) If the first two measurements are later found to have systematic error of $45''$ each and the last measurement is free of systematic error, determine, by systematic error propagation, the effect of these systematic error on the mean of the first two measurements and of the difference between the last measurement and that mean.

2.3 In Figure P2.3, the measured values for slope distance s and vertical angle θ are 1000.000 m and $15°00'00''$, respectively; the systematic errors for the distance s and the angle θ are 0.250 m and $50''$, respectively. Calculate the errors in the computed values of d and h by applying systematic error propagation rule.

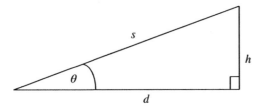

Figure P2.3

2.4 a) Given the elevations (above mean sea level) of two benchmarks HA and HB as 25.456 ± 0.002 m and 155.267 ± 0.005 m, respectively, the EDM slope distance (s) from HA to HB was measured as 1025.436 m with precision specification of 3 mm $\pm$ 2 ppm; and assuming the elevations of HA and HB, which are determined by differential leveling, are correlated by 0.5 (referring to Figure P2.4), estimate the horizontal distance (y_1) and the slope angle (y_2) from HA to HB and their corresponding standard deviations (with appropriate units).

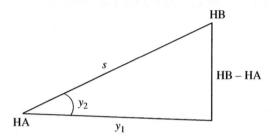

Figure P2.4

b) Calculate the correlation between the estimated horizontal distance and slope angle.
c) If there is a systematic error of +0.400 m on the elevation of benchmark HB (none on HA) and a systematic error of +0.050 m in the slope distance measurement, determine, using systematic error propagation method, the amount of correction to be applied to the calculated horizontal distance and the slope angle (providing appropriate units).

2.5 The Figure P2.5 shows a small rectangular lot divided into two rectangular areas A_1 and A_2 (that is, all angles are perfectly known).

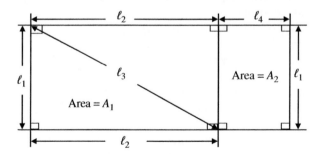

Figure P2.5

After correcting for possible systematic errors followed by the least squares adjustment of the measurements, the adjusted measurements with their corresponding standard deviations are as follows: $\hat{\ell}_2 = 30.455$ m ± 0.003, $\hat{\ell}_3 = 36.266 \pm 0.004$ m, and $\hat{\ell}_4 = 10.155 \pm 0.001$ m. If the correlation between $\hat{\ell}_2$ and $\hat{\ell}_3$ is 0.75, compute the adjusted areas of the two lots and their corresponding standard deviations and the correlation coefficient between areas A_1 and A_2.

2.6 For visible and near-infrared radiation and neglecting the effects of water vapor pressure, the formula for computing the refractive index, n, in an EDM can be determined by

$$n - 1 = \frac{0.269\,578[n_0 - 1]}{273.15 + t} p$$

where n_0 is the constant refractive index set in the EDM, t is the temperature in °C, p is the pressure in mbar, and n is the realistic refractive index to be determined. The EDM has a set constant value, $n_0 = 1.000\,294\,497$, and the average temperature and pressure during the measurements are expected to be +30 °C and 950 mb, respectively. Assuming the standard deviation of measuring temperature, $\sigma_t = 1.0$ °C, what would be the largest value of σ_p so that the error in σ_n will not be more than 2 ppm?

2.7 Given the variance–covariance matrix of the measurement vector $\ell = \begin{bmatrix} \ell_1 \\ \ell_2 \end{bmatrix}$

$$C_\ell = \begin{bmatrix} \dfrac{1}{2} & \dfrac{1}{4} \\ \dfrac{1}{4} & \dfrac{1}{2} \end{bmatrix}$$

and two functions of ℓ, $x = 4\ell_2$ and $y = 2\ell_1 + \ell_2$, use heuristic rule approach to determine the following:
a) The covariance between x and y (or C_{xy}).
b) The covariance between x and ℓ (or $C_{x\ell}$).
c) The covariance between y and ℓ (or $C_{y\ell}$).

2.8 a) In throwing a shot put from an origin A to any point B in an Olympic competition, the traditional surveying technique with robotic total station was used. In the design (referring to Figure P2.8), the distance (b) from the origin A to the robotic total station setup point C is measured as 9.005 m and fixed. The total station observables when a competitor throws the shot put are the angle C and distance "a" in order to calculate distance "c." For a competitor, the measured distance and angle are $a = 27.544$ m and $C = 40°48'20''$, respectively. If the standard error of each of the measured sides is 3 mm ± 2 ppm and the angular precision is 2″, determine the distance c and its standard deviation (note that angle A is not necessarily 90°).

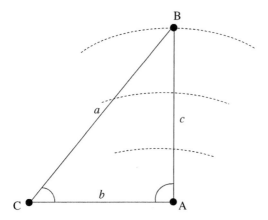

Figure P2.8

b) In reviewing if the competitor in Question (a) has broken the Olympic record, the surveyor found out that he had forgotten to subtract prism constant of 30 mm from each distance measurement and to subtract the initial direction measurement of 45″ from the angle measurement. Determine, using systematic error propagation method, the amount of correction to be applied to the calculated distance "c" in Question (a).

2.9 Given two distance measurements that are independent and have standard deviations $\sigma_1 = 0.20$ m and $\sigma_2 = 0.15$, respectively,
   a) Calculate the standard deviations of the sum and difference of the two measurements.
   b) Calculate the correlation between the sum and the difference.

3

Statistical Distributions and Hypothesis Tests

CHAPTER MENU

OBJECTIVES

After studying this chapter, you should be able to:

1) Define some basic statistical and probability terms for geomatics inferences and analysis.
2) Apply statistical distributions, such as normal, Student's t-, chi-square, and Fisher's F-distributions to solve geomatics problems.
3) Define significance and confidence levels, type I and II errors, and power of test in statistical decision-making.
4) Formulate null and alternative hypotheses.
5) Perform hypothesis testing in univariate cases.

Understanding Least Squares Estimation and Geomatics Data Analysis, First Edition.
John Olusegun Ogundare.
© 2019 John Wiley & Sons, Inc. Published 2019 by John Wiley & Sons, Inc.
Companion website: www.wiley.com/go/ogundare/Understanding-lse-and-gda

3.1 Introduction

This chapter discusses the important elements of statistical distributions commonly used in geomatics and explains how statistical problems in geomatics are solved, including how statistical decisions are made based on statistical hypothesis tests. As an introduction, the relevant statistical terms are first explained including how the commonly used terms such as statistics as a subject, statistics as an observable, the concepts of probability, and statistical distributions (or probability functions) are related. It is common in geomatics to confuse statistics as a subject with statistics as random observables. *Statistics* as a field allows one to make conclusions about data sets; it is a science that involves studying different aspects of a given data set, which may include the methods and procedures for performing the following:

- Collecting and analyzing the data sets (sampling distribution).
- Presenting and summarizing the data sets (frequency distribution).
- Deriving information from the data sets in order to use them for activities that may include prediction and decision-making (statistical inferencing).
- Validating the data sets against some given standards (hypothesis testing).

In statistics, the laws of chance (probability) are applied in order to obtain estimates or to make inferences based on a given set of observed data. The measure of uncertainty associated with the activities involved in statistics or a means of quantifying the chance that an event will occur is known as the *probability*. For example, the probability is 100% or 1 if it is certain that an event will occur; the probability of occurrence of the event will be zero if it is certain that it will not occur. Probability tends to predict and describe averages that would have been realized if the experiments were repeated infinite times. In another words, probability provides the limit of frequency of occurrence when the number of repetitions of an experiment approaches infinity. One of the commonly used procedures for estimating probabilities of events occurring is to use *probability functions or probability density functions* (pdfs). A probability function can be used to calculate the probability value associated with the occurrence of an event, or they can be used to determine the occurrence of an event associated with a probability value. Different types of probability functions discussed in this chapter are collectively referred to as *statistical distributions*.

The term *population* is used in this book to represent all possible measurements of a given random observable; it is a complete set of data that can be used to determine with absolute certainty the average value of the observable. Decisions and conclusions based on the population are free of any risk. The risk of incorrect decisions and conclusions is always present when a subset (a *sample*) of the population, which is representative of that population, is used. Since a conclusion about a *population* based solely on a *sample* cannot be made with certainty, the language of probability is usually employed in any statement of such a conclusion.

The usual quantities for summarizing or describing some aspects of an entire population are the population mean (μ) and the population standard deviation (σ), which are often referred to as *population parameters*. A parameter in statistical term is a constant numerical value that describes a population; it is usually unknown and not computed, but can be inferred from the sample. Any quantity obtained from a sample for the purpose of estimating a population parameter is called a *sample statistic* or simply *statistic*. A sample statistic can also be defined as a function of one or more random observables (or variables), which does not depend upon any unknown population parameter; if population parameters are involved in the function, their values must be already known. In general, the term statistic, which should not be confused with statistics as a science, is used to represent a random observable or its numerical value. It should also be emphasized that the values obtained from statistics as a function are themselves referred to as statistics. *Estimators* are subsets of statistics. The goals of statistics are to summarize information in a sample and sometimes to perform statistical tests, while estimators are mainly for determining the values of parameters. If a statistic is able to estimate a parameter of a population, it is also called an estimator. The examples of statistics (as values) are *test statistics* used in hypothesis testing, such as *t-score, z-score, F-statistic,* and *chi-square statistic*. These test statistics, however, are not estimators since they are not used for estimating any parameter. For example, the sample mean is a statistic as well as an estimator since it is often used to estimate the true population mean.

3.2 Probability Functions

Probability functions are mathematical models that are formulated with their validity already tested by experiments. They are otherwise referred to as *stochastic models*, which are functions of some random *observables (or variables)*. Being a function of some random observables, a stochastic model itself invariably becomes a random observable. A random observable (or variable) is a physical quantity, which can be assigned values through measurement or observation, such as distance, angle, height, etc. A random observable that can assume countable possible number of values is referred to as *discrete random observable*; the one that can assume infinite possible number of values is a *continuous random observable*. If the possible values of a random observable are arranged in an increasing order of magnitude with their associated probabilities, a *probability function* or *probability distribution function* ($f(x)$) is obtained. A probability function $f(x)$ is such that when one of the values that the random observable can assume is input, it produces an associated probability value as an output. This function describes the probability that a given value will occur. Some of the properties of a valid probability function are that its output is always greater than or equal to zero and the sum of all its possible output will tend to unity (or 1). If the random observable is a discrete type, the associated probability function is usually known as *probability mass function*. The probability mass functions specifically refer to discrete probability distributions in which the function $f(x)$ can take a discrete value (x) of a random variable (such as head or tail in tossing a coin) and provide its exact probability $[p(x)]$ of its occurrence, usually between the values of 0 and 1. A continuous random observable type, which can assume values between two different values in an interval from negative infinity to positive infinity (but not at a single point as in discrete type), will have a probability function known as *probability density function* (pdf) or simply as *density function*. Another type of probability function is *cumulative distribution function* (cdf), which gives the probability of a random observable taking a value less than or equal to a given value x. This type applies to both discrete and continuous random observables (variables).

It is typical in probability tradition to represent random observable (or variable) in capital letters, such as X, and its value in small letter, such as x. In this case, the probability mass function of a random variable X taking a value x can be expressed as $P(X = x)$, while the cdf of a random variable X taking a value less than or equal to x will be expressed as $P(X \leq x)$. Since cdf has important applications in statistical hypothesis testing, it will be explored further in this chapter. Also, normal distribution (or Gaussian distribution), which is one of the most important examples of a continuous probability distribution, will be discussed in Section 3.6.

3.2.1 Normal Probability Distributions and Density Functions

In geomatics, observables are random and their observations are considered continuous, i.e. each time an observable is measured, a different numerical value is possible for the observable, and usually the observable can assume any value within a certain interval. For example, a distance observable measured three times may assume the values 12.255, 12.250, and 12.245 m; considering these sample measurements, it can be seen that the distance observable is likely to assume any value within or outside the interval 12.245 and 12.555 m. This observable is then considered a continuous random variable since it can assume any value within an interval. From the principles of probability and statistics, a given continuous random observable X will have its probability distribution completely described by its *pdf* and its *cdf*. In geomatics, the observables, as continuous random variables, are assumed to have populations that are normally distributed with the errors involved assumed satisfying Gaussian theory of errors.

The geomatics observations can be represented and analyzed graphically or numerically. Graphical representation will involve creating histograms (showing relative frequencies of different values of the repeated measurement of an observable), which will lead to probability distribution of errors in repeated measurements. When systematic errors have been removed from survey observations, the random errors remaining shows a characteristic bell shape called the *normal error distribution curve* or the *pdf* of a normal random variable. The density function is a smoothed-out function of a histogram that describes the relative probabilities (or relative frequencies) of the values taken by the given observable. If the probability density around a value is large, it means the random observable is likely to be close to that value. For infinite number of values of the random observable with equal probability, this will theoretically mean that the probability of zero will be assumed by each value so that one cannot be thinking of pdf as providing probability values, but relative probabilities. When the density function is high, it means the associated value is relatively more probable, and when it is low, the value is relatively less probable. Probability density provides the probability of being within a very small interval containing the given value. The area under the density function represents the sum of all the probabilities of the values of the observable, and this is equal to 1; the areas under the density function between values on the x-axis give the probabilities of lying in those intervals. The equation of the normal distribution curve, also known as normal pdf, is given as follows:

$$f(x) = \frac{1}{\sigma\sqrt{2\pi}} e^{-\frac{1}{2}\left(\frac{x-\mu}{\sigma}\right)^2} \tag{3.1}$$

where $f(x)$ represents the probability of the occurrence of $(x - \mu)$ (i.e. the probability that the observable value x deviates by $x - \mu$ value from the central position μ of the distribution), σ is the true standard deviation (or spread of the distribution) of the measurement x, e is the base of the natural logarithms with approximate value of 2.718 28, π is the familiar mathematical constant with approximate value of 3.141 59, μ is the true value of x, $\dfrac{1}{\sigma\sqrt{2\pi}}$ is a factor to ensure that the total area under the distribution curve is equal to one (making the function a probability function), and the half in the exponent is to ensure that the distribution has unit variance (and also unit standard deviation) in the case of standard normal distribution. Although the $f(x)$ axis represents the frequencies of the values of the variable, the areas under the curve (representing the probabilities) are more important than the frequencies. This normal distribution is therefore a two-parameter family of curves with the parameters as μ and σ. Since the true value of random error should be zero ($\mu = 0$), then Equation (3.1), when used in relation to random errors, will become

$$f(x) = \frac{1}{\sigma\sqrt{2\pi}} e^{-\frac{1}{2}\left(\frac{x}{\sigma}\right)^2} \tag{3.2}$$

A standard normal distribution ($f(z)$) is obtained when Equation (3.1) is used in relation to a standardized random observable Z whose value can be represented by

$$z = \left(\frac{x - \mu}{\sigma}\right) \tag{3.3}$$

where the standard normal distribution ($f(z)$) sets σ to 1, giving, from Equation (3.1), the following:

$$f(z) = \frac{1}{\sqrt{2\pi}} e^{-\frac{1}{2}(z)^2} \tag{3.4}$$

Some of the important properties of z are that it has a population mean (μ) of zero and a population variance (σ^2) of 1 and it is dimensionless. The values of z are sometimes referred to as *standard scores* (or *z-scores*). Note that the z-score in Equation (3.3) represents how many standard deviations the observable value x is away from the population mean, μ. The graphical representation of the standard normal distribution is given in Figure 3.1. As it can be seen in the figure, the z-axis values are given in terms of how many standard deviations the $x-\mu$ observable value is away from the position where $x-\mu$ assumes zero value. The numerical values of the *pdf* in Equation (3.4) are usually provided in the statistical tables for standardized normal distribution for any given value of z.

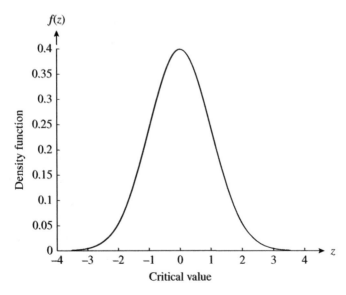

Figure 3.1 Plot of standard normal distribution function.

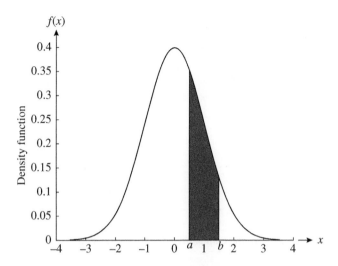

Figure 3.2 Probability of a random variable having its value in the interval between *a* and *b*.

In Figure 3.1, it can be seen that the density function is about 0.4 for a central value of 0, and for value of −1 or 1, its probability density is about 0.2. This means that a value close to 0 approximately is twice as probable as a value very close to −1 or 1. Note that the density at 0 is not the probability of getting 0, because density is not itself probability. The area (not the value of the function itself) under the probability function in Equation (3.1) (or the integral of the function), between two distinct points ($x = a$ and $x = b$), provides the probability for that interval as shown in Figure 3.2. This can be interpreted that the probability of the random variable X having its value in the interval between two points a and b is the shaded area under the density function from a to b, expressed mathematically as follows:

$$P(a \leq X \leq b) = \int_{a}^{b} f(x)\,dx \tag{3.5}$$

The area of the function over the whole population of values (from negative infinity to positive infinity) must be equal to one. The cdf $F(x)$ of a random variable X is defined in relation to the pdf for a variable value $x = a$ as follows:

$$F(x = a) = P(X \leq a) = \int_{-\infty}^{a} f(x)\,dx \tag{3.6}$$

Equation (3.6) means that the area ($F(a)$) under the pdf ($f(x)$) from negative infinity up to $x = a$ is the probability that the value of X will be at most equal

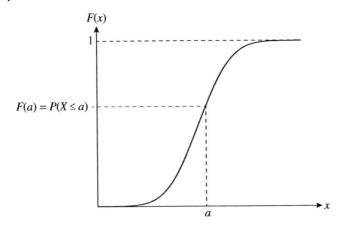

Figure 3.3 A graphical representation of the cumulative distribution function.

to *a*. From Equation (3.6), the mathematical relationship between *f(x)* and *F(x)* can be given as follows:

$$f(x) = \frac{\mathrm{d}(F(x))}{\mathrm{d}x} \tag{3.7}$$

The graphical representation of the cdf (*F(x)*) expressed by Equation (3.6) is given in Figure 3.3. In practice, *pdf* cannot be precisely known since it will require an exhaustive sampling of a whole population. Usually, estimates of the pdf are made based on finite samples with their associated uncertainties. The most popular estimators of probability density are the histograms. An example of a well-known pdf is based on the normal distribution curve (with parameters as the mean μ and the standard deviation σ), whose pdf is given for a variable value *x* by Equation (3.1).

The area under the standard normal pdf (the integral of the pdf) of a random variable (such as random error) over a given region as shown in Figure 3.4, for example, describes the relative likelihood for this random variable to occur within the region. The distribution of random errors (or standardized errors or standardized residuals) as shown in Figure 3.4 has the following properties:

- The frequency of occurrence (or probability) of a random error of a given magnitude depends on the magnitude of the error. For example, the probability of an error of magnitude -1σ or $+1\sigma$ (where σ is the standard deviation of measurement) as shown in Figure 3.4 is 68.27%. This means that for every 100 measurements, there is a probability that 68.27 of the measurements will

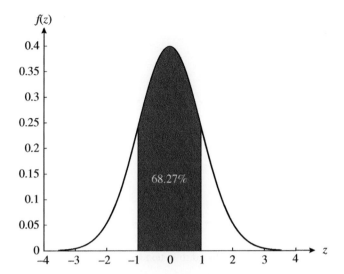

Figure 3.4 Area under standard normal distribution probability function.

have error of magnitude $\pm 1\sigma$. This value can be determined using the cdf on the standardized random variable Z, expressed as

$$P(-1 \leq Z \leq 1) = P(Z \leq 1) - P(Z \leq -1) \tag{3.8}$$

with $P(-1 \leq Z \leq 1)$ representing the probability of a random variable Z taking a value between plus or minus one standard deviation, $P(Z \leq 1)$ the probability of a random variable Z taking a value less than plus one standard deviation, and $P(Z \leq -1)$ the probability of a random variable Z taking a value less than minus one standard deviation. Equation (3.8) is further represented in Equation (3.9) and Figure 3.5.

$$P(-1 \leq Z \leq 1) = F(1) - F(-1) \tag{3.9}$$

Using the standard normal cumulative distribution, e.g. Microsoft Excel 2013 software routine NORM.S.DIST, Equation (3.9) can be rewritten as

$$P(-1 \leq Z \leq 1) = NORM.S.DIST(1, TRUE) - NORM.S.DIST(-1, TRUE) \tag{3.10}$$

giving the following:

$$P(-1 \leq Z \leq 1) = 0.841\ 34 - 0.158\ 655 \tag{3.11}$$

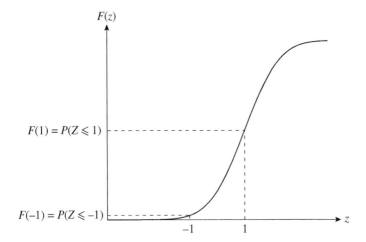

Figure 3.5 Area under standard normal cumulative distribution function.

which is 0.6827 or 68.27%. Similarly, the probability of an error of magnitude -3σ or $+3\sigma$ is 0.9973 or 99.73%, which can be given from the standard normal cumulative distribution in Figure 3.5 as follows:

$$P(-3 \le Z \le 3) = 0.998\ 65 - 0.001\ 35 \tag{3.12}$$

- Large errors occur less frequently than small ones. For example, in Figure 3.4, there is $(1-0.6827)$ or 31.73% probability of occurrence of errors greater than 1σ. This means that the larger the error, the less likely it is to occur in a measurement. The pdf in Equation (3.4) and Figure 3.4 show that the pdf approaches zero value as z (standardized value) increases in magnitude, meaning that great errors are extremely improbable.
- The error distribution can be positive and negative of the same magnitude. From Figure 3.4, positive and negative errors of the same magnitude are equally likely about the zero value.

In geomatics, observations and their residuals (or random errors) are considered to be normally distributed with the density function expressed by Equation (3.1). It is also common in geomatics to associate the maximum acceptable error in measurement with a range of uncertainty that has a very high probability level of occurrence, such as $\pm 3\sigma$ (or the error with probability of 99.73% that it will occur). This maximum acceptable error in measurement is often referred to as *measurement tolerance* or *survey tolerance*. In some geomatics projects, measurement tolerance may be specified at 95% probability level; if the tolerance is not specified, however, it will be safe to use 99.73% probability level (especially at the design stage of a project).

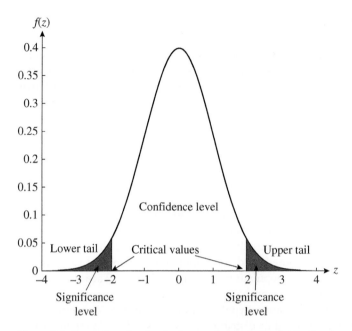

Figure 3.6 Properties of a standard normal probability distribution function.

If the probability of a variable Z taking a value less than an unknown value z is known (e.g. 95% probability level), the value of z can be estimated depending on whether two tails or one tail of the probability function is involved. The *tail* of a distribution usually refers to the region of the distribution that is far away from the mean (which is at the center of the distribution). A point where a tail starts when heading for positive infinity or where it ends when starting from negative infinity is known as the *critical value*. The areas under the tails from the critical values heading for negative and positive infinities are known as the *significance level* (α). The area of the region at the middle of the distribution contained by the lower and upper critical values is known as the *confidence level*. The tails, critical values, confidence level, and significance level of a standard normal distribution are illustrated in Figure 3.6. For example, given the probability of Z taking a value less than unknown value z as 95%, the value of z can be determined using the inverse standard normal probability function (NORM.S.INV (probability)) in Microsoft Excel 2013 software routine) as follows. For two-tailed problem (with significance level of 0.025 for lower and upper tails each), NORM.S.INV(0.025) is −1.96 as the critical value at the lower tail and NORM.S. INV(0.975) is 1.96 as the critical value at the upper tail. In the case of one-tailed problem (with 0.05 for either of the tails), NORM.S.INV(0.05) is −1.64 and 1.64 for the upper tail region. Thus, it can be seen that the critical values are symmetric about the mean with one of them being the negative of the other.

3.3 Sampling Distribution

In practice one is usually interested in examining and drawing valid conclusions about a population based on a sample. The process of collecting samples with the main objective of selecting that part of the population that is representative of the entire population is called *sampling*. A population can be finite or infinite depending on the *population size* (the number of items in the population). The process of choosing a sample is very important since the reliability of conclusions made concerning a population depends on whether the sample chosen sufficiently represents the population. If each member of the population has the same chance of being chosen during sampling, the sample chosen will be considered as *random sample*.

A population is considered known when the probability distribution (or density function) of the associated random observable is known. For example, if a random observable is normally distributed, then the population is considered to be normally distributed, and the population parameters such as the population mean (μ) and the population standard deviation (σ) are also considered known. In a case where the probability distribution of the population is not well known, one may make some hypotheses or assumptions concerning the general behavior of the probability distribution (or density function) like assuming normal distribution; if one or both values of μ and σ are unknown, one might need to estimate them. Random samples from the population are taken and used to obtain values for the population parameters.

One of the important problems of sampling theory is how to form a proper sample statistic that will best estimate a given population parameter. If all possible data sets (samples) of size n are randomly drawn from a given population containing all the possible data values, one can calculate a statistic (mean, z-score, t-score, or chi-square statistic) for each sample; the probability distribution of all the statistics of all the possible samples is called a *sampling distribution*. For an illustration, if samples of a given size are repeatedly taken from a population with the average of each sample calculated, the distribution of these averages is called sampling distribution of the sample mean. Similarly, the distribution of z-scores is known as *z-distribution* (or *standard normal distribution*) and the distribution of t-statistics is known as *t-distribution* (or *Student's t-distribution*). A distribution may be normal if the underlying population is normal and may be close to normal even when the population distribution is not according to what is known as *central limit theorem*. The theorem states that the sample mean will approach a normal distribution as the sample size increases (even if the observable itself is not normally distributed) and that if a random observable is the sum of some other random observables with associated pdf (which may be different from normal distribution), then that random variable tends to assume normal distribution. In practice, the sample size of 25 or 30 is considered large enough to assume that the sample mean will be approximately normally distributed. A knowledge of sampling distribution is important in inference statistics since it is needed to estimate parameters of the underlying population distributions. Sampling distributions provide probability distributions as well as measures of precision for estimated quantities.

3.3.1 Student's t-Distribution

Student's t-distribution is a pdf that depends on the degrees of freedom (df). The pdf can be given as

$$f(t) = \frac{\Gamma\left(\dfrac{df+1}{2}\right)}{\Gamma\left(\dfrac{df}{2}\right)\sqrt{\pi df}}\left(1+\frac{t^2}{df}\right)^{-(df+1)/2} \qquad \text{for } -\infty < t < \infty \qquad (3.13)$$

where $\Gamma(\cdot)$ is a gamma function. The curve associated with Student's t-distribution closely approximates normal curve when the degrees of freedom involved is equal to or greater than 30. The comparison of Student's t-distributions at three different degrees of freedom (5, 15, and 30) is given in Figure 3.7. In Figure 3.8, the normal distribution is compared with Student's t-distribution with the number of degrees of freedom as 15.

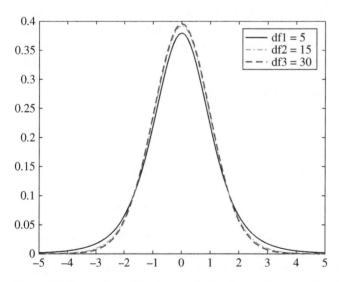

Figure 3.7 Comparison of Student's t-distributions at degrees of freedom 5, 15, and 30.

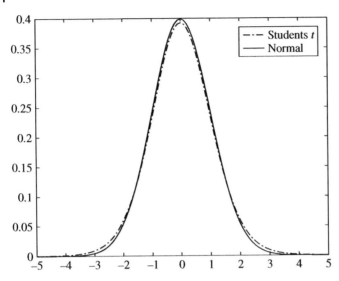

Figure 3.8 Comparison of standard normal and Student's t-distributions (for df = 15).

It can be seen in Figure 3.8 that normal and Student's t-distributions are very close, even at the low number of degrees of freedom of 15; the similarity of the distributions in Figure 3.8 is identical to the similarity of the t-distributions with degrees of freedom of 15 and 30 in Figure 3.7. For example, the probability of a random observable T taking a value less than $t = -1$ at the degrees of freedom of 15 can be determined using the cumulative Student's t-distribution (e.g. using the Microsoft Excel 2013 software routine T.DIST(x, df, TRUE)), which gives 0.17 for T.DIST(-1, 15, TRUE); for a value of $t = 1$, the cumulative probability is 0.83 using T.DIST(1, 15, TRUE). The probability of taking a value between $t = -1$ and $t = 1$ is 0.83 minus 0.17 (or 0.66 or 66%); it is 68% for standard normal distribution as shown in Figure 3.4. For the inverse case, if the probability of a random observable T taking a value less than an unknown value t is 95%, the value of t can be determined using the inverse standard Student's t-distribution function (T.INV(probability, df) in Microsoft Excel 2013 software routine) as follows. For two-tailed problem (with significance level of 0.025 for lower and upper tails each), T.INV(0.025, 15) for lower tail is -2.13 and T.INV(0.975, 15) for upper tail is 2.13; the values for standard normal distributions are -1.96 and 1.96 as discussed in Section 3.3.1. As in the case of normal distribution, it can be seen that the critical values are symmetric about the mean with one of them being the negative of the other.

3.3.2 Chi-square and Fisher's F-distributions

From the sampling distribution, the χ^2 *statistic (or chi-square statistic)* for the sample variance (s^2) can be given as

$$\chi^2 = \frac{(\mathrm{df})s^2}{\sigma^2} \tag{3.14}$$

where df is the number of degrees of freedom and σ is the population standard deviation. (Refer to Figure 3.9 and Table B.3 for chi-square values with p levels referring to the lower tail areas.) For example, the chi-square value ($\chi^2_{\mathrm{df}=4,\,p=1-0.005}$) has significance level, α as 0.005, and the p level as 0.995 with the number of degrees of freedom (df) as 4. This chi-square value can be determined using the Microsoft Excel 2013 software routine CHISQ.INV(0.995,4), which gives 14.86. The lower tail area value ($\chi^2_{\mathrm{df}=4,\,\alpha=0.005}$) having p equals α can be obtained using the routine CHISQ.INV(0.005,4), which gives 0.21.

The relationship between chi-square value (e.g. $\chi^2_{\mathrm{df}=4,\,p=1-0.005}$) and the level of significance α (which is the area from $\chi^2_{\mathrm{df}=4,\,p=1-0.005}$ to positive infinity ∞) is given by Equation (3.15):

$$\alpha = \int_{\chi^2_{\mathrm{df}=4,\,p=1-0.005}}^{\infty} \chi^2(x)\,dx \tag{3.15}$$

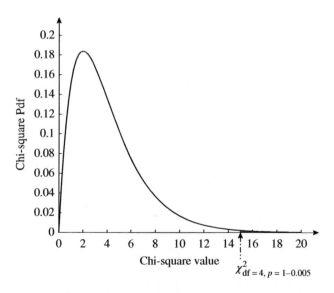

Figure 3.9 Typical chi-square distribution with critical values for degrees of freedom of 4 (with $\alpha = 0.005$ in the upper tail area).

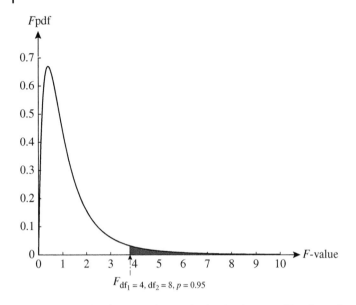

Figure 3.10 *F*-distribution with critical value for degrees of freedom of 4 and 8 (with $\alpha = 0.05$ in the upper tail area).

The other related density function is Fisher's *F*-distribution with two degrees of freedom (df_1 and df_2) involved. The *F*-distribution curve for the degrees of freedom 4 and 8 is shown in Figure 3.10. For example, the F-distribution value for significance level of 0.05 in the upper tail area (p equals 1 minus 0.05) with the degrees of freedom $df_1 = 4$ and $df_2 = 8$ can be determined using the Microsoft Excel 2013 software routine F.INV(0.95,4, 8), which gives 3.84. The lower tail area value (p equals α) can be obtained using the routine F.INV(0.05,4, 8), which gives 0.16.

The relationship between *F*-distribution and chi-square distribution values can be given as

$$F_{p,\,df_1,\,df_2} = \frac{\chi^2_{p,\,df_1}/df_1}{\chi^2_{p,\,df_2}/df_2} \tag{3.16}$$

where $F_{p,\,df_1,\,df_2}$ is the *F*-distribution value, $\chi^2_{p,\,df_1}$ and $\chi^2_{p,\,df_2}$ are the chi-square distribution values, df_1, df_2 are the degrees of freedom, $p = 1 - \alpha$, and α is the significance level. It should be noted also that the following relationship between chi-square and *F*-distributions is valid when the number of degrees of freedom for the denominator tends to infinity (∞):

$$F_{p,\,df_1,\,\infty} = \frac{\chi^2_{p,\,df_1}}{df_1} \tag{3.17}$$

The chi-square is also related to the z-score of the standard normal distribution when $df_1 = 1$, so that the following relationship is valid:

$$z_{1-\alpha/2} = \sqrt{\chi^2_{df=1,\,p}} \qquad (3.18)$$

where $z_{1-\alpha/2}$ is the z-score from the standard normal distribution, $p = 1 - \alpha$, and $|\cdot|$ is for absolute value of what is contained within the vertical bars. Note in Equation (3.18) that $\alpha/2$ is used in the standard normal distribution value, while α is used in the chi-square distribution value.

3.4 Joint Probability Function

The idea of probability functions involving only one random observable can be generalized to involve two or more random observables, producing what is commonly referred to *as joint probability function (joint density function)*. This type of probability function, for example, in the case involving two observables (x, y), represents a surface known as probability surface with the total volume bounded by this surface and the xy-plane being equal to 1; this surface is illustrated in Figure 3.11. The density function of n-dimensional multivariate (for multiple observables) normal distribution can be generalized from the univariate normal case for $n \times 1$ vector X of observations of several variables as

$$f(X) = \frac{1}{(2\pi)^{n/2}\sqrt{\det(\Sigma)}} e^{-\frac{1}{2}(X-\mu)^T \Sigma^{-1}(X-\mu)} \qquad (3.19)$$

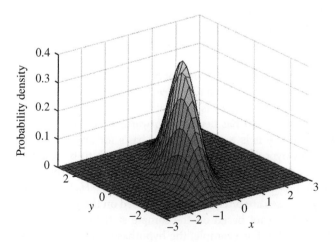

Figure 3.11 Bivariate normal probability density function.

where X is a vector of the measured values of the observables, Σ is $n \times n$ true covariance matrix of the vector of measured values, μ is the vector of the true values of the observables, n is the number of observables (not the number of observations) considered in the multivariate case, $(2\pi)^{n/2}\sqrt{\det(\Sigma)}$ is the normalizing constant that makes the volume under the multivariate normal distribution probability density equal to one, and $\det(\Sigma)$ is the determinant of the covariance matrix Σ. The multivariate normal distribution is a distribution for random vectors of observables, each observable of which has a univariate normal distribution. Figure 3.11 illustrates this with two-dimensional pdf, which is a volume that is bounded by a surface that is a function of x and y observables. The probability values of the density function are within the intervals $x = -3$ to $+3$ and $y = -3$ to $+3$. If the contour plot of the lines of constant probability density is projected onto the $x–y$ plane, a series of ellipses will be produced. The concepts of joint probability function are used in creating confidence error ellipses, which are discussed in Chapter 7 of this book.

3.5 Concepts of Statistical Hypothesis Tests

An assumption, guess, or statement that one makes about the unknown parameters of a probability distribution of a population or about the distribution itself is called *statistical hypothesis*. This assumption or guess may or may not be true. The statistical hypothesis can be made on whether or not a population has a particular probability distribution or if the mean or a variance belongs to a particular distribution. A decision-making procedure that consists of making statistical hypotheses and carrying out statistical tests to determine if these hypotheses are acceptable or not is known as *statistical hypothesis test* (or *test of significance*). Mathematical and stochastic models used in adjustments are based on some statistical hypotheses with statistical tests conducted to verify those hypotheses.

Two complementary hypotheses are commonly made and tested: *null hypothesis (or statistical hypothesis)*, H_0, and *alternative hypothesis*, H_A. Null hypothesis (or statistical hypothesis), H_0, is a specific statement (or a theory) about a property of population that has been put forward, either because it is believed to be true or because it is to be used as a basis for argument, but has not been proved. This hypothesis must be specific enough to be testable. The alternative hypothesis, H_A, is a statement of what a statistical hypothesis test is set up to establish. It is the statement to be accepted if the H_0 hypothesis fails. The H_0 and H_A hypotheses are often formulated based on the given values of the population parameters (not based on the sample statistics or the estimated values from the sample). For example, the hypotheses can be set up as $H_0: \mu = 0$ vs. $H_A: \mu > 0$ or $H_A: \mu < 0$ (but not as $H_0: \bar{x} = 0$ vs. $H_A: \bar{x} > 0$ or $H_0:$

$\bar{x} < 0$, since one cannot be making an assumption about a quantity like $\bar{x}$, which can be clearly determined from the sample). The commonly made statistical hypotheses (null hypotheses) in adjustments are:

- No gross errors (blunders) exist in the observations.
- Mathematical model formulated for the observations gives a correct relationship between the observations and the unknown parameters; systematic errors have been adequately taken care of.
- Chosen stochastic model for the observations (e.g. covariance matrix) appropriately describes the stochastic properties of the observations.

The outcome of a hypothesis test is to "reject H_0 in favor of H_A" or "do not reject H_0." The probability of rejecting H_0 hypothesis when it is correct is represented by the *significance level* α. The significance level is the extent that one is willing to take a risk of making wrong conclusion from a test and its value must be chosen before the test. It represents the total area from a particular value (*critical value*) up to the infinities. The probability of accepting H_0 when it is correct (resulting in making a correct judgment) is represented by the *confidence level* $(1 - \alpha)$. The confidence level describes the uncertainty associated with a particular sampling method. This means that if a particular sampling method is used to select different samples from a population and a different interval estimate is computed for each sample, some interval estimates would include the true population parameter and some would not. At 90% confidence level, for example, one would expect 90% of the interval estimates to include the population parameter. In this case, one may pose a question, "how confident is one in one's measurements or how many times out of 100 is one expecting one's sample mean to be close to the true mean?" The *critical value* for a hypothesis test is a threshold to which the value of the *test statistic* in a sample is compared to determine whether or not the null hypothesis is rejected. The null hypothesis is rejected if a test statistic falls into rejection regions (beyond the *critical values*) of the sampling distribution of the statistic; if the *rejection regions* become smaller in size (smaller significance level being chosen), it is less likely to reject the null hypothesis. *Test statistic* is a quantity calculated from a sample of data. Its value is used to decide whether or not the null hypothesis should be rejected in a hypothesis test. The possible test decision-making options are given in Table 3.1.

As can be seen in Table 3.1, two types of error of judgment are possible in hypothesis testing: *type I* and *type II* errors. *Type I error* is the error of rejecting H_0 hypothesis when it is in fact true, i.e. H_0 is wrongly rejected. The probability of type I error is the significance level (α) of the test; the significance level is the level at which a result is not likely to be due to random chance but something else. *Type II error* is the error of rejecting H_A hypothesis when it is in fact true; it occurs when the null hypothesis H_0 is not rejected when it is in fact false. The probability of a type II error is generally unknown, but is symbolized by β. The

Table 3.1 Possible test decisions with associated probability levels.

Reality	Test decision	Probability	Error in judgment
H_0 is true	H_0 accepted	$1 - \alpha$	OK (confidence level)
	H_0 rejected	α	Type I error
H_0 is false	H_0 accepted	β	Type II error
	H_0 rejected	$1 - \beta$	Ok (power of test)

probability of rejecting the false H_0 hypothesis is called the *power of a statistical hypothesis test* or simply the *power of test*, i.e. the power of the test to locate bad observation in a set of observations if the bad observation actually exists. Power of a statistical hypothesis test measures the test's ability to reject the null hypothesis when it is actually false, i.e. the ability to make a correct decision. In other words, the power of a hypothesis test is the probability of not committing a type II error. It is calculated by subtracting the probability of a type II error from one (or $1 - \beta$). The maximum power a test can have is 1, and the minimum is 0. Ideally, one would want a test to have high power that is close to 1. Generally, if one reduces the chance of type I error by reducing the size of the significance level α, the chance of making type II is automatically increased with β becoming bigger and the test becoming less powerful. Type I error is often considered to be more serious than type II error and therefore more important to avoid than a type II error. The hypothesis test procedure is often adjusted so that there is a guaranteed "low" probability of making type I error. This probability, however, is never as low as zero and its choice is usually arbitrary. In geomatics practice, significance level is usually chosen as $\alpha = 0.05$ (or 5%) or $\alpha = 0.01$ (or 1%), so that the corresponding confidence levels $(1 - \alpha)$ become 0.95 (or 95%) and 0.99 (or 99%), respectively. The significance level of 0.05 is considered "moderately significant" (taking a risk of being wrong 5 times per 100 trials), and 0.01 is "highly significant" (more stringent level, but providing more convincing or less risky conclusion, since one is only taking a risk of being wrong one time per 100 trials).

3.6 Tests of Statistical Hypotheses

Tests of hypotheses (or tests of significance) are procedures that enable one to decide whether to accept or reject hypotheses or determine whether observed samples differ significantly from expected results. During these tests, statistical decisions are made about populations based on sample data. Before arriving at a statistical decision, assumptions or guesses (statistical hypotheses) about populations are first made. Note that the type of hypothesis testing discussed in this section is based on the null hypothesis (H_0) probability distribution in which it is assumed that H_0 is true (with an error of judgment of α); the test does not include testing whether a distribution based on the alternative hypothesis is true (so that the probability β is not considered). Depending on the type of test, different assumptions may be made about the distribution of random variables that are being sampled. Whatever assumptions that are made, however, must be found to be consistent with the chosen test and also appropriate when interpreting the results of the test. The general steps for statistical hypothesis testing can be given as follows:

i) Formulate the appropriate hypotheses, providing H_0 and H_A. H_0 will specify the value of the population parameter, while H_A is the theory that the investigator believes is correct. Some statistical books will express the null hypothesis as a strict equality and specify a unique value for the parameter in the null hypothesis, this book recognizes that H_0 and H_A should be all-inclusive, but will adopt strict equality for null hypotheses for simplicity. The null hypotheses, however, can be modified appropriately to make the hypotheses all-inclusive if desired.

ii) Select the significance level (e.g. $\alpha = 0.10$) and extract the critical values based on the assumed statistical distribution of the population from which the sample data are taken.

iii) Give the appropriate test statistic. The test statistic should be chosen depending on the conditions satisfied by the sample data.

iv) Make decision. The decision is to "reject H_0 in favor of H_A" or "do not reject H_0." The rejection region usually identifies the values of the test statistic that support the alternative hypothesis.

v) Interpret the result and provide an appropriate conclusion. This is necessary so that the outcome of the test is presented in a manner that is easily understood by the client. This should go beyond just "reject" or "do not reject" the null hypothesis, but should relate the statistical result to the problem solved.

In formulating H_0 and H_A hypotheses, it must be decided whether the test will be *one tailed* or *two tailed*. The tails of a distribution are the regions away from the mean of the distribution from the critical points to the infinities; they are the critical regions of the distribution with their areas being equal to the level of significance (α). After selecting the α level for a test, the two-tailed test assigns $\alpha/2$ to test the hypothesis in one direction and the other $\alpha/2$ to test the hypothesis in the other direction so that the overall significance for the test is α. This type of test should be done if there is a possible relationship in both directions, especially if the effect in one direction cannot be neglected or considered irrelevant. In the case of one-tailed test, all of the chosen α for the test will be assigned to test the hypothesis in only the one direction of interest. This type of test is for testing relationship in one direction while disregarding the possibility of a relationship in the other direction. This test should be done if the effect of a test in the other direction is negligible or irrelevant. One-tailed test, however, is more powerful in detecting an effect in one direction by ignoring the effect in the other direction.

3.6.1 Test of Hypothesis on a Single Population Mean

Testing a hypothesis on a single population mean (μ) is a case where one has to decide if the population mean is equal to a known value (δ). In this test it is required to find if the sample mean ($\bar{x}$) is consistent with the population mean, i.e. if $\mu = \delta$, then $\bar{x}$ is consistent with μ. The concept of hypothesis testing discussed in Section 3.6 can be applied to testing hypotheses on a single population mean as follows:

i) Formulate the appropriate hypotheses for one-tailed test or two-tailed test, depending on what is of interest, as shown in Table 3.2. One-tailed test will be appropriate if one is interested in testing if the population mean is significantly greater than or significantly less than the sample mean, but not both. A two-tailed test should be done if one is interested in knowing if the population mean is significantly greater than or significantly less than the sample mean in the same single test. In a test, however, it is usually easier to identify the alternative hypothesis first. As can be seen in Table 3.2, when formulating the hypotheses, the alternative hypothesis is used to identify whether the test is one tailed and which tail is involved. The H_0 for one-tailed test in the table can be made all-inclusive by modifying it to $H_0: \mu \geq \delta$ or $H_0: \mu \leq \delta$, whichever is appropriate.

ii) Select the significance level (e.g. $\alpha = 0.10$) and extract the critical values from the appropriate statistical distribution functions. This step assumes that the sampled data are independently drawn from a normal distribution, so that the commonly used distributions are the standard normal and Student's t-distributions. Since these distributions are symmetric about zero, their values in one tail are the same (but with opposite sign) as the corresponding ones in the other tail.

Table 3.2 Hypotheses on a single population mean.

	One-tailed test	Two-tailed test
Null hypothesis	$H_0: \mu = \delta$	$H_0: \mu = \delta$
Alternative hypothesis	$H_A: \mu < \delta$ or $H_A: \mu > \delta$	$H_A: \mu \neq \delta$

iii) Give the appropriate test statistic. The commonly used test statistics for testing the population means are the z- and t-statistics, leading to z- and t-tests, respectively. The z-test or t-test can be performed depending on the conditions satisfied by the sample data. For example, for sample size $n \leq 30$ with the population standard deviation (σ) unknown, the *t-test* should be done using the following statistic:

$$t = \frac{\bar{x} - \mu}{SE} \tag{3.20}$$

where $SE = s/\sqrt{n}$ is the standard deviation of the mean computed from the sample standard deviation s (where $\bar{x}$ and s are from independent samples (Vanicek and Krakiwsky 1986)) with the t-statistic having Student's t-distribution with the number of degrees of freedom as $n - 1$. If $\bar{x}$ and s are calculated from the same sample, *tau distribution* by Pope (1976) should be used instead of t-distribution (tau distribution sample is provided in Appendix C). For $n > 30$ or in the case where the population standard deviation (σ) is known, the following *z-test* statistic should be used:

$$z = \frac{\bar{x} - \mu}{SE} \tag{3.21}$$

where $SE = \sigma/\sqrt{n}$ with the z-statistic having standard normal distribution with the population mean of zero and the population standard deviation of 1.

iv) Make decision. Usually, two-tailed test is commonly performed to test means of measurements. The decision is to "reject H_0 in favor of H_A" or "do not reject H_0." Typical decisions that can be made with regard to hypothesis test for a single population mean are given in Table 3.3, where $t_{1-\alpha, df}$ and $t_{1-\alpha/2, df}$ are Student's t-distribution critical values and $z_{1-\alpha}$ and $z_{1-\alpha/2}$ are the standard normal distribution critical values. In the table, the null hypothesis H_0: $\mu_1 - \mu_2 = 0$ should be rejected under the conditions stated for different alternative hypotheses. The one-tailed and two-tailed regions of normal distributions are illustrated in Figure 3.12.

For example, in Table 3.3, if z (or t) is less than the given critical value when H_A: $\mu < \delta$ (in a one-tailed test), then H_0: $\mu = \delta$ should be rejected.

Table 3.3 Decisions in hypothesis test on a single population mean.

Alternative hypothesis	Reject H_0: $\mu = \delta$ (for z-statistic)	Reject H_0: $\mu = \delta$ (for t-statistic)
H_A: $\mu < \delta$	$z > z_\alpha$ for $n > 30$	$t < t_\alpha$ for $n \leq 30$
H_A: $\mu > \delta$	$z > z_{1-\alpha}$ for $n > 30$	$t < t_{1-\alpha}$ for $n \leq 30$
H_A: $\mu \neq \delta$ two tailed	$z > z_{1-\alpha/2}$ or $z > z_{\alpha/2}$ for $n > 30$	$t < t_{1-\alpha/2}$ or $t < t_{\alpha/2}$ for $n \leq 30$

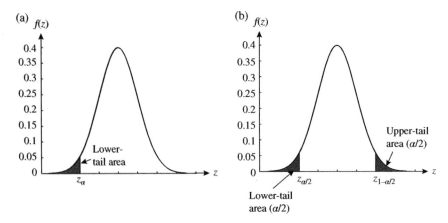

Figure 3.12 (a) One-tailed and (b) two-tailed regions of normal distributions.

In the case of two-tailed test, if the value of z (or t) satisfies the two given conditions when $H_A : \mu \neq \delta$, then the $H_0 : \mu = \delta$ should be rejected at the given significance level (e.g. $\alpha = 0.10$ level). Note that if one is accepting the H_0, it should be remembered that one is accepting it against the alternative hypothesis H_A. Generally, two-tailed test should be performed when testing the mean measurements since both the population mean being significantly greater than and significantly less than the sample mean are both equally relevant.

v) Interpret the result and provide a conclusion. This step should state if measurement being tested should be accepted at the given significance level or not.

Example 3.1 A baseline of calibrated length (μ) 1153.00 m is measured 5 times. Each measurement is independent and made with the same precision. The sample mean ($\bar{x}$) and sample standard deviation (s) are calculated from the measurements as $\bar{x} = 1153.39$ m and $s = 0.06$ m, respectively. Test at the 10% significant level if the measured distance is significantly different from the calibrated distance and also if the measured distance is significantly greater than the calibrated value.

Solution (Part 1):

$$n = 5, \quad \bar{x} = 1153.39 \text{ m} \quad s = 0.06 \text{ m} \quad \mu = 1153.00 \text{ m}$$

i) State the H_0 and H_A hypotheses. The term "significantly different" suggests that two directions of the test must be evaluated as stated in the alternative hypothesis, i.e. two-tailed test:

$$H_0 : \mu = 1153.00 \quad \text{vs.} \quad H_A : \mu \neq 1153.00$$

ii) Select the significance level – $\alpha = 0.10$ – and the critical values. Since the sample size ($n = 5$) involved is smaller than 30 and the population standard deviation (σ) is not provided, the Student's *t-distribution* should be used in extracting the critical values. In order to obtain the Student t values, one can use the Microsoft Excel 2013 software routine T.INV with the cumulative probability of $1-0.05$ or 0.95 (for $\alpha = 0.10/2$ being the desired significant level for a tail) and the degrees of freedom as 4; the calculated values are $t_{1-0.10/2, \ df \ = \ 5-1} = 2.132$ and $t_{0.10/2, \ df \ = \ 5-1} = -2.132$. Table B.2 can also be used.

iii) Give the test statistic from Equation (3.20) for unknown σ or $n \leq 30$, where SE is the standard deviation of the mean:

$$t = \frac{\bar{x} - \mu}{SE}$$

$$SE = \frac{s}{\sqrt{n}} \quad \rightarrow \quad SE = \frac{0.06}{\sqrt{5}} = 0.0268$$

$$t = \frac{1153.39 - 1153.00}{0.0268} = 14.552$$

iv) Decision for two-tailed test is to reject H_0: $\mu = 1153.00$ if one or both of the following conditions are satisfied:

$$t > t_{1-\alpha/2} \quad \text{or} \quad t < t_{\alpha/2}? \rightarrow 14.552 > 2.132?$$

Since one of the conditions ($14.552 > 2.132$) is satisfied, reject the H_0: $\mu = 1153.00$ at 10% significance level and accept the alternative hypothesis that H_A: $\mu \neq 1153.00$.

v) Based on the outcome of this test, the measured distance is significantly different from the calibrated distance at 10% significance level.

Solution (Part 2):

Testing if the measured distance is significantly greater than the calibrated value

$$n = 5, \quad \bar{x} = 1153.39 \text{ m} \quad s = 0.06 \text{ m} \quad \mu = 1153.00 \text{ m}$$

i) State the H_0 and H_A hypotheses to test if measurement is significantly greater; if the measurement is significantly greater, it means the corresponding population mean should be greater than the given value (the H_A hypothesis):

$$H_0 : \mu = 1153.00 \quad \text{vs.} \quad H_A : \mu > 1153.00$$

ii) Select the significance level, $\alpha = 0.10$ for one-tailed test and determine the corresponding t critical value using the Microsoft Excel 2013 software routine T.INV with the cumulative probability for $\alpha = 0.10$ being the desired significant level for the upper tail and the degrees of freedom as 4; the calculated value should be 1.533.

iii) Give the test statistic from part 1 for σ unknown or $n < 31$: $t = 14.552$.

iv) Decision for one-tailed test is as follows: Reject H_0 if $t > t_{1-\alpha, \, df} \rightarrow 14.552 > 1.533$? Since 14.552 is greater than 1.533, the H_0: $\mu = 1153.00$ is rejected at 10% significance level.

v) Based on the outcome of this test, the measured distance is significantly greater than the calibrated distance at 10% significance level.

3.6.2 Test of Hypothesis on Difference of Two Population Means

This is a case where one is trying to decide if two population means (μ_1 and μ_2) are equal or the two population means are not statistically different. For example, if two survey crews independently determined two sample means as $\bar{x}_1$ and $\bar{x}_2$ with the corresponding standard deviations as s_1 and s_2, respectively, one may want to decide if $\bar{x}_1$ and $\bar{x}_2$ are statistically equal. The hypotheses can be formulated as follows:

Null hypothesis : $H_0 : \mu_1 - \mu_2 = 0$ or $H_0 : \mu_1 = \mu_2$

Possible alternative hypotheses : $H_A : \mu_1 - \mu_2 \neq 0$ (or $H_A : \mu_1 \neq \mu_2$) or

$H_A : \mu_1 - \mu_2 < 0$ or $H_A : \mu_1 - \mu_2 > 0$

The t-statistic for this type of test for a case where the sample sizes, n_1 or $n_2 \leq 30$ and the population standard deviation (σ) unknown, can be given as

$$t = \frac{(\bar{x}_1 - \bar{x}_2) - (\mu_1 - \mu_2)}{SE} \tag{3.22}$$

where for small samples (n_1, $n_2 \leq 30$) having equal population variances ($\sigma_1^2 = \sigma_2^2$), the pooled variance is used to estimate SE as

$$SE = \sqrt{\frac{(n_1 - 1) s_1^2 + (n_2 - 1) s_2^2}{n_1 + n_2 - 2}} \tag{3.23}$$

with the number of degrees of freedom, df $= n_1 + n_2 - 2$. For a case where the sample sizes, n_1, $n_2 > 30$, the following test statistic is used:

$$z = \frac{(\bar{x}_1 - \bar{x}_2) - (\mu_1 - \mu_2)}{SE} \tag{3.24}$$

with

$$SE = \sqrt{\frac{s_1^2}{n_1} + \frac{s_2^2}{n_2}} \tag{3.25}$$

Table 3.4 Decisions of test on difference between population means.

Alternative hypothesis	Reject $H_0: \mu_1 - \mu_2 = 0$ (for z-statistic)	Reject $H_0: \mu_1 - \mu_2 = 0$ (for t-statistic)
$H_A: \mu_1 - \mu_2 < 0$	$z < z_\alpha$ for n_1 or $n_2 > 30$	$t < t_\alpha$ for n_1 or $n_2 \leq 30$
$H_A: \mu_1 - \mu_2 > 0$	$z < z_{1-\alpha}$ for n_1 or $n_2 > 30$	$t > t_{1-\alpha}$ for n_1 or $n_2 \leq 30$
$H_A: \mu_1 - \mu_2 \neq 0$ two tailed	$z > z_{1-\alpha/2}$ or $z < z_{\alpha/2}$ for n_1 or $n_2 > 30$	$t > t_{1-\alpha/2}$ or $t > t_{\alpha/2}$ for n_1 or $n_2 \leq 30$

or

$$SE = \sqrt{\frac{\sigma_1^2}{n_1} + \frac{\sigma_2^2}{n_2}} \qquad (3.26)$$

or using the propagated error for the difference $(\bar{x}_1 - \bar{x}_2)$, where σ_1^2 and σ_2^2 are the population variances for the two samples. From the H_0: $\mu_1 - \mu_2 = 0$, Equations (3.22) and (3.24) will be reduced to the following, respectively:

$$t = \frac{(\bar{x}_1 - \bar{x}_2)}{SE} \qquad (3.27)$$

and

$$z = \frac{(\bar{x}_1 - \bar{x}_2)}{SE} \qquad (3.28)$$

The null hypothesis H_0: $\mu_1 - \mu_2 = 0$ should be rejected under the conditions stated for different alternative hypotheses in Table 3.4.

Example 3.2 An angle was measured in six sets at a particular atmospheric condition using a particular survey instrument. Considering the mean of a set as a single measurement, the mean angle of all the six sets was determined as 115°45′37″ with the standard deviation of measurement determined from the sample as 4″. This angle was measured again in five sets with the same instrument, but different observer and at different atmospheric condition with the mean as 115°45′45″ and the new standard deviation determined from the sample as 6″. Determine at 95% confidence level if the two means are statistically the same, assuming the population variances of the two samples of measurements are equal.

Solution:

Given $\bar{x}_1 = 115°45′37″$, $s_1 = 4″$, $n_1 = 6$; $\bar{x}_2 = 115°45′45″$, $s_2 = 6″$, $n_2 = 5$.

Hypotheses: $H_0 : \mu_1 - \mu_2 = 0$ vs. $H_A : \mu_1 - \mu_2 \neq 0$

The *t*-statistic for the sample sizes n_1 or $n_2 \leq 30$ and the population standard deviation (σ) unknown use Equation (3.22) with SE given by Equation (3.23) as

$$SE = \sqrt{\frac{(6-1)4^2 + (5-1)6^2}{6+5-2}} \quad \text{or} \quad 4.99$$

$$t = \frac{\left(115°45'37'' - 115°45'37''\right) - 0}{4.99} \quad \text{or} \quad -1.603$$

The significance level is $\alpha = 0.05$; degrees of freedom is df = 9 (from df = n_1 + n_2 − 2); the critical values for two-tailed tests are obtained using Microsoft Excel 2013 software routine T.INV(0.025, 9) so that $t_{\alpha/2} = -2.262$ and the symmetric value is $t_{1-\alpha/2} = 2.262$. From Table 3.4, reject the H_0: $\mu_1 - \mu_2 = 0$ if the following conditions are satisfied: $t > t_{1-\alpha/2}$ or $t < t_{\alpha/2}$. Since −1.603 is not greater than 2.262 and −1.603 is not less than −2.262, H_0: $\mu_1 - \mu_2 = 0$ is not rejected at 95% confidence level. It can be concluded at 95% confidence level that the two means are statistically the same.

Example 3.3 Two survey crews independently determined the elevation of a single benchmark as $\bar{x}_1 = 105.254$ m ± 0.007 m and $\bar{x}_2 = 105.249$ m ± 0.004 m based on their leveling runs from different starting points and along different routes. If the standard deviations of the elevations are considered well estimated, determine if $\bar{x}_1$ is significantly greater than $\bar{x}_2$ at 90% confidence level.

Solution:

Given $\bar{x}_1 = 105.254$ m, $s_{\bar{x}_1} = 0.007$ m; $\bar{x}_2 = 105.249$ m, $s_{\bar{x}_2} = 0.004$ m.

The term "significantly greater than" forms the basis for the alternative hypothesis.

Hypotheses: H_0: $\mu_1 - \mu_2 = 0$ vs. H_A: $\mu_1 - \mu_2 > 0$
The SE for the $\bar{x}_1 - \bar{x}_2$ is propagated as

$$SE = \sqrt{s_{\bar{x}_1}^2 + s_{\bar{x}_2}^2} \text{ giving } \pm 0.0081 \text{ m}$$

For standard deviations of elevations well known, the statistic in Equation (3.28) can be used:

$$z = \frac{(105.254 - 105.245)}{0.0081} = 1.111$$

The critical value is based on the significance level, $\alpha = 0.10$; using Microsoft Excel 2013 software routine NORM.S.INV(0.90) gives $z_{1-\alpha} = 1.282$. According to Table 3.4, the H_0: $\mu_1 - \mu_2 = 0$ should be rejected if $z > z_{1-\alpha}$ is satisfied. Since 1.111 is not greater than 1.282, the H_0 hypothesis is not rejected, and it can be concluded at 90% confidence level that $\bar{x}_1$ is not significantly greater than $\bar{x}_2$.

3.6.3 Test of Measurements Against the Means

It is usually desired in geomatics to test individual measurement in a data set against the mean for the purpose of determining if the measurement is consistent with the rest of the measurements in that data set. If the measurement tested is not consistent with the rest, it is considered as an *outlier*. The test can be done *out of context* or *in context*, where out-of-context test ignores the influence of the other measurements in the choice of significance level while in-context test does not. The difference is in the choice of significance level (α_0): If $\alpha_0 = \alpha$ is used for out-of-context test, then $\alpha_0 = \alpha/n$ will be used for in-context test, where n is the number of measurements in the sample. According to *Bonferroni inequality*, if the significance level for joint probability of n equally likely events is α, the significance level for individual event (with consideration for possible dependence of events) should be greater than or equal to α/n. In this case, the joint probability of the n events occurring should be greater than or equal to $1 - \alpha$. The in-context and out-of-context tests are typically two tailed, and the null hypothesis is like

H_0 : *The measurement has normal pdf (i.e. it is normally distributed).*

Considering the two-tailed test further, it can be shown from Equations (3.27) and (3.28) that the expected critical value of the difference between the two sample means at $(1 - \alpha)$ confidence level will be

$$|\bar{x}_1 - \bar{x}_2| = (\text{SE})k_p \tag{3.29}$$

$$k_p = t_{p = 1 - \alpha_0/2, \text{ df}} \text{ or } k_p = z_{p = 1 - \alpha_0/2} \tag{3.30}$$

where k_p can be considered as a factor for scaling the standard deviation to the given probability (p) level. Following the procedure in Equations (3.29) and (3.30), the expected marginal values that must be satisfied by the differences between the measurements and their means can be given as follows:

i) If the population mean (μ) and the population standard deviation ($\text{SE} = \sigma$) are known, the difference between a measurement ℓ_i and its mean will be statistically insignificant at α significance level if the following is satisfied:

$$|\ell_i - \mu| < \sigma \times k_p \tag{3.31}$$

where $k_p = z_{p = 1 - \alpha_0/2}$. If the population mean (μ) is known but the population standard deviation (σ) is unknown but estimated as $\text{SE} = s$ using μ and the sample data (excluding the measurement ℓ_i being tested), Equation (3.29) will be modified as follows:

$$|\ell_i - \mu| < s \times k_p \tag{3.32}$$

where $k_p = t_{p = 1 - \alpha_0/2, \text{ df}}$ and n is the total number of measurements in the sample, including the one being tested.

ii) If the population standard deviation (σ) is known and the population mean (μ) is unknown but estimated as $\bar{\ell}$ and used in determining the sample standard deviation (s), the difference between a measurement ℓ_i and its mean will be statistically insignificant at α significance level if the following is satisfied:

$$|\ell_i - \bar{\ell}| < \left(\frac{n-1}{n}\right)^{1/2} \sigma \times k_p \tag{3.33}$$

where $SE = ((n-1)/n)^{1/2} \times \sigma$, $k_p = z_{p=1-\alpha_0/2}$, and n is the number of measurements in the sample including the one being tested. This relates to a case where a well-calibrated instrument with known standard deviation (σ) is being used to collect the measurements. If the population standard deviation (σ) and the population mean (μ) are both unknown and the sample standard deviation (s) and the sample mean ($\bar{\ell}$) are estimated from the same sample, the following must be satisfied:

$$|\ell_i - \bar{\ell}| < \left(\frac{n-1}{n}\right)^{1/2} s \times k_p \tag{3.34}$$

where $SE = ((n-1)/n)^{1/2} \times s$, $k_p = \tau_{\alpha_0/2,\, n-1}$, and $\tau_{\alpha_0/2,\, n-1}$ is the tau probability density (Pope 1976) with $n-1$ degrees of freedom and n is the number of measurements in the sample including the one being tested.

The out-of-context or in-context testing of the measurements will be done as above by replacing the significance level $\alpha_0 = \alpha$ or $\alpha_0 = \alpha/n$, respectively. Note that in Equations (3.33) and (3.34) the standard error of the residual for the ℓ_i measurement is $SE = ((n-1)/n)^{1/2} \times \sigma$ or $SE = ((n-1)/n)^{1/2} \times s$ with the measurement being tested included in determining the mean; if the measurement ℓ_i being tested is not included in the mean (i.e. the mean is based on $n-1$ previous measurements with n as the number of measurements including the one excluded), the standard error of the residual will be $SE = (n/(n-1))^{1/2} \times \sigma$ or $SE = (n/(n-1))^{1/2} \times s$ with the number of degrees of freedom as $n-2$.

Example 3.4 An angle is to be measured in six sets with a condition that each set must be compared with the mean of the previous sets of measurements; if the mean of the previous sets is significantly different from the mean of current set, the current set of measurements will be rejected and a new set of measurements must be made. If the mean of five sets of measurements of the angle is 106°55′33″ with the standard deviation of each measurement (considered as the mean of a set) assumed well known and given as 5″, determine if the difference between this mean and the current measured value of 106°55′40″ is significant at 95% confidence level. (Assume each set of measurements constitutes one single measurement.)

Solution:

$n = 6$, $\sigma = 5''$, $\alpha = 0.05$

Equation (3.33) should be used since the standard deviation of measurements is considered well known; the standard deviation of the difference (or SE) is then given as

$$\sigma_{\ell_i - \bar{\ell}} = \left(\frac{n}{n-1}\right)^{1/2} \sigma$$

Using Equation (3.33) and $z_{1 - \alpha/2} = 1.96$ (assuming out-of-context testing),

$$\left|106°\,55'40'' - 106°\,55'33''\right| < \left(\frac{n}{n-1}\right)^{1/2} \sigma \times z_{1-\alpha/2}$$

$$7'' < \left(\frac{6}{5}\right)^{1/2} 5 \times 1.96 \rightarrow 7'' < 10.7''?$$

Since the condition $7'' < 10.7''$ is true, it can be concluded at 95% confidence level that the difference between the mean and the given measurement is not significant and the measurement can be accepted.

3.6.4 Test of Hypothesis on a Population Variance

This is a case where one is to decide if the sample standard deviation (s) compares with the published precision (or population standard deviation), $\sigma = \sigma_0$. The hypotheses can be formulated as follows:

Null hypothesis: H_0: $\sigma^2 = \sigma_0^2$
Possible alternative hypotheses: H_A: $\sigma^2 > \sigma_0^2$ or H_A: $\sigma^2 < \sigma_0^2$ or H_A: $\sigma^2 \neq \sigma_0^2$

The test statistic for this type of test is the χ^2 statistic (or chi-square statistic) given as

$$\chi^2 = \frac{(\mathrm{df})s^2}{\sigma^2} \tag{3.35}$$

The number of degrees of freedom (df) depends on how the sample standard deviation (s) is calculated; if the sample standard deviation is calculated using the population mean (μ), df $= n$, where n is the sample size; if the sample standard deviation is calculated using the sample mean ($\bar{x}$), df $= n - 1$. If $\chi^2_{p_1, \mathrm{df}}$ and $\chi^2_{p_2, \mathrm{df}}$ are the critical values for one-tailed and two-tailed tests, respectively, from the chi-square distribution curve, the H_0: $\sigma^2 = \sigma_0^2$ should be rejected if the conditions stated under the different alternative hypotheses in Table 3.5 are satisfied. The one-tailed and two-tailed regions of chi-square distributions

Table 3.5 Decisions of test on a population variance.

Alternative hypothesis	Reject H_0: $\sigma^2 = \sigma_0^2$
H_A: $\sigma^2 < \sigma_0^2$	$\chi^2 > \chi^2_{p_1 = \alpha, df}$ (for α in lower tail)
H_A: $\sigma^2 > \sigma_0^2$	$\chi^2 > \chi^2_{p_1 = 1-\alpha, df}$ (for α in upper tail)
H_A: $\sigma^2 \neq \sigma_0^2$ two tailed	$\chi^2 < \chi^2_{p_2 = \alpha/2, df}$ (for $\alpha/2$ in lower tail) or $\chi^2 > \chi^2_{p_2 = 1-\alpha/2, df}$ (for $\alpha/2$ in upper tail)

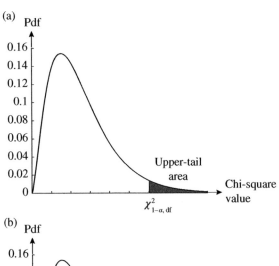

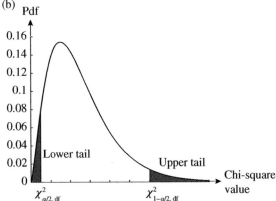

Figure 3.13 (a) One-tailed and (b) two-tailed regions of chi-square distributions.

are illustrated in Figure 3.13. In Table 3.5, $p_1 = \alpha$ for α being in the lower tail area or $p_1 = 1 - \alpha$ for α being in the upper tail area in the case of one-tailed test, and $p_2 = \alpha/2$ for $\alpha/2$ being in the lower tail area or $p_2 = 1 - \alpha/2$ for $\alpha/2$ being in the upper tail area in the case of two-tailed test

Considering one-tailed test further, it can be shown from Equation (3.35) for the case of H_A: $\sigma^2 > \sigma_0^2$ (for α in the upper tail area) that the expected marginal value of the sample standard deviation at $(1 - \alpha)$ confidence level will be

$$s_{1-\alpha} = \sqrt{\frac{\chi^2_{p_1 = 1-\alpha,\,df}\sigma^2}{df}} \tag{3.36}$$

where $s_{1-\alpha}$ is the marginal standard deviation (or the standard deviation that is being tested against the given standard deviation σ). Usually, the sample standard deviation (s) must be less than or equal to this marginal standard deviation in order to accept that the sample standard deviation (s) compares with the published value (σ) according to the one-tailed hypothesis test formulated above. This can be formulated from Equation (3.36), for one-tailed test, that H_0: $\sigma^2 = \sigma_0^2$ should be accepted if the following condition is satisfied:

$$s \leq \sqrt{\frac{\chi^2_{p_1 = 1-\alpha,\,df}\sigma^2}{df}} \tag{3.37}$$

If Equation (3.37) is not satisfied, then H_A: $\sigma^2 > \sigma_0^2$ should be accepted, which means that the population standard deviation is actually larger than the value given or that the sample standard deviation is larger than the population standard deviation. Strictly speaking, one-tailed test should be performed when testing a sample standard deviation, since one may not be interested in whether the sample standard deviation is smaller (or more precise) than the population standard deviation.

Example 3.5 If the standard deviation of a horizontal direction measurement with a theodolite is provided by the manufacturer as $\sigma = 2''$, check if the experimental standard deviation (observed in both face left and face right positions) $s = 2.3''$ is consistent with (not significantly larger than) the manufacturer's value at 95% confidence level. Assume that the number of degrees of freedom for the determination of s is 32.

Solution:

Given $\sigma = 2''$ and $s = 2.3''$, $\alpha = 0.05$, and $v = 32$; the statement "not significantly larger than" suggests what the alternative hypothesis should look like, i.e. one-tailed test.

One-tailed test: H_0: $\sigma^2 = (2)^2$ vs. H_A: $\sigma^2 > (2)^2$

The test statistic is $s \leq \sqrt{\dfrac{\chi^2_{p_1 = 1-\alpha,\,df}\sigma^2}{df}}$

Using the Microsoft Excel 2013 software routine CHISQ.INV(0.95,32) for α in the upper tail area gives the value 46.19.

$$s \leq \sqrt{\frac{\chi^2_{0.95,32} \times 2^2}{32}} = \sqrt{\frac{46.19 \times 4}{32}}$$

$$2.3'' \leq 2.4''$$

Since the above condition is fulfilled, the null hypothesis stating that the empirically determined standard deviation, $s = 2.3''$, is equal to the manufacturer's value, $\sigma = 2''$, is not rejected at the confidence level of 95%.

3.6.5 Test of Hypothesis on Two Population Variances

This section deals with testing if two experimental standard deviations, s_1 and s_2, as determined from two different samples of measurements belong to the same population ($\sigma_1 = \sigma_2$) at the confidence level, $1 - \alpha$. The two samples may be considered different if (a) the samples are collected using the same instrument but different observers, (b) the samples are collected using different instruments with the same observer, or (c) the samples are collected at different times using the same instrument with the same observer. The statistical tests can be expressed as follows:

Null hypothesis: H_0: $\sigma_1^2 = \sigma_2^2$
Possible alternative hypothesis: H_A: $\sigma_1^2 < \sigma_2^2$ or H_A: $\sigma_1^2 > \sigma_2^2$ or H_A: $\sigma_1^2 \neq \sigma_2^2$

The test statistic can be given as

$$F = \frac{s_1^2}{s_2^2} \tag{3.38}$$

which is distributed as F-distribution. If F_{p_1, df_1, df_2} and F_{p_2, df_1, df_2} (with df_1 and df_2 as the numbers of degrees of freedom in determining s_1^2 and s_2^2, respectively) are the critical values from the F-distribution curve, the H_0: $\sigma^2 = \sigma_0^2$ should be rejected if the conditions stated under the different alternative hypotheses in Table 3.6 are satisfied. The one-tailed and two-tailed regions of the F-distributions are illustrated in Figure 3.14. In Table 3.6, $p_1 = \alpha$ for α being in the lower tail area or $p_1 = 1 - \alpha$ for α being in the upper tail area in the case of one-tailed test, and $p_2 = \alpha/2$ for $\alpha/2$ being in the lower tail area or $p_2 = 1 - \alpha/2$ for $\alpha/2$ being in the upper tail area in the case of two-tailed test.

For example, for two-tailed test, H_0: $\sigma_1^2 = \sigma_2^2$ is not rejected if the following condition is satisfied:

$$\frac{1}{F_{p_2 = 1-\alpha/2, \, df_2, \, df_1}} \leq \frac{s_1^2}{s_2^2} \leq F_{p_2 = 1-\alpha/2, \, df_1, \, df_2} \tag{3.39}$$

Table 3.6 Decisions of test on two population variances.

Alternative hypothesis	Reject H_0: $\sigma^2 = \sigma_0^2$
H_A: $\sigma_1^2 < \sigma_2^2$	$F < F_{p_1 = \alpha, df_1, df_2}$ (for α in lower tail)
H_A: $\sigma_1^2 > \sigma_2^2$	$F > F_{p_1 = 1-\alpha, df_1, df_2}$ (for α in upper tail)
H_A: $\sigma_1^2 \neq \sigma_2^2$ two tailed	$F < F_{p_2 = \alpha/2, df_1, df_2}$ (for $\alpha/2$ in lower tail) or $F > F_{p_2 = 1-\alpha/2, df_1, df_2}$ (for $\alpha/2$ in upper tail)

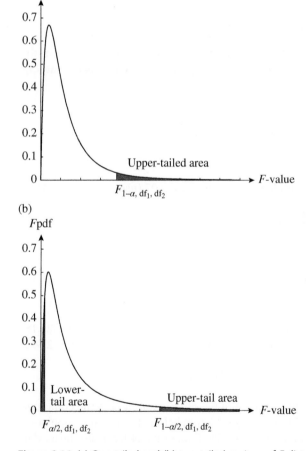

Figure 3.14 (a) One-tailed and (b) two-tailed regions of F-distributions.

where $F_{p_2 = \alpha/2, \, df_1, \, df_2}$ that is the same as $1/F_{p_2 = 1 - \alpha/2, \, df_2, \, df_1}$ is the Fisher distribution value that can be extracted from the F-distribution curve for $\alpha/2$ being in the upper tail area of the F-distribution curve (taking note of the flipping around of the degrees of freedom in the denominator). In this case,

$$F_{p_2 = \alpha/2, \, df_1, \, df_2} \leq \frac{s_1^2}{s_2^2} \leq F_{p_2 = 1 - \alpha/2, \, df_1, \, df_2} \tag{3.40}$$

Example 3.6 In order to determine that the precision of leveling equipment is appropriate for the intended measuring task, two calibration tests were carried out with two samples of measurements by the same instrument but different observers. The results of the tests are as follows:

Test 1: Computed standard deviation of instrument (s_1) = 2.5 mm, number of degrees of freedom (df_1) = 25.

Test 2: Computed standard deviation of instrument (s_2) = 2.0 mm, number of degrees of freedom (df_2) = 18.

Do the two computed standard deviations, s_1 and s_2, as determined from the two different samples of measurements belong to the same population at 95% confidence level?

Solution:

Given s_1 = 2.5 mm, s_2 = 2.0 mm, df_1 = 25, and df_2 = 18.

$$H_0 : \sigma_1^2 = \sigma_2^2 \quad H_A : \sigma_1^2 \neq \sigma_2^2$$

Evaluate the following confidence interval at $\alpha = 0.05$ (using Equation (3.39):

$$\frac{1}{F_{p_2 = 1 - \alpha/2, \, df_2, \, df_1}} \leq \frac{s_1^2}{s_2^2} \leq F_{p_2 = 1 - \alpha/2, \, df_1, \, df_2}$$

$$\frac{1}{F_{0.975, \, 18, 25}} \leq \frac{2.5^2}{2.0^2} \leq F_{0.975, 25, 18}$$

Note that $\dfrac{1}{F_{0.975, 18, 25}} = F_{0.025, \, 25, \, 18} = 0.428$. Using Microsoft Excel 2013 software routine for $\alpha/2$ in the lower tail area of F-distribution F.INV(0.025,25,18) gives 0.428, and for $\alpha/2$ in the upper tail area, F.INV(0.975,25,18) gives 2.491.

$$\frac{1}{2.338} \leq \frac{6.25}{4.0} \leq 2.491 \quad \text{or} \quad 0.43 \leq 1.56 \leq 2.49$$

Since the condition is fulfilled, the null hypothesis stating that the computed standard deviations s_1 = 2.5 mm and s_2 = 2.0 mm belong to the same population is not rejected at the confidence level of 95%. Equation (3.40) can also be used to solve this problem as follows:

$$F_{p_2 = \alpha/2,\, df_1,\, df_2} \leq \frac{s_1^2}{s_2^2} \leq F_{p_2 = 1-\alpha/2,\, df_1,\, df_2}$$

$$F_{0.025,25,18} \leq \frac{2.5^2}{2.0^2} \leq F_{0.975,25,18}$$

Using Microsoft Excel, F.INV(0.975,25,18) gives 2.491 as above, and F.INV (0.025,25,18) gives 0.43 so that the same result is obtained as

$$0.43 \leq 1.56 \leq 2.49$$

Problems

3.1 An angle is measured 10 times. Each measurement is independent and made with the same precision. The sample mean is $42°12'14.6''$ and the sample standard deviation is $3.7''$. Answer the following.

a) Test at a significance level of 5% the hypothesis that the population standard deviation σ of the measurements is $2.0''$ against the alternative that σ is not $2.0''$.

b) Test at a significance level of 5% the hypothesis that the population mean of the measurements is $42°12'16.0''$ against the alternative that it is not $42°12'16.0''$.

c) With regard to information provided in questions (a) and (b), if one of the sample measurements is $42°12'10.5''$, perform out-of-context test to determine if this measurement is an outlier at 99% confidence level, assuming the population mean and standard deviation are both known.

d) With regard to information provided in questions (a) and (b), if one of the sample measurements is $42°12'10.5''$, perform out-of-context test to determine if this measurement is an outlier at 99% confidence level, assuming the population mean is unknown, but the population standard deviation is known.

e) With regard to information provided in questions (a) and (b), if one of the sample measurements is $42°12'20.0''$, perform in-context test to determine if this measurement is an outlier at 98% confidence level, assuming the population mean and the population standard deviation are unknown.

3.2 The line between two survey markers P and Q was measured repeatedly by a survey crew A, and the adjusted distance obtained was 1500.030 m; the survey crew B obtained the adjusted distance for the same line as 1500.042. If the standard error of the adjusted distance by each crew is 4 mm (considered well known), determine if the expected critical value of the difference in the two distances has been exceeded at 80% confidence level.

Based on your result, are the two distances significantly different at 80% confidence level?

3.3 The standard deviation of measuring a 1000.000 m long baseline with the Leica TPS 1203 equipment is 1.8 mm (according to the manufacturer's specification). After calibrating the equipment on the 1000.000 m long baseline, the calculated standard deviation was 2.5 mm based on 15 measurements of the baseline. Determine, statistically at 95% confidence level, if the equipment is performing according to the manufacturer's specification.

3.4 A distance has been independently measured 4 times and its sample unit variance obtained from the adjustment $\hat{\sigma}_0^2$ is equal to 1.44 cm^2. If the a priori standard deviation σ_0 is 1.0 cm, conduct a statistical test to decide if the adjustment result is acceptable with a significance level of $\alpha = 5\%$.

3.5 The manufacturer's specification for Leica TPS 1203 equipment is quoted as 1 mm + 1.5 ppm for the standard deviation (ISO 17123-4) of distance measurement in standard (IR) mode. After calibrating the equipment on a baseline that is 1000.000 m long, the calculated standard deviation was 2.5 mm based on 20 measurements of the baseline. Would you be 95% confident in the manufacturer's specification? Explain your answer with a statistical proof.

4

Adjustment Methods and Concepts

OBJECTIVES

After studying this chapter, you should be able to:

1) Explain the differences among the traditional methods of adjustment (transit, compass, Crandall's) and the least squares method of adjustment, including their limitations.
2) Perform simple traverse network adjustment using traditional methods of adjustment (transit, compass, Crandall's).
3) Discuss the advantages of least squares method.
4) Explain the least squares criterion (or condition).
5) Explain different least squares adjustment model types and the least squares adjustment steps.
6) Explain datum definitions and different constraints in least squares adjustment, including their practical importance.

Understanding Least Squares Estimation and Geomatics Data Analysis, First Edition.
John Olusegun Ogundare.
© 2019 John Wiley & Sons, Inc. Published 2019 by John Wiley & Sons, Inc.
Companion website: www.wiley.com/go/ogundare/Understanding-lse-and-gda

4.1 Introduction

In the past, three traditional adjustment methods (transit rule, compass or Bowditch rule, and Crandall's rule) of adjusting measurements were commonly used in geomatics. Today, these adjustment methods are inadequate since different measurement techniques are now employed by surveyors to allow gross errors in measurements to be checked and to increase precision of measurements and the quantities derivable from those measurements. The modern measurement techniques produce redundant measurements, which are better evaluated using the method generally referred to as *least squares estimation* or *least squares adjustment*. Adjustment is usually considered as a method of removing *misclosures*, while estimation is associated with statistical evaluation. The method of least squares is both estimation and adjustment methods; it generally requires that measurements (or observations) be first represented mathematically before adjusting and statistically evaluating the measurements.

Adjustment of mathematical models (or the observations) is meaningful only if there are *redundant models* or *redundant observations* involved. There are redundant models in the system if there are more model equations than unknown parameters (the quantities being determined). The models are then said to be *overdetermined* or *redundant*. Overdetermination or redundancy in geomatics creates inconsistencies or misclosures in observations but has some advantages such as providing checks on gross errors in measurements, which can lead to a more precise and accurate estimation of the unknown parameters. The whole adjustment concept is basically about selecting an adjustment criterion (or condition) and following it to determine a new set of unique values for the observables or for the unknown parameters using the given redundant observations. The purpose of such a criterion is to help in determining when a solution is achieved. Many adjustment criteria (or conditions) are possible; some of them, which include those relating to *traditional methods of adjustment*, will be discussed in this chapter. However, it must be mentioned that the *least squares criterion* has become the most acceptable standard for adjusting measurements by the geomatics professionals. The least squares method computes the best-fit solution for a mathematical model (or a system of equations) with redundant measurements. The least squares method is considered the best adjustment method since it provides the least possible amount of corrections when adjusting the measurements.

4.2 Traditional Adjustment Methods

Property surveys to locate or establish property boundaries are traditionally based on traverse surveys. Three classes of traverse surveys are common: *open traverse*, in which the starting position is known with the position of the end point unknown; *closed traverse*, in which the positions of the start and end points are known; and *closed-loop traverse*, which is a form of closed traverse in which the start and end points are the same. Generally, traverse surveys require adjustments; otherwise, the errors in the traverse will accumulate and be assigned to the last traverse leg of the traverse network, which will be unjustifiable.

There are three traditional adjustment methods used in geomatics: *transit rule, compass or Bowditch rule*, and *Crandall's rule*. These methods are commonly used with respect to closed traverses that represent parcels of land in subdivision plans. Each of these methods requires that the traverse angular measurements component be geometrically adjusted first prior to adjusting the whole traverse. These methods, however, will not accomplish the desired adjustment results when a network of closed traverses with common stations is adjusted. Moreover, the methods usually ignore the quality of observations in their adjustment procedures. This means that the surveyor who purchases an accurate surveying instrument will have to ignore the accuracy achieved with such an instrument when adjusting the measurements acquired with the instrument, which will be unreasonable.

In traditional methods of adjustment, the term *latitude* is used to mean change in northing value (ΔN), which can be expressed as

$$\Delta N = S \times \cos\beta \tag{4.1}$$

where S is the length of a given traverse leg and β is the bearing of the traverse leg (or the course of the traverse). The term *departure* is used to mean change in easting value (ΔE), which can be expressed as

$$\Delta E = S \times \sin\beta \tag{4.2}$$

The error of closure for the whole traverse is the length (linear error of closure) (LC) and direction (θ) connecting the computed and the known final traverse stations, which can be given as

$$LC = \sqrt{(\varepsilon_N)^2 + (\varepsilon_E)^2} \tag{4.3}$$

$$\theta = a\tan\left(\frac{\varepsilon_E}{\varepsilon_N}\right) \tag{4.4}$$

$$\varepsilon_N = \sum_{i=1}^{n} \Delta N_i \tag{4.5}$$

$$\varepsilon_E = \sum_{i=1}^{n} \Delta E_i \tag{4.6}$$

where ε_N and ε_E are the total closure errors of the traverse in the northing and the easting directions, respectively; LC is the linear error of closure; and n is the number of traverse legs measured in the traverse. The ratio of LC to the total length (D) of the traverse can be given as

$$RC = \frac{LC}{D} \tag{4.7}$$

Equations (4.1)–(4.7) are applied similarly in all of the three traditional adjustment methods. The main differences are in how the corrections to the changes in easting and northing coordinates are calculated. These differences are discussed in the following subsections.

4.2.1 Transit Rule Method of Adjustment

The transit rule method is an empirical method of adjustment that has no theoretical basis. It was initially developed for adjusting random errors of closure in closed traverse surveys based on an assumption that transit and tape equipment are used. The method assumes the angles measured with transit are more precise than the corresponding distances measured with tape. This method of adjustment, although not commonly used, computes an adjustment or correction ($c_{\Delta N_i}$) to the change in northing value (ΔN_i) for a given traverse leg i, as

$$c_{\Delta N_i} = -\varepsilon_N \left(\frac{|\Delta N_i|}{\sum_i^n |\Delta N_i|} \right) \tag{4.8}$$

where $|\Delta N_i|$ is the absolute value of the change in northing value for the given traverse leg i. Similarly, the correction ($c_{\Delta E_i}$) to be applied to the change in easting value (ΔE_i) for the traverse leg i can be given as

$$c_{\Delta E_i} = -\varepsilon_E \left(\frac{|\Delta E_i|}{\sum_i^n |\Delta E_i|} \right) \tag{4.9}$$

where $|\Delta E_i|$ is the absolute value of the change in easting value for the given traverse leg i. The adjusted coordinate changes for a given traverse leg i are given as

$$\Delta \hat{N}_i = \Delta N_i + c_{\Delta N_i} \tag{4.10}$$

$$\Delta \hat{E}_i = \Delta E_i + c_{\Delta E_i} \tag{4.11}$$

where $\Delta \hat{N}_i$ and $\Delta \hat{E}_i$ are the adjusted changes in northing and easting values for the given traverse leg i. The adjusted coordinates for a given traverse point k are given as

$$\hat{N}_k = \hat{N}_{k-1} + \Delta \hat{N}_{k-1 \to k} \tag{4.12}$$

$$\hat{E}_k = \hat{E}_{k-1} + \Delta \hat{E}_{k-1 \to k} \tag{4.13}$$

where $\hat{N}_k$ and $\hat{E}_k$ are the adjusted northing and easting coordinate values, respectively, for the current traverse point k; $\hat{N}_{k-1}$ and $\hat{E}_{k-1}$ are the previously adjusted northing and easting coordinate values, respectively, for the traverse point $k-1$ preceding the current traverse point k; and $\Delta\hat{N}_{k-1\to k}$ and $\Delta\hat{E}_{k-1\to k}$ are the adjusted changes in northing and easting values for the traverse leg from point $k-1$ to point k. Example 4.1 is given to illustrate the computational steps for the transit adjustment method.

Example 4.1 The loop traverse network shown in Figure 4.1 is a simple network, which is common in property survey. The initial bearing of line P–Q in the figure is fixed as $79°37'22''$. Given in Table 4.1 are the distance and interior angle measurements of the network made with a total station instrument, which has a standard deviation for horizontal angle measurement as $\pm3''$ according to ISO 17123-3 standard and the standard deviation of distance measurement as 2 mm + 2 ppm according to ISO 17123-4. The angular error was computed as $\pm4''$ for a set of horizontal angular measurement at each station; the horizontal distance standard deviations are derived from the distance specification with

Figure 4.1 Sample closed-loop traverse network.

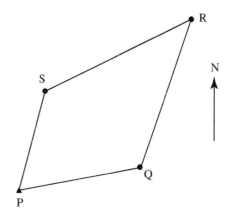

Table 4.1 Traverse measurements.

From–at–to	Interior angle	Distance (at–to)
P–Q–R	$119°44'50'' \pm 4''$	$223.600 \text{ m} \pm 2.7 \text{ mm}$
Q–R–S	$45°00'10'' \pm 4$	$237.184 \text{ m} \pm 2.7 \text{ mm}$
R–S–P	$130°35'50'' \pm 4''$	$145.775 \text{ m} \pm 2.6 \text{ mm}$
S–P–Q	$64°39'20'' \pm 4''$	$180.278 \text{ m} \pm 2.6 \text{ mm}$

the assumption that the instrument and targets can be centered to an accuracy of ±0.8 mm each. Adjust the measurements by transit method.

Solution:

The traverse computations are given in Tables 4.2 and 4.3, where the symbols used are based on Equations (4.1)–(4.13). The total closure errors in the easting and northing directions are given as sums in columns (5) and (6) of Table 4.2 as $\varepsilon_E = -0.0119$ m and $\varepsilon_N = -0.0022$ m, respectively; the total length (D) of traverse is 786.837 m; the linear error of closure (LC) based on Equation (4.3) is 0.0121 m; direction of closure (θ) using Equation (4.4) is 79°39′; and the ratio of LC to D using Equation (4.7) is 1 : 64 975, which is better than the usually acceptable ratio 1 : 5 000 for property surveys.

Table 4.2 Departure and latitude computations.

Station (from) (1)	Adjusted angle (2)	β (3)	S (m) (4)	ΔE (5)	ΔN (6)	Station (to) (7)
						P
P		79°37′22″	180.278	177.3292	32.4731	Q
Q	119°44′48″	19°22′10″	223.600	74.1587	210.9442	R
R	45°00′07″	244°22′17″	237.184	−213.8490	−102.5906	S
S	130°35′48″	194°58′05″	145.775	−37.6508	−140.8289	P
P	64°39′17″	79°37′22″				Q
Sum	360°00′00″		786.837	−0.0119	−0.0022	

Table 4.3 Adjustments and adjusted coordinates by transit method.

Station (from) (1)	$c_{\Delta E}$ (2)	$c_{\Delta N}$ (3)	$\Delta \hat{E}$ (4)	$\Delta \hat{N}$ (5)	$\hat{E}_k$ (6)	$\hat{N}_k$ (7)	Station (to) (8)
					1000.0000	1150.0000	P
P	0.0042	0.0001	177.3334	32.4733	1177.3334	1182.4733	Q
Q	0.0018	0.0009	74.1605	210.9451	1251.4939	1393.4184	R
R	0.0051	0.0005	−213.8440	−102.5902	1037.6499	1290.8282	S
S	0.0009	0.0006	−37.6499	−140.8282	1000.0000	1150.0000	P
Sum	0.0119	0.0022	0.0000	0.0000			

4.2.2 Compass (Bowditch) Rule Method

The compass (otherwise, known as Bowditch) rule method was invented in 1773–1838 by the American navigator known as Nathaniel Bowditch. The method was originally designed to statistically adjust random errors of closure in closed traverses measured with compass and tape equipment. The method assumes that the bearing and distance measurements of a traverse leg are uncorrelated and the standard deviations (s_d) of the distance measurements and those of the corresponding bearing measurements (s_α) are consistent or equivalent, i.e. $s_d = d(s_\alpha)$, where d is the associated distance. The other statistical assumption of this method is that the most likely error in a distance measurement will be proportional to the square root of that distance measurement (Sprinsky 1987; Crandall 1901). Since modern survey equipment provide flexible and different types of measurement precisions, the compass rule adjustment assumptions have become unnecessary.

The compass rule method, however, determines adjustments or corrections to coordinate changes as follows. For a given total closure error of a traverse in the northing direction as ε_N, the correction ($c_{\Delta N_i}$) to be applied to the change in northing value (ΔN_i) for the traverse leg i can be given as

$$c_{\Delta N_i} = -\varepsilon_N \left(\frac{S_i}{D}\right) \tag{4.14}$$

where S_i is the length of the given traverse leg i and D is the total length of the traverse. Similarly, the correction ($c_{\Delta E_i}$) to be applied to the change in easting value (ΔE_i) for the traverse leg i can be given as

$$c_{\Delta E_i} = -\varepsilon_E \left(\frac{S_i}{D}\right) \tag{4.15}$$

where ε_E is the total closure error of the traverse in the easting direction. Again, the adjusted coordinate changes for a given traverse leg i are given by Equations (4.10) and (4.11); and the adjusted northing coordinate ($\hat{N}_k$) and the adjusted easting coordinate ($\hat{E}_k$) are calculated from Equations (4.12) and (4.13), respectively. Example 4.2 is given to illustrate the computational steps for the compass rule method of adjustment.

Example 4.2 Adjust the loop traverse network in Example 4.1 by compass rule method.

Solution:

The traverse computations given in Table 4.2 are the same for compass rule method. The computations in Table 4.4 are based on Equations (4.14)–(4.15) and (4.12)–(4.13).

Table 4.4 Adjustments and adjusted coordinates by compass rule method.

Station (from) (1)	$c_{\Delta E}$ (2)	$c_{\Delta N}$ (3)	$\Delta \hat{E}$ (4)	$\Delta \hat{N}$ (5)	$\hat{E}_k$ (6)	$\hat{N}_k$ (7)	Station (to) (8)
					1000.0000	1150.0000	P
P	0.0027	0.0005	177.3320	32.4736	1177.3320	1182.4736	Q
Q	0.0034	0.0006	74.1621	210.9448	1251.4941	1393.4184	R
R	0.0036	0.0007	−213.8455	−102.5900	1037.6486	1290.8284	S
S	0.0022	0.0004	−37.6486	−140.8284	1000.0000	1150.0000	P
Sum	0.0119	0.0022	0.0000	0.0000			

4.2.3 Crandall's Rule Method

Crandall's rule method of adjustment was developed by Crandall (1901) to reflect the instrumentation of that time by applying the principle of minimum variance. The method simulates the least squares method of adjustment of closure errors in closed traverse networks and assumes that angle (or azimuth) measurements in traverse networks are more accurate than distance measurements. In this case, the method attributes the closure errors to distance measurements, distributes the closure error corrections only to distances, and leaves the azimuth (angle) measurements unadjusted. This assumption of errorless azimuth (angle) measurements, however, is incorrect for modern surveying equipment, such as total station. The corrections $c_{\Delta N_i}$ and $c_{\Delta E_i}$ to be applied to coordinate changes using this method can be expressed as

$$c_{\Delta N_i} = A(\Delta N_i)^2 + B(\Delta N_i \times \Delta E_i) \tag{4.16}$$

$$c_{\Delta E_i} = A(\Delta N_i \times \Delta E_i) + B(\Delta E_i)^2 \tag{4.17}$$

$$B = \frac{\left[\varepsilon_N \times \left(\sum_i^n \Delta N_i \times \Delta E_i\right) - \varepsilon_E \times \left(\sum_i^n (\Delta N_i)^2\right)\right]}{\left[\left(\sum_i^n (\Delta E_i)^2\right)\left(\sum_i^n (\Delta N_i)^2\right) - \left(\sum_i^n \Delta N_i \times \Delta E_i\right)^2\right]} \tag{4.18}$$

$$A = \frac{-\left[B\left(\sum_i^n \Delta N_i \times \Delta E_i\right) + \varepsilon_N\right]}{\sum_i^n (\Delta N_i)^2} \tag{4.19}$$

Again, the adjusted coordinate changes for a given traverse leg i are given by the Equations (4.10) and (4.11); and the adjusted northing coordinate $(\hat{N}_k)$ and the adjusted easting coordinate $(\hat{E}_k)$ are calculated from Equations (4.12) and (4.13), respectively. Example 4.3 is given to illustrate the computational steps for Crandall's rule method of adjustment.

Example 4.3   Adjust the loop traverse network in Example 4.1 by Crandall's rule method.

Solution:

The traverse computations given in Table 4.2 are the same for Crandall's rule method. The computations in Table 4.5 are based on Equations (4.16)–(4.19) and (4.12)–(4.13). From Equation (4.18), B is calculated as 1.987 46E-7, and A is calculated from Equation (4.19) as $-9.869\,016$E-8.

Table 4.5 Adjustments and adjusted coordinates by Crandall's rule method.

Station (from) (1)	$c_{\Delta E}$ (2)	$c_{\Delta N}$ (3)	$\Delta\hat{E}$ (4)	$\Delta\hat{N}$ (5)	$\hat{E}_k$ (6)	$\hat{N}_k$ (7)	Station (to) (8)
					1000.0000	1150.0000	P
P	0.0057	0.0010	177.3349	32.4742	1177.3349	1182.4742	Q
Q	−0.0005	−0.0013	74.1583	210.9429	1251.4932	1393.4171	R
R	0.0069	0.0033	−213.8421	−102.5873	1037.6511	1290.8298	S
S	−0.0002	−0.0009	−37.6511	−140.8298	1000.0000	1150.0000	P
Sum	0.0119	0.0022	0.0000	0.0000			

4.3 The Method of Least Squares

Today, the traditional methods of adjusting measurements are inadequate for precision surveys involving large redundant measurements (to check gross errors in measurements), different measurement techniques (such as in triangulation and trilateration surveys), and variable measurement precisions. For examples, the traditional methods clearly show their weaknesses when adjusting traverse networks with cross ties and in post-processed global navigation satellite system (GNSS) network adjustment where many observations are involved. The adjustment method that is able to accommodate all of these properties and evaluate redundant measurements at the same time is the least squares method, which is generally referred to as *least squares estimation* or *least squares adjustment*. The method of least squares provides organized and mathematically logical treatment of observation random errors and inconsistencies in mathematical models. It uses a certain error distribution approach in applying corrections to observations. The main features of least squares adjustment are to adjust the observations so as to eliminate possible misclosures and to keep changes to the observations as small as possible (according to what is known as the *least squares criterion*).

The concepts of least squares method have evolved over several years. The following is the summary of some of the important aspects of the history of the method (Merriman 1877; Cooper 1987). In fact, the background to the method actually started with the various studies and investigations carried out by the astronomers and scientists, such as the following:

- In 1788, *Laplace* (French scientist) used the so-called method of averages to combine different data collected under the same conditions.
- *Boscovich* in 1757 and Laplace in 1799 suggested the use of the method of least absolute deviation to combine different data collected under different conditions.
- In 1795, 1806, and 1809, *Legendre* (French mathematician) and *Gauss* (German scientist) came up with the concept of optimization of the sum of squares of corrections to measurements. The name "method of least squares" was first published by Legendre (1806), but the detailed description of the method was provided by Gauss (1809), who claimed that he had independently derived the method of least squares since 1795. *Legendre* and *Gauss*, however, are understood to have approached the method of least squares differently:
  - *Legendre* used mathematical models to eliminate misclosures and provided numerical consistency of observations connected by models. He demonstrated the procedure by fitting linear equations to some observations. In this case, he treated least squares method as a pure numerical estimation procedure with no probability distribution considered.
  - *Gauss* assumed normality of measurements to produce results with the highest probability of being true. He discovered that observational errors follow the law of normal probability distribution function, which he used to provide numerical and statistical properties of the unknown parameters. In his publication in 1809, Gauss used the method of least squares to calculate the orbit of celestial bodies and connected the method with the principles of probability and normal distribution.
  - *Gauss* (1809) and *Laplace* (1811) specified a mathematical model for the probability density of observations. They also proved that observational errors follow the law of normal probability distribution function with the mean of repeated measurements as the most probable value of the unknown. This method is commonly known as the *maximum likelihood estimator method*.

- In general, around 1800, Legendre, Gauss, and Laplace were able to fit functional models to data through different types of least squares methods.

Real solution approach of inconsistencies caused by redundant measurements is actually understood to have been provided by *Legendre* and *Gauss*, and they are considered as the founders or fathers of the method of least squares. The method of least squares, however, became attractive to the geodesists sometimes later. The geodesists employed the method in determining the best-fitting ellipsoid based on inconsistent meridian arc measurements. Some of the geodesists who made their contributions in one way or another to the method of least squares are as follows:

- *Adrian* (American surveyor) (1808) independently formulated the principle of least squares analysis, but did not actually refer to it by that name. He also suggested the use of normal distribution for describing errors in measurements as far back as 1808, even before Gauss.
- *Bessel* (German astronomer and geodesist) in 1837 solved the problem of least squares adjustment of parametric models by determining the elements of the figure of the Earth (such as flattening, polar and equatorial radii, etc.) based on the available geodetic coordinates. The parametric model adjustment method has become the most useful adjustment procedure today.
- *Helmert* (German geodesist) in 1907 determined the coefficients of linear transformation by parametric method of least squares known as *Helmert's transformation*.
- *Bjerhammar* (Swedish geodesist) (1951) contributed to the simplification of least squares method with the use of matrix notations.

4.3.1 Least Squares Criterion

To illustrate some of the features of least squares method, consider a resection problem in Figure 4.2 in which the coordinates (x, y) of point P are to be determined. In this problem, the horizontal angles (α, β, γ) and horizontal distances (d_1, d_2, d_3) with their associated standard deviations are measured from point P to three control points A, B, and C whose coordinates are well known.

The six measurements in Figure 4.2 are to be adjusted to make them fit together in determining the two unknown coordinates (x, y) of point P. In this case, the sum of the angles at point P about the horizon must be equal to $360°$, and the corrections to the distance measurements must be as small as possible; all of these conditions are to be satisfied simultaneously. The corrections (or residuals) to be applied to the measurements depend on how well the measurements fit together. One approach in solving for the coordinates of point P is to guess the coordinates of point P using the coordinates of points A, B, and C as

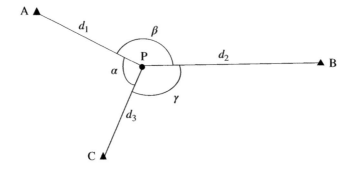

Figure 4.2 Resection of point P.

guide, then calculate independently the measurements, and use them to determine the residuals. The method of least squares iteratively corrects initial guessed coordinates so that the weighted sum of corrections to the measurements is a minimum value. This is to say that the method of least squares is designed to keep changes (or adjustments) to the original measurements as small as possible. This leads to what is known as the *least squares criterion* or the *least squares condition*. The general *least squares criterion* or *least squares condition* states that the sum of weighted squares of corrections to measurements is a minimum. This can be expressed mathematically as

$$v^T P v = \text{minimum} \tag{4.20}$$

where v is the vector of observation residuals (corrections to be applied to measurements after adjustment or the differences between the actual measurements and the values predicted by the model) and P is the weight matrix of the observations derived from the standard deviations of the observations. If the weight matrix is diagonal (with off-diagonal elements as zeros), the least squares criterion in Equation (4.20) can be rewritten as

$$v_1^2 p_{11} + v_2^2 p_{22} + v_3^2 p_{33} + \cdots + v_n^2 p_{nn} = \text{minimum} \tag{4.21}$$

where $P = \text{diag}(p_{11}, p_{22}, p_{33}, ..., p_{nn})$ is the weight matrix of the measurements and $v = (v_1, v_2, v_3, ..., v_n)^T$ is the vector of residuals, which are to be determined. In other words, the least squares method provides the way of finding the most probable value of a measured quantity through the use of least squares criterion. In order to demonstrate the application of the least squares criterion, let an observable ℓ be measured repeatedly n times with the vector of measurements as $\ell = (\ell_1, \ell_2, \ell_3, ..., \ell_n)^T$; the weight matrix of the measurements as $P = \text{diag}(p_{11}, p_{22}, p_{33}, ..., p_{nn})$, assuming different means were used to measure the observable so that each measurement has different weight; the mean value of the observations as $\bar{x}$ (unknown to be determined); and the vector of residuals as

$v = (v_1, v_2, v_3, ..., v_n)^T$ (also unknown to be determined). The only quantities known for the observable are the measurements and their weights. The least squares criterion can be used to determine the unknowns as follows. The least squares criterion is formulated as in Equation (4.21) as follows:

$$v_1^2 p_{11} + v_2^2 p_{22} + v_3^2 p_{33} + \cdots + v_n^2 p_{nn} = \text{minimum} \tag{4.22}$$

The minimum of Equation (4.22) is found by setting the partial derivatives of the equation (with respect to each observation residual) to zero as follows:

$$\frac{\partial\left(v_1^2 p_{11}\right)}{\partial v_1} + \frac{\partial\left(v_2^2 p_{22}\right)}{\partial v_2} + \frac{\partial\left(v_3^2 p_{33}\right)}{\partial v_3} + \cdots + \frac{\partial\left(v_n^2 p_{nn}\right)}{\partial v_n} = 0 \tag{4.23}$$

or

$$2v_1 p_{11} + 2v_2 p_{22} + 2v_3 p_{33} + \cdots + 2v_n p_{nn} = 0 \tag{4.24}$$

Remembering that a residual (v_i) is the difference between the measurement (ℓ_i) and its mean value $(\bar{x})$, expressed as $v_i = (\bar{x} - \ell_i)$, Equation (4.24) can be rewritten as

$$(\bar{x} - \ell_1)p_{11} + (\bar{x} - \ell_2)p_{22} + (\bar{x} - \ell_3)p_{33} + \cdots + (\bar{x} - \ell_n)p_{nn} = 0 \tag{4.25}$$

Equation (4.25) can be solved to give

$$\bar{x} = \frac{\ell_1 p_{11} + \ell_2 p_{22} + \ell_3 p_{33} + \cdots + \ell_n p_{nn}}{p_{11} + p_{22} + p_{33} + \cdots + p_{nn}} \tag{4.26}$$

Equation (4.26), which is simply the weighted mean of the measurements, is the least squares solution for the measurements ℓ having a weight matrix P. It can be stated therefore that least squares estimate is simply a weighted average of the observations for the unknown parameter. In this case, all the observations somehow contribute to the "best" estimated parameter. The weighted mean is the most probable value in this problem. The ordinary least squares method is referred to as the *best linear unbiased estimator (BLUE)* of parameters in relation to *Gauss–Markov model*. A Gauss–Markov model is a linear regression model in which uncorrelated errors have zero expectation and equal variances. The term "best" means that the lowest possible mean squared error of the estimated parameter is achieved. It is also known that when a weighted sum of squared residuals is minimized, as in Equation (4.21), the estimated parameter is BLUE if each weight is equal to the reciprocal of the variance of the measurement. If the errors are correlated, the resulting estimator is also BLUE if the weight matrix is equal to the inverse of the variance–covariance matrix of the observations, as in the case of Equation (4.20). The relationships between standard deviations of measurements, the variance–covariance matrix, the cofactor matrix, and the weights matrix were discussed in Chapter 1. From that chapter, it can be seen that weight matrix P of observations can be given as

$$P = \sigma_0^2 C_\ell^{-1} \tag{4.27}$$

where $C_\ell = \sigma_0^2 Q$ is the variance–covariance matrix of the observation vector ℓ (with the squared standard deviations or variances, of observations as the diagonal elements), σ_0^2 is the a priori variance factor of unit weight (which is always taken as the general population variance with a value of 1), and Q is the cofactor matrix of the observation vector.

4.4 Least Squares Adjustment Model Types

The concepts of survey observables, adjustments, functional models, and stochastic models were discussed in Chapter 1. Functional models relate survey observables (such as distances, directions, azimuths, angles, and elevation differences) with the unknown parameters such as coordinates of survey network points. Before any adjustment can be done, a functional model relating observables to parameters must be formulated, or a functional model relating observables to each other must be identified and formulated.

The need for adjustment of models is based on redundancy (or overdetermination) of models; there will be no need for adjustment if the models are not overdetermined. Depending on how these observables relate with the unknown parameters or with each other, the mathematical models formed can be referred to as *parametric, conditional,* or *general*. Parametric models have one observation (ℓ_i) per equation, which is a function of some or all of the unknown parameters (x). Since the number of equations is the same as the number of observations, the number of degrees of freedom (redundancy) is the number of equations (or observations) minus the number of parameters. The conditional models geometrically relate only the observations (ℓ) together without involving any parameters in the equations. Since the number of independent equations involved may be less than the number of observations, the number of degrees of freedom for conditional model adjustment is the same as the number of independent model equations formulated. The general models somehow have the observations (ℓ) intertwined with the parameters (x) in the same equations; the number of equations involved minus the number of parameters is the number of degrees of freedom. It should be mentioned, however, that the number of equations involved in general model adjustment is not necessarily the same as the number of observations, as in the case of parametric model adjustment; it is usually less.

The special case of general models in which each unknown parameter is a function of observations is referred to as *special model* in this book. This type of model was formulated for error propagation problems in Chapter 2. The relationships between the different models are shown in Figure 4.3. Examples of each model are given in appropriate chapters dealing with parametric, conditional, and general least squares adjustment methods. The convention that will be followed in the following least squares adjustment methods is such that the symbols representing unadjusted observations and the unadjusted parameters will have no hats, such as ℓ for observations (quantities measured) and x as unadjusted parameters (quantities that are not measured but whose values are to be estimated); the adjusted observations will be represented with hats by $\hat{\ell}$, and the adjusted parameters will be represented with hats by $\hat{x}$. It should be emphasized that whenever each type of the different models (parametric, conditional, general) is being formulated, the order of laying out the observations and the parameters on either side of the equality sign in the equations is very important. The idea behind using adjusted quantities (those represented with "hat" on them) in model equation formulation is to indicate that those

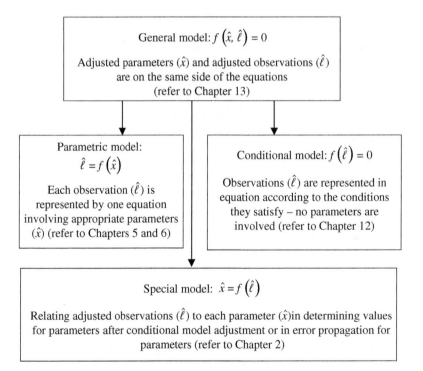

Figure 4.3 Relationships between main mathematical models used in geomatics.

quantities are going to be adjusted and the equations so formed are complete and consistent, i.e. they will not produce any misclosures.

Since matrix algebra is convenient in representing model equations in least squares estimation, it will be used throughout this book. As a form of review, introductory matrices are presented in Chapter 1, and some important matrix lemmas are provided in the Appendix E. It should be mentioned here that whenever the mathematical model equations are being formulated, the numerical values of the observations should not be used immediately, but the appropriate symbols representing those measurements, such as $\hat{\ell}_1, \hat{\ell}_2, \Delta\hat{h}_1, \Delta\hat{h}_2$. It should be remembered also that at the equation formulation stage, one is not solving the equations but just presenting them symbolically. It should also be noted that, at this stage, what is important is how to identify unknown parameters and observations. The unknown parameters (x) are quantities one wants to determine, and the observations (ℓ) are the measurements or quantities provided with their standard deviations or covariance matrix that are to be used for determining the unknown parameters.

4.5 Least Squares Adjustment Steps

After survey measurements have been collected, the measurements must be corrected for some physical conditions such as mark-to-mark reductions, refraction corrections to zenith angles, corrections due to deflections of the vertical to convert astronomic azimuth to geodetic, scale correction, etc. After these reductions and corrections have been applied to the measurements, *corrected observations* are obtained. Corrected observations must not be confused with *adjusted observations*. Adjusted observations are corrected observations whose random errors have been minimized through adjustment process involving redundant observations; the adjusted observations produce the best values of the unknown quantities. The reduced and corrected observations are evaluated using the least squares tool as illustrated in Figure 4.4.

Basically, the least squares method requires the following steps as stated in Figure 4.4:

- Formulating some suitable mathematical models (parametric, conditional, or general) and stochastic model (see Chapters 5, 12, and 13).
- Incorporating the models into the least squares criterion (forming what is known as variation function) (see Sections 5.5, 12.3, and 13.4).
- Finding the partial derivatives of the *least squares criterion* or the variation function (formulated from the least squares criterion), setting the partial derivatives to zero, and solving them. The "minimum" of the least squares criterion or of the variation function is found by setting the partial derivatives of the least squares criterion or of the variation function to zero (see Sections 5.5–5.7, 12.3, and 13.5).

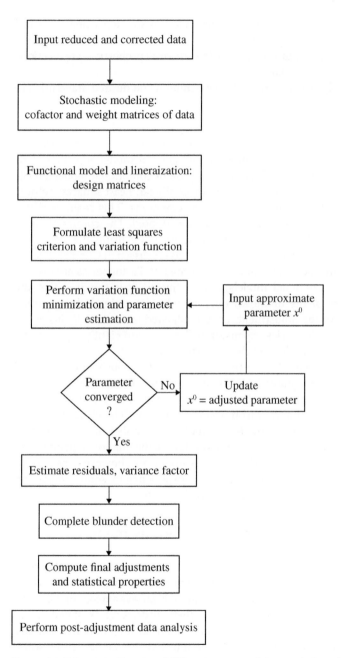

Figure 4.4 Schematic diagram of different aspects of data analysis and least squares estimation.

- Detecting possible blunders that may degrade the solution using statistical testing approach (see Chapter 11).
- Computing the final solutions and their statistical properties (see Sections 5.7–5.8, 12.4, and 13.6–13.8).
- Evaluating the adjustment results to determine if the desired standards are satisfied.

The starting point of least squares adjustment is the formulation of functional mathematical models relating observations and the unknown quantities (the parameters) as in the cases of parametric and general model adjustments or relating observations to themselves as in the case of conditional model adjustment. Functional models in geomatics can be nonlinear. Least squares method, however, are generally performed with linear equations. This requires that a means of linearizing nonlinear models must be adopted. The commonly used means of linearizing the equations in geomatics is by using the zero- and first-order terms of the Taylor series expansion as discussed in Chapter 2. Since only the linear terms (zero- and first-order terms) of Taylor series are used in approximating the functional models to be adjusted, it is required that the approximate values of the unknown quantities be determined close enough to the end results of the adjustment, i.e. the adjusted quantities. If this is not the case, then the Taylor series expansion will not be sufficiently representing the original functional model and the corrected values of the parameters from the adjustment must be used again (now as a new set of approximate or initial values) in the new least squares adjustment procedure. This procedure may be repeated as many times as may be found necessary until the final solution becomes insignificantly different from the previous one. The closer the initial values are to the final solution, the fewer will be the number of iterations necessary. This is to say that the choice of initial values of the unknown quantities to be determined in least squares adjustment is important; however, there is no special way of choosing them. Usually, some experience or some simple and quick computational means may be used to choose them.

4.6 Network Datum Definition and Adjustments

A *network datum* is a reference or a base network to which another network of measurements is attached through adjustment process. In an adjustment where coordinates of points are to be determined, there is always a need for a coordinate system, otherwise known as datum. In other words, it can be stated that in adjusting a survey network in which coordinates of network points are desired, a coordinate system (a datum) must first be attached to the survey network. Attaching a datum to a survey network is the same as defining the *datum parameters* for the network. The least squares adjustment of the survey network will only be possible if the network datum parameters are defined. The elements of the datum that must be known prior to it being used are known as *datum para-meters*. Typical parameters of a datum are its *spatial location, orientation,* and *scale*.

In geomatics, one is usually concerned with positioning or determining positions (or coordinates) of points on or near the Earth's surface. Positions or coordinates of points, unlike the measurable quantities like distances and angles, need to be specified with regard to some datum parameters. The number of datum parameters required in an adjustment depends on the type of positioning or coordinates to be determined. A common datum for three-dimensional (3-D) spatial positioning is usually a right-hand orthogonal 3-D (X, Y, Z) coordinate system that has a well-defined location or origin ($x = 0$, $y = 0$, $z = 0$); the orientation of its X-, Y-, and Z-coordinate axes are known with regard to the Earth; and the unit of length (or scale) to be used with the coordinate system is known. These seven parameters (three translations or x, y, z coordinates of the origin, three orientation elements of the coordinate axes, and a unit of length or a scale for the coordinate system) are known as *datum parameters* for 3-D positioning problems. The datum for a 3-D network is defined practically by specifying values for the seven datum parameters as follows:

- Specify three coordinates (x, y, z) of at least one of the network points to define the location of the network in space.
- Specify three spatial azimuths of the X-, Y- and Z-coordinate axes of the network; or use horizontal and zenith angles, if measured, to provide two rotation parameters with one azimuth completing the remaining rotation parameter.
- Measure the spatial distance of at least one of the lines in the network to provide the required scale for the network.
- Alternatively, fixing three coordinates (x, y, z) of a point with height difference measurements and azimuth measurements are sufficient; or fixing the coordinates (x, y, z) of two points and the height of another point can be sufficient in 3-D positioning.

If two-dimensional (2-D) coordinates (x, y) are to be determined in an adjustment, the number of datum parameters will be four: two translations or x, y coordinates of the origin, one orientation element of the two coordinate axes, and a unit of length or scale for the coordinate system. The datum for a 2-D network is defined practically by specifying values for the four datum parameters as follows:

- Specify two coordinates (x, y) of at least one of the network points to define the location or origin of the network on a map projection plane.
- Specify the azimuths of the X- and Y-coordinate axes of the network; the azimuth or grid bearing of a line in the network will provide the needed orientation parameter; horizontal angles, however, do not provide any value for the orientation parameter for 2-D network.

- Measure the horizontal distance of at least one of the lines in the network to provide the needed scale for the network.
- Alternatively, fixing the coordinates (x, y) of two points can be sufficient for 2-D adjustment.

If one-dimensional (1-D) coordinates (elevations or heights of points) are to be determined in an adjustment, the number of datum parameters required will be one: elevation or height of one point. The datum for a 1-D network (such as differential leveling network) is defined if the elevation of one benchmark is specified (with the height difference measurements providing the needed scale). Note that 1-D problems require no orientation parameter. Usually, height difference, if measured between two points of a network, will provide two rotation and one scale parameters if the network is 3-D and one scale parameter if the network is 2-D.

In summary, any adjustment of survey network that just satisfy the above minimum number of datum parameters is known as *minimum constraint* adjustment.

4.6.1 Datum Defect and Configuration Defect

The datum parameters that are not accounted for in the adjustment of network measurements constitute what is usually referred to as *datum defects* or *defects in datum*. A datum defect is the amount by which the datum parameters are unknown or are missing. It is closely connected with the datum parameters that are undefined through observations. For example, in 2-D traverse survey, in which distances and angles are usually measured, if no bearing of a line is known, there will be 3 datum defects as a result of no bearing and no fixed or constrained (x, y) coordinates. If the traverse is now done with a bearing of a line known, there will be 2 datum defects for lack of fixed or constrained (x, y) coordinates. One of the consequences of datum defects is that they make it impossible to solve least squares adjustment of survey network where coordinates are the unknown parameters.

Configuration defects arise when the number or the choice of observations are not sufficient to uniquely determine the coordinates of given network points, e.g. triangulation networks have datum defects as well as configuration defects, due to missing information about scale; the configuration defect in this case can be removed by measuring one distance; if datum parameters are defined for the network, then it will be free from the datum defects. Sometimes, a datum may be defined for a network, but the network may still have configuration defect, e.g. if a side shot is made to a point with only a distance measurement without an angle measurement to connect the side shot to the main network, it will be impossible to determine the coordinates of that point using only the distance measurement; this network is said to have configuration defect. Points that are disconnected from the main network will create configuration defects for the network.

4.7 Constraints in Adjustment

Constraint equations are the equations needed to complete a datum definition. This is necessary since the datum for the unknown coordinates to be determined is incompletely defined by the measurements. The type and the number of constraint equations needed depends on the number of *datum parameters* that are undefined by the network measurements being adjusted, i.e. the number of datum defects. These defects are to be taken care of by formulating appropriate constraint equations. The number of constraint equations required in an adjustment depends on the type of positioning or coordinates to be determined and also on the number of datum parameters taken care of by survey measurements. Generally, some of the datum parameters taken care of by survey measurements are as follows:

- *Scale parameter*: This is taken care of by either horizontal distance measurements, mark-to-mark distance measurements from GPS surveys, slope distance measurements, or height difference measurements.
- *Position parameters*: These are taken care of by providing appropriate coordinates (1-D, 2-D, or 3-D) of one point depending on whether 1-D, 2-D, or 3-D coordinates are being determined in the adjustment.
- *Orientation parameters*: These are taken care of by measuring or fixing an azimuth (or bearing) of a line for 2-D (x, y) positioning. For 3-D (x, y, z) determinations, the orientation parameters will be defined by measuring an azimuth and horizontal and vertical (zenith) angles; an azimuth, horizontal angles, and height differences; azimuths and height differences; or an azimuth, horizontal distances, and height differences.

One can fix a coordinate by including an appropriate equation that specifies the value to be assigned to the coordinate, such as

$$x + v_x = \hat{x} \tag{4.28}$$

$$y + v_y = \hat{y} \tag{4.29}$$

where the measured coordinates (x, y) with their corrections (v_x, v_y) are expressed as functions of the adjusted coordinates $(\hat{x}, \hat{y})$ themselves. When coordinates are expressed mathematically as in Equations (4.28) and (4.29), the coordinates are directly observed, forming the *constraint equations*. In this case, whenever the parametric equations are formed, it is necessary to add a set of constraint equations that will complete the definition of the coordinate datum. These constraint equations force those parameters involved to obey the rules defined by the equations. The other way of applying these constraint equations in adjustment is to leave them out when forming the first design matrix; in this case the number of parameters to be determined in the adjustment is reduced. This approach is adopted throughout this book.

4.7.1 Minimal Constraint Adjustments

When the number and types of constraint equations are just sufficient to define a given datum, the adjustment resulting from such a system will be called *minimal constraint adjustment*. This type of adjustment will provide adjusted coordinates that are solely dependent on the observations and the arbitrary choice of datum parameters. Least squares adjustment will be impossible without at least minimal constraint equations imposed. In practice, a survey network is said to have minimum constraints if it satisfies any of the following:

1) In 1-D *leveling network* (where only elevation differences are measured), the elevation of one point in the network can be fixed (not subject to changes) while all of the remaining network points are free to change by the adjustment process.

2) In 2-D *triangulation network* (where only angles are measured), the coordinates of two points in the network can be fixed (not subject to changes) while all of the remaining network points are free to change (not fixed). In this case, the coordinates of one of the two points are to define the location of the network; the bearing and the calculated distance between the two points are to provide the needed orientation (bearing) and scale to complete the datum definition.

3) In 2-D *trilateration network* (where only the distances are measured), the coordinates of one point in the network can be fixed (to define the location of network); one coordinate (*x* or *y*) of another network point can be fixed to complete the datum if the approximate coordinates of the other network points are "well known." In this case, the distance measurements will provide the scale and the fixed coordinates, and the approximate coordinates of the network points will define the azimuth orientation of the network.

4) In a *combined 2-D triangulation and 2-D trilateration network* (known also as *2-D triangulateration network*) in which an azimuth of a line is measured, the coordinates of one point in the network can be fixed (to define the location of network); the azimuth and the angle measurements will define the orientation of the network, while the distance measurements will provide the needed scale for the network to complete the datum definition.

5) In *2-D traverse network* in which angles and distances are measured at traverse stations, the datum is defined similarly to the cases of 2-D trilateration network and a combined 2-D triangulation and 2-D trilateration network.

6) In *3-D traverse network* in which horizontal angles, zenith angles, slope distances, and elevation differences are measured, the datum is defined by fixing 3-D coordinates of one of the network points and the azimuth of a line.

Fixing any other coordinates in any of the above networks will *overconstrain* the network, but adding more measurements such as angles, distances, and even azimuths will not overconstrain the network. Adjusting a network with the above specified minimum constraints is referred to as minimum constraint adjustment. For example, GPS-only solution (in which the coordinates (x, y, z) of a reference point are known with the baseline vectors (dx, dy, dz) measured) is minimally constrained solution; the coordinates of the reference point provide the location parameter, and the baseline vector provides both the orientations and the scale parameters. It should also be remembered that if a datum is not established for a network, it will be impossible to adjust the network by least squares method no matter the number of redundant observations involved.

4.7.2 Overconstrained and Weight-Constrained Adjustments

In minimum constraint adjustment, the published coordinates (if not fixed in the adjustment) may still move in one direction (which is normally unacceptable). It might be preferable to make sure that the existing control network stays fixed by constraining (or holding fixed) their coordinates. This is common in practice, assuming the quality of the control network is far superior to the quality of new observations being added. When a network is constrained to two or more control stations, it is said to be *overconstrained*. An overconstrained network is forced to accommodate itself (the measurements), the previously established control (errors in previous survey), and the new survey. In this network, any errors in the fixed parameters will deform the adjusted observations and the adjusted parameters. In other words, when more constraint equations are used in an adjustment than just what are needed, the adjustment is said to be *constrained adjustment* or what is commonly known as *overconstrained adjustment*. In this case, the resulting adjusted quantities are constrained to be in the proper coordinate system and also to fit the additional constraints. The error ellipses derived for the new points, however, will appear too optimistic in the sense that the errors in the control points are not considered. Every time a new survey is adjusted into an existing coordinate system by using existing control points, an *overconstrained (or simply constrained) adjustment* is performed.

Constrained adjustments have some advantages over minimal constraint adjustment, which include the following:

- Placing new surveys into some larger coordinate system by providing the datum (origin and orientation) for the new survey.
- Providing the means of detecting blunders in the new survey; this makes the resulting adjusted observations from constrained adjustment to be more

accurate than the observed values and the adjusted coordinates to be more accurate than those from minimal constraint adjustment.

- Controlling or dampening the buildup of the effect of observational (accidental) errors on the adjusted coordinates.

Some of the disadvantages of constrained adjustments include the following:

- The control points fixed (with their coordinates assumed to be correct and errorless) for the adjustment may be inadequate in the sense that they may lead to misclosures that may be much larger than expected. This is usually the case since misclosures in such adjustments are often associated with errors in the new surveys (which may be wrong) rather than in the fixed control points. The consequence of this association is that the difference between observed and adjusted values of observations may be much larger than can be explained by observational (accidental) error.
- Every constrained adjustment in which the control points are held fixed (having coordinates with zero variances) is *suboptimal*. This means that least squares adjustments using a weight matrix that is not inversely proportional to the covariance matrix of observations do not have the minimum variance property (they are suboptimal). Note that the coordinates of the fixed control points are pseudo-observations, but their covariance matrices are not used in weighting them since they have zero values as their elements. For an example, a highly accurate GPS survey may be distorted or degraded when fitted to lower-grade control points. In this case, the covariance matrix of the adjusted observations may not be less than that of the real observations as it should be. By fixing control points, the adjusted values of observations will likely become worse than the real observations. For example, the effects of errors in fixed control points must be considered when fitting GPS vectors that are accurate to 1 ppm into fixed control points based on the original NAD83 network that is accurate to about 1 : 300 000. This means that the covariances of the fixed control points must be considered in the adjustment so as to account for the effects of the errors of the fixed control points.

Constrained adjustments are often said to distort observations since they produce adjusted observations that are more uncertain or are likely to have greater errors than the real observations. For this reason, when surveyors process, for example, their GPS baseline vectors, they usually prefer minimal constraint adjustment to constrained adjustment since minimal constraint adjustments do not distort measurements after adjustments. The most correct approach for adjusting survey measurements is to perform weight-constrained adjustment, which considers both the errors in the new surveys and in the control points. This approach, however, has a problem of including the coordinates of the control points for adjustment, thereby changing their values after the adjustment. Since surveyors do not want coordinates of control points to keep on changing in any adjustment, the most acceptable option for surveyors is still to keep control points fixed (defining the datum) and to adjust new surveys relative to them.

The type of adjustment to be performed on survey stations of a new survey depends on the types of constraints imposed on the survey station coordinates. Survey station coordinates can be defined as *fixed (or absolutely constrained)*, *weight constrained*, or *free*. Fixed (or absolutely constrained) station coordinates will have coordinates that will remain the same after adjustment since their coordinates are not open to any adjustment. These stations are said to have coordinates with infinite weights or a cofactor matrix with zero values as its elements. When survey stations are fixed in order to define a datum, those fixed stations are considered as constituting a *zero variance reference base*. In this case, the covariance matrix of the estimated coordinates will depend on the choice of the zero variance reference base (datum). It is possible, however, using similarity transformations to transform a given cofactor matrix from one datum or zero variance reference base to another on a different zero variance reference base. The transformation steps are discussed later in Chapter 14. If the initial coordinates of a survey station (before adjustment) are associated with some standard deviations, the station is said to be a *weight-constrained station*. This means the weighted station coordinates are not known well enough to be held fixed so that they are open to adjustment to the extent limited by the associated standard deviations of the station coordinates; these station coordinates can be used as pseudo-measurements with their standard deviations as weighting factor. The standard deviations of the weighted points are to improve after the adjustment and may as well improve other parameters through correlation. The weighted stations adjustment is also known as *Bayesian least squares*.

Usually, a least squares estimation based on initial coordinates that are associated with a cofactor matrix (*weighted constraints*) will give adjustment results that are almost identical to those by *minimum constraint* adjustment; the adjusted values based on this method is somewhere between the values derived using the overconstrained method and those based on minimally constrained method. This can be interpreted to mean that previously estimated data in an adjustment do not significantly distort new measurements. In this case, the geometry (relative position of points) is defined by the measurements, and the weight constraints are not allowed to affect the geometry. When coordinates of survey stations are unknown and are open to unlimited or unconstrained adjustment, the stations are referred to as *free stations*.

4.7.3 Adjustment Constraints Examples

Circle the letters corresponding to the most appropriate answers for the following multiple-choice questions.

1. Which of the following networks is not minimally constrained?
   (a) A 2-D traverse network with only one network point fixed with an azimuth of one of the legs given.
   (b) A 2-D horizontal network with all of the possible distances measured with only one fixed network point and one known bearing.
   (c) A 2-D horizontal network with all of the possible distances measured with only the center of gravity of the network points fixed.
   (d) A 2-D network in which two points are fixed with all of the possible distances measured.
   Answer: (d) A 2-D network in which two points are fixed with all of the possible distances measured. As discussed in Section 4.7.1, item (v), in a traverse survey, distances and angles at the setup stations are measured. The distances provide scale, leaving 3 datum parameters (two coordinates and an azimuth) to be fixed; all the other answers will provide minimal constraints with item (c) being a different type of minimal constraint discussed in Chapter 14.

2. In a 2-D network survey, all of the distances and the included angles were measured without a known point. What is the number of datum deficiencies of the network?
   (a) 1
   (b) 2
   (c) 3
   (d) 4
   Answer: (c) 3. The number of datum deficiencies of the 2-D network is 3; in Section 4.7.1, item (iv), angles do not contribute to datum; the distance measurements fix only the scale, leaving two coordinates and one azimuth parameters to be fixed.

3. In a 2-D traverse survey, all of the distances and the included angles were measured with the coordinates of the starting point known. What is the number of datum deficiencies of the network?
   (a) 0
   (b) 1
   (c) 2
   (d) 3
   Answer: (b) 1. The number of datum deficiencies of the network is 1; this question is similar to question (2) except that the starting two coordinates are now provided, leaving only one azimuth parameter to be defined.

4. Which of the following is a weighted constraint point?
   (a) A fixed network point
   (b) A point with zero standard deviation
   (c) A point with nonzero standard deviation
   (d) A point that is open to unlimited adjustment

Answer: (c) A point with nonzero standard deviation. The weighted constraint point is a point with nonzero standard deviations, which can be used to weight the coordinates as pseudo-observations as discussed in Section 4.7.2.

5. All of the possible distances in a horizontal survey network of five points were measured. If no other measurements were made and if the coordinates of all the network points are only approximately known, how many constraint equations will be formed for the minimal constraint least squares adjustment of the network?

(a) 1
(b) 2
(c) 3
(d) 4

Answer: (c) 3. The number of constraint equations that can be formed for the minimal constraint least squares adjustment of the network is 3. This is a trilateration network (discussed in Section 4.7.1, item (iii)) with distances defining the scale, leaving two coordinates and an azimuth to be fixed. Three equations representing these defects must be formulated as constraint equations in the least squares adjustment.

6. All of the possible total station directions in a horizontal survey network of five points were measured. If no other measurements were made and if the coordinates of all the network points are only approximately known, how many constraint equations will be formed for the minimal constraint least squares adjustment of the network?

(a) 1
(b) 2
(c) 3
(d) 4

Answer: (d) 4. The number of constraint equations that can be formed for the minimal constraint least squares adjustment of the network is 4. Note that the total station directions are arbitrary directions and are not the same as azimuths or bearings. This problem is similar to a triangulation network in Section 4.7.1, item (ii), where no datum parameters are fixed with the measurements; the number of datum parameters to be fixed for the 2-D network is 4.

7. Which of the following will not distort observations after basic least squares adjustment?

(a) Minimum constraint adjustment
(b) Overconstrained adjustment
(c) Constrained adjustment
(d) Unconstrained adjustment

Answer: (a) Minimum constraint adjustment. Minimum constraint adjustment will not distort observations after basic least squares adjustment

(Section 4.7.2). Note that unconstrained adjustment is impossible with the basic least squares method.

8. What is the possible number of datum defects in a simple three-dimensional GPS survey?
   (a) 1
   (b) 2
   (c) 3
   (d) 4
   Answer: (c) 3. The possible number of datum defects in a simple three-dimensional GPS survey is 3. From Section 4.7.1, the (x, y, z) coordinates of the base station are required to define the datum; the scale and orientation are provided by the baseline vectors.
9. Which of the following constraints adjustments will change the coordinates of the control points after adjustments?
   (a) When the control points are used with their corresponding standard deviations
   (b) When the control points are used to overconstrain the adjustment
   (c) When the control points are used as zero variance reference base
   (d) When the control points are fixed
   Answer: (a) When the control points are used with their corresponding standard deviations. Using control points with their standard deviations make them pseudo-measurements, which will be adjusted by the least squares procedure.

4.8 Comparison of Different Adjustment Methods

In recent times, geomatics professionals have become interested in how traditional traverse adjustment methods compare with the method of least squares adjustment. The well-known traditional adjustment methods in surveying are the transit, Bowditch, and Crandall's methods. Another traditional traverse survey adjustment algorithm referred to as Tuttle's method was discussed and compared with the Bowditch and Crandall's procedures in Sprinsky (1987). Although the traditional methods of adjustment are inadequate in adjusting high precision survey networks involving trilateration and triangulation networks, they are still very attractive to land surveyors who specialize in property boundary establishment. This section uses two examples (Examples 4.4 and 4.5) to explain the assumptions and relationships among the traditional adjustment methods and the least squares adjustment method. It is expected that geomatics professionals will appreciate the limitations of the different adjustment methods and how the methods are related through the given examples. Moreover, geomatics professionals are encouraged to verify any adjustment results obtained through commercial software by performing routine traverse computations manually by hand rather than just accepting values provided by some software packages. The adjustment results based on transit, Crandall's, Bowditch, and the least squares methods are presented, and some aspects of the traditional methods are compared with the least squares adjustment method. In order to quantify the effect of any adjustment based on any method, the adjusted observations (distances and angles) are analyzed rather than the plane coordinates computed from those observations. The corrections to the observations should be based on the quality of the observations; large corrections to high quality observations indicate a possible systematic error in the observations. The steps for quantifying the effect of any adjustment can be given as follows:

- Calculate the coordinates from the field measurements.
- Calculate angles and distances by back-computations using the calculated coordinates.
- Compare the back-computed quantities with the actual measurements to see how far they deviate from the actual measurements.

Example 4.4   Adjust the loop traverse network described in Example 4.1 by the method of least squares adjustment (using any commercial least squares adjustment software package), and compare the results of the adjustment with those based on transit method (Example 4.1), compass rule method (Example 4.2), and Crandall's rule method (Example 4.3) by analyzing the residuals computed from each of the methods.

Solution:

The traditional adjustments were performed using the Microsoft Excel 2013 program (refer to Examples 4.1–4.3). The least squares adjustment was done using the least squares adjustment software known as MicroSurvey STAR∗NET Pro; the STAR∗NET software code for the adjustment is listed in Table 4.6. The adjustment gives the number of redundant observations as 3 and the a posteriori standard factor of unit weight as 1.532 with the chi-squares test passing at 95% confidence level. The adjusted coordinates of the traverse points are given in Table 4.7.

The residual components of the adjustments by four methods (least squares, transit, compass, and Crandall) of the closed-loop traverse network in Figure 4.1 and Table 4.1 are presented in Tables 4.8–4.10. The distance residuals are given in Table 4.8, while the angle residuals are given in Table 4.9. Given in Table 4.10 are the sums of weighted squares of residuals, which are based on the computed residuals from the four methods of adjustment with the variances (derived from the standard deviations in Table 4.1) of the measurements used as weighting factors. It can be seen in the table that the least squares method has the least

Table 4.6 MicroSurvey STAR*NET 8 code for the 2-D closed-loop traverse adjustment.

```
# Least Squares adjustment of the sample closed-loop
# traverse network
.2D
.ORDER EN FromAtTo
.UNITS Meters DMS
# Given coordinates
C P 1000.000 1150.000 ! !
# Bearing measurement
B P-Q 79-37-22 !
# Angle measurements
A P-Q-R 119-44-50 4
A Q-R-S 45-00-10 4
A R-S-P 130-35-50 4
A S-P-Q 64-39-20 4
# Distance measurements
D Q-R 223.600
D R-S 237.184
D S-P 145.775
D P-Q 180.278
```

Table 4.7 Adjusted coordinates of the closed-loop traverse network based on least squares adjustment.

Station	Easting (m)	Northing (m)
P	1000.0000	1150.0000
Q	1177.3322	1182.4737
R	1251.4939	1393.4186
S	1037.6483	1290.8282

Table 4.8 Computed distance residuals from the four adjustment methods.

Traverse leg	Least squares (mm)	Transit (mm)	Compass (mm)	Crandall (mm)
QR	1.6	1.5	1.7	−1.4
RS	−3.2	−4.8	−3.5	−7.7
SP	−1.3	−0.8	−1.0	0.9
PQ	3.1	4.2	2.8	5.8

Table 4.9 Computed angle residuals from the four adjustment methods.

Traverse point	Least squares (arc sec)	Transit (arc sec)	Compass (arc sec)	Crandall (arc sec)	Pre-adjust correction (arc sec)
Q	0.3	−1.5	0.8	−2.0	−2.0
R	−6.4	−5.8	−6.6	−3.0	−3.0
S	−4.3	−1.4	−4.0	−2.0	−2.0
P	0.3	−1.3	−0.1	−3.0	−3.0

Table 4.10 Sum of squares of weighted residuals.

Method	Value
Transit	8.535
Compass	7.146
Crandall's	15.032
Least squares	7.040

sum of weighted squares of residuals of 7.040, which is very close to that of the compass rule method (7.146). Note that the maximum number of redundancies in a closed-loop traverse survey is three, which makes traverse survey technique less reliable compared with triangulation and trilateration networks, which can have large and variable numbers of redundancies. Note also that the traditional methods of adjustment are inadequate in adjusting triangulation and trilateration networks, which are commonly used in high precision surveys.

In Table 4.10, the sum of weighted squares of residuals for Crandall's method will be larger than what was computed if the angular misclosure of the traverse were not first accounted for before the adjustment. The error of closure of this traverse before adjusting the angles is 1 : 45 000; after pre-adjusting the angles to satisfy the traverse geometry, the error of the traverse closure becomes 1 : 65 000. These errors of closures are quite within the usually acceptable 1 : 5000 accuracy specification for property survey. It can be seen from Tables 4.8 and 4.9 that all the residuals are acceptable since none of them is greater than thrice the standard deviation of measurement limit for the residuals to be considered as outliers.

The graphical representations of the computed residuals are given in Figures 4.5 and 4.6. In Figure 4.5, it can be seen that Crandall's method adjusted the distances the most, especially the first traverse leg P–Q and the traverse leg

(R–S) before the last leg closing the loop. The adjustment of the distances is identical in compass and least squares methods. In Figure 4.6, Crandall's method produces the least adjustments to the angles; it can be seen in Table 4.9 and Figure 4.6 that only the pre-adjustment angular corrections are provided as residuals. Again, the compass and least squares methods produce consistent adjustments as shown in Figure 4.6.

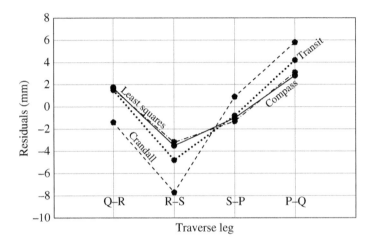

Figure 4.5 Plot of distance residuals from four adjustment methods.

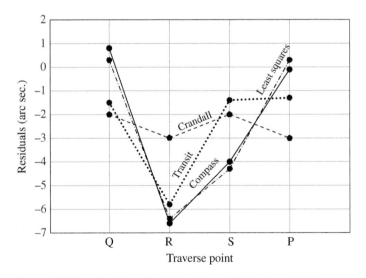

Figure 4.6 Plot of angle residuals from four adjustment methods.

Example 4.5 In a heavily vegetated area, the labor involved in brushing traverse lines may make it impractical to have a smooth closed-loop traverse. In this situation an alternate point is established between the section corners, producing alternate-point traverse network as shown in Figure 4.7 and Table 4.11. In this traverse, the main traverse corners are P–Q–S–T–U–W–P, but points Q and S and points U and W are not intervisible so that an alternate point

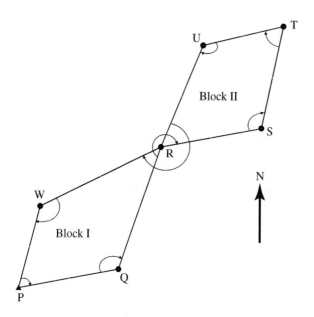

Figure 4.7 An alternate-point traverse network.

Table 4.11 Field measurements.

Obs. no.	From	At	To	Angle	Distance (m) (at–to)
1	P	Q	R	119°44′54″	223.603
2	Q	R	S	240°15′30″	180.276
3	R	S	T	112°36′56″	180.265
4	S	T	U	64°39′25″	145.755
5	T	U	R	125°50′27″	190.398
6	U	R	W	221°37′54″	237.184
7	R	W	P	130°35′50″	145.775
8	W	P	Q	64°39′23″	180.276

R has to be established. In establishing this alternative point, two blocks of network (Block I and Block II) are created as shown in Figure 4.7. In the network, assume the initial bearing of line P–Q is fixed at 79°37′22″. In Table 4.11, the total station instrument used is assumed having a standard deviation for horizontal angle measurement as ±3″ according to ISO 17123-3 standard and the standard deviation of distance measurement as ±3 mm. The angular error was computed as ±4″ for a set of horizontal angular measurement at each station. Adjust the measurements by transit, compass rule, Crandall's rule, and least squares methods, and analyze the residuals computed from each of the methods.

Solution:

The traverse computations for departures and latitudes are given in Table 4.12; the total closure errors in the easting and northing directions are given as sums in columns (5) and (6) as ε_E = −0.0053 m and ε_N = −0.0206 m, respectively; the total length (D) of traverse is 1483.532 m; the linear error of closure (LC) based on Equation (4.3) is 0.0213 m; direction of closure (θ) using Equation (4.4) is 14°32′; and the ratio of LC to D using Equation (4.7) is 1 : 69 673. The adjustments of the measurements by transit, compass rule, and Crandall's rule methods are given in Tables 4.13–4.15, respectively. Note in each method that points R and V represent the same point and their mean coordinates are adopted for the point.

Table 4.12 Departure and latitude computations.

Station (from) (1)	Adjusted angle (2)	β (3)	S (m) (4)	ΔE (5)	ΔN (6)	Station (to) (7)
						P
P		79°37′22″	180.276	177.3273	32.4728	Q
Q	119°44′52″	19°22′14″	223.603	74.1638	210.9456	R
R	240°15′28″	79°37′42″	180.276	177.3304	32.4556	S
S	112°36′53″	12°14′35″	180.265	38.2269	176.1652	T
T	64°39′22″	256°53′57″	145.755	−141.9614	−33.0376	U
U	125°50′25″	202°44′22″	190.398	−73.5966	−175.5988	V
V	221°37′52″	244°22′14″	237.184	−213.8476	−102.5937	W
W	130°35′47″	194°58′01″	145.775	−37.6481	−140.8296	P
P	64°39′21″	79°37′22″				Q
Sum	1080°00′00″		1483.5320	−0.0053	−0.0206	

Table 4.13 Adjustments and adjusted coordinates by transit method.

Station (from) (1)	$c_{\Delta E}$ (2)	$c_{\Delta N}$ (3)	$\Delta \hat{E}$ (4)	$\Delta \hat{N}$ (5)	$\hat{E}_k$ (6)	$\hat{N}_k$ (7)	Station (to) (8)
					1000.0000	1150.0000	P
P	0.0010	0.0007	177.3283	32.4735	1177.3283	1182.4735	Q
Q	0.0004	0.0048	74.1643	210.9504	1251.4925	1393.4239	R
R	0.0010	0.0007	177.3314	32.4563	1428.8239	1425.8802	S
S	0.0002	0.0040	38.2271	176.1692	1467.0510	1602.0494	T
T	0.0008	0.0008	−141.9606	−33.0369	1325.0904	1569.0125	U
U	0.0004	0.0040	−73.5962	−175.5948	1251.4942	1393.4178	V
V	0.0012	0.0023	−213.8463	−102.5914	1037.6479	1290.8264	W
W	0.0002	0.0032	−37.6479	−140.8264	1000.0000	1150.0000	P
Sum	0.0053	0.0206	0.0000	0.0000			

Table 4.14 Adjustments and adjusted coordinates by compass rule method.

Station (from) (1)	$c_{\Delta E}$ (2)	$c_{\Delta N}$ (3)	$\Delta \hat{E}$ (4)	$\Delta \hat{N}$ (5)	$\hat{E}_k$ (6)	$\hat{N}_k$ (7)	Station (to) (8)
					1000.0000	1150.0000	P
P	0.0006	0.0025	177.3279	32.4753	1177.3279	1182.4753	Q
Q	0.0008	0.0031	74.1646	210.9487	1251.4925	1393.4239	R
R	0.0006	0.0025	177.3311	32.4581	1428.8236	1425.8820	S
S	0.0006	0.0025	38.2275	176.1677	1467.0511	1602.0497	T
T	0.0005	0.0020	−141.9609	−33.0356	1325.0902	1569.0141	U
U	0.0007	0.0026	−73.5960	−175.5961	1251.4943	1393.4180	V
V	0.0009	0.0033	−213.8467	−102.5904	1037.6476	1290.8276	W
W	0.0005	0.0020	−37.6476	−140.8276	1000.0000	1150.0000	P
Sum	0.0053	0.0206	0.0000	0.0000			

Table 4.15 Adjustments and adjusted coordinates by Crandall's rule method.

Station (from) (1)	$c_{\Delta E}$ (2)	$c_{\Delta N}$ (3)	$\Delta \hat{E}$ (4)	$\Delta \hat{N}$ (5)	$\hat{E}_k$ (6)	$\hat{N}_k$ (7)	Station (to) (8)
					1000.0000	1150.0000	P
P	−0.0009	−0.0002	177.3263	32.4726	1177.3263	1182.4726	Q
Q	0.0025	0.0072	74.1663	210.9527	1251.4927	1393.4253	R
R	−0.0009	−0.0002	177.3295	32.4554	1428.8221	1425.8807	S
S	0.0011	0.0053	38.2280	176.1705	1467.0501	1602.0512	T
T	−0.0004	−0.0001	−141.9618	−33.0377	1325.0883	1569.0135	U
U	0.0020	0.0048	−73.5946	−175.5940	1251.4937	1393.4195	V
V	0.0011	0.0005	−213.8464	−102.5932	1037.6472	1290.8263	W
W	0.0009	0.0033	−37.6472	−140.8263	1000.0000	1150.0000	P
Sum	0.0053	0.0206	0.0000	0.0000			

The mean coordinates for the point represented by R and V in transit method are $\hat{E}$ = 1251.4934 m, $\hat{N}$ = 1393.4208 m.

The mean coordinates for the point represented by R and V in compass rule adjustment method are $\hat{E}$ = 1251.4934 m, $\hat{N}$ = 1393.4210 m.

From Equation (4.18), B is calculated as −6.36702E-08 and A is calculated from Equation (4.19) as 1.831E-07. The mean coordinates for the point represented by R and V in Crandall's rule method are $\hat{E}$ = 1251.4932 m, $\hat{N}$ = 1393.4224 m.

The least squares adjustment was done using the least squares adjustment software known as MicroSurvey STAR∗NET Pro; the STAR∗NET software code for the adjustment is listed in Table 4.16. The adjustment gives the number of redundant observations as 5 and the a posteriori standard factor of unit weight as 1.477 with the chi-squares test passing at 95% confidence level. The adjusted coordinates of the traverse points are given in Table 4.17. In Tables 4.16 and 4.17, the coordinates of point R are uniquely determined unlike in the traditional methods in which point R is also represented as point V in the return traverse.

The residual components of the adjustments of the measurements in Table 4.11 by the traditional methods and the least squares methods are given in Tables 4.18 and 4.19 and also displayed graphically in Figures 4.8 and 4.9. The computed distance residuals for the four methods of adjustment are given in Table 4.18 and Figure 4.8. In the figure, it can be seen that compass and least squares methods produce adjustments to distance measurements in an identical manner and Crandall's method consistently produces largest corrections to the measurements. Figure 4.9 shows that the compass rule method produces the

Table 4.16 MicroSurvey STAR*NET 8 code for the 2-D alternate-point traverse adjustment.

```
# Least Squares adjustment of the alternate point
# traverse network
.2D
.ORDER EN FromAtTo
.UNITS Meters DMS
# Given coordinates
C P 1000.000 1150.000 ! !
# Bearing measurement
B P-Q 79-37-22 !
# Angle measurements
A P-Q-R 119-44-54 4
A Q-R-S 240-15-30 4
A R-S-T 112-36-56 4
A S-T-U 64-39-25 4
A T-U-R 125-50-27 4
A U-R-W 221-37-54 4
A R-W-P 130-35-50 4
A W-P-Q 64-39-23 4
# Distance measurements
D Q-R 223.603 0.003
D R-S 180.276 0.003
D S-T 180.265 0.003
D T-U 145.755 0.003
D U-R 190.398 0.003
D R-W 237.184 0.003
D W-P 145.775 0.003
D P-Q 180.276 0.003
```

Table 4.17 Adjusted coordinates of the alternate-point traverse network based on least squares adjustment.

Station	Easting (m)	Northing (m)
P	1000.0000	1150.0000
Q	1177.3294	1182.4732
R	1251.4941	1393.4198
S	1428.8252	1425.8779
T	1467.0540	1602.0468
U	1325.0930	1569.0134
W	1037.6478	1290.8285

Table 4.18 Distance residuals from the four methods of adjustment.

Traverse leg	Transit	Compass	Crandall	Least squares
QR	2.1	0.6	5.0	1.3
RS	0.8	0.8	−0.9	1.2
ST	4.0	2.6	5.4	4.0
TU	−1.0	−1.0	0.4	−1.4
UV	−6.3	−5.1	−7.7	−3.9
VW	−1.6	−1.7	−0.4	−2.2
WP	−3.2	−2.1	−3.4	−1.2
PQ	1.1	1.1	−1.0	2.2

Table 4.19 Comparison of the angle residuals from the four adjustment methods.

Traverse point	Transit (arc sec)	Compass (arc sec)	Crandall's (arc sec)	Least squares (arc sec)	Pre-adjust correction (arc sec)
Q	−0.8	2.1	−0.7	−1.6	−2.0
R	−6.8	−9.6	−6.7	−5.1	−2.0
S	0.5	3.3	0.4	0.9	−3.0
T	−1.5	−0.5	−3.0	1.5	−3.0
U	0.6	−2.1	−0.3	−3.3	−2.0
R	−6.7	−4.9	−6.2	−5.1	−2.0
W	−0.8	−2.6	−0.5	−4.4	−3.0
P	−3.5	−4.7	−2.0	−2.0	−2.0

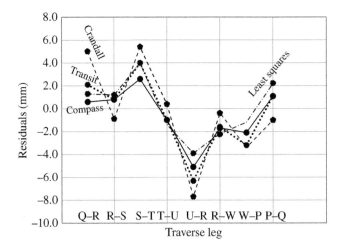

Figure 4.8 Plot of distance residuals from four adjustment methods.

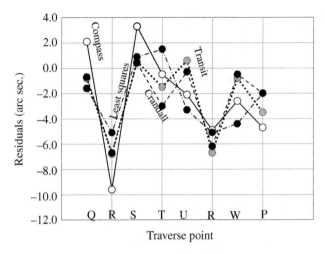

Figure 4.9 Comparison of angle residuals from four methods of adjustment.

Table 4.20 Sum of squares of weighted residuals.

Method	Value
Transit	15.155
Compass	15.174
Crandall's	20.240
Least squares	10.915

largest adjustments to angle measurements at the starting point P and the next three points Q, R, and S. Crandall's method provides the least adjustments to angle measurements in the Block I of the traverse (at points P, Q, and W) excluding only the alternate point R that is common to Block I and Block II. The least squares method produces the least adjustment to the angle measurement at the alternate point R.

The sums of weighted squares of residuals for the four methods are given in Table 4.20. These sums are based on the computed residuals from the four methods of adjustments with the variances of the measurements used as weighting factors. The variances are derived from the standard deviations of measurements of ±4″ for angle measurements and ±3 mm for distance measurements. As can be seen in Table 4.20, the sum of the weighted squares of residuals based on the least squares method is 10.915, which is still the least when compared with the three traditional adjustment methods. The sum of weighted squares of residuals for Crandall's method will be larger than what was computed if the angular misclosure of the traverse were not first accounted for before the adjustment. The error of closure of this traverse before adjusting the angles is 1 : 54 000; after pre-adjusting the angles to satisfy the traverse geometry, the error of the traverse closure becomes 1 : 70 000. This is quite within the usually acceptable 1 : 5000 accuracy specification for property survey. It can be seen from Tables 4.18 and 4.19 that all the residuals are acceptable since none of them is greater than thrice the standard deviation of measurement limit for the residuals to be considered as outliers.

4.8.1 General Discussions

From Table 4.9, it can be seen that the pre-adjustment corrections are the same as the residuals from Crandall's method. This is consistent with the idea that Crandall's method does not adjust angles (bearings); the residuals are actually due to the pre-adjustment corrections applied to the angles. This is different in the case of alternate-point traverse network adjustment as shown in Table 4.19. It can be seen from Table 4.19 that the residuals from Crandall's method are different from the pre-adjustment corrections to the angles, which seems to suggest that Crandall's method adjusted the bearings; this is because of the common point R. If point R, however, is treated as two different points in the adjustment, the residuals will be the same as the pre-adjustment corrections. It can be seen that those points connected to point R, such as Q, S, U, and W, have different residuals, but unconnected points T and P have residuals that are equal to their pre-adjustment corrections. Generally, from the two given examples, it can be seen that maximum distance adjustment expected over an average of 200 m is within ±10 mm in traditional method (in Crandall's method) and is within ±4 mm using least squares method. The angular adjustments are within ±10″ in traditional method (in compass method) and within ±7″ using least squares method. The adjusted coordinates from the three traditional methods vary from those of least squares within ±5 mm. Although the results from the three traditional adjustment methods are comparable with those of least squares method, there are a number of significant advantages of least squares method over the traditional methods, which include the following:

1) It has an advantage over the other methods since it is based on the theory of mathematical probability, which makes it more rigorous. It also provides the smallest corrections to measurements after fitting observations together with zero misclosure.
2) It allows all observations to be appropriately weighted using their expected precisions; the other methods do not allow this.
3) The same procedure is followed in solving many different types of geomatics problems.

4) It enables complete statistical analyses to be made after adjustments, which includes determining the expected precisions of adjusted quantities. Since current survey accuracy standards are based on the concepts of least squares estimation, the surveyors are only able to properly test their observations for compliance with the accuracy standards by adjusting their data using the method of least squares estimation.

5) GNSS and geographic information systems (GISs) rely heavily on rigorous adjustment and statistical analysis procedures of the least squares estimation methods.

6) It can be used to perform pre-survey planning, such as in pre-analysis and network design.

The disadvantages of ordinary least squares method are generally with regard to other adjustment methods, such as *least squares collocation*. In this case, ordinary least squares method can be considered as having oversmoothing properties when applied to prediction or interpolation problems. The ordinary least squares method (also known as L_2 norm) may also be weak compared with other estimation method like L_1 norm method (also known as robust alternative to least squares, which minimizes the sum of absolute values of residuals or weighted residuals) (Huber 1964; Barrodale and Roberts 1974; Harvey 1993; Bektas and Sisman 2010) when dealing with observations that are likely to be affected by outliers. In this case, no particular adjustment method can be considered best for all applications. The following general comments can be made with regard to errors of closure of traditional traverse survey:

1) The ratio of the error of closure to the length of traverse has been useful in indicating the existence of some blunders and has been an indication of the accuracy of the traverse. However, there is no guarantee that all measurements are free from blunders or the geometry of the traverse is such that all stations are within a desired relative accuracy to each other.

2) The ratio of the error of closure to the length of traverse (e.g. 1 : 5000) has traditionally been used to determine the accuracy of legal survey work. This ratio has limitations for very large surveys and surveys involving very short lines. In very long survey traverses, the allowable error may be misleading and, in some cases, may conceal a blunder. For a traverse length of 100 m, for example, the expected linear closure will be 0.02 m, which may be difficult to meet on the basis of 1/5000 error closure specification.

3) In GIS and land information system where information from several data sets are usually integrated, it is necessary to know positional accuracy. So, it is no longer sufficient to assess accuracy of a legal survey in terms of the ratio of the error of closure to the length of the traverse. Moreover, there is no simple correlation between relative closure ratio accuracies and 95% radial positional accuracies.

Problems

4.1 A leveling net of five points (A, 1, 2, 3, B) shown in Figure P4.1 was measured in a leveling survey. The total number of elevation differences measured between the points is 7. The elevation of point A is known as 100.000 m with standard deviation of 0.006 m; the elevation of point B is known and fixed.

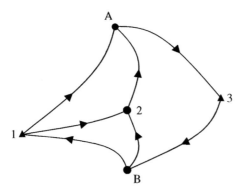

Figure P4.1 (Not to scale.)

Use the above information to answer the following, giving your reasons for each of them:
a) What is the number of datum defect(s)?
b) What is the number of free points?
c) Give the weight constraint point.
d) Give the absolutely constrained point.
e) Which of the points will have zero weights?
f) Which of the points will have the highest (infinite) weight?

4.2 A horizontal triangular survey network (A–B–C) in Figure P4.2 was measured for the purpose of determining the coordinates of a point C (x_C, y_C). The network measurements include two distances; two angles; a bearing measured in a previous survey; the coordinates (and their standard deviations) of point B being determined from the previous GPS survey; and point A being fixed and errorless. Answer the following, giving your reasons.

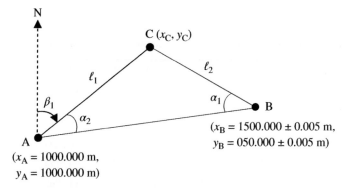

Figure P4.2 (Not to scale.)

a) What is the number of datum defect(s)?
b) List the free coordinates in the network
c) List the weight constraint coordinate(s) in the network.
d) Give the absolutely constrained point in the network.
e) Is the network overconstrained? Why?

4.3 A small two-dimensional network is to be adjusted by the parametric method of least squares. The measurements include nine distances, four directions observed at point A, and three directions observed at point 2.

a) If the network is to be adjusted without fixing any of the network points, how many datum defects are contained in the network? List the types of parameters needed to just fix the datum problem.
b) How many unknown parameters does the network contain if a minimum constrained adjustment is to be performed? Give details of how you obtain your answer.
c) How many parameters are there if the network is fully constrained with fixed stations A and B?

4.4 Repeat Example 4.5 using the following field measurements and compare the adjusted coordinates of point R from the four adjustment methods. Assume the initial bearing of line P–Q is fixed at $106°41'25''$.

Field measurements.

Obs. no.	From	At	To	Angle	Distance (m) (at–to)
1	P	Q	R	119°44′45″	223.590
2	Q	R	S	240°15′35″	180.272
3	R	S	T	112°36′45″	180.262
4	S	T	U	64°39′20″	145.750
5	T	U	R	125°50′30″	190.382
6	U	R	W	221°38′00″	237.178
7	R	W	P	130°35′45″	145.765
8	W	P	Q	64°39′25″	180.278

5

Parametric Least Squares Adjustment

Model Formulation

CHAPTER MENU

Understanding Least Squares Estimation and Geomatics Data Analysis, First Edition.
John Olusegun Ogundare.
© 2019 John Wiley & Sons, Inc. Published 2019 by John Wiley & Sons, Inc.
Companion website: www.wiley.com/go/ogundare/Understanding-lse-and-gda

---

OBJECTIVES

After studying this chapter, you should be able to:

1) Formulate parametric model equations relating typical geomatics observables.
2) Derive basic parametric least squares adjustment models, including linearized models, variation functions, normal equations, and solution equations.
3) Explain parametric least squares adjustment steps.
4) Apply the laws of variance–covariance propagation to determine the stochastic models of adjusted quantities, such as adjusted parameters, adjusted observations, and observation residuals.
5) Formulate weight constraint parametric least squares adjustment models, including the solution equations and the associated stochastic models.

5.1 Parametric Model Equation Formulation

Parametric model is a model relating some adjusted observations $(\hat{\ell})$ with some unknown (adjusted) parameters $(\hat{x})$. This can be symbolically represented as $\hat{\ell} = f(\hat{x})$. In this type of model, each of the measurements will produce one equation involving some or all of the specified parameters with no other observations involved in the equation. For example, 10 measurements will produce 10 independent equations with observation (the subject) appearing only once in each equation. The parametric equations are formulated so that when approximate values of the unknown parameters are substituted into the equations, the derived or calculated observations are obtained; the differences between the original observations and these derived observations are the misclosures. The *redundancy of parametric model equations* or the *number of degrees of freedom* of an adjustment is determined as the number of model equations formed minus the number of unknown parameters involved in the equations. Since the number of parametric model equations is always the same as the number of observations, the number of degrees of freedom can also be determined as the number of observations minus the number of unknown parameters. The typical quantities measured (the observables) in surveying are slope (or horizontal) distances, horizontal directions (or angles), vertical (or zenith) angles, azimuths (or bearings), elevation differences, coordinate differences (in Global Navigation Satellite Systems [GNSS] surveys), etc. The unknown parameters are typically coordinates of network points, which are linked together with the observations in form of parametric models. Take note of the following in formulating parametric model equations:

1) The important elements to be manipulated in the equations are the parameters (x) and observations (ℓ). Identify what constitutes the parameters (x) and what constitutes the observations (ℓ) from the question. Use suitable

Figure 5.1 Measurement of distance, azimuth, and total station direction observables.

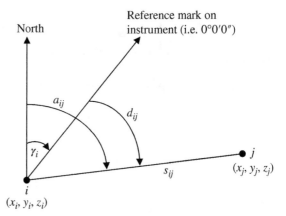

symbols to represent each of the observations and the parameters, and formulate the equations using the symbols.

2) Arrange those elements representing the parameters (x) and the observations (ℓ) in the equations ($\hat{\ell} = f(\hat{x})$) such that each observation is made a subject of an equation involving only parameters and possible constant values. Each of the already identified parameters and observations should be represented, for now, in symbols while deferring the use of their numerical values till when the equations are being solved.

The following shows how the observables and the parameters (i.e. coordinates) are linked together mathematically. The formulated equations for distance, azimuth, and total station (horizontal) direction observables are given with regard to Figure 5.1.

5.1.1 Distance Observable

For a measured EDM slope distance (s_{ij}) (from station i to station j in Figure 5.1) after meteorological and instrumental corrections have been applied, the distance observation (parametric) equation can be given as

$$s_{ij} = \sqrt{\left(x_j - x_i\right)^2 + \left(y_j - y_i\right)^2 + \left(z_j - z_i\right)^2} \tag{5.1}$$

where ($x_i,\ y_i,\ z_i$) and ($x_j,\ y_j,\ z_j$) are the three-dimensional coordinates of the instrument station i and target station j, respectively. For two-dimensional coordinate system, the z components will be discarded.

5.1.2 Azimuth and Horizontal (Total Station) Direction Observables

For the measured raw azimuth (by star shot or by the use of gyro equipment on the line i to j in Figure 5.1) reduced to the map projection coordinate system (α_{ij}), the parametric equation can be given as

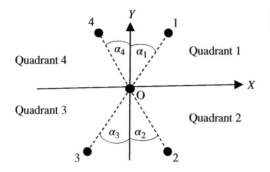

Figure 5.2 Quadrant analysis for bearing calculation.

$$\alpha_{ij} = \text{atan}\left[\frac{(x_j - x_i)}{(y_j - y_i)}\right] + [\text{Quadrant analysis}] \tag{5.2}$$

where *atan()* is used to represent arc tangent or tangent inverse trigonometric function. The quadrant analysis is illustrated using Figure 5.2. In the figure, the positive grid easting is X, and the positive grid northing is Y with four possible quadrants (quadrant 1, quadrant 2, quadrant 3, and quadrant 4). Assume the measured bearings O–1, O–2, O–3 and O–4 are located in the four different quadrants as shown in Figure 5.2; the parametric equations of the measured bearings can be formulated as follows.

In the following formulated equations, atan () represents arctangent of or Tangent inverse of; this symbol is mostly used in this book. Using atan () will produce the angles α_1, α_2, α_3, α_4 shown in Figure 5.2 depending on the change in easting ($dx = x_i - x_O$) from point O and the change in northing ($dy = y_i - y_O$) from point O with i representing the station sighted to (such as $i = 1, 2, 3, 4$). These angles can be determined using atan () as follows:

1) If dx is positive and dy is positive, locating line O–1 in quadrant 1 (giving positive angle α_1: bearing O–1 (β_{O-1}) is the same as α_1):

$$\alpha_1 = \text{atan}\left(\frac{dx}{dy}\right) \tag{5.3}$$

$$\beta_{O-1} = \alpha_1 \tag{5.4}$$

2) If dx is positive and dy is negative, locating line O–2 in quadrant 2 (giving negative angle α_2: bearing O–2 (β_{O-2}) is the same as $\alpha_2 + 180$):

$$\alpha_2 = \text{atan}\left(\frac{dx}{dy}\right) \tag{5.5}$$

$$\beta_{O-2} = \alpha_2 + 180 \tag{5.6}$$

3) If dx is negative and dy is negative, locating line O–3 in quadrant 3 (giving positive angle α_3: bearing O–3 (β_{O-3}) is the same as $\alpha_3 + 180$):

$$\alpha_3 = \text{atan}\left(\frac{dx}{dy}\right) \tag{5.7}$$

$$\beta_{O-3} = \alpha_3 + 180 \tag{5.8}$$

4) If dx is negative and dy is positive, locating line O–4 in quadrant 4 (giving negative angle α_4: bearing O–4 (β_{O-4}) is the same as $\alpha_4 + 360$):

$$\alpha_4 = \text{atan}\left(\frac{dx}{dy}\right) \tag{5.9}$$

$$\beta_{O-4} = \alpha_4 + 180 \tag{5.10}$$

For the measured raw horizontal (total station) direction observations (d_{ij}) reduced to the map projection coordinate system and measured relative to the "zero" of the horizontal circle, the parametric equation can be given as

$$d_{ij} = \text{atan}\left(\frac{(x_j - x_i)}{(y_j - y_i)}\right) - \gamma_i + [\text{Quadrant analysis}] \tag{5.11}$$

where γ_i is the orientation unknown (parameter) at the station i. The orientation parameter term in Equation (5.11) needs a little more emphasis so that azimuth (bearing) and direction measurements are not misunderstood as meaning the same thing. Total station (or theodolite) instrument directions are just arbitrary direction readings given by the instrument with respect to the zero reading of the instrument; they are not the same as azimuths (or bearings), which are given with respect to the direction of the north (the direction of the north being their reference zero point). The difference between the zero point reading of the total station and the zero point for azimuth (or bearing) measurement is the *unknown orientation parameter*. This unknown orientation parameter for the total station direction reading on a line is the amount of angle by which the total station direction measurement on that line must be adjusted in order to make that total station direction measurement identical to the azimuth or bearing of that line. This value is not determined directly in a setup, but it is considered as an unknown orientation parameter (γ), which makes it a parameter to be determined in an adjustment. Since no one is usually interested in knowing the numerical value of an orientation parameter, the orientation parameter is otherwise referred to as *nuisance parameter*. It is required in the model equations in order to obtain correct equations, but the knowledge of its value is not required. Note that there will always be one unknown orientation parameter per instrument setup for all directions having a common total station zero point reading; if the total station zero point orientation in space changes (as a result of releveling and recentering of the instrument over another point or over the same point), another unknown orientation parameter will be associated with the new setup. For every total station setup involving direction measurement, there will be one unknown orientation parameter for every setup station. For example, if the total station directions are made at four setup stations (or the zero of the instrument is reset four times at a station), there will be four unknown orientation parameters for the four sets of directions made at the four stations (or on the same station). Note that orientation parameters are involved when angles between two lines are measured, as discussed in Section 5.1.3.

The main purpose of relating total station direction to azimuth (or bearing) is because there is no direct equation for total station direction measurement; this, therefore, requires that the corresponding azimuth (or bearing) equation of the line be derived indirectly with the orientation parameter introduced to reduce the azimuth to the total station direction measurement. For example, *the total station direction of a line is equal to the azimuth of that line minus the unknown orientation parameter for that setup*, as given in Equation (5.11). Note that in choosing the unknown orientation parameter, one does not need to first visualize how the value will relate to the azimuth; all that will be required is to choose arbitrary symbol for the parameter. Moreover, consistency in applying the orientation parameter in the model equations is important. The concept of unknown orientation parameter is illustrated in Examples 5.5 and 5.6.

5.1.3 Horizontal Angle Observable

The measured horizontal angle θ_{ijk} can be given as the difference between two direction measurements (d_{ik} and d_{ij}) as shown in Figure 5.3.

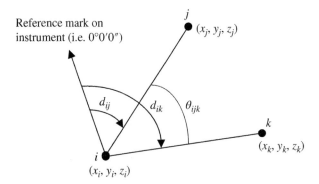

Figure 5.3 Measurement of a horizontal angle.

For the measured raw angle (θ_{ijk}) reduced to the map projection coordinate system, the parametric equation can be given as

$$\theta_{ijk} = \operatorname{atan}\left[\frac{(x_k - x_i)}{(y_k - y_i)}\right] - \operatorname{atan}\left[\frac{(x_j - x_i)}{(y_j - y_i)}\right] \tag{5.12}$$

where point i is the instrument station and points j and k are the target points. It can be seen from Equation (5.12) that the orientation parameter (γ) is not involved in angle measurements.

5.1.4 Zenith Angle Observable

For the measured raw zenith angle (Z_{ij}) reduced to the local plane for line i to j in Figure 5.3, the parametric equation (in three-dimensional case) can be given as

$$Z_{ij} = \operatorname{atan}\left[\frac{\sqrt{(x_j - x_i)^2 + (y_j - y_i)^2}}{(z_j - z_i)}\right] \tag{5.13}$$

5.1.5 Coordinate Difference Observable

The GNSS coordinate differences (in three-dimensional geocentric coordinate system) for a baseline i to j in Figure 5.3 produce the following parametric equations:

$$\begin{aligned}
\Delta x_{ij} &= x_j - x_i \\
\Delta y_{ij} &= y_j - y_i \\
\Delta z_{ij} &= z_j - z_i
\end{aligned} \tag{5.14}$$

where Δx_{ij}, Δy_{ij}, and Δz_{ij} are the measured baseline vectors between points i and j.

5.1.6 Elevation Difference Observable

The elevation difference measurement (Δh_{ij}) between points i and j in Figure 5.3 (based on differential leveling procedure) produces the following parametric equation:

$$\Delta h_{ij} = h_j - h_i \tag{5.15}$$

where h_i and h_j are the heights of points i and j above a given datum, respectively. The elevation difference (Δh_{ij}) between points i and j in Figure 5.3 (based on trigonometric leveling procedure with total station equipment, assuming the distance between points i and j is short and ignoring the earth curvature and atmospheric refraction corrections) is not measured directly but calculated (as unknown parameter, x) as follows:

$$\Delta h_{ij} = \text{HI} - \text{HT} + s_{ij}\cos z_{ij} \qquad (5.16)$$

where HI is the measured height of total station instrument at point i, HT is the measured height of target at point j, s_{ij} is the measured slope distance, and z_{ij} is the measured zenith angle. Since the quantities measured in Equation (5.16) are $\ell^T = [\text{HI} \ \text{HT} \ s_{ij} \ z_{ij}]$ and the unknown parameter, $x = [\Delta h_{ij}]$, then Equation (5.16) is a special form of general model equation $\hat{x} = f(\hat{\ell})$. If the equation is rearranged, it forms the general model equation, $f(\hat{x}, \hat{\ell}) = 0$, as follows:

$$\Delta \hat{h}_{ij} - \text{H}\hat{\text{I}} + \text{H}\hat{\text{T}} - \hat{s}_{ij}\cos\hat{z}_{ij} = 0 \qquad (5.17)$$

5.2 Typical Parametric Model Equations

Typical parametric model equations are provided in the examples given in this section.

Example 5.1 Given the following two-parameter transformation problem relating local photo coordinates (e, n) *system* with global photogrammetric model coordinates (E, N) *system* and the corresponding coordinates of points 1 and 2 (referring to Table 5.1 and Figure 5.4):

Table 5.1 Coordinates of points 1 and 2.

Point	Local system (photo)		Global system (model)	
i	e_i (cm)	n_i (cm)	E_i (cm)	N_i (cm)
1	0.0	1.0	−2.1	1.1
2	1.0	0.0	1.0	2.0

$$E_i = ae_i - bn_i \qquad (5.18)$$

$$N_i = an_i + be_i \qquad (5.19)$$

If the (E, N) coordinates of points 1 and 2 are considered as measurements, (e, n) coordinates of points 1 and 2 are precisely known values, and the quantities a and b are the unknown to be determined, write down the appropriate mathematical model ($\hat{\ell} = f(\hat{x})$) for the problem.

Figure 5.4 Relationship between the two coordinate systems.

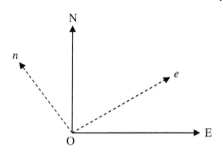

Solution:

Parametric model should look like $\hat{\ell} = f(\hat{x})$. Take note that in this problem, (e, n) coordinates of points 1 and 2 are precisely known values (considered as constants). The steps given in Section 5.1 are followed in formulating the parametric equations as follows:

1) Since the (E, N) coordinates of points 1 and 2 under the global system are considered as the only measurements with the other given coordinates as constants, there will only be four observations. The unknown parameters (x) stated in the question are a and b. At this stage, the observations are represented by the symbols $\ell = [E_1 \ \ N_1 \ \ E_2 \ \ N_2]^T$, where E_1 and N_1 are the easting and northing (in the global system) for point 1.
2) Looking at the representative Equations (5.18) and (5.19), it can be seen that each element of the observation vector (ℓ) is a stand-alone subject equated to an expression involving the elements (a, b) of the parameters (x) with some constant quantities e_i and n_i. The equations are already in the required arrangement $\hat{\ell} = f(\hat{x})$ with each point associated with two equations (Equations (5.18) and (5.19)). The final parametric model equations may be written for all the points as

$$\hat{E}_1 = \hat{a}e_1 - \hat{b}n_1 \tag{5.20}$$

$$\hat{N}_1 = \hat{a}n_1 + \hat{b}e_1 \tag{5.21}$$

$$\hat{E}_2 = \hat{a}e_2 - \hat{b}n_2 \tag{5.22}$$

$$\hat{N}_2 = \hat{a}n_2 + \hat{b}e_2 \tag{5.23}$$

where (e_1, n_1) and (e_2, n_2) are, respectively, the precisely known coordinates of points 1 and 2 that will remain unadjusted (considered as constant quantities); $(\hat{E}_1, \hat{N}_1)$ and $(\hat{E}_2, \hat{N}_2)$ represent the adjusted model coordinates of points 1 and 2, respectively; and $(\hat{a}, \hat{b})$ represent the adjusted parameters (unknowns). Note that (e_1, n_1) and (e_2, n_2) do not have any hats added to them since they are constant quantities.

Example 5.2 The horizontal coordinates (x, y) of point PT2 are to be determined in a horizontal traverse survey in Figure 5.5. The measured quantities are the azimuth Az, distances d_1 and d_2, and the angle (α) at point PT2.

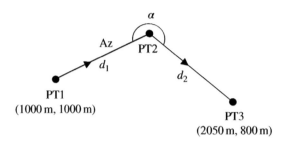

Figure 5.5 Horizontal traverse survey (not to scale).

If the coordinates of points PT1 and PT3 are fixed, formulate the relationships $(\hat{\ell} = f(\hat{x}))$ between the observations and the unknown parameters.

Solution:

The parametric model to be formulated should look like $\hat{\ell} = f(\hat{x})$.

The unknown parameters in this problem are x and y of point PT2. Following the steps given in Example 5.1 and the quadrant analysis procedure in Section 5.1.2, the required parametric model can be formulated as follows:

1) The measured quantities are the azimuth Az, distances d_1 and d_2, and angle α; they constitute the observations (ℓ), which are four in number. At this stage, the observations are represented by the symbols $\ell = [\text{Az} \ d_1 \ d_2 \ \alpha]^T$. The parameters are the horizontal coordinates (x, y) of point PT2 that are to be determined. The coordinates of points PT1 and PT3 are given, but they are not to be adjusted; so, they are considered as constant quantities.

2) From the understanding of the type of model $(\hat{\ell} = f(\hat{x}))$ to be formulated, each element of the observation vector (ℓ) must stand alone as a subject equated to an expression involving the parameters (x, y) and some constant quantities (coordinates of points PT1 and PT3). The formulated are Equations (5.24)–(5.27).

Since there are four observations, four corresponding equations are possible as follows. For the bearing Az of line PT1–PT2, it can be seen from Figure 5.5 that the line is in the first quadrant (referring to Figure 5.2). The corresponding equation for Az can be given as

$$A\hat{z} = \text{atan} \left[\frac{(\hat{x} - 1000)}{(\hat{y} - 1000)} \right] \tag{5.24}$$

For each distance observation, the inverse of the two corresponding coordinates using the usual Pythagoras theorem gives

$$\hat{d}_1 = \sqrt{(\hat{x} - 1000)^2 + (\hat{y} - 1000)^2} \tag{5.25}$$

$$\hat{d}_2 = \sqrt{(2050 - \hat{x})^2 + (800 - \hat{y})^2} \tag{5.26}$$

Using the concept that an angle is the difference between two bearings, the measured angle α will be equal to the bearing of line PT2–PT3 minus the bearing of line PT2–PT1 plus 360° to make the bearing a positive value:

$$\text{Bearing of line PT2} - \text{PT1 (in quadrant 3)} = \text{atan}\left[\frac{(1000 - \hat{x})}{(1000 - \hat{y})}\right] + 180$$

$$\text{Bearing of line PT2} - \text{PT3 (in quadrant 2)} = \text{atan}\left[\frac{(2050 - \hat{x})}{(800 - \hat{y})}\right] + 180$$

$$\hat{\alpha} = \text{atan}\left[\frac{(2050 - \hat{x})}{(800 - \hat{y})}\right] - \text{atan}\left[\frac{(1000 - \hat{x})}{(1000 - \hat{y})}\right] + 360° \tag{5.27}$$

where $(\hat{x}, \hat{y})$ are the adjusted coordinates (parameters) of point PT2, $(\hat{d}_1, \hat{d}_2)$ are the adjusted distances, A$\hat{z}$ is the adjusted bearing, and $\hat{\alpha}$ is the adjusted angle.

Example 5.3 Consider a leveling network shown in Figure 5.6 with the directions of the arrows indicating the directions of leveling runs. Points

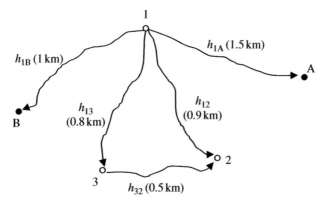

Figure 5.6 Leveling network.

A and B are the given benchmarks whose elevations are well known as H_A and H_B, respectively; the elevations H_1, H_2, and H_3 of points 1, 2, and 3, respectively, are to be determined. The leveled height differences and the approximate distances between the corresponding points are measured and given as shown in the figure. Formulate the parametric model equations ($\hat{\ell} = f(\hat{x})$) for this problem.

Solution:

The parametric model equations should look like $\hat{\ell} = f(\hat{x})$. Using the steps given in Example 5.1, the required parametric model can be formulated as follows:

1) The leveled height differences and the approximate distances between the corresponding points are measured. Since approximate distances are not used directly in deriving elevations of points in differential leveling surveys, the only observations (ℓ) to be considered are the measured elevation differences (h_{1B}, h_{13}, h_{12}, h_{1A}, h_{32}). As usual, the observations are represented by symbols $\ell = [h_{1B} \ h_{13} \ h_{12} \ h_{1A} \ h_{32}]^T$. The elevations of given benchmarks (HA and HB) are considered as constants since they are not to change in values in an adjustment and they are not parameters and not observations. The elevations H_1, H_2, H_3 are to be determined, so they are the parameters (x).

2) In formulating the model ($\hat{\ell} = f(\hat{x})$), each element of the observation vector (ℓ) must stand alone as a subject equated to an expression involving the elements (H_1, H_2, H_3) of the parameters (x) and the elevations of points A and B as constant quantities. For five observations, the following five parametric equations are possible:

$$\hat{h}_{1B} = H_B - \hat{H}_1 \tag{5.28}$$

$$\hat{h}_{13} = \hat{H}_3 - \hat{H}_1 \tag{5.29}$$

$$\hat{h}_{12} = \hat{H}_2 - \hat{H}_1 \tag{5.30}$$

$$\hat{h}_{1A} = H_A - \hat{H}_1 \tag{5.31}$$

$$\hat{h}_{32} = \hat{H}_2 - \hat{H}_3 \tag{5.32}$$

It can be seen in Equations (5.28) and (5.31) that H_A and H_B have no hats on them since they are not parameters and not observations.

Example 5.4 The angle θ and distance S_{AB} in the traverse network shown in Figure 5.7 were measured. In the traverse, two points A and C are fixed. Formulate the parametric model equations for determining the coordinates (x, y) of point B.

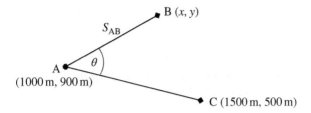

Figure 5.7 Traverse survey with points A and C fixed.

Solution:

The parametric model equations should look like $\hat{\ell} = f(\hat{x})$. Using the steps given in Example 5.2 and the relation that angle θ is equal to the bearing of line AC minus the bearing of line AB, the Equation (5.33) can be formulated; the Pythagoras theorem is applied in order to obtain Equation (5.34):

$$\hat{\theta} = \operatorname{atan}\left(\frac{1500 - 1000}{500 - 900}\right) - \operatorname{atan}\left(\frac{\hat{x} - 1000}{\hat{y} - 900}\right) + 180° \tag{5.33}$$

$$\hat{S}_{AB} = \sqrt{(\hat{x} - 1000)^2 + (\hat{y} - 900)^2} \tag{5.34}$$

where $\hat{S}_{AB}$ is the adjusted distance and $(\hat{x}, \hat{y})$ are the adjusted coordinates (parameters) of point B.

Example 5.5 Point O (x, y) in Figure 5.8 is to be resected from two given control points A (x_A, y_A) and B (x_B, y_B). The azimuth O–A was measured as $\ell_1 = 40°$; and the total station directions O–A and O–B were measured as $\ell_2 = 10°$ and

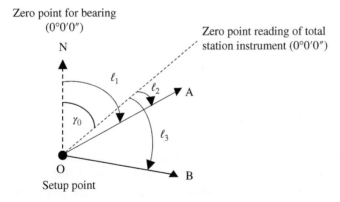

Figure 5.8 Resection problem to illustrate the difference between azimuth and direction measurements.

$\ell_3 = 65°$, respectively. Formulate the parametric model equations $(\hat{\ell} = f(\hat{x}))$ for this problem.

Solution:

This example illustrates the *difference between azimuth and total station direction (the orientation parameter)*. Based on the concept of orientation parameters discussed in Section 5.1.2 and the steps given in Example 5.1, the parametric model equations $(\hat{\ell} = f(\hat{x}))$ can be formulated as follows:

1) From the question, the azimuth O–A (or ℓ_1) and the total station directions O–A (ℓ_2) and O–B (ℓ_3) were measured, making them the observations (ℓ) for this problem. The observations are to be represented by the symbols $\ell = [\ell_1 \ \ell_2 \ \ell_3]^T$. The coordinates (x, y) of point O are the main parameters to be determined. Since the total station directions are also measured from one setup, there is one additional unknown orientation parameter (γ_0) required for the model formulation to be complete. The total parameters in this problem are $x = [x \ y \ \gamma_0]^T$.

2) In formulating the parametric model equations $(\hat{\ell} = f(\hat{x}))$, each element of the observation vector (ℓ) must stand alone as a subject equated to an expression involving the elements of the parameters $x = [x \ y \ \gamma_0]^T$ and the constant quantities (that may be required in quadrant analysis). The parametric equations are formulated as follows:

   • For azimuth measurement of line O–A (in first quadrant):

$$\hat{\ell}_1 = \text{atan}\left[\frac{(x_A - \hat{x})}{(y_A - \hat{y})}\right] \tag{5.35}$$

   • For total station direction measurement of line O–A, which is equal to the azimuth of line O–A (which is in first quadrant) minus the unknown orientation parameter (γ_0), can be given as

$$\hat{\ell}_2 = \text{atan}\left[\frac{(x_A - \hat{x})}{(y_A - \hat{y})}\right] - \hat{\gamma}_0 \tag{5.36}$$

   • Similarly, for the total station direction measurement of line O–B:

$$\hat{\ell}_3 = \text{atan}\left[\frac{(x_B - \hat{x})}{(y_B - \hat{y})}\right] - \hat{\gamma}_0 + 180° \tag{5.37}$$

As can be seen in Equation (5.36), if the unknown orientation parameter ($\hat{\gamma}_0$) is not applied, the equation will exactly be the same as that of the adjusted bearing ($\hat{\ell}_1$); this should not be so since the bearing of the given line is approximately $\ell_1 = 40°$ and the total station direction measurement of the same line is $\ell_2 = 10°$. It can be seen that the unknown orientation parameter

$(\hat{\gamma}_0)$ in this case will be close to $30°$. It is advisable to always keep the orientation parameter as unknown in the least squares adjustment so that its best estimate can be determined as part of the unknown parameters. In the above problem, the number of observations is 3 (i.e. ℓ_1, ℓ_2, ℓ_3), and the number of unknown parameters is 3 (i.e. x, y, $\hat{\gamma}_0$). The steps discussed in this example must be followed in formulating equations (or models) involving total station direction measurements.

Example 5.6 Points 1 and 2 in Figure 5.9 are to be coordinated by intersection from two control points A (x_A, y_A) and B (x_B, y_B). The total station direction measurements from point A are ℓ_1, ℓ_2, and ℓ_3; the total station direction measurements at point B are ℓ_4, ℓ_5, and ℓ_6; the measured distances from point A are ℓ_7 and ℓ_8; and the measured distances from point B are ℓ_9 and ℓ_{10}, as shown in the figure. Formulate the parametric model equations $(\hat{\ell} = f(\hat{x}))$ for this problem.

Figure 5.9 Intersection problem.

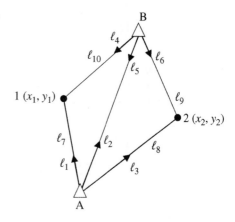

Solution:

This example is to further explain how to introduce unknown orientation parameters into parametric model equations. Using the steps given in Example 5.5, the following parametric model equations can be formulated:

1) From the question, the observations (ℓ) are the total station directions at A (ℓ_1, ℓ_2, ℓ_3), the total station directions at B (ℓ_4, ℓ_5, ℓ_6), and the distances $(\ell_7, \ell_8, \ell_9, \ell_{10})$, which can be given as $\ell = [\ell_1 \; \ell_2 \; \ell_3 \; \ell_4 \; \ell_5 \; \ell_6 \; \ell_7 \; \ell_8 \; \ell_9 \; \ell_{10}]^T$. The main parameters are the coordinates (x_A, y_A) and (x_B, y_B) to be determined directly. Since there are two total station setup points (points A and B) where direction measurements are made, there will be two unknown

orientation parameters (γ_A and γ_B) corresponding to those setup points. The total parameters in this problem can be given as $x^T = [x_A \ y_A \ x_B \ y_B \ \gamma_A \ \gamma_B]$.

2) In formulating the parametric model equations ($\hat{\ell} = f(\hat{x})$), each element of the observation vector (ℓ) must stand alone as a subject equated to an expression involving the elements of the parameters (x) and some constant quantities (that may be required in quadrant analysis). Given 10 observations, the number of parametric model equations possible is 10. They are formulated as follows:

- At point A, assume the instrument zero is on the left-hand side of line A-1. Use the concept that each observation of a line is equal to the azimuth of that line minus orientation parameter at that point:

$$\hat{\ell}_1 = \text{atan}\left[\frac{(\hat{x}_1 - x_A)}{(\hat{y}_1 - y_A)}\right] + 360 - \hat{\gamma}_A \tag{5.38}$$

$$\hat{\ell}_2 = \text{atan}\left[\frac{(\hat{x}_B - x_A)}{(\hat{y}_B - y_A)}\right] + 360 - \hat{\gamma}_A \tag{5.39}$$

$$\hat{\ell}_3 = \text{atan}\left[\frac{(\hat{x}_2 - x_A)}{(\hat{y}_2 - y_A)}\right] + 360 - \hat{\gamma}_A \tag{5.40}$$

- At point B, assume the instrument zero is on the left-hand side of line B-2. Use the concept that each observation of a line is equal to the azimuth of that line minus orientation parameter at that point:

$$\hat{\ell}_4 = \text{atan}\left[\frac{(\hat{x}_1 - x_B)}{(\hat{y}_1 - y_B)}\right] + 180 - \hat{\gamma}_B \tag{5.41}$$

$$\hat{\ell}_5 = \text{atan}\left[\frac{(x_A - x_B)}{(y_A - y_B)}\right] + 180 - \hat{\gamma}_B \tag{5.42}$$

$$\hat{\ell}_6 = \text{atan}\left[\frac{(\hat{x}_2 - x_B)}{(\hat{y}_2 - y_B)}\right] + 180 - \hat{\gamma}_B \tag{5.43}$$

- Use the Pythagoras theorem to determine distances from the corresponding coordinates:

$$\hat{\ell}_7 = \sqrt{(\hat{x}_1 - x_A)^2 + (\hat{y}_1 - y_A)^2} \tag{5.44}$$

$$\hat{\ell}_8 = \sqrt{(\hat{x}_2 - x_A)^2 + (\hat{y}_2 - y_A)^2} \tag{5.45}$$

$$\hat{\ell}_9 = \sqrt{(\hat{x}_2 - x_B)^2 + (\hat{y}_2 - y_B)^2} \tag{5.46}$$

$$\hat{\ell}_{10} = \sqrt{(\hat{x}_1 - x_B)^2 + (\hat{y}_1 - y_B)^2} \tag{5.47}$$

5.3 Basic Adjustment Model Formulation

The parametric model equations are expressed symbolically as $\hat{\ell} = f(\hat{x})$, meaning that one adjusted observation in a vector $(\hat{\ell})$ of adjusted observations is functionally equal to a combination of adjusted parameters $(\hat{x})$. The parametric least squares model is also known as the *Gauss–Markov model* or the *observation equation model*. In deriving the parametric model adjustment steps, the following *guidelines* may be useful:

1) All symbols or variables used in formulating a particular parametric model equation must be unique; no two variables must represent the same quantity in the same model. In another words, two different variables in a model must not have the same meaning. This is necessary in order to avoid confusing the meanings of symbols or variables used in the same model.
2) Every symbol or variable used in a model must be clearly defined (usually after they have just been used) in order to increase the clarity of the steps being taken.
3) Adjustment model derivation must flow logically from one step to another with a brief and concise explanation of every step, including an explanation of the purpose of the step and why the step is necessary.
4) Every derived equation must be properly numbered so that it can be cited later using that number. The numbering convention must be consistent with the convention used in this book.

The formulation of the least squares parametric adjustment models consists of the following general steps, which are developed further in the following sections:

i) *Formulating the parametric model equations* $[\hat{\ell} = f(\hat{x})]$ as discussed in Sections 5.1 and 5.2. Note that the covariance matrices (or weight matrices) declared with the formulated model are relevant in formulating the variation functions later.
ii) *Linearizing the model equations.* Since model equations may be nonlinear, they should be linearized by Taylor series expansion (refer to Section 5.4).
iii) *Deriving a new set of equations known as variation function.* This is done by incorporating the linearized parametric model equations into the least squares criterion (refer to Section 5.5).
iv) *Deriving the normal equation system.* This is done by determining the partial derivatives of the variation function with respect to the unknowns in the function and setting the partial derivatives to zero to produce what is commonly referred to as a system of normal equations (refer to Section 5.6).
v) *Solving the normal equation system.* The solution of the normal equations provides the adjusted values for the unknown quantities (refer to Section 5.7)

5.4   Linearization of Parametric Model Equations

The linearization of parametric model equations are based on Taylor series expansion discussed in Chapter 2. Refer to Equations (E.22)–(E.29) for review of partial derivatives of matrix-related and vector-related functions. In this section, the parametric model with and without nuisance parameters will be linearized. Remember also that least squares estimation is iterative due to the practical need to linearize the functional model by the Taylor series expansion. This iterative solution is necessary since the initial (approximate) values of the unknown parameters are hardly obtained to sufficient level of accuracy for the series expansion to be valid. After the initial parameter values are input into the first adjustment, the subsequent iterative solutions are used as new initial values, and the adjustment procedure is repeated until a particular solution is insignificantly different from the previous one; at this point, the adjustment is said to have converged to a final solution. This final solution is then considered as containing the final values of the unknown parameters.

5.4.1   Linearization of Parametric Model Without Nuisance Parameter

Given the following parametric model, which contains no nuisance parameters (Equation (5.48)),

$$\hat{\ell} = f(\hat{x}) \quad P \tag{5.48}$$

where f is a vector of mathematical models, $\hat{x}$ is a vector of unknown parameters, $\hat{\ell}$ is a vector of observations, and P is the weight matrix of observations. A particular attention must be given to the weight matrix P, which must be explicitly used in any derivation involving the given function; this is the main significance of declaring it with the function. Let Equation (5.48) be rewritten as

$$\ell + v = f\left(x^0 + \delta\right) \tag{5.49}$$

where v is the observation residual vector, ℓ is a vector of original observations, x^0 is a vector of approximate values of the parameters, and δ is a vector of corrections to the approximate parameters. By performing Taylor series expansion on Equation (5.49), the following linearized form of the parametric model is obtained:

$$v = f\left(x^0\right) + \frac{\partial f}{\partial x}\delta - \ell \tag{5.50}$$

The linearized Equation (5.50) can be given in simpler terms as follows:

$$v = w + A\delta \tag{5.51}$$

where $w = f(x^0) - \ell$ is the misclosure vector and $A = \partial f / \partial x$ is the Jacobian matrix, also known as the first design matrix. It must be emphasized that Equation (5.51) is produced just for the purpose of facilitating possible further operations on the linearized Equation (5.50). The partial derivatives needed for the linearization of typical observables are given in Appendix D as follows: azimuth observable (Section D.1), total station observable (Section D.2), horizontal angle observable (Section D.3), distance observable (Section D.4), and zenith angle observable (Section D.5). The examples of linearized models are given as follows.

Example 5.7   Linearize Equations (5.20)–(5.23) in Example 5.1 and present it in form of Equation (5.51).

Solution:

For the Equations (5.20)–(5.23), the linearized model can be given as Equation (5.51)where parameter vector is $x = \begin{bmatrix} a \\ b \end{bmatrix}$; given the vector of approximate parameters $x^0 = \begin{bmatrix} a^0 \\ b^0 \end{bmatrix}$, the first design matrix (evaluated at x^0) can be given as

$$A = \left.\frac{\partial f}{\partial x}\right|_{x^0} = \begin{bmatrix} \dfrac{\partial \hat{E}_1}{\partial \hat{a}} & \dfrac{\partial \hat{E}_1}{\partial \hat{b}} \\[2ex] \dfrac{\partial \hat{N}_1}{\partial \hat{a}} & \dfrac{\partial \hat{N}_1}{\partial \hat{b}} \\[2ex] \dfrac{\partial \hat{E}_2}{\partial \hat{a}} & \dfrac{\partial \hat{E}_2}{\partial \hat{b}} \\[2ex] \dfrac{\partial \hat{N}_2}{\partial \hat{a}} & \dfrac{\partial \hat{N}_2}{\partial \hat{b}} \end{bmatrix}_{x^0} \qquad A = \begin{bmatrix} e_1 & -n_1 \\ n_1 & e_1 \\ e_2 & -n_2 \\ n_2 & e_2 \end{bmatrix}$$

The misclosure vector (w) evaluated at x^0, the vector of residuals (v), and the vector of corrections to the approximate parameters (δ) can be given, respectively, as

$$w = \begin{bmatrix} a^0 e_1 - b^0 n_1 - E_1 \\ a^0 n_1 + b^0 e_1 - N_1 \\ a^0 e_2 - b^0 n_2 - E_2 \\ a^0 n_2 + b^0 e_2 - N_2 \end{bmatrix}, \quad v = \begin{bmatrix} v_1 \\ v_2 \\ v_3 \\ v_4 \end{bmatrix}, \quad \delta = \begin{bmatrix} \delta a \\ \delta b \end{bmatrix}$$

The linearized model of Equations (5.20)–(5.23) (evaluated at x^0) can be given as

$$
\begin{bmatrix} v_1 \\ v_2 \\ v_3 \\ v_4 \end{bmatrix} = \begin{bmatrix} a^0 e_1 - b^0 n_1 - E_1 \\ a^0 n_1 + b^0 e_1 - N_1 \\ a^0 e_2 - b^0 n_2 - E_2 \\ a^0 n_2 + b^0 e_2 - N_2 \end{bmatrix} + \begin{bmatrix} e_1 & -n_1 \\ n_1 & e_1 \\ e_2 & -n_2 \\ n_2 & e_2 \end{bmatrix} \begin{bmatrix} \delta a \\ \delta b \end{bmatrix}
$$

Example 5.8 Linearize Equations (5.24)–(5.27) in Example 5.2 and present it in the form of Equation (5.51), referring to the partial derivatives for azimuth observable in Section D.1, distance observable in Section D.4, and horizontal angle observable in Section D.3.

Solution:

For the Equations (5.24)–(5.27), the linearized model can be given as Equation (5.51) where parameter vector is $p = \begin{bmatrix} x \\ y \end{bmatrix}$; given the vector of approximate parameters $p^0 = \begin{bmatrix} x^0 \\ y^0 \end{bmatrix}$, the first design matrix (evaluated at p^0) can be given as

$$
A = \left. \frac{\partial f}{\partial \hat{x}} \right|_{p^0} = \begin{bmatrix} \dfrac{\partial A\hat{z}}{\partial \hat{x}} & \dfrac{\partial A\hat{z}}{\partial \hat{y}} \\[2mm] \dfrac{\partial \hat{d}_1}{\partial \hat{x}} & \dfrac{\partial \hat{d}_1}{\partial \hat{y}} \\[2mm] \dfrac{\partial \hat{d}_2}{\partial \hat{a}} & \dfrac{\partial \hat{d}_2}{\partial \hat{b}} \\[2mm] \dfrac{\partial \hat{\alpha}}{\partial \hat{x}} & \dfrac{\partial \hat{\alpha}}{\partial \hat{y}} \end{bmatrix}_{x^0}
$$

where

$$
\left. \frac{\partial A\hat{z}}{\partial \hat{x}} \right|_{p^0} = \frac{(y^0 - 1000)}{(y^0 - 1000)^2 + (x^0 - 1000)^2}
$$

$$
\left. \frac{\partial A\hat{z}}{\partial \hat{y}} \right|_{p^0} = \frac{-(x^0 - 1000)}{(y^0 - 1000)^2 + (x^0 - 1000)^2}
$$

$$\left.\frac{\partial \hat{d}_1}{\partial \hat{x}}\right|_{p^0} = \frac{(x^0 - 1000)}{\sqrt{(y^0 - 1000)^2 + (x^0 - 1000)^2}}$$

$$\left.\frac{\partial \hat{d}_1}{\partial \hat{y}}\right|_{p^0} = \frac{(y^0 - 1000)}{\sqrt{(y^0 - 1000)^2 + (x^0 - 1000)^2}}$$

$$\left.\frac{\partial \hat{d}_2}{\partial \hat{x}}\right|_{p^0} = \frac{-(2050 - x^0)}{\sqrt{(800 - y^0)^2 + (2050 - x^0)^2}}$$

$$\left.\frac{\partial \hat{d}_2}{\partial \hat{y}}\right|_{p^0} = \frac{-(800 - y^0)}{\sqrt{(800 - y^0)^2 + (2050 - x^0)^2}}$$

$$\left.\frac{\partial \hat{a}}{\partial \hat{x}}\right|_{p^0} = \frac{-(800 - y^0)}{(800 - y^0)^2 + (2050 - x^0)^2} - \frac{(1000 - y^0)}{(1000 - y^0)^2 + (1000 - x^0)^2}$$

$$\left.\frac{\partial \hat{a}}{\partial \hat{x}}\right|_{p^0} = \frac{(2050 - x^0)}{(800 - y^0)^2 + (2050 - x^0)^2} + \frac{(1000 - x^0)}{(1000 - y^0)^2 + (1000 - x^0)^2}$$

The misclosure vector (w) evaluated at x^0, the vector of residuals (v), and the vector of corrections to the approximate parameters (δ) can be given, respectively, as

$$w = \begin{bmatrix} Az^0 - Az \\ d_1^0 - d_1 \\ d_2^0 - d_2 \\ a^0 - a \end{bmatrix}, \quad v = \begin{bmatrix} v_1 \\ v_2 \\ v_3 \\ v_4 \end{bmatrix}, \quad \delta = \begin{bmatrix} \delta x \\ \delta y \end{bmatrix}$$

where $\begin{bmatrix} Az^0 & d_1^0 & d_2^0 & a^0 \end{bmatrix}^T$ is a vector of derived observations using the approximate values of the parameters into Equations (5.24)–(5.27), respectively.

Example 5.9 Linearize Equations (5.28)–(5.32) in leveling Example 5.3 and present it in the form of Equation (5.51).

Solution:

For the Equations (5.28)–(5.32), the linearized model can be given as Equation (5.51) where parameter vector is $x = \begin{bmatrix} H_1 & H_2 & H_3 \end{bmatrix}^T$; given the vector of approximate parameters $x^0 = \begin{bmatrix} H_1^0 & H_2^0 & H_3^0 \end{bmatrix}^T$, the first design matrix (evaluated at x^0) can be given as

$$A = \frac{\partial f}{\partial \hat{x}}\bigg|_{x^0} = \begin{bmatrix} \dfrac{\partial \hat{h}_{1B}}{\partial \hat{H}_1} & \dfrac{\partial \hat{h}_{1B}}{\partial \hat{H}_2} & \dfrac{\partial \hat{h}_{1B}}{\partial \hat{H}_3} \\[2ex] \dfrac{\partial \hat{h}_{13}}{\partial \hat{H}_1} & \dfrac{\partial \hat{h}_{13}}{\partial \hat{H}_2} & \dfrac{\partial \hat{h}_{13}}{\partial \hat{H}_3} \\[2ex] \dfrac{\partial \hat{h}_{12}}{\partial \hat{H}_1} & \dfrac{\partial \hat{h}_{12}}{\partial \hat{H}_2} & \dfrac{\partial \hat{h}_{12}}{\partial \hat{H}_3} \\[2ex] \dfrac{\partial \hat{h}_{1A}}{\partial \hat{H}_1} & \dfrac{\partial \hat{h}_{1A}}{\partial \hat{H}_2} & \dfrac{\partial \hat{h}_{1A}}{\partial \hat{H}_3} \\[2ex] \dfrac{\partial \hat{h}_{32}}{\partial \hat{H}_1} & \dfrac{\partial \hat{h}_{32}}{\partial \hat{H}_2} & \dfrac{\partial \hat{h}_{32}}{\partial \hat{H}_3} \end{bmatrix}_{x^0} \qquad A = \begin{bmatrix} -1 & 0 & 0 \\ -1 & 0 & 1 \\ -1 & 1 & 0 \\ -1 & 0 & 0 \\ 0 & 1 & -1 \end{bmatrix}$$

The misclosure vector (w) evaluated at x^0, the vector of residuals (v), and the vector of corrections to the approximate parameters (δ) can be given, respectively, as

$$w = \begin{bmatrix} H_B - H_1^0 - h_{1B} \\ H_3^0 - H_1^0 - h_{13} \\ H_2^0 - H_1^0 - h_{12} \\ H_A - H_1^0 - h_{1A} \\ H_2^0 - H_3^0 - h_{32} \end{bmatrix}, \quad v = \begin{bmatrix} v_1 \\ v_2 \\ v_3 \\ v_4 \\ v_5 \end{bmatrix}, \quad \delta = \begin{bmatrix} \delta H_1 \\ \delta H_2 \\ \delta H_3 \end{bmatrix}$$

The linearized model of Equations ((5.28)–(5.32)) (evaluated at x^0) can be given as

$$\begin{bmatrix} v_1 \\ v_2 \\ v_3 \\ v_4 \\ v_5 \end{bmatrix} = \begin{bmatrix} H_B - H_1^0 - h_{1B} \\ H_3^0 - H_1^0 - h_{13} \\ H_2^0 - H_1^0 - h_{12} \\ H_A - H_1^0 - h_{1A} \\ H_2^0 - H_3^0 - h_{32} \end{bmatrix} + \begin{bmatrix} -1 & 0 & 0 \\ -1 & 0 & 1 \\ -1 & 1 & 0 \\ -1 & 0 & 0 \\ 0 & 1 & -1 \end{bmatrix} \begin{bmatrix} \delta H_1 \\ \delta H_2 \\ \delta H_3 \end{bmatrix}$$

5.4.2 Linearization of Parametric Model with Nuisance Parameter

For an illustration of the concepts of linearization by Taylor series expansion, let a parametric model be given by

$$\hat{\ell} = f(\hat{x}_1, \hat{x}_2) \tag{5.52}$$

where $\hat{\ell}$ is a vector of adjusted observations, $\hat{x}_1$ is a vector of adjusted coordinates, and $\hat{x}_2$ is a vector of adjusted orientation parameters (or nuisance parameters). The Taylor series expansion (up to the first order) of the model can be given as follows:

$$\hat{\ell} = f(x_1, x_2) + \frac{\partial f}{\partial x_1}(\delta_1) + \frac{\partial f}{\partial x_2}(\delta_2) \tag{5.53}$$

where Equation (5.53) is the approximate linear form of the given function $f(\hat{x}_1, \hat{x}_2)$ or $\hat{\ell}$ in Equation (5.52); $f(x_1, x_2)$, the zero-order term of Taylor series, is the original function evaluated at the given initial numerical values of the coordinates (x_1) and the orientation parameters (x_2); δ_1 is a vector of corrections to be applied to the initial values of coordinates (x_1); δ_2 is a vector of corrections to be applied to the initial values of the orientation parameters (x_2); $\partial f/\partial x_1$ is a Jacobian matrix containing the partial derivatives with respect to the coordinates; and $\partial f/\partial x_1$ is a Jacobian matrix of the partial derivatives with respect to the orientation parameters.

A typical survey observation that is represented in form of Equation (5.52) is total station direction observation (d_{ij}) from station i to j, reduced to the map projection coordinate system. This observation is expressed in explicit mathematical form in Section 5.1.2 as

$$\hat{d}_{ij} = \text{atan}\left[\frac{(\hat{x}_j - \hat{x}_i)}{(\hat{y}_j - \hat{y}_i)}\right] - \hat{\gamma}_i \tag{5.54}$$

where $\hat{\gamma}_i$ is the adjusted orientation parameter at the station i and $(\hat{x}_i, \hat{y}_i)$ and $(\hat{x}_j, \hat{y}_j)$ are the adjusted horizontal coordinates of points i and j, respectively. Relating Equation (5.52) to Equation (5.54),

$$\hat{\ell} = \hat{d}_{ij}$$

$$\hat{x}_1 = \left(\hat{x}_i, \hat{y}_i, \hat{x}_j, \hat{y}_j\right)^T$$

$$\hat{x}_2 = \hat{\gamma}_i$$

$$f(\hat{x}_1, \hat{x}_2) = \text{atan}\left[\frac{(\hat{x}_j - \hat{x}_i)}{(\hat{y}_j - \hat{y}_i)}\right] - \hat{\gamma}_i$$

The Taylor series expansion of Equation (5.54) with regard to Equation (5.53) will be

$$\hat{d}_{ij} = \text{atan}\left[\frac{(x_j^0 - x_i^0)}{(y_j^0 - y_i^0)}\right] - \gamma_i^0 + A_1\begin{bmatrix}\delta x_i \\ \delta y_i \\ \delta x_j \\ \delta y_j\end{bmatrix} + A_2[\delta\gamma_i] \tag{5.55}$$

where

$$f(x_1, x_2) = \text{atan} \left[\frac{\left(x_j^0 - x_i^0 \right)}{\left(y_j^0 - y_i^0 \right)} \right] - \gamma_i^0 \text{ corresponding zero order of Taylor series}$$

evaluated at the vector of approximate values of the parameters,

$$x^0 = \begin{bmatrix} x_i^0 & y_i^0 & x_j^0 & y_j^0 & \gamma_i^0 \end{bmatrix}^T, \quad A_1 = \frac{\partial f}{\partial \hat{x}_1}\Big|_{x^0} = \begin{bmatrix} \dfrac{\partial \hat{d}_{ij}}{\partial \hat{x}_i} & \dfrac{\partial \hat{d}_{ij}}{\partial \hat{y}_i} & \dfrac{\partial \hat{d}_{ij}}{\partial \hat{x}_j} & \dfrac{\partial \hat{d}_{ij}}{\partial \hat{y}_j} \end{bmatrix}_{x^0}, \quad A_2 = \frac{\partial f}{\partial \hat{x}_2}\Big|_{x^0} =$$

$$\begin{bmatrix} \dfrac{\partial \hat{d}_{ij}}{\partial \hat{\gamma}_i} \end{bmatrix}_{x^0}, \quad \delta_1 = \begin{bmatrix} \delta x_i \\ \delta y_i \\ \delta x_j \\ \delta y_j \end{bmatrix}, \text{ and } \delta_2 = [\delta \gamma_i]; \text{ the Jacobian matrices } A_1 \text{ and } A_2 \text{ are known}$$

with regard to parametric least squares adjustment as the first design matrices. The elements of the matrices A_1 and A_2 are given in Section D.2.

5.5 Derivation of Variation Function

Two approaches can be followed in deriving the variation function from the linearized parametric model given in Equation (5.51): *direct approach* and *Lagrangian approach*. The Lagrangian approach seems to be more flexible when solving "minimum-related problem" as in the case of least squares criterion. The choice of either of the two approaches, however, is a matter of convenience since they both end up producing the same solution vector.

5.5.1 Derivation of Variation Function Using Direct Approach

Variation function is the quadratic form of the linearized parametric model given in Equation (5.51) except that it is now "correlated" with the *least squares criterion*. In Equation (5.48), the weight matrix P is clearly associated with the given model, meaning that P must be explicitly used in the variation function. Variation function based on Equation (5.51) is derived using direct approach as follows. The least squares criterion for this type of problem (referring to Section 4.2.1) can be given as

$$v^T P v = \text{minimum} \tag{5.56}$$

The variation function (φ) in Equation (5.57) is obtained by imposing the least squares criterion (Equation (5.56)) on the linearized model (5.51). This is done by substituting the linearized model directly into the least squares criterion as follows:

$$\varphi = v^T P v = (A\delta + w)^T P(A\delta + w) = \text{minimum} \tag{5.57}$$

In expanded form, the variation function (Equation (5.57)) can be given as

$$\varphi = \delta^T A^T P A \delta + \delta^T A^T P w + w^T P A \delta + w^T P w = \text{minimum} \tag{5.58}$$

Following the rules of basic matrix operations in Section 1.6.3, it can be shown that $(A\delta + w)^T = \delta^T A^T + w^T$ (noting that A and δ are now switched around). If the covariance matrix of observations (C_ℓ) is provided in Equation (5.48) instead of the given weight matrix P, the Equation (5.48) will be modified to look like the following model:

$$\hat{\ell} = f(\hat{x}) \; C_\ell \tag{5.59}$$

where P is now replaced with the covariance matrix of the observations C_ℓ, which is now clearly associated with the given model. This means that C_ℓ must be explicitly used in the least squares criterion and the variation function. By substituting $P = \sigma_0^2 C_\ell^{-1}$ (and setting the a priori variance factor of unit weight, $\sigma_0^2 = 1$) into Equations (5.56) and (5.58), the corresponding least squares criterion and the variation function are obtained as Equations (5.60) and (5.61), respectively:

$$v^T C_\ell^{-1} v = \text{minimum} \tag{5.60}$$

$$\varphi = \delta^T A^T C_\ell^{-1} A \delta + 2\delta^T A^T C_\ell^{-1} w + w^T C_\ell^{-1} w = \text{minimum} \tag{5.61}$$

5.5.2 Derivation of Variation Function Using Lagrangian Approach

The variation function based on Equation (5.51) associating with the model in Equation (5.48) and the least squares criterion in Equation (5.56) is derived as follows using Lagrangian approach. The Lagrangian approach is more flexible than the direct approach especially when the least squares criterion must be imposed on many linearized models; in this case, all that one needs to do is to multiply each linearized model by a different Lagrangian multiplier (k) and subtract each product from (or add each product to) the criterion function to produce a variation function. Imposing the least squares criterion (Equation (5.56)) on the linearized model (Equation (5.51)) using a vector of Lagrangian multipliers k (also known as a vector of correlates) gives the variation function by the Lagrangian approach as follows:

$$\varphi = v^T P v - 2k^T (A\delta + w - v) = \text{minimum} \tag{5.62}$$

If the variation function is based on Equation (5.51) associating with the model in Equation (5.59) and the least squares criterion in Equation (5.60), the following will be obtained:

$$\varphi = v^T C_\ell^{-1} v - 2k^T (A\delta + w - v) = \text{minimum} \tag{5.63}$$

Equations (5.62) and (5.63) are the *variation functions based on Lagrangian* approach. As it can be seen in the two equations, the only difference is the use of P in Equation (5.62) and C_ℓ^{-1} in Equation (5.63).

5.6 Derivation of Normal Equation System

The least squares normal equation system is derived from the expanded variation function given in Equation (5.58) or Equation (5.61) depending on whether C_ℓ or P is given in the case of direct approach or Equation (5.62) or Equation (5.63) depending on whether C_ℓ or P is given in the case of Lagrangian approach. The partial derivatives of the variation function must be found and set equal to zero as follows in order to obtain the normal equation system.

5.6.1 Normal Equations Based on Direct Approach Variation Function

The normal equation system can be derived from the variation functions obtained based on the direct approach, as follows. For example, consider the variation function in Equation (5.58), find its partial derivatives (referring to Equations (1.43)–(1.50)) with respect to the unknown in the function (i.e. δ), and set the resulting equation to zero as follows:

$$\frac{\partial \varphi}{\partial \delta} = 2\delta^T A^T PA + 2w^T PA = 0 \tag{5.64}$$

With regard to the partial derivatives done on Equation (5.58) leading to result in Equation (5.64), the following matrix rules should be noted, which can be confirmed from basic matrix algebra books:

- Partial derivatives of $\delta^T A^T PA\delta$ with respect to δ give $2\delta^T A^T PA$ (since there are two δ involved in the term with one being the transpose of the other).
- The two terms $\delta^T A^T Pw + w^T PA\delta$ will give the same numerical values if numbers are used in the vectors and matrices involved; they will sum up to $2w^T PA\delta$ or $2\delta^T A^T Pw$ since $w^T PA\delta = \delta^T A^T Pw$. For convenience, the partial derivative of $2w^T PA\delta$ with respect to δ is considered, giving $2w^T PA$.

The least squares *normal equation system* is obtained by rearranging Equation (5.64) as follows (dividing the equation by 2 and then transposing it):

$$(A^T PA)\delta + A^T Pw = 0 \tag{5.65}$$

If the variation function in Equation (5.61) is used in the partial derivatives in Equation (5.64), the following normal equation system will be obtained:

$$\left(A^T C_\ell^{-1} A\right)\delta + A^T C_\ell^{-1} w = 0 \tag{5.66}$$

5.6.2 Normal Equations Based on Lagrangian Approach Variation Function

The least squares normal equation system can be derived from the variation functions obtained based on the Lagrangian approach in Section 5.5.2 as follows. For example, by finding the partial derivatives of the variation function in Equation (5.63) with respect to the unknown in the function (i.e. v, δ, and k), the following are obtained (referring to Equations (1.43)–(1.50) for partial derivatives of matrices):

$$\frac{\partial \varphi}{\partial v} = 2v^T C_\ell^{-1} + 2k^T = 0 \tag{5.67}$$

$$\frac{\partial \varphi}{\partial \delta} = -2k^T A = 0 \tag{5.68}$$

$$\frac{\partial \varphi}{\partial k^T} = 2(A\delta + w - v) = 0 \tag{5.69}$$

Equations (5.67)–(5.69) can be rewritten (dividing by 2 or −2) as follows:

$$v^T C_\ell^{-1} + k^T = 0 \tag{5.70}$$

$$k^T A = 0 \tag{5.71}$$

$$A\delta + w - v = 0 \tag{5.72}$$

Equations (5.70)–(5.72) are the least squares normal equation system (in the most expanded form) based on the Lagrangian approach. If the variation function in Equation (5.62) is used in the partial derivatives in Equations (5.67)–(5.69), the only difference in the normal equation system will be C_ℓ^{-1} being changed to P in Equation (5.70).

5.7 Derivation of Parametric Least Squares Solution

The least squares solution can be derived from the normal equation system given in Section 5.6.1 in the case of direct approach or in Section 5.6.2 in the case of the Lagrangian approach.

5.7.1 Least Squares Solution from Direct Approach Normal Equations

The least squares solution can be derived from the normal equation system given in Equation (5.65) by solving directly for the unknown (δ) as follows:

$$\delta = -\left(A^T P A\right)^{-1} A^T P w \tag{5.73}$$

or from Equation (5.66), giving

$$\delta = -\left(A^T C_\ell^{-1} A\right)^{-1} A^T C_\ell^{-1} w \tag{5.74}$$

Equations (5.73) and (5.74) are the solution vectors for the unknown corrections to be applied to the approximate values of the parameters (x^0). The adjusted parameters ($\hat{x}$) can be given as

$$\hat{x} = x^0 + \delta \tag{5.75}$$

From the linearized parametric model Equation (5.51), the observation residual vector (v) can be obtained as

$$v = w + A\delta \tag{5.76}$$

with the adjusted observations given as follows:

$$\hat{\ell} = \ell + v \tag{5.77}$$

Generally, least squares method hides errors when minimizing the corrections (residuals) to the observations by spreading the errors around.

5.7.2 Least Squares Solution from Lagrangian Approach
Normal Equations

The least squares solution can also be derived by simultaneously solving for δ from the Lagrangian approach normal equation system in Equations (5.70)–(5.72). By post-multiplying Equation (5.70) by matrix A and setting $K^T A = 0$ according to Equation (5.71), the following is obtained:

$$v^T C_\ell^{-1} A = 0 \tag{5.78}$$

Transposing Equation (5.78) gives

$$A^T C_\ell^{-1} v = 0 \tag{5.79}$$

Substituting v from Equation (5.72) into Equation (5.79) gives the following:

$$A^T C_\ell^{-1} A\delta + A^T C_\ell^{-1} w = 0 \tag{5.80}$$

Equation (5.80) is the normal equation system (in condensed form) based on Lagrangian approach. The solution of Equation (5.80) gives

$$\delta = -\left(A^T C_\ell^{-1} A\right)^{-1} A^T C_\ell^{-1} w \tag{5.81}$$

Substituting $P = C_\ell^{-1}$ into Equation (5.81) gives

$$\delta = -\left(A^T P A\right)^{-1} A^T P w \tag{5.82}$$

Equations (5.81) and (5.82) are the solution vectors for the unknown corrections to be applied to the approximate values of the parameters (x^0). These equations are the same as those derived from the direct approach in Equations (5.73) and (5.74). This is to say that the end results are the same whichever approach is used. In summary, the useful equations in the above derivations for the parametric least squares adjustment procedure can be given as follows:

$$\delta = -\left(A^T P A\right)^{-1} A^T P w \tag{5.83}$$

or

$$\delta = -N^{-1} u \tag{5.84}$$

$$N = A^T P A \tag{5.85}$$

$$u = A^T P w \tag{5.86}$$

$$A = \frac{\partial f}{\partial x} \tag{5.87}$$

$$w = f\left(x^0\right) - \ell \tag{5.88}$$

$$\hat{x} = x^0 + \delta \tag{5.89}$$

$$\hat{\ell} = \ell + v \tag{5.90}$$

$$v = w + A\delta \tag{5.91}$$

where δ is the vector of corrections to the vector of approximate parameters x^0, N is usually referred to as the matrix of the coefficients of normal equations, A is the first design matrix, w is the misclosure vector, $\hat{x}$ is the vector of adjusted parameters, $\hat{\ell}$ is the vector of adjusted observations, v is the vector of residuals or corrections to be applied to the original observations ℓ, and P is the weight matrix of the observations ℓ.

5.8 Stochastic Models of Parametric Adjustment

The relationship between the variance–covariance matrix (C_ℓ) of some observation vector ℓ and the cofactor matrix (Q_ℓ) of the observation vector can be given as

$$C_\ell = \sigma_0^2 Q_\ell \tag{5.92}$$

where σ_0^2 is a priori variance factor of unit weight (which is always taken as the general population variance with a value of 1). Since $\sigma_0^2 = 1$ in Equation (5.92), then, at the start of the least squares adjustment, it is assumed that the variance–covariance matrix of observations is the same as the cofactor matrix of the observations. With this assumption, the weight matrix (P) of observations is taken to be the inverse of the variance–covariance matrix of observations, given as follows:

$$P = C_\ell^{-1} = Q_\ell^{-1} \tag{5.93}$$

Since the observations ℓ and its variance–covariance matrix C_ℓ are used in computing the unknown quantities, such as the adjusted parameters ($\hat{x}$), the adjusted observations ($\hat{\ell}$), and the observation residuals (v), it is obvious that the errors in those observations will propagate into these quantities also. The concepts of variance–covariance propagation (Chapter 2) will be used as follows to derive the cofactor matrix of the adjusted parameters ($Q_{\hat{x}}$), the cofactor matrix of the adjusted observation ($Q_{\hat{\ell}}$), and the cofactor matrix of the observation residuals (Q_v).

5.8.1 Derivation of Cofactor Matrix of Adjusted Parameters

The vector of least squares adjusted parameters can be given from Equations (5.83), (5.88), and (5.89) as

$$\hat{x} = x^0 - \left(A^T PA\right)^{-1} A^T P\left[f\left(x^0\right) - \ell\right] \tag{5.94}$$

Equation (5.94) is a form of $\hat{x} = f(\ell)$ with the usual propagated cofactor matrix (based on variance–covariance propagation rules) given as

$$Q_{\hat{x}} = JQ_\ell J^T \tag{5.95}$$

where Q_ℓ is the cofactor matrix of the observations and J is the Jacobian matrix given as follows:

$$J = \frac{\partial \hat{x}}{\partial \ell} = \left(A^T PA\right)^{-1} A^T P \tag{5.96}$$

taking note that the partial derivatives with respect to ℓ of all the other terms (except ℓ) will be zero. Substituting Equation (5.96) into Equation (5.95) gives the following:

$$Q_{\hat{x}} = \left(A^T PA\right)^{-1} A^T P Q_\ell PA \left(A^T PA\right)^{-1} \tag{5.97}$$

Since, by definition, $P = Q_\ell^{-1}$, then $PQ_\ell = I$ (an identity matrix with all the principal diagonal elements as one and all the remaining elements as zero), so that Equation (5.97) can be given as

$$Q_{\hat{x}} = \left(A^T PA\right)^{-1} \left(A^T PA\right) \left(A^T PA\right)^{-1} \tag{5.98}$$

Similarly, $(A^T PA)(A^T PA)^{-1} = I$, so that Equation (5.98) can further be reduced to the following:

$$Q_{\hat{x}} = \left(A^T PA\right)^{-1} \tag{5.99}$$

Equation (5.99) is the cofactor matrix of the adjusted parameters ($\hat{x}$); this matrix can actually be deduced directly, by inspection, from the parametric least squares solution vector given in Equation (5.83).

5.8.2 Derivation of Cofactor Matrix of Adjusted Observations

The vector of least squares adjusted observations can be given from Equations (5.88), (5.90), and (5.91) as

$$\hat{\ell} = \ell + A\delta + f\left(x^0\right) - \ell \tag{5.100}$$

From Equations (5.83), (5.88), and (5.100), the following equation is obtained:

$$\hat{\ell} = -A\left(A^T PA\right)^{-1} A^T P\left[f\left(x^0\right) - \ell\right] + f\left(x^0\right) \tag{5.101}$$

Equation (5.101) is a form of $\hat{\ell} = f(\ell)$ with the usual propagated cofactor matrix (based on variance–covariance propagation rules) given as

$$Q_{\hat{\ell}} = JQ_\ell J^T \tag{5.102}$$

where Q_ℓ is the cofactor matrix of the observations and J is the Jacobian matrix given as follows:

$$J = \frac{\partial \hat{\ell}}{\partial \ell} = A\left(A^T PA\right)^{-1} A^T P \tag{5.103}$$

taking note that the partial derivatives with respect to ℓ of all the other terms (except ℓ) will be zero. Substituting Equation (5.103) into Equation (5.102) gives the following:

$$Q_{\hat{\ell}} = A\left(A^T PA\right)^{-1} A^T P Q_\ell PA\left(A^T PA\right)^{-1} A^T \tag{5.104}$$

Similarly in Equation (5.104), $PQ_\ell = I$, so that Equation (5.104) can be given as

$$Q_{\hat{\ell}} = A\left(A^T PA\right)^{-1} \left(A^T PA\right)\left(A^T PA\right)^{-1} A^T \tag{5.105}$$

Similarly, $(A^T PA)(A^T PA)^{-1} = I$, so that Equation (5.105) can further be reduced to the following:

$$Q_{\hat{\ell}} = A\left(A^T PA\right)^{-1} A^T \tag{5.106}$$

Equation (5.106) is the cofactor matrix of the adjusted observations.

5.8.3 Derivation of Cofactor Matrix of Observation Residuals

The vector of observation residuals can be given from Equations (5.83), (5.88), and (5.91) as

$$v = -A(A^TPA)^{-1}A^TP[f(x^0) - \ell] + f(x^0) - \ell \tag{5.107}$$

Equation (5.107) is a form of $v = f(\ell)$ with the usual propagated cofactor matrix (based on variance–covariance propagation rules) given as

$$Q_v = JQ_\ell J^T \tag{5.108}$$

where Q_ℓ is the cofactor matrix of the observations and J is the Jacobian matrix given as follows:

$$J = \frac{\partial v}{\partial \ell} = A(A^TPA)^{-1}A^TP - I \tag{5.109}$$

taking note that the partial derivative with respect to ℓ for the last term in Equation (5.107) is an identity matrix I since ℓ is a vector quantity. Substituting Equation (5.109) into Equation (5.108) gives the following:

$$Q_v = \left[A(A^TPA)^{-1} - I\right]Q_\ell\left[PA(A^TPA)^{-1}A^T - I\right] \tag{5.110}$$

Equation (5.110) in expanded form can be given as

$$Q_v = A(A^TPA)^{-1}A^TPQ_\ell PA(A^TPA)^{-1}A^T - A(A^TPA)^{-1}A^TPQ_\ell$$
$$- Q_\ell PA(A^TPA)^{-1}A^T + Q_\ell \tag{5.111}$$

Taking $PQ_\ell = I$ and $Q_\ell P = I$ in Equation (5.111) reduces the equation to the following:

$$Q_v = A(A^TPA)^{-1}(A^TPA)(A^TPA)^{-1}A^T - A(A^TPA)^{-1}A^T$$
$$- A(A^TPA)^{-1}A^T + Q_\ell \tag{5.112}$$

Taking $(A^TPA)(A^TPA)^{-1} = I$ and simplifying Equation (5.112) further produces the following:

$$Q_v = A(A^TPA)^{-1}A^T - A(A^TPA)^{-1}A^T - A(A^TPA)^{-1}A^T + Q_\ell \tag{5.113}$$

which simplifies further to

$$Q_v = Q_\ell - A(A^TPA)^{-1}A^T \tag{5.114}$$

Equation (5.113) is the cofactor matrix of the observation residuals (v). The variance–covariance matrices of the adjusted quantities can be given from Equations (5.99), (5.106), and (5.114), respectively, as follows:

$$C_{\hat{x}} = s_0^2 \left(A^T P A\right)^{-1} \tag{5.115}$$

$$C_{\hat{\ell}} = s_0^2 \left[A\left(A^T P A\right)^{-1} A^T\right] \tag{5.116}$$

$$C_v = s_0^2 Q_\ell - s_0^2 \left[A\left(A^T P A\right)^{-1} A^T\right] \tag{5.117}$$

where $C_{\hat{x}}$ is the covariance matrix of the adjusted parameters, $C_{\hat{\ell}}$ is the covariance matrix of the adjusted observations, C_v is the covariance matrix of the observation residuals, and s_0^2 is *a posteriori variance factor (APVF) of unit weight*, which is calculated after the least squares adjustment as follows:

$$s_0^2 = \frac{v^T P v}{n - u} \tag{5.118}$$

where the redundancy is expressed as $n - u$, n is the number of parametric equations, u is the number of unknown parameters (which should not be confused with the similar symbol used in Equations (5.84) and (5.86) such as $u = A^T P w$), v is the vector of observation residuals, and P is the weight matrix of observations. In Equation (5.117), it can be seen that s_0^2 is used to scale the cofactor matrix (Q_ℓ) of the original observations to give a more acceptable covariance matrix of the observations; this is a form of indirectly calibrating the instrument used in making the measurements. The APVF (s_0^2) is an indicator (for the whole network) as to the consistency of the adjusted network based on the a priori predicted errors. The predicted errors are represented by the standard deviations of the observations. If on average, each observation is adjusted by an amount larger than its predicted error, the APVF will tend to be greater than 1; if on average, each observation is adjusted by an amount less than its predicted error, the APVF will tend to be less than 1. In either case, one may not be correctly estimating the quality of one's measurements or one may have a few bad measurements in survey (especially those measurements receiving large adjustment). It should be remembered that if the a priori variance factor σ_0^2 is well known, it should be used in Equations (5.115)–(5.117); when unknown its value would be set to 1 before the adjustment in order to determine the weight matrix (P) of observations. From Equations (5.115) to (5.117), it can be seen that the cofactor matrix of adjusted parameters ($Q_{\hat{x}}$) is the same as N^{-1} with the cofactor of the unadjusted observations (Q_ℓ) given as P^{-1}. From Equations (5.116) and (5.117), the covariance matrix of the observation residuals can be rewritten as

$$C_v = \hat{C}_\ell - C_{\hat{\ell}} \tag{5.119}$$

where $C_{\hat{\ell}}$ (given in Equation (5.116)) is the covariance matrix of the adjusted observations, which should not be confused with the scaled covariance matrix of observations ($\hat{C}_\ell$) of the original observations ℓ, expressed as

$$\hat{C}_\ell = s_0^2 Q_\ell \qquad (5.120)$$

Equation (5.120) should also not be confused with $C_\ell = \sigma_0^2 Q_\ell$ (with $\sigma_0^2 = 1$), which is the unscaled covariance matrix of the original observations, commonly used at the beginning of a least squares adjustment. Actually, Equation (5.120) and $C_\ell = \sigma_0^2 Q_\ell$ will be identical if $s_0^2 = \sigma_0^2$. By rearranging Equation (5.119), the following is obtained:

$$C_{\hat{\ell}} = \hat{C}_\ell - C_v \qquad (5.121)$$

By interpreting Equation (5.121), it can be understood that the covariance matrix of the adjusted observation ($C_{\hat{\ell}}$) is always smaller than the scaled covariance matrix of the original observations ($\hat{C}_\ell$) by the amount of the covariance matrix of the residuals (C_v). This means that least squares adjustment method improves precisions of original observations. For example, in minimal constraint adjustment (discussed in Chapter 4) in which a control point is held fixed for an adjustment, the covariance matrix of the adjusted parameters given by Equation (5.115) gives the uncertainty in the adjusted coordinates that is due to the uncertainties of the new observations. This covariance matrix indicates how well the coordinates of the new survey points are known relative to the fixed control point, but not how well they are known relative to the datum as a whole. The covariance matrix of the adjusted observations (Equation (5.116)) is true for all minimal constrained adjustment, irrespective of how the coordinate system is defined and irrespective of the uncertainty of the fixed control point. However, the coordinates obtained for the free points may be affected by the errors in the fixed control point, even though the adjusted observations are free of errors. The idea that adjusted observations from minimal constraint adjustments are not affected by errors in the fixed control point is the sense in which minimal constraint adjustments are considered "free adjustments."

5.8.4 Effects of Variance Factor Variation on Adjustments

In some of the least squares adjustment software packages today, the a priori variance factor of unit weight can be changed from the usual value of one to some other numbers, especially in the process of blunder detection. This is usually done in order to make the APVF of unit weight equal to one. The common problem with the users of such software packages is understanding how the changed a priori variance factor actually affects the adjusted quantities apart from changing the APVF of unit weight (s_0^2). This is commonly the case in the least squares adjustment of GNSS measurements in which variances of baseline vectors are usually too optimistic and sometimes approach zero values. When those baseline vectors and their covariance matrices are used in subsequent network adjustment, the computed APVF of unit weight for the adjustment becomes extremely too large that the covariance matrices of the input baseline vectors have to be scaled by a factor. For example, consider a case in which the a priori variance factor of unit weight ($\sigma_0^2 = 1$) is changed to a constant factor $1/k$ (i.e. $\sigma_0^2 = 1/k$). It can be shown mathematically, using Equations (5.81), (5.82), (5.99), (5.115), and (5.118), that the adjusted vector of parameters $\hat{x}$ and its covariance matrix $C_{\hat{x}}$ will remain unchanged. This is to say that when the initial covariance matrix (C_ℓ) of measurements is multiplied by a factor k (i.e. $Q_\ell = kC_\ell$), this will not affect $\hat{x}$ and $C_{\hat{x}}$ after the adjustment; hence the covariance matrix of the adjusted parameters will not be affected by the choice of a priori variance factor. The cofactor matrix of the adjusted parameters ($Q_{\hat{x}}$), however, will be increased by a factor k, and the APVF of unit weight (s_0^2) will be reduced by a factor k. Similarly, using Equations (5.92), (5.114), (5.117), and (5.118), it can be shown mathematically that the cofactor matrix of residuals will be scaled by a factor of k, but the residuals and the covariance matrix of the residuals will remain unchanged when $\sigma_0^2 = 1/k$.

5.9 Weight-constrained Adjustment Model Formulation

A survey station is said to be *weight constrained* if the coordinates of the station are associated with some standard deviations or some uncertainty. This is to say that the coordinates of this station are open to adjustment to the extent limited by the associated standard deviations of the station coordinates. In a least squares estimation, these station coordinates can be considered as pseudo-measurements and their standard deviations as weighting factors. Usually, a least squares estimation based on prior coordinates and an associated a priori covariance matrix C_x (or *weight constraints*, P_x) will give adjusted values that are almost identical to those based on *minimum constraint* adjustment. The solution model for weight-constrained adjusted is derived in this section. Example 5.10 gives the derivation of weight-constrained adjustment procedure in symbolic form. Assume a survey network is composed of two sets of parameters (x_1, x_2) with the previously estimated parameters as x_1 with a covariance matrix C_{x_1} associated with them. If a new adjustment is performed on this survey network, the parameters x_1 will be considered as both measurements as well as parameters (having *dual nature*) in the new adjustment. If x_1 and x_2 have n_1 and n_2 elements, respectively, the weight matrix P_x to be formulated for the parameters in the new adjustment will have matrix size ($n_1 + n_2$) by ($n_1 + n_2$) with the inverse of C_{x_1} occupying the space corresponding to x_1, and all other elements of the matrix P_x set to zero. This is a typical example of an adjustment problem involving weight constraints on a subset of network points. It is also possible to have weight constraints on all of the given network points. This may be the case when a whole network previously measured (probably imprecisely using lower-grade equipment) in a survey (first survey) is again measured with higher-precision equipment in another survey (second survey). The adjusted network coordinates and their covariance matrices from the first survey will serve as pseudo-measurements and weight constraints (P_x) for the adjustment of the measurements in the second survey.

Example 5.10 Given the following parametric model with weight constraint (P_x) on the parameter:

$$\hat{\ell} = f(\hat{x}) \quad P \quad P_x \tag{5.122}$$

where f is a vector of mathematical model, x is a vector of unknown parameters, ℓ is a vector of observations, and P is the weight matrix of the observations.

a) Formulate the variation function based on Lagrangian approach.

Solution:

Rewrite Equation (5.122) in terms of residual vectors v, vector of adjusted observation $\hat{\ell}$, vector of approximate value of parameters (x^0), and vector of corrections to approximate parameters δ:

$$\ell + v = f\left(x^0 + \delta\right) \tag{5.123}$$

Linearize Equation (5.123) by Taylor series expansion:

$$v = f\left(x^0\right) + \frac{\partial f}{\partial x}\delta - \ell \tag{5.124}$$

or

$$v = w + A\delta \tag{5.125}$$

where $w = f(x^0) - \ell$ is the misclosure vector and $A = \partial f/\partial x$ is the first design matrix. Equation (5.124) or Equation (5.125) is the linearized form of the model given in Equation (5.122).

Since the weight matrix of observations (P) and the weight matrix of the parameters (P_x) are directly associated with the given functional model in Equation (5.122), they should be used directly in formulating the least squares criterion as follows:

$$v^T P v + \delta^T P_x \delta = \text{minimum} \tag{5.126}$$

where the weight matrix of the parameter is $P_x = C_x^{-1}$ and that of the observation is $P = C_\ell^{-1}$ with the a priori variance factor of unit weight $\sigma_0^2 = 1$. Impose the least squares criterion on the linearized model

Equation (5.125), modified as $(w + A\delta - v = 0)$. The following *variation function* is then obtained:

$$\varphi = v^T P v + \delta^T P_x \delta - 2k^T (A\delta + w - v) = \text{minimum} \qquad (5.127)$$

where k is a vector of correlates (or unknown constant values).

b) Derive the most expanded form of the least squares normal equation system.

Solution:

Find the partial derivatives of φ in Equation (5.127) with respect to the unknown quantities v, δ, k^T and set their values to zero as follows:

$$\frac{\partial \varphi}{\partial v} = 2v^T P + 2k^T = 0 \qquad (5.128)$$

$$\frac{\partial \varphi}{\partial \delta} = 2\delta^T P_x - 2k^T A = 0 \qquad (5.129)$$

$$\frac{\partial \varphi}{\partial k^T} = -2(A\delta + w - v) = 0 \qquad (5.130)$$

The most expanded least squares normal equation system is given by simplifying Equations (5.128)–(5.130) as follows:

$$v^T P + k^T = 0 \qquad (5.131)$$
$$\delta^T P_x - k^T A = 0 \qquad (5.132)$$
$$A\delta + w - v = 0 \qquad (5.133)$$

c) Derive the least squares solution of the normal equation system.

Solution:

From Equation (5.131), transpose the equation and rearrange to give

$$v = -P^{-1}k \qquad (5.134)$$

Substitute Equation (5.134) into Equation (5.133) to obtain

$$A\delta + w + P^{-1}k = 0 \qquad (5.135)$$

Rearrange the equation and solve for k as follows:

$$k = -P(A\delta + w) \qquad (5.136)$$

Transpose Equation (5.132), and then substitute Equation (5.136) into it, giving

$$P_x\delta + A^T P(A\delta + w) = 0 \qquad (5.137)$$

Solve for the unknown δ in Equation (5.137):

$$\delta = -\left(P_x + A^T P A\right)^{-1} A^T P w \tag{5.138}$$

The adjusted parameters can be determined from

$$\hat{x} = x^0 + \delta \tag{5.139}$$

By using Equations (5.138), (5.136), and (5.134), the residual vector v can be determined, so that the adjusted observations can be obtained as

$$\hat{\ell} = \ell + v \tag{5.140}$$

5.9.1 Stochastic Model for Weight-constrained Adjusted Parameters

The adjusted parameters from the weight-constrained adjustment can be given from Equations (5.138) and (5.139) as follows:

$$\hat{x} = x^0 - \left(P_x + A^T P A\right)^{-1} A^T P\left[f\left(x^0\right) - \ell\right] \tag{5.141}$$

where the misclosure vector is $w = [f(x^0) - \ell]$. It is assumed in Equation (5.141) that the weight matrix P_x has a full matrix size of all the parameters to be adjusted (including the unknown parameters) and the elements of the weight matrix can be derived from the a priori cofactor matrix (Q_x) of the parameters; the parameters with unknown values can be assigned zero weights instead of deriving them directly from their cofactors. By variance–covariance propagation laws on Equation (5.141), assuming the cofactor matrix of parameters (Q_x) with "zero elements" for the unknown parameters, the cofactor matrix of the adjusted parameters ($Q_{\hat{x}}$) can be given as

$$Q_{\hat{x}} = J \begin{bmatrix} Q_x & 0 \\ 0 & Q_\ell \end{bmatrix} J^T \tag{5.142}$$

where Q_ℓ is the cofactor of the measurements and J is the Jacobian matrix of the Equation (5.141) with respect to the vector of parameters (x) and of the measurements (ℓ), given as follows:

$$J = \left[\frac{\partial \hat{x}}{\partial x} \quad \frac{\partial \hat{x}}{\partial \ell}\right] \tag{5.143}$$

with

$$\frac{\partial \hat{x}}{\partial x} = I - \left(P_x + A^T P A\right)^{-1} A^T P A \tag{5.144}$$

$$\frac{\partial \hat{x}}{\partial \ell} = \left(P_x + A^T P A\right)^{-1} A^T P \tag{5.145}$$

Take note from Equation (5.144) that $\partial f(x^0)/\partial x = A$, the first design matrix, and I is an identity matrix with all the principal diagonal elements as ones and the off-diagonal elements are zeros. The expanded form of Equation (5.142) can be given as

$$Q_{\hat{x}} = Q_x - Q_x NW - WNQ_x + WNQ_x NW + WNW \tag{5.146}$$

where

$$W = \left(P_x + A^T PA\right)^{-1} \tag{5.147}$$

$$N = \left(A^T PA\right) \tag{5.148}$$

Equation (5.146) can be shown to be equivalent to the following:

$$Q_{\hat{x}} = \left(P_x + A^T PA\right)^{-1} \tag{5.149}$$

The variance–covariance matrix of the adjusted parameters can be given as

$$C_{\hat{x}} = s_0^2 Q_{\hat{x}} \tag{5.150}$$

where the APVF of unit s_0^2 can be given as

$$s_0^2 = \frac{v^T Pv + \delta^T P_x \delta}{n_d - u_d} \tag{5.151}$$

with n_d as the number of direct (or actual) observations and u_d as the number of direct parameters (those parameters whose values are not associated with any a priori standard deviations).

5.9.2 Stochastic Model for Weight-constrained Adjusted Observations

The adjusted observations can be given from Equations (5.133), (5.138), and (5.140) with the misclosure vector $w = f(x^0) - \ell$ as follows:

$$\hat{\ell} = \ell - A\left(P_x + A^T PA\right)^{-1} A^T P\left[f\left(x^0\right) - \ell\right] + \left[f\left(x^0\right) - \ell\right] \tag{5.152}$$

or in an expanded form as

$$\hat{\ell} = -A\left(P_x + A^T PA\right)^{-1} A^T P\left[f\left(x^0\right)\right] + A\left(P_x + A^T PA\right)^{-1} A^T P\ell + f\left(x^0\right) \tag{5.153}$$

It is assumed in Equation (5.153) that the weight matrix P_x has a full matrix size of all the parameters to be adjusted (including the unknown parameters) and the elements of the weight matrix can be derived from the a priori cofactor matrix (Q_x) of the parameters; the parameters with unknown values can be assigned zero weights instead of deriving them directly from their cofactors.

By variance–covariance propagation laws on Equation (5.153), assuming the cofactor matrix of parameters (Q_x) with "zero elements" for the unknown parameters, the cofactor matrix of the adjusted parameters $(Q_{\hat{x}})$ can be given as

$$Q_{\hat{l}} = J \begin{bmatrix} Q_x & 0 \\ 0 & Q_l \end{bmatrix} J^T \qquad (5.154)$$

where Q_l is the cofactor of the measurements and J is the Jacobian matrix of the Equation (5.153) with respect to the vector of parameters (x) and of the measurements (l), given as follows:

$$J = \begin{bmatrix} \dfrac{\partial \hat{l}}{\partial x} & \dfrac{\partial \hat{l}}{\partial l} \end{bmatrix} \qquad (5.155)$$

with

$$\frac{\partial \hat{l}}{\partial x} = -A \left(P_x + A^T PA \right)^{-1} A^T PA + A \qquad (5.156)$$

$$\frac{\partial \hat{l}}{\partial l} = A \left(P_x + A^T PA \right)^{-1} A^T P \qquad (5.157)$$

Take note from Equation (5.156) that $\partial f(x^0)/\partial x = A$, the first design matrix, and I is an identity matrix with all the principal diagonal elements as ones and the off-diagonal elements are zeros. The expanded form of Equation (5.154) can be given as

$$Q_{\hat{l}} = A Q_x A^T - A Q_x NWA^T - A WNQ_x A^T + A WNQ_x NWA^T + A WNWA^T \qquad (5.158)$$

where W and N are as previously defined in Equations (5.147) and (5.148). Equation (5.158) can be shown to be equivalent to the following:

$$Q_{\hat{l}} = A \left(P_x + A^T PA \right)^{-1} A^T \qquad (5.159)$$

The variance–covariance matrix of the adjusted observations can be given as

$$C_{\hat{l}} = s_0^2 Q_{\hat{l}} \qquad (5.160)$$

Problems

5.1 In a three-point resection problem shown in Figure P5.1, the coordinates of point P (x, y) are to be determined using control points A (x_A, y_A), B (x_B, y_B), and C (x_C, y_C). The total station direction measurements from the setup point P are l_1, l_2, and l_3; the angle measurements at point B are

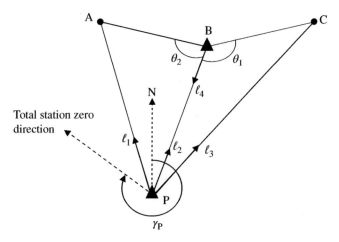

Figure P5.1 (**Note:** Lines BN and PN are the directions of the north.)

θ_1 and θ_2; distance measurement from point B to P is ℓ_4; and the orientation parameter at the setup point P is as shown on the figure. Formulate the parametric model equations ($\hat{\ell} = f(\hat{x})$) for this problem.

5.2 Given the following linearized parametric model

$$r = Jy + m \quad C_\lambda$$

where J is the first design matrix with respect to the vector of parameters x, y is a vector of corrections to the approximate parameters (x^0), m is the misclosure vector, C_λ is a covariance matrix of observation vector λ, and r is a vector of residuals. Answer the following questions using the given symbols.

a) Give the least squares criterion and the variation function (based on the Lagrangian approach) for this problem in terms of the symbols used in this question.

b) Derive the least squares normal equation system and the solution vector (y) for this problem in terms of the symbols used in this question (clearly explaining every step of your derivations).

5.3 a) Referring to Section 5.5 of the book, assume the a priori variance factor of unit weight (σ_0^2) is not equal to one but equal to an unknown constant factor k (i.e. $\sigma_0^2 = k$). Show mathematically, using Equations (5.73), (5.74), (5.99), (5.115), and (5.118), how the parametric model solution vector and its cofactor and covariance matrices will be affected by the unknown constant factor k. (Full marks will not be awarded except all the logical steps and explanations, including substitutions into formulas, are clearly provided.)

b) Referring to Section 5.5 of the book, assume the a priori variance factor of unit weight (σ_0^2) is not equal to one but equal to an unknown constant factor k (i.e. $\sigma_0^2 = k$). Show mathematically, using Equations (5.92), (5.114), (5.117), and (5.118), how the cofactor and covariance matrices of residuals will be affected by the unknown constant factor k. (Full marks will not be awarded except all the logical steps and explanations, including substitutions into formulas, are clearly provided.)

5.4 Given the following mathematical models

$$\hat{\lambda} = f(\hat{y})\ \ C_\lambda$$

where f is a vector of mathematical models, $\hat{y}$ is a vector of adjusted parameters, $\hat{\lambda}$ is a vector of adjusted observations, and C_λ is a covariance matrix of observations. Answer the following questions, defining every other symbol used in your derivations.

a) Derive (as completely as possible with full explanation of your steps) the linearized parametric model.

b) Give the least squares criterion and the variation function (based on the Lagrangian approach) for this problem in terms of the symbols used in this question.

5.5 Given the following mathematical models

$$\hat{\ell} = f(\hat{x}_1, \hat{x}_2, \hat{x}_3)\ \ P\ \ P_{x_1}\ \ P_{x_3}$$

where f is a vector of mathematical models, $\hat{x}_1$, $\hat{x}_2$, and $\hat{x}_3$ are vectors of adjusted parameters, $\hat{\ell}$ is a vector of adjusted observations, P is a weight matrix of observations, P_{x_1} is the weight matrix of parameters $\hat{x}_1$, and P_{x_3} is the weight matrix of parameters $\hat{x}_3$. Answer the following questions, defining every other symbol used in your derivations.

a) Derive (as completely as possible with full explanation of your steps) the linearized model.

b) Give the least squares criterion for this problem.

c) Derive the variation function (based on the Lagrangian approach).

d) Derive the least squares normal equation system by minimizing the variation function.

6

Parametric Least Squares Adjustment

Applications

CHAPTER MENU

OBJECTIVES

After studying this chapter, you should be able to:

1) Formulate parametric model equations for various geomatics problems.
2) Apply the concepts of parametric least squares adjustment to various geomatics problems, which include differential leveling, station adjustment, traverse, triangulation, trilateration, resection, and curve fitting.
3) Apply the laws of variance–covariance propagation to determine the variance–covariance matrices of the adjusted quantities, such as adjusted parameters, adjusted observations, and observation residuals.
4) Adjust geomatics problems associated with weight constraints.

Understanding Least Squares Estimation and Geomatics Data Analysis, First Edition.
John Olusegun Ogundare.
© 2019 John Wiley & Sons, Inc. Published 2019 by John Wiley & Sons, Inc.
Companion website: www.wiley.com/go/ogundare/Understanding-lse-and-gda

6.1   Introduction

The method of least squares is meaningful if the number (n) of parametric model equations is greater than the number (u) of parameters. This means that least squares adjustment by parametric method is possible if the redundancy (degrees of freedom) given as ($n - u$) is greater than zero. Mathematically, this means that the least squares normal equations are *singular* if the redundancy is less than zero, assuming the datum is already defined for the adjustment as discussed in Chapter 4. The *singularity* of normal equations in least squares adjustment also means that the inverse of the matrix of the coefficients of normal equations (N) in the solution steps given in Section 5.7 will be impossible. Note that the greater the redundancy, the more reliable are the results of the adjustment.

In this chapter, some examples of least squares adjustment by parametric method are given to illustrate the steps involved in the parametric least squares adjustment. Continuing from the mathematical derivations given in Chapter 5, the summary of the least squares steps for the adjustment of parametric model equations can be given as follows:

1) Formulate the parametric model equations ($\hat{\ell} = f(\hat{x})$). In this model formulation, each observation (ℓ_i) will be translated into an equation involving only some or all (depending on the type of observable) of the unknown parameters (x_i) and possible constant values using symbols for observations and parameters (refer to Chapter 5 on how to formulate the equations).
2) Form the first design matrix (A), which is the matrix of the partial derivatives of the parametric equations with respect to the unknown parameters. The size of this matrix will be equal to the number of equations or number of observations (as the number of rows) by the number of parameters (as the number of columns).
3) Choose approximate values for the parameters (x^0); these values are initial values that are considered to be close enough to what their final values should be. Usually, if this is not the case, the adjusted values based on these initial values can be used again as approximate values in another round of adjustment, in an iterative way, until the final adjusted values are not significantly different from the previous approximate (or initial) values. In this case, the adjustment solution is said to have converged. Least squares adjustment may have to be repeated as many times as possible until the difference between the final solution and the previous one converges to a specific tolerance. The best way to obtain approximate coordinate parameters is to calculate the coordinates from selected observations.

4) Substitute the approximate values (x^0) into the parametric equations formulated in step 1 so as to obtain the so-called simulated values of the observations $(\ell^0 = f(x^0))$; use these simulated values to obtain the misclosure vector $(w = f(x^0) - \ell)$ as the difference between the simulated observations (ℓ^0) and the actual observations (ℓ).

5) Form the cofactor matrix (Q) of the observations (assuming the variance factor, $\sigma_0^2 = 1$) by positioning squares of standard deviations of observations along the principal diagonal of the matrix (in the order in which the parametric equations are formulated in step 1). The inverse of this cofactor matrix is the weight matrix of the observations (P).

6) Form the following matrix products using the already determined quantities: $N = A^T P A$ and $u = A^T P w$. Note that A-matrix is evaluated using the approximate values (x^0) of the parameters and N is a square matrix with its size being equal to the number of unknown parameters.

7) Solve the normal equation system to obtain the corrections (δ) to the approximate parameters (x^0) as $\delta = -N^{-1}u$.

8) Determine the values of the adjusted parameters as $\hat{x} = x^0 + \delta$.

9) If required, determine the residuals (v) of the observations and the adjusted observations $(\hat{\ell})$ as $v = A\delta + w$ and $\hat{\ell} = \ell + v$, respectively.

6.2 Basic Parametric Adjustment Examples

Basic examples are given in this section to illustrate the parametric least squares adjustment steps formulated in Chapter 5 and summarized in Section 6.2. In some of the examples, hats are ignored in their model formulations, which is acceptable for brevity since their solutions are of main interest.

6.2.1 Leveling Adjustment

Determine the adjusted elevations h_A and h_B of points A and B, respectively, by the least squares parametric adjustment of the field leveling notes in Table 6.1, which relates to the leveling network in Figure 6.1. Assume the elevations of points FH1 and FH2 are known and fixed as $h_1 = 100.000$ m and $h_2 = 99.729$ m, respectively.

Solution:

Known elevations of FH1 and FH2 define the datum.

$$\text{Adjusted parameters: } \hat{x} = \begin{bmatrix} \hat{h}_A \\ \hat{h}_B \end{bmatrix}; \text{ adjusted observations: } \hat{\ell} = \begin{bmatrix} d\hat{h}_1 \\ d\hat{h}_2 \\ d\hat{h}_3 \end{bmatrix}.$$

Table 6.1 Differential leveling field notes.

Leg *i*	BS	FS	Elevation difference dh$_i$ (m)	Standard deviation (m)
1	FH1	BMA	−7.341	0.008
2	BMA	BMB	2.495	0.005
3	BMB	FH2	5.107	0.004

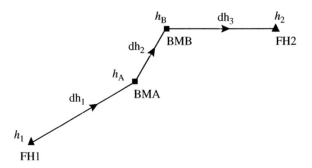

Figure 6.1 Sample differential leveling network.

Determine approximate elevations of point A and B using any direction of leveling:

$$x^0 = \begin{bmatrix} h_A^0 = 92.659 \\ h_B^0 = 95.154 \end{bmatrix}$$

Form the parametric model (remember from Chapter 5): $\hat{\ell} = f(\hat{x})$

$$d\hat{h}_1 = \hat{h}_A - 100 \qquad (6.1)$$

$$d\hat{h}_2 = \hat{h}_B - \hat{h}_A \qquad (6.2)$$

$$d\hat{h}_3 = 99.729 - \hat{h}_B \qquad (6.3)$$

Find the A-matrix from Equations (6.1) to (6.3) (with hats ignored for brevity):

$$A = \frac{\partial \ell}{\partial x} = \begin{bmatrix} \dfrac{\partial dh_1}{\partial h_A} & \dfrac{\partial dh_1}{\partial h_B} \\ \dfrac{\partial dh_2}{\partial h_A} & \dfrac{\partial dh_2}{\partial h_B} \\ \dfrac{\partial dh_3}{\partial h_A} & \dfrac{\partial dh_2}{\partial h_B} \end{bmatrix} \rightarrow A = \begin{bmatrix} 1 & 0 \\ -1 & 1 \\ 0 & -1 \end{bmatrix}$$

As can be seen in the A-matrix above, the number of columns corresponds with the number of unknown parameters (h_A and h_B) in that order, and the number of rows corresponds with the number of equations. The size of the A-matrix, therefore, is always the same as the number of equations (rows) by the number of parameters (columns). Form the misclosure vector w by substituting the approximate values of the parameters x^0 and the original observations ℓ into Equations (6.1)–(6.3):

$$w = f(x^0) - \ell \quad \ell = \begin{bmatrix} dh_1 = -7.341 \\ dh_2 = 2.495 \\ dh_3 = 5.107 \end{bmatrix}$$

The $f(x^0)$ function from above gives the following:

$$dh_1^0 = 92.659 - 100 \quad \rightarrow \quad -7.341$$
$$dh_2^0 = 95.154 - 92.659 \quad \rightarrow \quad 2.495$$
$$dh_3^0 = 99.729 - 95.154 \quad \rightarrow \quad 4.575$$

$$f(x^0) = \begin{bmatrix} -7.341 \\ 2.495 \\ 4.575 \end{bmatrix}$$

$$w = f(x^0) - \ell \rightarrow \begin{bmatrix} -7.341 \\ 2.495 \\ 4.575 \end{bmatrix} - \begin{bmatrix} -7.341 \\ 2.495 \\ 5.107 \end{bmatrix} \rightarrow w = \begin{bmatrix} 0.000 \\ 0.000 \\ -0.532 \end{bmatrix}$$

Form the weight matrix (P) of the elevation differences using the given standard deviations of the elevation differences:

$$P = \begin{bmatrix} 1/(0.008)^2 & 0 & 0 \\ 0 & 1/(0.005)^2 & 0 \\ 0 & 0 & 1/(0.004)^2 \end{bmatrix}$$

Form the following products:

$$u = A^T P w = \begin{bmatrix} 0.000 \\ 33\,250 \end{bmatrix} \quad N = A^T P A = \begin{bmatrix} 55\,625 & -40\,000 \\ -40\,000 & 102\,500 \end{bmatrix}$$

Find the solution vector:

$$\delta = \begin{bmatrix} \delta_{h_A} \\ \delta_{h_B} \end{bmatrix} = -N^{-1}u$$

$$\delta = \begin{bmatrix} \delta_{h_A} \\ \delta_{h_B} \end{bmatrix} = \begin{bmatrix} -0.324 \\ -0.451 \end{bmatrix}$$

The adjusted elevations of BMA and BMB are given as follows: $\hat{x} = x^0 + \delta$

$$\hat{h}_A = h_A^0 + \delta_{h_A} = 92.659 - 0.324 \rightarrow 92.335 \, \text{m}$$

$$\hat{h}_B = h_B^0 + \delta_{h_B} = 95.154 - 0.451 \rightarrow 94.703 \, \text{m}$$

Calculate the observation residuals: $v = A\delta + w$

$$v = A\delta + w = \begin{bmatrix} 1 & 0 \\ -1 & 1 \\ 0 & -1 \end{bmatrix} \begin{bmatrix} -0.324 \\ -0.451 \end{bmatrix} + \begin{bmatrix} 0.000 \\ 0.000 \\ -0.532 \end{bmatrix} \quad v = \begin{bmatrix} -0.324 \\ -0.127 \\ -0.081 \end{bmatrix}$$

Calculate the adjusted observations as follows: $\hat{\ell} = \ell + v$

$$d\hat{h}_1 = dh_1 + v_1 \quad \rightarrow \quad -7.341 - 0.324 = -7.665 \, \text{m}$$

$$d\hat{h}_2 = dh_2 + v_2 \quad \rightarrow \quad 2.495 - 0.127 = 2.368 \, \text{m}$$

$$d\hat{h}_3 = dh_3 + v_3 \quad \rightarrow \quad 5.107 - 0.081 = 4.494 \, \text{m}$$

The adjusted observations are $\hat{\ell} = \begin{bmatrix} d\hat{h}_1 \\ d\hat{h}_2 \\ d\hat{h}_3 \end{bmatrix} = \begin{bmatrix} -7.665 \\ 2.368 \\ 4.494 \end{bmatrix}$ m.

The MATLAB code for the least squares adjusted of the leveling network in Figure 6.1 is given in Table 6.2.

Table 6.2 MATLAB code for the least squares adjustment of the leveling survey.

```
>> % use syms to define the variables hA and hB
>> syms hA hB
>> % create the parametric Equations (6.1) to (6.3)
>> dh1=hA-100;
>> dh2=hB-hA;
>> dh3=99.729-hB;
>> % Determine the Jacobian of equations wrt hA and hB
>> J=jacobian([dh1;dh2;dh3],[hA hB]);
>> hA=92.659;hB=95.154;
>> % evaluate Jacobian as the A matrix, computed
observations as L0 and given observations as L
>> A=eval(J);
>> L0=eval([dh1;dh2;dh3]);
>> L=[-7.341;2.495;5.107];
>> % determine misclosure w, weight matrix P and solution d
>> w=L0-L;
>> P=diag([1/0.008^2 1/0.005^2 1/0.004^2]);
>> u=A'*P*w;
>> N=A'*P*A;
>> d=-inv(N)*u
d =
   -0.3243
   -0.4509
>> % determine adjusted parameters
>> hAa=hA+d(1)
hAa =
   92.3347
>> hBa=hB+d(2)
hBa =
   94.7031
>> % determine residuals v and adjusted observations La
>> v=A*d+w;
>> La=L0+v
La =
   -7.6653
    2.3683
    4.4939
>>
```

Example 6.1   Figure 6.2 shows a leveling network with the height difference measurements Δh_{ij} given in Table 6.3. Assume that all observations have the same standard deviation $\sigma_{\Delta h_{ij}} = 0.05$ m and points P1 and P5 are fixed with known elevations of $H_1 = 107.50$ m and $H_5 = 101.00$ m, respectively. Perform parametric least squares adjustment on the leveling network and compute the following:

a) Adjusted elevations of points P2, P3, and P4.
b) Observation residuals.
c) Adjusted height differences.

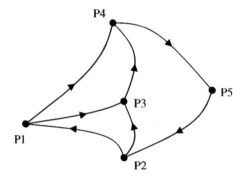

P4

P3

P5

P1

P2

Figure 6.2 Leveling net.

Table 6.3 Leveling net.

No.	i	j	Δh_{ij}
	P2	P1	1.34
	P1	P3	−5.00
	P1	P4	−2.25
	P2	P3	−3.68
	P5	P2	5.10
	P3	P4	2.70
	P4	P5	−4.13

Solution:

a) Adjusted elevations of points P2, P3, and P4 (note that hats are ignored for brevity):
    All the observations have the same standard deviations, $\sigma_{\Delta h_{ij}} = 0.05$ m, and points P1 and P2 are fixed with $H_1 = 107.50$ m and $H_5 = 101.00$ m. The parametric model equations ($\ell = f(x)$) are formulated as follows:

$$\Delta h_{21} = H_1 - H_2$$
$$\Delta h_{13} = H_3 - H_1$$
$$\Delta h_{14} = H_4 - H_1$$
$$\Delta h_{23} = H_3 - H_2$$
$$\Delta h_{52} = H_2 - H_5$$
$$\Delta h_{34} = H_4 - H_3$$
$$\Delta h_{45} = H_5 - H_4$$

The given observations (ℓ) and vector of parameters (x) are as follows:

$$\ell = \begin{bmatrix} \Delta h_{21} \\ \Delta h_{13} \\ \Delta h_{14} \\ \Delta h_{23} \\ \Delta h_{52} \\ \Delta h_{34} \\ \Delta h_{45} \end{bmatrix} \quad \text{or} \quad \ell = \begin{bmatrix} 1.34 \\ -5.00 \\ -2.25 \\ -3.68 \\ 5.10 \\ 2.70 \\ -4.13 \end{bmatrix} \quad x = \begin{bmatrix} H_2 \\ H_3 \\ H_4 \end{bmatrix}$$

Approximate values of parameters x^0:

$$x^0 = \begin{bmatrix} 107.50 - 1.34 \\ -5.00 + 107.50 \\ -2.25 + 107.50 \end{bmatrix} \quad \text{or} \quad x^0 = \begin{bmatrix} 106.16 \\ 102.50 \\ 105.25 \end{bmatrix}$$

The first design matrix A and the weight matrix P (a diagonal matrix) are given as follows:

$$A = \begin{bmatrix} \dfrac{\partial \Delta h_{21}}{\partial H_2} & \dfrac{\partial \Delta h_{21}}{\partial H_3} & \dfrac{\partial \Delta h_{21}}{\partial H_4} \\ \dfrac{\partial \Delta h_{13}}{\partial H_2} & \dfrac{\partial \Delta h_{13}}{\partial H_3} & \dfrac{\partial \Delta h_{13}}{\partial H_4} \\ \dfrac{\partial \Delta h_{14}}{\partial H_2} & \dfrac{\partial \Delta h_{14}}{\partial H_3} & \dfrac{\partial \Delta h_{14}}{\partial H_4} \\ \dfrac{\partial \Delta h_{23}}{\partial H_2} & \dfrac{\partial \Delta h_{23}}{\partial H_3} & \dfrac{\partial \Delta h_{23}}{\partial H_4} \\ \dfrac{\partial \Delta h_{52}}{\partial H_2} & \dfrac{\partial \Delta h_{52}}{\partial H_3} & \dfrac{\partial \Delta h_{52}}{\partial H_4} \\ \dfrac{\partial \Delta h_{34}}{\partial H_2} & \dfrac{\partial \Delta h_{34}}{\partial H_3} & \dfrac{\partial \Delta h_{34}}{\partial H_4} \\ \dfrac{\partial \Delta h_{45}}{\partial H_2} & \dfrac{\partial \Delta h_{45}}{\partial H_3} & \dfrac{\partial \Delta h_{45}}{\partial H_4} \end{bmatrix} \quad \text{or} \quad A = \begin{bmatrix} -1 & 0 & 0 \\ 0 & 1 & 0 \\ 0 & 0 & 1 \\ -1 & 1 & 0 \\ 1 & 0 & 0 \\ 0 & -1 & 1 \\ 0 & 0 & -1 \end{bmatrix}$$

$$P = \text{diag}[400 \ 400 \ 400 \ 400 \ 400 \ 400 \ 400]$$

Matrix of the coefficients of normal equations ($N = A^T PA$) and its inverse:

$$N = \begin{bmatrix} 1200 & -400 & 0 \\ -400 & 1200 & -400 \\ 0 & -400 & 1200 \end{bmatrix} \quad N^{-1} = \begin{bmatrix} 0.000\ 952 & 0.000\ 357 & 0.000\ 119 \\ 0.000\ 357 & 0.001\ 0714 & 0.000\ 357 \\ 0.000\ 119 & 0.000\ 357 & 0.000\ 952 \end{bmatrix}$$

The misclosure vector w and the product $A^T Pw$:

$$w = \begin{bmatrix} 0.00 \\ 0.00 \\ 0.00 \\ 0.02 \\ 0.06 \\ 0.05 \\ -0.12 \end{bmatrix} \quad A^T Pw = \begin{bmatrix} 16.0 \\ -12.0 \\ 68.0 \end{bmatrix}$$

Corrections to the approximate elevations:

$$\delta = -\left(A^T PA\right)^{-1} A^T Pw = \begin{bmatrix} -0.019 \\ -0.017 \\ -0.062 \end{bmatrix}$$

Adjusted parameters:

$$\hat{x} = x^0 + \delta = \begin{bmatrix} 106.16 \\ 102.50 \\ 105.25 \end{bmatrix} + \begin{bmatrix} -0.019 \\ -0.017 \\ -0.062 \end{bmatrix} \rightarrow \hat{x} = \begin{bmatrix} 106.141 \\ 102.483 \\ 105.188 \end{bmatrix}$$

b) Observation residuals:

$$v = A\delta + w = \begin{bmatrix} 0.019 \\ -0.017 \\ -0.062 \\ 0.002 \\ -0.019 \\ -0.045 \\ 0.062 \end{bmatrix} + \begin{bmatrix} 0.00 \\ 0.00 \\ 0.00 \\ 0.02 \\ 0.06 \\ 0.05 \\ -0.12 \end{bmatrix} \rightarrow \begin{bmatrix} 0.019 \\ -0.017 \\ -0.062 \\ 0.022 \\ 0.041 \\ 0.005 \\ -0.058 \end{bmatrix}$$

c) Adjusted height differences:

$$
\hat{\ell} = \ell + v =
\begin{bmatrix}
1.34 \\
-5.00 \\
-2.25 \\
-3.68 \\
5.10 \\
2.70 \\
-4.13
\end{bmatrix}
+
\begin{bmatrix}
0.019 \\
-0.017 \\
-0.062 \\
0.022 \\
0.041 \\
0.005 \\
-0.058
\end{bmatrix}
\rightarrow
\begin{bmatrix}
1.359 \\
-5.017 \\
-2.312 \\
-3.658 \\
5.141 \\
2.705 \\
-4.188
\end{bmatrix}
$$

6.2.2 Station Adjustment

Station adjustment is a procedure that consists of finding the adjusted values of angles measured at a survey station so that they are consistent with the associated geometry. The following example is given to illustrate the simplest form of station adjustment, which can be extended to other forms of station adjustments as illustrated in the following examples. For example, an angle was measured repeatedly over two nights. The arithmetic means of the measurements made in the two nights $(\bar{\ell}_1, \bar{\ell}_2)$ with the seconds given only and the numbers of repetitions (n_1, n_2) of measurements for each night are as follows:

Night 1 : $\bar{\ell}_1 = 40''$ $n_1 = 3$
Night 2 : $\bar{\ell}_2 = 45''$ $n_2 = 4$

Determine the least squares adjusted angle for the two nights using parametric least squares adjustment method with the numbers of repetitions considered as weights.

Solution:

Consider Figure 6.3 for the illustration of the problem in this example. The means of the measurements of the same angle (θ) are given as $\bar{\ell}_1$ and $\bar{\ell}_2$; there is now a need to identify what to refer to as parameters since the problem is not to be formulated as conditional model equations (equations involving only observations). Since directions are used in obtaining angles, let the two directions involved for this angle θ be d_1 (the backsight direction) and d_2 (the foresight direction). Since the backsight direction reading is usually set to zero for convenience, set d_1 to $0°$ and d_2 as the unknown direction (parameter) to be determined by the method of parametric least squares adjustment.

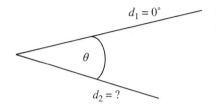

Figure 6.3 Approximate locations of total station directions.

Note that this approach can be extended to involve many angles measured about a point, as shown in the following examples.

Let the parameter be $x = [d_2]$:

$$\text{Observations}: \ell = \begin{bmatrix} \bar{\ell}_1 \\ \bar{\ell}_2 \end{bmatrix}$$

Formulate the parametric equations ($\ell = f(x)$) as follows. Since there are two independent observations involved, there will be two equations formed as follows (one equation for each observation with hats ignored for brevity):

$$\bar{\ell}_1 = d_2 - 0.0 \tag{6.4}$$

$$\bar{\ell}_2 = d_2 - 0.0 \tag{6.5}$$

The first design matrix (A) can be formed by finding the partial derivatives of Equations (6.4) and (6.5) with respect to the unknown parameter d_2 as follows:

$$A = \frac{\partial f}{\partial x} = \begin{bmatrix} \dfrac{\partial \bar{\ell}_1}{\partial d_2} \\ \dfrac{\partial \bar{\ell}_2}{\partial d_2} \end{bmatrix} \rightarrow A = \frac{\partial f}{\partial x} = \begin{bmatrix} 1 \\ 1 \end{bmatrix}$$

Let x^0 (approximate value of parameter) = 40″; substitute this value into Equations (6.4) and (6.5) to obtain

$$f(x^0) = \begin{bmatrix} 40 - 0 \\ 40 - 0 \end{bmatrix} \rightarrow \begin{bmatrix} 40 \\ 40 \end{bmatrix}$$

Form the misclosure vector w:

$$w = f(x^0) - \ell = \begin{bmatrix} 40 - 40 \\ 40 - 45 \end{bmatrix} \rightarrow w = \begin{bmatrix} 0 \\ -5 \end{bmatrix}$$

Since the question requires that the number of repetitions be taken as corresponding to the weight of each mean measurement, the weight matrix (*P*) of the observations can be given as

$$P = \begin{bmatrix} 3 & 0 \\ 0 & 4 \end{bmatrix}$$

Other products are as follows:

$$N = A^T P A \rightarrow 7$$
$$u = A^T P w \rightarrow -20$$

Vector of corrections to the approximate parameter:

$$\delta = -N^{-1}u \rightarrow \left(\frac{1}{7}\right) \times 20 = +2.86''$$

The adjusted direction $\hat{d}_2$:

$$\hat{d}_2 = x^0 + \delta$$

$$\hat{d}_2 = 40 + 2.86 \rightarrow 42.86''$$

The residual vector *v*: $v = A\delta + w$

$$v = \begin{bmatrix} 1 \\ 1 \end{bmatrix}[2.86] + \begin{bmatrix} 0 \\ -5 \end{bmatrix} \rightarrow v = \begin{bmatrix} 2.86 \\ -2.14 \end{bmatrix}$$

The adjusted angle: $\hat{\ell} = \ell + v$

$$\hat{\ell} = \begin{bmatrix} 40 \\ 45 \end{bmatrix} + \begin{bmatrix} 2.86 \\ -2.14 \end{bmatrix} \rightarrow \begin{bmatrix} 42.86 \\ 42.86 \end{bmatrix}$$

Adjusted angle $\theta = 42.86''$

Example 6.2   Three horizontal angles are measured with a total station instrument around the horizon at point A as shown in Figure 6.4. The measurements and their standard deviations are given in Table 6.4. Adjust the measurements using parametric least squares adjustment.

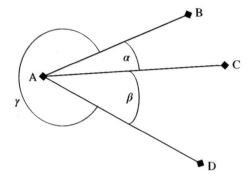

Figure 6.4 Horizontal angle measurements around the horizon.

Table 6.4 Horizontal angle measurements around the horizon.

Angle	Measurement	Standard deviation
α	42°12′13″	3″
β	59°56′15″	2″
γ	257°51′35″	4″

Solution:

If the total station direction A–B is taken as fixed value 0°00′00″, the other directions (A–C and A–D) are considered as the unknown parameters (x), and the measured horizontal angles are the observables (ℓ). Let the unknown parameters be A–C $= d_1$ and A–D $= d_2$. The parametric equations $(\hat{\ell} = f(\hat{x}))$ for the problem can be given as follows:

$$\hat{\alpha} = \hat{d}_1 - 0 \tag{6.6}$$

$$\hat{\beta} = \hat{d}_2 - \hat{d}_1 \tag{6.7}$$

$$\hat{\gamma} = 360° - \hat{d}_2 \tag{6.8}$$

A-matrix is derived from Equations (6.6)–(6.8) (with hats ignored for brevity) as follows:

$$A = \begin{bmatrix} \dfrac{\partial\alpha}{\partial d_1} & \dfrac{\partial\alpha}{\partial d_2} \\[1.5ex] \dfrac{\partial\beta}{\partial d_1} & \dfrac{\partial\beta}{\partial d_2} \\[1.5ex] \dfrac{\partial\gamma}{\partial d_1} & \dfrac{\partial\gamma}{\partial d_2} \end{bmatrix} \rightarrow A = \begin{bmatrix} 1 & 0 \\ -1 & 1 \\ 0 & -1 \end{bmatrix}$$

Use the given angle measurements to determine the initial directions (approximate parameters, x^0), and use them in Equations (6.6)–(6.8) to simulate the observations; the following are then obtained:

$$x^0 = \begin{bmatrix} 42°12'13'' \\ 102°8'28'' \end{bmatrix} \quad f(x^0) = \begin{bmatrix} 42°12'13'' \\ 59°56'15'' \\ 257°51'32'' \end{bmatrix}$$

Misclosure vector:

$$w = f(x^0) - \ell = \begin{bmatrix} 42°12'13'' \\ 59°56'15'' \\ 257°51'32'' \end{bmatrix} - \begin{bmatrix} 42°12'13'' \\ 59°56'15'' \\ 257°51'35'' \end{bmatrix}$$

$$w = \begin{bmatrix} 0 \\ 0 \\ -3'' \end{bmatrix}$$

Determine the weight matrix (P) using the given standard deviations (in arc seconds, since all observations are in angular units):

$$P = \begin{bmatrix} \frac{1}{9} & & \\ & \frac{1}{4} & \\ & & \frac{1}{16} \end{bmatrix} \quad \text{or} \quad P = \begin{bmatrix} 0.1111 & & \\ & 0.2500 & \\ & & 0.0625 \end{bmatrix}$$

Form the matrix of coefficients of normal equations (N) and its inverse:

$$N = A^T PA = \begin{bmatrix} 0.3611 & -0.2500 \\ -0.2500 & 0.3125 \end{bmatrix} \quad N^{-1} = \begin{bmatrix} 6.2069 & 4.9655 \\ 4.9655 & 7.1724 \end{bmatrix}$$

Parametric least squares adjusted parameters:

$$u = A^T Pw = \begin{bmatrix} 0.0 \\ 0.1875 \end{bmatrix}$$

Correction to approximate parameter:

$$\delta = -N^{-1} u = \begin{bmatrix} -0.93 \\ -1.34 \end{bmatrix}$$

Adjusted parameter: $\hat{x} = x^0 + \delta$

$$\hat{x} = \begin{bmatrix} 42°12'13'' \\ 102°8'28'' \end{bmatrix} + \begin{bmatrix} -0.93 \\ -1.34 \end{bmatrix} \rightarrow \begin{bmatrix} 42°12'12.1'' \\ 102°8'26.7'' \end{bmatrix}$$

Residual vector and the adjusted observations:
Residual vector: $v = A\delta + w$

$$v = \begin{bmatrix} 1 & 0 \\ -1 & 1 \\ 0 & -1 \end{bmatrix} \begin{bmatrix} -0.93 \\ -1.34 \end{bmatrix} + \begin{bmatrix} 0.0 \\ 0.0 \\ -3.0 \end{bmatrix} \rightarrow \begin{bmatrix} -0.93 \\ -0.41 \\ -1.66 \end{bmatrix}$$

Adjusted observations, which are the adjusted angle measurements: $\hat{\ell} = \ell + v$

$$\hat{\ell} = \begin{bmatrix} 42°12'13'' \\ 59°56'15'' \\ 257°51'32'' \end{bmatrix} + \begin{bmatrix} -0.93 \\ -0.41 \\ -1.66 \end{bmatrix} \rightarrow \begin{bmatrix} 42°12'12.1'' \\ 59°56'14.6'' \\ 257°51'30.3'' \end{bmatrix}$$

Example 6.3 Consider Table 6.5 and Figure 6.5 in which four sets of direction measurements are made from a theodolite station P to three target stations Q, R, and S. Each set consists of two measurements (made in face I and face II positions of the theodolite) with releveling and recentering and changing of zero graduation direction of theodolite made between sets. Determine the parametric least squares adjusted direction measurement to each target.

Table 6.5 Direction measurements from station P to stations Q, R, and S.

Set no.	Station sighted	Face I	Face II
1	Q	0°00'00''	180°00'05''
	R	37°30'27''	217°30'21''
	S	74°13'42''	254°13'34''
2	Q	90°00'00''	270°00'08''
	R	127°30'32''	307°30'28''
	S	164°13'48''	344°13'42''
3	Q	180°00'00''	0°00'15''
	R	217°30'26''	37°30'26''
	S	254°13'36''	74°13'40''
4	Q	270°00'00''	90°00'12''
	R	307°30'28''	127°30'30''
	S	344°13'45''	164°13'41''

Figure 6.5 Direction measurements (not to scale).

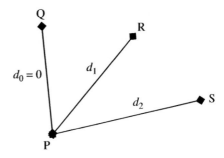

Solution:

Let the unknown orientation parameters for all the four setups (one per set) be $x_1 = \begin{bmatrix} \delta_0 \\ \delta_1 \\ \delta_2 \\ \delta_3 \end{bmatrix}$.

Given that the first direction is fixed and set to zero, let the unknown direction parameters be $x_2 = \begin{bmatrix} d_1 \\ d_2 \end{bmatrix}$; combined parameters, $x = \begin{bmatrix} \delta_0 & \delta_1 & \delta_2 & \delta_3 & d_1 & d_2 \end{bmatrix}^T$.

Observation vector: $\ell = \begin{bmatrix} \ell_1 & \ell_2 & \ell_3 & \ell_4 & \ell_5 & \ell_6 & \ell_7 & \ell_8 & \ell_9 & \ell_{10} & \ell_{11} & \ell_{12} \end{bmatrix}^T$.

The parametric equations are formulated as follows: $\hat{\ell} = f(\hat{x})$

Since there are 12 independent observations involved, there will be 12 equations formed as follows (one equation for each set, the mean of face I and face II measurements):

$$\hat{\ell}_1 = \hat{\delta}_0 + d_0$$

$$\hat{\ell}_2 = \hat{\delta}_0 + \hat{d}_1 \tag{6.9}$$

$$\hat{\ell}_3 = \hat{\delta}_0 + \hat{d}_2$$

$$\hat{\ell}_4 = \hat{\delta}_1 + d_0$$

$$\hat{\ell}_5 = \hat{\delta}_1 + \hat{d}_1 \tag{6.10}$$

$$\hat{\ell}_6 = \hat{\delta}_1 + \hat{d}_2$$

$$\hat{\ell}_7 = \hat{\delta}_2 + d_0$$

$$\hat{\ell}_8 = \hat{\delta}_2 + \hat{d}_1 \tag{6.11}$$

$$\hat{\ell}_9 = \hat{\delta}_2 + \hat{d}_2$$

$$\hat{\ell}_{10} = \hat{\delta}_3 + \hat{d}_0$$

$$\hat{\ell}_{11} = \hat{\delta}_3 + \hat{d}_1 \qquad\qquad (6.12)$$

$$\hat{\ell}_{12} = \hat{\delta}_3 + \hat{d}_2$$

Approximate parameters:

$$x^0 = \begin{bmatrix} \delta_0^0 \\ \delta_1^0 \\ \delta_2^0 \\ \delta_3^0 \\ d_1^0 \\ d_2^0 \end{bmatrix} = \begin{bmatrix} 0° \\ 90° \\ 180° \\ 270° \\ 37°30'27'' \\ 74°13'42'' \end{bmatrix}$$

The first design matrix A can be formed by finding the partial derivatives of Equations (6.9)–(6.12) (ignoring the hats for brevity) with respect to the unknown parameter vector x as follows:

$$A = \frac{\partial \ell}{\partial x} = \begin{bmatrix} 1 & 0 & 0 & 0 & 0 & 0 \\ 1 & 0 & 0 & 0 & 1 & 0 \\ 1 & 0 & 0 & 0 & 0 & 1 \\ 0 & 1 & 0 & 0 & 0 & 0 \\ 0 & 1 & 0 & 0 & 1 & 0 \\ 0 & 1 & 0 & 0 & 0 & 1 \\ 0 & 0 & 1 & 0 & 0 & 0 \\ 0 & 0 & 1 & 0 & 1 & 0 \\ 0 & 0 & 1 & 0 & 0 & 1 \\ 0 & 0 & 0 & 1 & 0 & 0 \\ 0 & 0 & 0 & 1 & 1 & 0 \\ 0 & 0 & 0 & 1 & 0 & 1 \end{bmatrix}$$

The $A^T A$ product and its inverse (assuming equal weights of unity for all the observations) can be given as follows:

$$A^T A = \begin{bmatrix} 3 & 0 & 0 & 0 & 1 & 1 \\ 0 & 3 & 0 & 0 & 1 & 1 \\ 0 & 0 & 3 & 0 & 1 & 1 \\ 0 & 0 & 0 & 3 & 1 & 1 \\ 1 & 1 & 1 & 1 & 4 & 0 \\ 1 & 1 & 1 & 1 & 0 & 4 \end{bmatrix} \quad (A^T A)^{-1} = \begin{bmatrix} 0.5 & 0.16 & 0.16 & 0.16 & -0.25 & -0.25 \\ 0.16 & 0.5 & 0.16 & 0.16 & -0.25 & -0.25 \\ 0.16 & 0.16 & 0.5 & 0.16 & -0.25 & -0.25 \\ 0.16 & 0.16 & 0.16 & 0.5 & -0.25 & -0.25 \\ -0.25 & -0.25 & -0.25 & -0.25 & 0.5 & 0.25 \\ -0.25 & -0.25 & -0.25 & -0.25 & 0.25 & 0.5 \end{bmatrix}$$

Form the misclosure vector w and the product $A^T w$ as follows:

$$w = f(x^0) - \ell = \begin{bmatrix} -2.5'' \\ 3.0'' \\ 4.0'' \\ -4.0'' \\ -3.0'' \\ -3.0'' \\ -7.5'' \\ 1.0'' \\ 4.0'' \\ -6.0'' \\ -2.0'' \\ -1.0'' \end{bmatrix} \quad A^T w = \begin{bmatrix} 4.5'' \\ -10.0'' \\ -2.5'' \\ -9.0'' \\ -1.0'' \\ 4.0'' \end{bmatrix}$$

The vector of corrections (δ) to the approximate parameters is solved for as follows:

$$\delta = -(A^T A)^{-1} A^T w = \begin{bmatrix} 2.1'' \\ 6.9'' \\ 4.4'' \\ 6.6'' \\ -4.8'' \\ -6.0'' \end{bmatrix}$$

The adjusted direction parameters (excluding the orientation parameters): $\hat{x} = x^0 + \delta$

$$\hat{d}_1 = d_1^0 + \hat{d}_1 \quad \rightarrow \quad \hat{d}_1 = 37°30'27'' - 4.8'' \ (\text{or } 37°30'22.2'')$$
$$\hat{d}_2 = d_2^0 + \hat{d}_2 \quad \rightarrow \quad \hat{d}_2 = 74°13'42'' - 6.0'' (\text{or } 74°13'36.0'')$$

As can be seen in the above results, the adjusted direction values are identical to those obtained based on the approach used in Example 1.3. Hence, that adjustment approach is based on the least squares method.

6.2.3 Traverse Adjustment

A simple traverse adjustment is used in this section to illustrate how to adjust a traverse network using parametric least squares method. The approach can be extended to a more complex traverse network with the steps given in this section applicable in the same way. For illustration, the 2-D fitted traverse network in Figure 6.6 with field measurements given in Table 6.6 is to be adjusted using parametric least squares adjustment method. The measurements consist of two angles, two bearings, and three distances with their standard deviations provided. The coordinates of the control points B and E (to be fixed in the adjustment) and the approximate coordinates of the unknown points C and D to be determined are given in Table 6.7. Compute the adjusted coordinates of points C and D.

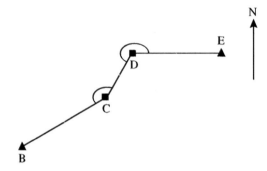

N **Figure 6.6** Two-dimensional traverse net.

Table 6.6 Field measurements.

Symbol	At	From	To	Measurement	Standard deviation
ℓ_1	C	B	D	149°59′45″	10″
ℓ_2	D	C	E	240°01′00″	10″
ℓ_3		D	E	90°00′00″	2″
ℓ_4		B	C	59°59′15″	2″
ℓ_5		B	C	199.880 m	3 mm + 2 ppm
ℓ_6		C	D	99.900 m	0.005 m
ℓ_7		D	E	177.000 m	0.005 m

Table 6.7 Coordinates of control points and approximate coordinates of unknown points.

Station	X (m)	Y (m)
B	1000.000	1000.000
E	1400.000	1186.500
C	1173	1100
D	1223	1186

Parametric model equations ($\hat{\ell} = f(\hat{x})$) are formulated in the order in which the observations are given in Table 6.6 as follows:

$$\hat{\ell}_1 = \text{atan}\left[\frac{(\hat{X}_D - \hat{X}_C)}{(\hat{Y}_D - \hat{Y}_C)}\right] - \text{atan}\left[\frac{(X_B - \hat{X}_C)}{(Y_B - \hat{Y}_C)}\right] \tag{6.13}$$

$$\hat{\ell}_2 = \text{atan}\left[\frac{(X_E - \hat{X}_D)}{(Y_E - \hat{Y}_D)}\right] + 180° - \text{atan}\left[\frac{(\hat{X}_C - \hat{X}_D)}{(\hat{Y}_C - \hat{Y}_D)}\right] \tag{6.14}$$

$$\hat{\ell}_3 = \text{atan}\left[\frac{(X_E - \hat{X}_D)}{(Y_E - \hat{Y}_D)}\right] \tag{6.15}$$

$$\hat{\ell}_4 = \text{atan}\left[\frac{(\hat{X}_C - X_B)}{(\hat{Y}_C - Y_B)}\right] \tag{6.16}$$

$$\hat{\ell}_5 = \left[(\hat{X}_C - X_B)^2 + (\hat{Y}_C - Y_B)^2\right]^{1/2} \tag{6.17}$$

$$\hat{\ell}_6 = \left[(\hat{X}_D - \hat{X}_C)^2 + (\hat{Y}_D - \hat{Y}_C)^2\right]^{1/2} \tag{6.18}$$

$$\hat{\ell}_7 = \left[(X_E - \hat{X}_D)^2 + (Y_E - \hat{Y}_D)^2\right]^{1/2} \tag{6.19}$$

Substitute the coordinate values in Table 6.7 into Equations (6.13)–(6.19) to obtain the simulated (or derived) observations $\ell^0 = f(x^0)$ (with the angular values in radians). The derived observations ($f(x^0)$) and the actual observations (ℓ) are given, respectively, as follows:

$$f(x^0) = \begin{bmatrix} 2.621\ 535\ 53 \\ 4.182\ 936\ 86 \\ 1.567\ 971\ 48 \\ 1.046\ 684\ 4 \\ 199.822\ 4 \\ 99.478\ 6 \\ 177.000\ 7 \end{bmatrix} \quad \ell = \begin{bmatrix} 2.617\ 921\ 156 \\ 4.189\ 081\ 093 \\ 1.570\ 796\ 327 \\ 1.046\ 979\ 39 \\ 199.880 \\ 99.900 \\ 177.000 \end{bmatrix} \tag{6.20}$$

The misclosure vector is given as

$$w = f(x^0) - \ell = \begin{bmatrix} 3.614\ 375\text{E}-3 \\ -6.144\ 235\text{E}-3 \\ -2.824\ 851\text{E}-3 \\ -2.949\ 914\text{E}-4 \\ -5.757\ 884\text{E}-2 \\ -4.213\ 591\text{E}-1 \\ 7.062\ 133\text{E}-4 \end{bmatrix}$$

The diagonal values of the weight matrix (P) of the observations using the given standard deviations (with the standard deviations of angles converted to radians) are given as follows:

$$P = \text{diag}[4.254\text{E}8 \quad 4.254\text{E}8 \quad 1.064\text{E}10 \quad 1.064\text{E}10 \quad 8.651\text{E}10 \quad 4.0\text{E}4 \quad 4.0\text{E}4]$$

First design (A) matrix, which is the Jacobian matrix of Equations (6.13)–(6.19) (ignoring the hats for brevity) with respect to the unknown parameters, $x = [X_C \ Y_C \ X_D \ Y_D]^T$, is obtained as follows. In the following partial derivatives, the derived values of the measured distances ($\ell_5^0, \ell_6^0, \ell_7^0$) are obtained from $f(x^0)$ given in Equation (6.20). The calculated values are $\ell_5^0 = 199.8224$ m, $\ell_6^0 = 99.4786$ m, and $\ell_7^0 = 177.0007$ m:

$$\frac{\partial \ell_1}{\partial X_C} = \frac{-(Y_D - Y_C)}{(\ell_6^0)^2} + \frac{(Y_B - Y_C)}{(\ell_5^0)^2} = -0.011\ 195;$$

$$\frac{\partial \ell_1}{\partial Y_C} = \frac{(X_D - X_C)}{(\ell_6^0)^2} - \frac{(X_B - X_C)}{(\ell_5^0)^2} = 0.009\ 385\ 2;$$

$$\frac{\partial \ell_1}{\partial X_D} = \frac{(Y_D - Y_C)}{(\ell_6^0)^2} = 0.008\ 690\ 4; \quad \frac{\partial \ell_1}{\partial Y_D} = \frac{-(X_D - X_C)}{(\ell_6^0)^2} = -0.005\ 052\ 5;$$

$$\frac{\partial \ell_2}{\partial X_C} = \frac{-(Y_C - Y_D)}{(\ell_6^0)^2} = 0.008\ 690\ 4; \quad \frac{\partial \ell_2}{\partial Y_C} = \frac{(X_C - X_D)}{(\ell_6^0)^2} = -0.005\ 052\ 55;$$

$$\frac{\partial \ell_2}{\partial X_D} = \frac{-(Y_E - Y_D)}{(\ell_7^0)^2} + \frac{(Y_C - Y_D)}{(\ell_6^0)^2} = -0.008\ 706\ 3;$$

$$\frac{\partial \ell_2}{\partial Y_D} = \frac{(X_E - X_D)}{(\ell_7^0)^2} - \frac{(X_C - X_D)}{(\ell_6^0)^2} = 0.010\ 702\ 2;$$

$$\frac{\partial \ell_3}{\partial X_C} = 0.0 \quad \frac{\partial \ell_3}{\partial Y_C} = 0.0$$

$$\frac{\partial \ell_3}{\partial X_D} = \frac{-(Y_E - Y_D)}{\left(\ell_7^0\right)^2} = -0.000\ 015\ 96 \quad \frac{\partial \ell_3}{\partial Y_D} = \frac{(X_E - X_D)}{\left(\ell_7^0\right)^2} = 0.005\ 649\ 7;$$

$$\frac{\partial \ell_4}{\partial X_C} = \frac{(Y_C - Y_B)}{\left(\ell_5^0\right)^2} = 0.002\ 504\ 45; \quad \frac{\partial \ell_4}{\partial Y_C} = \frac{-(X_C - X_B)}{\left(\ell_5^0\right)^2} = -0.004\ 332\ 7;$$

$$\frac{\partial \ell_4}{\partial X_D} = 0.0; \quad \frac{\partial \ell_4}{\partial Y_D} = 0.0$$

$$\frac{\partial \ell_5}{\partial X_C} = \frac{(X_C - X_B)}{\ell_5^0} = 0.865\ 768\ 7; \quad \frac{\partial \ell_5}{\partial Y_C} = \frac{(Y_C - Y_B)}{\ell_5^0} = 0.500\ 444\ 3;$$

$$\frac{\partial \ell_5}{\partial X_D} = 0.0; \quad \frac{\partial \ell_5}{\partial Y_D} = 0.0$$

$$\frac{\partial \ell_6}{\partial X_C} = \frac{-(X_D - X_C)}{\ell_6^0} = -0.502\ 620\ 46; \quad \frac{\partial \ell_6}{\partial Y_C} = \frac{-(Y_D - Y_C)}{\ell_6^0} = -0.864\ 507\ 19;$$

$$\frac{\partial \ell_6}{\partial X_D} = \frac{(X_D - X_C)}{\ell_6^0} = 0.502\ 620\ 5; \quad \frac{\partial \ell_6}{\partial Y_D} = \frac{(Y_D - Y_C)}{\ell_6^0} = 0.864\ 507\ 2;$$

$$\frac{\partial \ell_7}{\partial X_C} = 0.0; \quad \frac{\partial \ell_7}{\partial Y_C} = 0.0$$

$$\frac{\partial \ell_7}{\partial X_D} = \frac{-(X_E - X_D)}{\ell_7^0} = -1.000; \quad \frac{\partial \ell_7}{\partial Y_D} = \frac{-(Y_E - Y_D)}{\ell_7^0} = -0.002\ 825$$

The A-matrix is given as follows:

$$A = \begin{bmatrix} -0.011\ 195 & 0.009\ 385 & 0.008\ 690 & -0.005\ 053 \\ 0.008\ 690 & -0.005\ 053 & -0.008\ 706 & 0.010\ 702 \\ 0.0 & 0.0 & -0.000\ 016 & 0.005\ 650 \\ 0.002\ 504 & -0.004\ 333 & 0.0 & 0.0 \\ 0.865\ 769 & 0.500\ 444 & 0.0 & 0.0 \\ -0.502\ 620 & -0.864\ 507 & 0.502\ 621 & 0.864\ 507 \\ 0.0 & 0.0 & -1.000\ 0 & -0.002\ 825 \end{bmatrix}$$

The matrix of coefficients of normal equations (N) is given as

$$N = A^T PA = \begin{bmatrix} 2.2711 & -1.2394 & -0.8369 & 0.4625 \\ -1.2394 & 2.9956 & 0.3604 & -0.7308 \\ -0.8369 & 0.3604 & 1.1449 & -0.4179 \\ 0.4625 & -0.7308 & -0.4179 & 4.2899 \end{bmatrix} \times E+5$$

$$N^{-1} = \begin{bmatrix} 7.528 & 2.567 & 4.726 & 0.086 \\ 2.567 & 4.4526 & 0.6747 & 0.5474 \\ 4.726 & 0.6747 & 12.2689 & 0.8010 \\ 0.086 & 0.5474 & 0.8010 & 2.493 \end{bmatrix} \times E-6$$

$$u = A^T Pw = \begin{bmatrix} -0.43630 \\ 0.53312 \\ 0.28103 \\ -2.2007 \end{bmatrix} \times E+5$$

The vector of corrections to the approximate coordinates is given as

$$\delta = -N^{-1}u = \begin{bmatrix} 0.0777 \\ -0.0239 \\ 0.0016 \\ 0.5007 \end{bmatrix}$$

The adjusted coordinates of points C and D are

$$\hat{x} = x^0 + \delta = \begin{bmatrix} 1173 \\ 1100 \\ 1223 \\ 1186 \end{bmatrix} + \begin{bmatrix} 0.0777 \\ -0.0239 \\ 0.0016 \\ 0.5007 \end{bmatrix} = \begin{bmatrix} 1173.0777 \\ 1099.9761 \\ 1223.0016 \\ 1186.5007 \end{bmatrix} \text{ m}$$

The adjusted coordinates of points C and D from $\hat{x}$ are given in Table 6.8.

Table 6.8 Adjusted coordinates of points C and D.

Station	X (m)	Y (m)
C	1173.0777	1099.9761
D	1223.0016	1186.5007

Example 6.4 In order to determine the coordinates of point B (x, y) in Figure 6.7, the following measurements were made: the angle θ as $60°00'00''$ $\pm 5''$, the distance S_{AB} as 89.50 ± 0.01 m, and the distance S_{CB} as 601.00 ± 0.05 m.

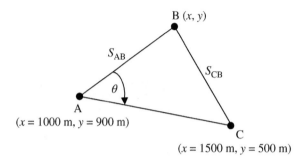

Figure 6.7 A simple traverse network.

a) Formulate the functional model $(\hat{\ell} = f(\hat{x}))$ relating the measurements with the unknown coordinates of point B.

Solution:

$$\text{Parameter} : x = \begin{bmatrix} x \\ y \end{bmatrix} \quad \text{Observations} : \ell = \begin{bmatrix} \theta \\ S_{AB} \\ S_{CB} \end{bmatrix}$$

The parametric model equations $(\hat{\ell} = f(\hat{x}))$ are formulated as follows:

$$\hat{\theta} = \operatorname{atan}\left(\frac{1500-1000}{500-900}\right) + 180° - a\tan\left(\frac{\hat{x}-1000}{\hat{y}-900}\right) \tag{6.21}$$

$$\hat{S}_{AB} = \sqrt{(\hat{x}-1000)^2 + (\hat{y}-900)^2} \tag{6.22}$$

$$\hat{S}_{CB} = \sqrt{(\hat{x}-1500)^2 + (\hat{y}-500)^2} \tag{6.23}$$

b) Form the first design matrix, A, assuming the approximate coordinates (x^0) of point B are $x = 1083$ m and $y = 933$ m.

Solution: (Note: The following A-matrix is evaluated at the approximate coordinate values of point B)

$$A = \frac{\partial f}{\partial x} \begin{bmatrix} \dfrac{-(\hat{y}-900)}{(\hat{x}-1000)^2 + (\hat{y}-900)^2} & \dfrac{(\hat{x}-1000)}{\sqrt{(\hat{x}-1000)^2 + (\hat{y}-900)^2}} \\[4mm] \dfrac{(\hat{x}-1000)}{\sqrt{(\hat{x}-1000)^2 + (\hat{y}-900)^2}} & \dfrac{(\hat{y}-900)}{\sqrt{(\hat{x}-1000)^2 + (\hat{y}-900)^2}} \\[4mm] \dfrac{(\hat{x}-1500)}{\sqrt{(\hat{x}-1500)^2 + (\hat{y}-500)^2}} & \dfrac{(\hat{y}-500)}{\sqrt{(\hat{x}-1500)^2 + (\hat{y}-500)^2}} \end{bmatrix} \qquad (6.24)$$

$$A = \begin{bmatrix} -4.136\ 38\mathrm{E}{-3} & 1.040\ 36\mathrm{E}{-2} \\ 0.929\ 247 & 0.369\ 460 \\ -0.693\ 674 & 0.720\ 289 \end{bmatrix}$$

The MATLAB code for calculating the *A*-matrix in Equation (6.24) is given in Table 6.9.

Table 6.9 MATLAB code for calculating the *A*-matrix in the traverse.

```
>> % Use syms to define variables x and y
>> syms x y
>> % form parametric Equations (6.21) to (6.23)
>> theta =atan((1500-1000)/(500-900))+pi-atan((x-1000)/
(y-900));
>> SAB=sqrt((x-1000)^2+(y-900)^2);
>> SCB=sqrt((x-1500)^2+(y-500)^2);
>> % do the Jacobian of the equations and evaluate as A matrix
>> J=jacobian([theta;SAB;SCB],[x y]);
>> format long
>> x=1083.0;y=933.0;
>> A=eval(J);
>>
```

c) Using the approximate coordinates of point B $\left(x^0 = \begin{bmatrix} x = 1083 \\ y = 933 \end{bmatrix}\right)$, form the vector of the values of the functions of the approximate parameters, $f(x^0)$.

Solution (substitute approximate values into Equations (6.21)–(6.23)):

$$\theta^0 = 128°39'35'' - 68°19'04'' \rightarrow 60°20'31'' (1.053\ 165\ 61\ \text{rad}) \qquad (6.25)$$

$$S_{AB}^0 = \sqrt{(x - 1000)^2 + (y - 900)^2} \rightarrow 89.320\ \text{m} \qquad (6.26)$$

$$S_{CB}^0 = \sqrt{(x - 1500)^2 + (y - 500)^2} \rightarrow 601.147\ \text{m} \qquad (6.27)$$

$$f(x^0) = \begin{bmatrix} 1.053\ 165\ 61\ \text{rad} \\ 89.320\ \text{m} \\ 601.147\ \text{m} \end{bmatrix} \quad \ell = \begin{bmatrix} 1.047\ 197\ 551\ 19\ \text{rad} \\ 89.50\ \text{m} \\ 601.10\ \text{m} \end{bmatrix}$$

Misclosure vector: $w = f(x^0) - \ell$

$$w = \begin{bmatrix} 0.005\ 968\ 058\ 81\ \text{rad} \\ -0.18\ \text{m} \\ 0.047\ \text{m} \end{bmatrix} \qquad (6.28)$$

The MATLAB code for calculating the misclosure in Equation (6.28) is given in Table 6.10. This code is used together with the code in Table 6.9.

Table 6.10 MATLAB code to calculate the misclosure vector *w*.

```
>> % Continue from the code in Table 6.9 and evaluate computed
observations L0 and misclosure w
>> L0=eval([theta;SAB;SCB])
L0 =
   1.0e+02 *
   0.010531683202522
   0.893196506934504
   6.011472365402673
>> L=[60*pi/180; 89.50; 601.00];
>> w=L0-L
w =
   0.005970769055632
  -0.180349306549573
   0.147236540267272
>>
```

d)   Calculate the weight matrix of the observations.

   Solution:

$$P = C_\ell^{-1} = \begin{bmatrix} \dfrac{1}{5.876\ 107\ 634\ 7\mathrm{E}-10} & 0 & 0 \\ 0 & \dfrac{1}{0.0001} & 0 \\ 0 & 0 & \dfrac{1}{0.0025} \end{bmatrix}$$

$$P = C_\ell^{-1} = \begin{bmatrix} 1\ 701\ 806\ 811.84 & 0 & 0 \\ 0 & 10\ 000 & 0 \\ 0 & 0 & 400 \end{bmatrix}$$

   The MATLAB code for calculating the weight matrix is given in Table 6.11.

Table 6.11 MATLAB code for weight matrix *P*.

```
>> % Form the weight matrix P
>> P=inv(diag([(5/206265)^2 (.01)^2
(0.05)^2]))
P =
   1.0e+09 *
   1.701810009000000                                0
0
               0    0.000010000000000
0
               0                0
0.000000400000000
>>
```

e)   Calculate the adjusted values of the unknown parameters.

   Solution:

   Calculate the matrix of coefficients of normal equations (*N*) and its inverse:

$$N = A^T P A = \begin{bmatrix} 37\ 944.774\ 374\ 1 & -70\ 000.928\ 353\ 2 \\ -70\ 000.928\ 353\ 2 & 185\ 767.411\ 532 \end{bmatrix}$$

$$N^{-1} = \begin{bmatrix} 8.645\ 280\ 388\mathrm{E}-5 & 3.257\ 716\ 991\ 6\mathrm{E}-5 \\ 3.257\ 716\ 991\ 6\mathrm{E}-5 & 1.765\ 881\ 383\ 71\mathrm{E}-5 \end{bmatrix}$$

The other product:

$$u = A^T P w = \begin{bmatrix} -43\ 724.444\ 330\ 2 \\ 105\ 041.156\ 97 \end{bmatrix}$$

The vector of corrections (δ) to the approximate parameters:

$$\delta = -N^{-1}u = \begin{bmatrix} 0.3582 \\ -0.4305 \end{bmatrix} m$$

The adjusted coordinates of point B:

$$\hat{x} = x^0 + \delta = \begin{bmatrix} \hat{x} \\ \hat{y} \end{bmatrix} = \begin{bmatrix} 1083.000 \\ 933.000 \end{bmatrix} + \begin{bmatrix} 0.3582 \\ -0.4305 \end{bmatrix} = \begin{bmatrix} \hat{x} \\ \hat{y} \end{bmatrix} = \begin{bmatrix} 1083.358 \\ 932.570 \end{bmatrix} m$$

The MATLAB code for completing the calculation of every element in question (e) is given in Table 6.12.

Table 6.12 MATLAB code for the adjusted coordinates computation.

```
>> % Continuing from the codes in Tables 6.9 – 6.11 determine
solution vector d
>> N=A'*P*A
N =
  1.0e+05 *
  0.379447553016646   -0.700010525582975
 -0.700010525582975    1.857681061924442
>> u=A'*P*w
u =
  1.0e+05 *
 -0.437469242869248
  1.050883754955768
>> d=-inv(N)*u
d =
  0.358569563656848
 -0.430580524631959
>> % determine the adjusted parameters xa and ya
>> xa=x+d(1)
xa =
     1.083358569563657e+03
>> ya=y+d(2)
ya =
     9.325694194753680e+02
>>
```

f) Calculate the adjusted observations.

Solution:

The residual vector:

$$v = A\delta + w = \begin{bmatrix} -5.960\ 053\ 256\ 25\text{E}-3 \\ 0.173\ 77 \\ -0.558\ 5 \end{bmatrix} + \begin{bmatrix} +5.968\ 05\text{E}-3\ \text{rad} \\ -0.18\ \text{m} \\ +0.147\ \text{m} \end{bmatrix}$$

$$v = \begin{bmatrix} +7.996\ 743\ 75\text{E}-6\ \text{rad}(1.65'') \\ -0.006\ 2\ \text{m} \\ -0.411\ 5\ \text{m} \end{bmatrix}$$

The adjusted observations:

$$\hat{\ell} = \ell + v = \begin{bmatrix} 60°00'00'' \\ 89.500 \\ 601.000 \end{bmatrix} + \begin{bmatrix} +7.996\ 743\ 75\text{E}-6\ \text{rad}(1.65'') \\ -0.006\ 2\ \text{m} \\ -0.411\ 5\ \text{m} \end{bmatrix} \rightarrow \begin{bmatrix} 60°00'02'' \\ 89.494 \\ 600.589 \end{bmatrix}$$

The MATLAB code for computing every element under question (f) is given in Table 6.13.

Table 6.13 MATLAB code for adjusted observations computation.

```
>> % Continuing from Table 6.13 determine residuals v and
adjusted observations La
>> v=A*d +w
v =
   0.000007999044961
  -0.006231782019172
  -0.411636319473537
>> La=L+v
La =
   1.0e+02 *
   0.010472055502416 rad
   0.894937682179808
   6.005883636805264
>>
```

6.2.4 Triangulateration Adjustment

A triangulateration network is a network of triangles that combines triangulation and trilateration networks in a single network. In this case, all the angles in the triangles making up the network are measured as in triangulation survey, and the lengths of all the sides of the triangles are measured as in the case of trilateration survey. In order to illustrate how parametric least squares adjustment method can be used to adjust a triangulateration network, a simple triangle is used with the understanding that the procedure will be the same when a more complex network is involved. To illustrate the procedure, a simple triangle given in Figure 6.8, consisting of three angles and three distances, was measured and given in Table 6.14. For minimal constraint parametric adjustment, the coordinates of point A will be fixed, and the azimuth of line A–C is measured to provide orientation for the network. The coordinates of point A and the approximate coordinates of the unknown points B and C are given in Table 6.15. The adjusted coordinates of points B and C are to be determined by the method of parametric least squares adjustment.

Figure 6.8 Sample triangulateration network.

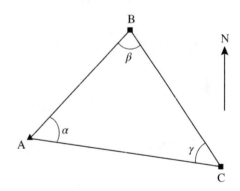

Table 6.14 Field measurements.

Symbol	At	From	To	Measurement	Standard deviation
ℓ_{AB}		A	B	533.808 m	0.003 m
ℓ_{BC}		B	C	592.215 m	0.003 m
ℓ_{CA}		C	A	710.095 m	0.003 m
α	A	B	C	54°39′50″	4″
β	B	C	A	78°00′03″	4″
γ	C	A	B	47°20′05″	4″
Br_{AC}		A	C	98°15′10″	5″

Table 6.15 Coordinates of control point A and approximate coordinates of unknown points B and C.

Station	X (m)	Y (m)
A	1000.000	1000.000
B	1367.720	1387.390
C	1702.150	898.160

The parametric model equations $(\hat{\ell} = f(\hat{x}))$ are formulated as follows in the order in which the observations are given in Table 6.14:

$$\hat{\ell}_{AB} = \left[\left(\hat{X}_B - X_A \right)^2 + \left(\hat{Y}_B - Y_A \right)^2 \right]^{1/2} \tag{6.29}$$

$$\hat{\ell}_{BC} = \left[\left(\hat{X}_C - \hat{X}_B \right)^2 + \left(\hat{Y}_C - \hat{Y}_B \right)^2 \right]^{1/2} \tag{6.30}$$

$$\hat{\ell}_{CA} = \left[\left(X_A - \hat{X}_C \right)^2 + \left(Y_A - \hat{Y}_C \right)^2 \right]^{1/2} \tag{6.31}$$

$$\hat{\alpha} = \text{atan} \left[\frac{\left(\hat{X}_C - X_A \right)}{\left(\hat{Y}_C - Y_A \right)} \right] + 180° - \text{atan} \left[\frac{\left(\hat{X}_B - X_A \right)}{\left(\hat{Y}_B - Y_A \right)} \right] \tag{6.32}$$

$$\hat{\beta} = \text{atan} \left[\frac{\left(X_A - \hat{X}_B \right)}{\left(Y_A - \hat{Y}_B \right)} \right] - \text{atan} \left[\frac{\left(\hat{X}_C - \hat{X}_B \right)}{\left(\hat{Y}_C - \hat{Y}_B \right)} \right] \tag{6.33}$$

$$\hat{\gamma} = a \tan \left[\frac{\left(\hat{X}_B - \hat{X}_C \right)}{\left(\hat{Y}_B - \hat{Y}_C \right)} \right] - a \tan \left[\frac{\left(X_A - \hat{X}_C \right)}{\left(Y_A - \hat{Y}_C \right)} \right] \tag{6.34}$$

$$\hat{B}r_{AC} = \text{atan} \left[\frac{\left(\hat{X}_C - X_A \right)}{\left(\hat{Y}_C - Y_A \right)} \right] + 180° \tag{6.35}$$

Substitute the values in Table 6.15 into Equations (6.29)–(6.35) to obtain the derived observations $\ell^0 = f(x^0)$ (with the angular values in radians). The derived observations $(f(x^0))$ and the actual observations (ℓ) are given, respectively, as follows:

$$f(x^0) = \begin{bmatrix} 534.124\ 5 \\ 592.612\ 4 \\ 709.497\ 0 \\ 0.955\ 477\ 29 \\ 1.358\ 978\ 49 \\ 0.827\ 136\ 88 \\ 1.714\ 832\ 15 \end{bmatrix} \quad \ell = \begin{bmatrix} 533.808 \\ 592.215 \\ 710.095 \\ 0.954\ 065 \\ 1.361\ 371 \\ 0.826\ 147 \\ 1.715\ 271 \end{bmatrix}$$

The misclosure vector is given as

$$w = f(x^0) - \ell = \begin{bmatrix} 0.316\ 527 \\ 0.397\ 367 \\ -0.597\ 989 \\ 1.412\ 442\text{E}-3 \\ -2.392\ 87\text{E}-3 \\ 9.901\ 27\text{E}-4 \\ -4.386\ 504\text{E}-4 \end{bmatrix}$$

The diagonal values of the weight matrix (P) of the observations using the given standard deviations (with the standard deviations of angles converted to radians) are given as follows:

$$P = \text{diag}[1.1\text{E}5 \quad 1.1\text{E}5 \quad 1.1\text{E}5 \quad 2.66\text{E}9 \quad 2.66\text{E}9 \quad 2.66\text{E}9 \quad 1.70\text{E}9]$$

The A-matrix of Equations (6.29)–(6.35) (ignoring hats for brevity) derived with respect to the unknown parameters, $x = [X_B \ Y_B \ X_C \ Y_C]^T$, is obtained as follows. In the following partial derivatives, the derived values of the measured distances (ℓ^0_{AB}, ℓ^0_{BC}, ℓ^0_{CA}) are obtained from the $f(x^0)$. The calculated values are $\ell^0_{AB} = 562.0595$ m, $\ell^0_{BC} = 623.8324$ m, and $\ell^0_{CA} = 746.8338$ m:

$$\frac{\partial \ell_{AB}}{\partial X_B} = \frac{(X_B - X_A)}{\ell^0_{AB}} = 0.688\ 454; \quad \frac{\partial \ell_{AB}}{\partial Y_B} = \frac{(Y_B - Y_A)}{\ell^0_{AB}} = 0.725\ 280$$

$$\frac{\partial \ell_{AB}}{\partial X_C} = 0.0; \quad \frac{\partial \ell_{AB}}{\partial Y_C} = 0.0$$

$$\frac{\partial \ell_{BC}}{\partial X_B} = \frac{-(X_C - X_B)}{\ell^0_{BC}} = -0.564\ 332; \quad \frac{\partial \ell_{BC}}{\partial Y_B} = \frac{-(Y_C - Y_B)}{\ell^0_{BC}} = 0.825\ 548$$

$$\frac{\partial \ell_{BC}}{\partial X_C} = \frac{(X_C - X_B)}{\ell^0_{BC}} = 0.564\ 332; \quad \frac{\partial \ell_{BC}}{\partial Y_C} = \frac{(Y_C - Y_B)}{\ell^0_{BC}} = -0.825\ 548$$

$$\frac{\partial \ell_{CA}}{\partial X_B} = 0.0; \quad \frac{\partial \ell_{CA}}{\partial Y_B} = 0.0;$$

$$\frac{\partial \ell_{CA}}{\partial X_C} = \frac{-(X_A - X_C)}{\ell^0_{CA}} = 0.989\ 645; \quad \frac{\partial \ell_{CA}}{\partial Y_C} = \frac{-(Y_A - Y_C)}{\ell^0_{CA}} = -0.143\ 538\ 3$$

$$\frac{\partial \alpha}{\partial X_B} = \frac{-(Y_B - Y_A)}{(\ell^0_{AB})^2} = -1.357\ 886\text{E}-3; \quad \frac{\partial \alpha}{\partial Y_B} = \frac{(X_B - X_A)}{(\ell^0_{AB})^2} = 1.288\ 939\text{E}-3;$$

$$\frac{\partial \alpha}{\partial X_C} = \frac{(Y_C - Y_A)}{\left(\ell_{CA}^0\right)^2} = -2.023\ 099E-4; \quad \frac{\partial \alpha}{\partial Y_C} = \frac{-(X_C - X_A)}{\left(\ell_{CA}^0\right)^2} = -1.394\ 854E-3;$$

$$\frac{\partial \beta}{\partial X_B} = \frac{-(Y_A - Y_B)}{\left(\ell_{AB}^0\right)^2} + \frac{(Y_C - Y_B)}{\left(\ell_{BC}^0\right)^2} = -3.517\ 973E-5;$$

$$\frac{\partial \beta}{\partial Y_B} = \frac{(X_A - X_B)}{\left(\ell_{AB}^0\right)^2} - \frac{(X_C - X_B)}{\left(\ell_{BC}^0\right)^2} = -2.241\ 217E-3;$$

$$\frac{\partial \beta}{\partial X_C} = \frac{-(Y_C - Y_B)}{\left(\ell_{BC}^0\right)^2} = 1.393\ 065\ 89E-3; \quad \frac{\partial \beta}{\partial Y_C} = \frac{(X_C - X_B)}{\left(\ell_{BC}^0\right)^2} = 9.522\ 781E-4;$$

$$\frac{\partial \gamma}{\partial X_B} = \frac{(Y_B - Y_C)}{\left(\ell_{BC}^0\right)^2} = 1.393\ 065\ 9E-3; \quad \frac{\partial \gamma}{\partial Y_B} = \frac{-(X_B - X_C)}{\left(\ell_{BC}^0\right)^2} = 9.522\ 781E-4;$$

$$\frac{\partial \gamma}{\partial X_C} = \frac{(Y_A - Y_C)}{\left(\ell_{AC}^0\right)^2} - \frac{(Y_B - Y_C)}{\left(\ell_{BC}^0\right)^2} = -1.190\ 755\ 9E-3;$$

$$\frac{\partial \gamma}{\partial Y_C} = \frac{-(X_A - X_C)}{\left(\ell_{CA}^0\right)^2} + \frac{(X_B - X_C)}{\left(\ell_{BC}^0\right)^2} = 4.425\ 759E-4;$$

$$\frac{\partial Br_{AC}}{\partial X_B} = 0.0; \quad \frac{\partial Br_{AC}}{\partial Y_B} = 0.0;$$

$$\frac{\partial Br_{AC}}{\partial X_C} = \frac{(Y_C - Y_A)}{\left(\ell_{CA}^0\right)^2} = -2.023\ 099\ 5E-4; \quad \frac{\partial Br_{AC}}{\partial Y_C} = \frac{-(X_C - X_A)}{\left(\ell_{CA}^0\right)^2} = -1.394\ 854E-3;$$

The A-matrix is given as follows:

$$A = \begin{bmatrix} 0.688\ 454 & 0.725\ 2803 & 0.0 & 0.0 \\ -0.564\ 332 & 0.825\ 548\ 1 & 0.564\ 331\ 8 & -0.825\ 548 \\ 0.0 & 0.0 & 0.989\ 644\ 8 & -0.143\ 538\ 3 \\ -1.357\ 886E-3 & 1.288\ 939E-3 & -2.023\ 10E-4 & -1.394\ 854E-3 \\ -3.517\ 973E-5 & -2.241\ 22E-3 & 1.393\ 066E-3 & 9.522\ 781E-4 \\ 1.393\ 065\ 9E-3 & 9.522\ 781E-4 & -1.190\ 756E-3 & 4.425\ 759E-4 \\ 0.0 & 0.0 & -2.023\ 10E-4 & -1.394\ 854E-3 \end{bmatrix}$$

Table 6.16 Adjusted coordinates of points B and C.

Station	X (m)	Y (m)
B	1368.2179	1386.4819
C	1702.6965	897.7663

The matrix of coefficients of normal equations is given as

$$
N = A^T PA = \begin{bmatrix}
9.811\ 53 & 0.279\ 86 & -3.919\ 6 & 5.835\ 16 \\
0.279\ 86 & 15.435\ 92 & 3.975\ 41 & -8.506\ 1 \\
-3.919\ 6 & 3.975\ 41 & 15.331\ 65 & -6.419\ 2 \\
5.835\ 16 & -8.506\ 1 & -6.419\ 2 & 8.943\ 16
\end{bmatrix} \times E + 4
$$

$$
u = A^T Pw = \begin{bmatrix}
-0.191\ 203 \\
8.356\ 607 \\
-5.344\ 661 \\
-3.600\ 39
\end{bmatrix} \times E + 4
$$

The vector of corrections to the approximate coordinates is given as

$$
\delta = -N^{-1}u = \begin{bmatrix}
0.4979 \\
-0.9081 \\
0.5465 \\
-0.3937
\end{bmatrix}
$$

The adjusted coordinates of points B and C are

$$
\hat{x} = x^0 + \delta = \begin{bmatrix}
1367.720 \\
1387.390 \\
1702.150 \\
898.160
\end{bmatrix} + \begin{bmatrix}
0.4979 \\
-0.9081 \\
0.5465 \\
-0.3937
\end{bmatrix} = \begin{bmatrix}
1368.2179 \\
1386.4819 \\
1702.6965 \\
897.7663
\end{bmatrix} m
$$

The adjusted coordinates from $\hat{x}$ are given in Table 6.16.

Example 6.5   In order to determine the least squares adjusted azimuths x_2 and x_3 in a triangular lot in Figure 6.9 based on a fixed azimuth $x_1 = 90°$, three angles $\alpha, \beta,$ and γ of the triangle are measured with their estimated standard deviations as shown in Table 6.17.

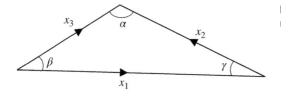

Figure 6.9 Sample triangulation network.

Table 6.17 Simple triangulation problem.

Angle	Measurement	Standard deviation
β	43°17′35″	6.7″
α	104°38′56″	9.9″
γ	32°03′14″	4.3″

Determine the least squares adjusted azimuths x_2 and x_3, assuming their approximate values as $x_2^0 = 302°03′14″$ and $x_3^0 = 46°42′10″$.

Solution (Note that hats are ignored in the equations for brevity):

Given parameters $x = \begin{bmatrix} x_2 \\ x_3 \end{bmatrix}$ and approximate values of parameters $x^0 = \begin{bmatrix} 302°03′14″ \\ 46°42′10″ \end{bmatrix}$, formulate the parametric model equations, $\ell = f(x)$:

$$\beta = 90 - x_3 \tag{6.36}$$

$$\alpha = 180 + x_3 - (x_2 - 180)$$

or

$$\alpha = x_3 - x_2 + 360 \tag{6.37}$$

$$\gamma = x_2 - 270 \tag{6.38}$$

From Equations (6.36)–(6.38), the following first design matrix (A) is obtained:

$$A = \frac{\partial f}{\partial x} = \begin{bmatrix} 0 & -1 \\ -1 & 1 \\ 1 & 0 \end{bmatrix}$$

The weight matrix (P) determined from the standard deviations of observations:

$$P = \begin{bmatrix} \dfrac{1}{44.89} & & \\ & \dfrac{1}{98.01} & \\ & & \dfrac{1}{18.49} \end{bmatrix}$$

$$N = A^{T}PA = \begin{bmatrix} 0.064\ 286\ 33 & -1.020\ 304\text{E}-2 \\ -1.020\ 304\text{E}-2 & 0.032\ 479\ 717 \end{bmatrix}$$

$$N = A^{T}PA = \begin{bmatrix} 16.371\ 652\ 518\ 7 & 5.142\ 921\ 494\ 53 \\ 5.142\ 921\ 494\ 53 & 32.404\ 021\ 314\ 8 \end{bmatrix}$$

The derived observations ($f(x^0)$) using the approximate values (x^0) of the parameters and the misclosure vector (w) are given as follows:

$$f(x^0) = \begin{bmatrix} 90° - 46°42'10'' \\ 46°42'10'' - 302°03'14'' + 360° \\ 302°03'14'' - 270° \end{bmatrix} = \begin{bmatrix} 43°17'50'' \\ 104°38'56'' \\ 32°03'14'' \end{bmatrix}$$

$$w = f(x^0) - \ell = \begin{bmatrix} 43°17'50'' \\ 104°38'56'' \\ 32°03'14'' \end{bmatrix} - \begin{bmatrix} 43°17'35'' \\ 104°38'56'' \\ 32°03'14'' \end{bmatrix} = \begin{bmatrix} 15'' \\ 0 \\ 0 \end{bmatrix}$$

Other product and the vector of corrections (δ) to the approximate parameters:

$$A^{T}Pw = \begin{bmatrix} 0.0 \\ -0.334\ 150\ 145 \end{bmatrix}$$

$$\delta = -N^{-1}A^{T}Pw = \begin{bmatrix} 1.718\ 51'' \\ 10.827\ 81'' \end{bmatrix}$$

Corrected azimuths (parameters):

$$\hat{x} = x^0 + \delta = \begin{bmatrix} 302°03'14'' \\ 46°42'10'' \end{bmatrix} + \begin{bmatrix} 1.718\ 51'' \\ 10.827\ 81'' \end{bmatrix}$$

$$\hat{x} = \begin{bmatrix} \hat{x}_2 \\ \hat{x}_3 \end{bmatrix} = \begin{bmatrix} 302°03'15.72'' \\ 46°42'20.83'' \end{bmatrix}$$

6.3 Stochastic Properties of Parametric Adjustment

The summary of the stochastic models for parametric least squares adjustment discussed in Chapter 5 is given in this section as follows. The models are used in the examples that follow.

1) At the beginning of a parametric least squares adjustment, the given standard deviations of the measurements are used to form the variance–covariance matrix (C_ℓ) of the observation vector ℓ. The variance–covariance matrix formed and its cofactor matrix (Q_ℓ) are related by the Equation (6.39) as

$$C_\ell = \sigma_0^2 Q_\ell \tag{6.39}$$

where $\sigma_0^2 = 1$ is the a priori variance factor of unit weight. In essence, this means that the variance–covariance matrix of observations and its cofactor are the same at the beginning of every adjustment.

2) During or after the least squares adjustment, the cofactor matrices of the adjusted parameters ($Q_{\hat{x}}$), the adjusted observations ($Q_{\hat{\ell}}$), and the residual vector (Q_v) are derived. These derived quantities are summarized by Equations (6.40), (6.41), and (6.41), respectively:

$$Q_{\hat{x}} = \left(A^T P A\right)^{-1} \tag{6.40}$$

$$Q_{\hat{\ell}} = A\left(A^T P A\right)^{-1} A^T \tag{6.41}$$

$$Q_v = Q_\ell - A\left(A^T P A\right)^{-1} A^T \tag{6.42}$$

where A is the first design matrix and P is the weight matrix of the observations given as $P = Q_\ell^{-1}$.

3) The following variance–covariance matrices of the adjusted quantities (Equations (6.43)–(6.45)) can be given from Equations (6.40), (6.41), and (6.42), respectively:

$$C_{\hat{x}} = s_0^2 \left(A^T P A\right)^{-1} \tag{6.43}$$

$$C_{\hat{\ell}} = s_0^2 A\left(A^T P A\right)^{-1} A^T \tag{6.44}$$

$$C_v = s_0^2 Q_\ell - s_0^2 A\left(A^T P A\right)^{-1} A^T \tag{6.45}$$

where $C_{\hat{x}}$ is the covariance matrix of the adjusted parameters, $C_{\hat{\ell}}$ is the covariance matrix of the adjusted observations, C_v is the covariance matrix o the observation residuals, and s_0^2 is the a posteriori variance factor (APVF) of unit weight, which is calculated after the least squares adjustment as

$$s_0^2 = \frac{v^T P v}{n - u} \tag{6.46}$$

with the redundancy expressed as $n - u$, n the number of parametric equations, u the number of independent parameters, v the vector of observation residuals, and P the weight matrix of observations.

4) The covariance matrix (C_v) of the observation residuals can be rewritten from Equation (6.45) as

$$C_v = \hat{C}_\ell - C_{\hat{\ell}} \tag{6.47}$$

where $\hat{C}_\ell$ is the scaled covariance matrix of the original observations ℓ, expressed as

$$\hat{C}_\ell = s_0^2 Q_\ell \tag{6.48}$$

with all the symbols as defined in step (3). The scaled covariance matrix $\hat{C}_\ell$ should not be confused with $C_\ell = \sigma_0^2 Q_\ell$ (with $\sigma_0^2 = 1$), which is the unscaled covariance matrix of the original observations, used at the beginning of the least squares adjustment. Actually, $\hat{C}_\ell = s_0^2 Q_\ell$ and $C_\ell = \sigma_0^2 Q_\ell$ will be identical when $\sigma_0^2 = s_0^2$.

6.4 Application of Stochastic Models

Example 6.6 Continuing from the Example 6.1, on leveling network adjustment, compute the following with regard to the network and the computations already done.

a) A posteriori variance–covariance factor: $s_0^2 = \frac{v^T P v}{n - u}$ where the number of parametric equations involved is $n = 7$ and the number of unknown parameters is $u = 3$.

Solution:

Using the residual vector (v) and the weight matrix (P) already computed in Example 6.1, the a posteriori variance factor of unit weight is computed as follows:

$$s_0^2 = \frac{4.019\ 047\ 620}{7 - 3} \longrightarrow 1.004\ 761\ 91$$
$$s_0 = 1.002\ 4$$

b) The standard deviations of the adjusted elevations of points P2, P3, and P4: $C_{\hat{x}} = s_0^2 N^{-1}$.

Solution:

The inverse of the matrix of coefficients of normal equations (N^{-1}) determined from Example 6.1 for the adjusted parameters, $\hat{x} = \begin{bmatrix} \hat{H}_2 & \hat{H}_3 & \hat{H}_4 \end{bmatrix}$, is as follows:

$$N^{-1} = \begin{bmatrix} 0.000\ 952 & 0.000\ 357 & 0.000\ 119 \\ 0.000\ 357 & 0.001\ 071 & 0.000\ 357 \\ 0.000\ 119 & 0.000\ 357 & 0.000\ 952 \end{bmatrix}$$

The covariance matrix of the adjusted parameters is

$$C_{\hat{x}} = s_0^2 N^{-1} = \begin{bmatrix} 0.000\ 957 & 0.000\ 359 & 0.000\ 120 \\ 0.000\ 359 & 0.001\ 077 & 0.000\ 359 \\ 0.000\ 120 & 0.000\ 359 & 0.000\ 957 \end{bmatrix}$$

and the corresponding standard deviations of the adjusted parameters are

$$\hat{\sigma}_{H_2} = \sqrt{0.000\ 957} \rightarrow 0.031 \text{ m}$$

$$\hat{\sigma}_{H_3} = \sqrt{0.001\ 077} \rightarrow 0.033 \text{ m}$$

$$\hat{\sigma}_{H_4} = \sqrt{0.000\ 957} \rightarrow 0.031 \text{ m}$$

c) The standard deviations of the adjusted observations: $C_{\hat{\ell}} = s_0^2 \left(AN^{-1}A^T \right)$.

Solution:

The adjusted observations:

$$\hat{\ell} = \begin{bmatrix} \Delta\hat{h}_{21} & \Delta\hat{h}_{13} & \Delta\hat{h}_{14} & \Delta\hat{h}_{23} & \Delta\hat{h}_{52} & \Delta\hat{h}_{34} & \Delta\hat{h}_{45} \end{bmatrix}^T$$

Note that the cofactor matrix of the adjusted observations, $Q_{\hat{\ell}} = AN^{-1}A^T$, and the covariance matrix of the adjusted observations, $C_{\hat{\ell}} = s_0^2 Q_{\hat{\ell}}$, where $s_0^2 = 1.004\ 761\ 91$. The following are determined:

$$Q_{\hat{\ell}} = \begin{bmatrix} 9.52\text{E}{-4} & -3.57\text{E}{-4} & -1.19\text{E}{-4} & 5.95\text{E}{-4} & -9.52\text{E}{-4} & 2.38\text{E}{-4} & 1.19\text{E}{-4} \\ -3.57\text{E}{-4} & 1.071\text{E}{-3} & 3.57\text{E}{-4} & 7.14\text{E}{-4} & 3.57\text{E}{-4} & -7.14\text{E}{-4} & -3.57\text{E}{-4} \\ -1.19\text{E}{-4} & 3.57\text{E}{-4} & 9.52\text{E}{-4} & 2.38\text{E}{-4} & 1.19\text{E}{-4} & 5.95\text{E}{-4} & -9.52\text{E}{-4} \\ 5.95\text{E}{-4} & 7.14\text{E}{-4} & 2.38\text{E}{-4} & 1.31\text{E}{-3} & -5.95\text{E}{-4} & -4.76\text{E}{-4} & -2.38\text{E}{-4} \\ -9.52\text{E}{-4} & 3.57\text{E}{-4} & 1.19\text{E}{-4} & -5.95\text{E}{-4} & 9.52\text{E}{-4} & -2.38\text{E}{-4} & -1.19\text{E}{-4} \\ 2.38\text{E}{-4} & -7.14\text{E}{-4} & 5.95\text{E}{-4} & -4.76\text{E}{-4} & -2.38\text{E}{-4} & 1.31\text{E}{-3} & -5.95\text{E}{-4} \\ 1.19\text{E}{-4} & -3.57\text{E}{-4} & -9.52\text{E}{-4} & -2.38\text{E}{-4} & -1.19\text{E}{-4} & -5.95\text{E}{-4} & 9.52\text{E}{-4} \end{bmatrix}$$

$$C_{\hat{\ell}} = s_0^2 Q_{\hat{\ell}} = \begin{bmatrix} 9.569E-4 & -3.588E-4 & -1.196E-4 & 5.981E-4 & -9.569E-4 & 2.392E-4 & 1.196E-4 \\ -3.588E-4 & 1.076E-3 & 3.588E-4 & 7.177E-4 & 3.588E-4 & -7.177E-4 & -3.588E-4 \\ -1.196E-4 & 3.588E-4 & 9.569E-4 & 2.392E-4 & 1.196E-4 & 5.981E-4 & -9.569E-4 \\ 5.981E-4 & 7.177E-4 & 2.392E-4 & 1.316E-3 & -5.981E-4 & -4.785E-4 & -2.392E-4 \\ -9.569E-4 & 3.588E-4 & 1.196E-4 & -5.981E-4 & 9.569E-4 & -2.392E-4 & -1.196E-4 \\ 2.392E-4 & -7.177E-4 & 5.981E-4 & -4.785E-4 & -2.392E-4 & 1.316E-3 & -5.981E-4 \\ 1.196E-4 & -3.588E-4 & -9.569E-4 & -2.392E-4 & -1.196E-4 & -5.981E-4 & 9.569E-4 \end{bmatrix}$$

The diagonal elements of the above matrix are the corresponding standard deviations of the adjusted observations:

$$\hat{\sigma}_{\Delta h_{21}} = \sqrt{9.569E-4} \ \rightarrow \ 0.031 \text{ m}$$

$$\hat{\sigma}_{\Delta h_{13}} = \sqrt{1.076E-3} \ \rightarrow \ 0.033 \text{ m}$$

$$\hat{\sigma}_{\Delta h_{14}} = \sqrt{9.569E-4} \ \rightarrow \ 0.031 \text{ m}$$

$$\hat{\sigma}_{\Delta h_{23}} = \sqrt{1.316E-3} \ \rightarrow \ 0.036 \text{ m}$$

$$\hat{\sigma}_{\Delta h_{52}} = \sqrt{9.569E-4} \ \rightarrow \ 0.031 \text{ m}$$

$$\hat{\sigma}_{\Delta h_{34}} = \sqrt{1.316E-3} \ \rightarrow \ 0.036 \text{ m}$$

$$\hat{\sigma}_{\Delta h_{45}} = \sqrt{9.569E-4} \ \rightarrow \ 0.031 \text{ m}$$

Example 6.7 Continuing from Example 6.3, on station adjustment, determine the standard deviation of a direction measurement.

Solution:

Using the vector of corrections (δ) to the approximate parameters (x^0), the first design matrix (A), and the misclosure vector (w) computed in Example 6.3, the residual vector (v) is calculated as follows:

$$v = A\delta + w$$

$$= [-0.42''\ \ 0.33''\ \ 0.08''\ \ 2.92''\ \ -0.83''\ \ -2.08''\ \ -3.08''\ \ 0.67''\ \ 2.42''\ \ 0.58''\ \ -0.17''\ \ -0.42'']^T$$

The variance factor of unit weight (which is the variance of a direction measurement) can be given as

$$s_0^2 = \frac{v^T v}{df} \ \rightarrow \ s_0^2 = \frac{30.1668}{12-6} = 5.0278$$

The standard deviation of one direction measurement is $s_0 = 2.2''$. As can be seen in the above results, the residuals and the standard deviation of one direction measurement are identical to those obtained based on the station adjustment approach in Example 1.3. This confirms that the station adjustment approach is based on the least squares adjustment method.

Example 6.8 Continuing from the traverse adjustment problem in Section 6.2.3, determine the standard deviations of the adjusted coordinates of points C and D.

Solution:

Using the vector of corrections (δ) to the approximate parameters (x^0), the first design matrix (A), and the misclosure vector (w) computed in Section 6.2.3, the residual vector (v) is calculated as follows:

$$v = A\delta + w = \begin{bmatrix} 4.595\ 39\text{E}-6 \\ -3.683\ 6\text{E}-6 \\ 3.953\ 24\text{E}-6 \\ 3.041\ 45\text{E}-6 \\ -2.247\ 0\text{E}-3 \\ -6.106\ 6\text{E}-3 \\ -2.320\ 2\text{E}-3 \end{bmatrix}$$

A posteriori variance factor of unit weight:

$$s_0^2 = \frac{v^T P v}{df} = \frac{2.423\ 09}{3} = 0.807\ 7$$

Covariance matrix of the adjusted coordinates of points C and D:

$$C_{\hat{x}} = s_0^2 N^{-1} = \begin{bmatrix} 6.080 & 2.073 & 3.817 & 0.069 \\ 2.073 & 3.596 & 0.545 & 0.442 \\ 3.817 & 0.545 & 9.910 & 0.647 \\ 0.069 & 0.442 & 0.647 & 2.014 \end{bmatrix} \times \text{E}-6$$

where $N = A^T P A$ was computed in Section 6.2.3. The adjusted coordinates for $\hat{x}$ computed in Section 6.2.3 and their standard deviations computed from $C_{\hat{x}}$ above are given in Table 6.18. Note that the standard deviation values given in the table had been scaled using the a posteriori standard factor of unit weight (s_0); if the a priori variance factor of unit weight $\sigma_0^2 = 1$ is considered well known, the scaling with s_0 will not be necessary.

Table 6.18 Adjusted coordinates with their standard deviations.

Station	X (m)	Standard deviation (m)	Y (m)	Standard deviation (m)
C	1173.0777	0.0025	1099.9761	0.0019
D	1223.0016	0.0031	1186.5007	0.0014

Example 6.9   Continuing from Section 6.2.4 on triangulateration adjustment, determine the standard deviations of the adjusted coordinates of points B and C.

Solution:

Using the vector of corrections (δ) to the approximate parameters (x^0), the first design matrix (A), and the misclosure vector (w) computed in Section 6.2.4, the residual vector (v) is calculated as follows:

$$v = A\delta + w = \begin{bmatrix} 6.523\ 97\text{E}-4 \\ 1.597\ 20\text{E}-4 \\ -6.245\ 44\text{E}-4 \\ 4.505\ 483\text{E}-6 \\ 1.128\ 285\text{E}-6 \\ -6.092\ 06\text{E}-5 \\ 0.000 \end{bmatrix}$$

A posteriori variance factor of unit weight:

$$s_0^2 = \frac{v^T P v}{df} = \frac{0.584\ 637}{7-4} = 0.194\ 9$$

Covariance matrix of the adjusted coordinates of points B and C:

$$C_{\hat{x}} = s_0^2 N^{-1} = \begin{bmatrix} 1.9535 & -1.6602 & -0.3787 & -3.1254 \\ -1.6602 & 1.6924 & 0.3777 & 2.9641 \\ -0.3787 & 0.3777 & 0.2662 & 0.7975 \\ -3.1254 & 2.9641 & 0.7975 & 5.6488 \end{bmatrix} \times \text{E}-5 \quad (6.49)$$

The adjusted coordinates obtained for $\hat{x}$ in Table 6.16 and their standard deviations computed from $C_{\hat{x}}$ in Equation (6.49) are given in Table 6.19. Note that the standard deviations given in the table had been scaled using the a posteriori standard factor of unit weight (s_0); if the a priori variance factor of unit weight $\sigma_0^2 = 1$ is considered well known, the scaling with s_0 will not be necessary.

Table 6.19 Adjusted coordinates with their standard deviations.

Station	X (m)	Standard deviation (m)	Y (m)	Standard deviation (m)
B	1368.2179	0.0044	1386.4819	0.0041
C	1702.6965	0.0016	897.7663	0.0075

Example 6.10 Continuing from Example 6.5, on triangulation problem, answer the following:

a) Determine the covariance matrix of the adjusted parameters.

Solution:

Using the vector of corrections (δ) to the approximate parameters (x^0), the first design matrix (A), and the misclosure vector (w) computed in Section 6.2.4, the residual vector (v) is calculated as follows:

$$v = A\delta + w \quad \rightarrow \quad v = \begin{bmatrix} 0 & -1 \\ -1 & 1 \\ 1 & 0 \end{bmatrix} \begin{bmatrix} 1.718\ 51'' \\ 10.827\ 81'' \end{bmatrix} + \begin{bmatrix} 15'' \\ 0 \\ 0 \end{bmatrix}$$

$$v = \begin{bmatrix} 4.172'' \\ 9.109'' \\ 1.719'' \end{bmatrix} \quad s_0^2 = \frac{v^T P v}{3-2}$$

$$s_0^2 = \frac{1.3941}{1} = 1.3941$$

The variance–covariance matrix of the adjusted parameters (azimuths $\hat{x}_2$ and $\hat{x}_3$):

$$C_{\hat{x}} = s_0^2 N^{-1} = 1.394\ 1 \begin{bmatrix} 16.371\ 652\ 518\ 7 & 5.142\ 921\ 494\ 53 \\ 5.142\ 921\ 494\ 53 & 32.404\ 021\ 314\ 8 \end{bmatrix}$$

$$C_{\hat{x}} = \begin{bmatrix} 22.823\ 720\ 78 & 7.169\ 746\ 856 \\ 7.169\ 746\ 856 & 45.174\ 446\ 115 \end{bmatrix}$$

b) Determine the covariance matrix of the adjusted observations and the standard deviations of the corresponding adjusted observations.

Solution:

Refer to the quantities determined in Example 6.5 and substitute them in the variance–covariance equation for the adjusted observations, given as

$$C_{\hat{\ell}} = s_0^2 A N^{-1} A^T$$

$$A = \begin{bmatrix} 0 & -1 \\ -1 & 1 \\ 1 & 0 \end{bmatrix} \quad N^{-1} = \begin{bmatrix} 16.371\ 652\ 518\ 7 & 5.142\ 921\ 494\ 53 \\ 5.142\ 921\ 494\ 53 & 32.404\ 021\ 314\ 8 \end{bmatrix}$$

Covariance matrix of the adjusted observations:

$$C_{\hat{\ell}} = 1.3941 \times \begin{bmatrix} 32.404\ 021\ 314\ 8 & -27.261\ 099\ 820\ 3 & -5.142\ 921\ 494\ 53 \\ -27.261\ 099\ 820\ 3 & 38.489\ 830\ 844\ 5 & -11.228\ 731\ 024\ 2 \\ -5.142\ 921\ 494\ 53 & -11.228\ 731\ 024\ 2 & 16.371\ 652\ 518\ 7 \end{bmatrix}$$

$$C_{\hat{\ell}} = \begin{bmatrix} 45.174\ 45 & -38.004\ 7 & -7.169\ 75 \\ -38.004\ 7 & 53.658\ 673 & -15.653\ 973 \\ -7.169\ 75 & -15.653\ 973 & 22.823\ 721 \end{bmatrix}$$

Standard deviations of the adjusted observations:

$$\sigma_\beta = \sqrt{45.174\ 45} = 6.7''$$

$$\sigma_\alpha = \sqrt{53.658\ 673} = 7.3''$$

$$\sigma_\gamma = \sqrt{22.823\ 721} = 4.8''$$

6.5 Resection Example

A point P (x_P, y_P) is to be resected from three control points A, B, and C as shown in Figure 6.10 with the coordinates of the control points A, B, and C as given in Table 6.20. The field measurements, given in Table 6.21, consist of three distances d_1, d_2, d_3; three bearings PA, PB, PC as α_{pa}, α_{pb}, α_{pc}, respectively; and an angle APB as θ. Form the parametric equations for this problem and compute the adjusted coordinates of point P, assuming the approximate coordinates of point P are $x_P^0 = 391.165$ m, $y_P^0 = 405.029$ m; each bearing has a standard deviation of $\pm 4''$; the angle is precise to $\pm 6''$; and the distances are precise to ± 0.002 m.

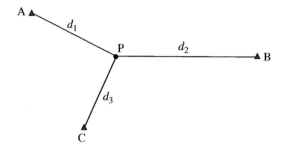

Figure 6.10 Resection of point P.

Table 6.20 Coordinates of control points.

Point	X (m)	Y (m)
A	316.682	505.015
B	500.005	400.040
C	356.804	310.032

Table 6.21 Field measurements.

At	To	Bearing (angle)	Distance (m) (At–To)
P	A	323°19′10″	124.679
P	B	92°37′15″	108.957
P	C	199°53′00″	101.015
Angle A-P-B		129°18′15″	—

Formulate the parametric model equations $\ell = f(x)$ (hats ignored) as follows:

$$\alpha_{pa} = \text{atan}\left(\frac{x_A - x_P}{y_A - y_P}\right) + 360 \tag{6.50}$$

$$\alpha_{pb} = \text{atan}\left(\frac{x_B - x_P}{y_B - y_P}\right) + 180 \tag{6.51}$$

$$\alpha_{pc} = \text{atan}\left(\frac{x_C - x_P}{y_C - y_P}\right) + 180 \tag{6.52}$$

$$\theta = \alpha_{pb} - \alpha_{pa} + 360° = \text{atan}\left(\frac{x_B - x_P}{y_B - y_P}\right) + 180° - \left[\text{atan}\left(\frac{x_A - x_P}{y_A - y_P}\right) + 360\right] + 360°$$

or

$$\theta = \text{atan}\left(\frac{x_B - x_P}{y_B - y_P}\right) - \text{atan}\left(\frac{316.682 - x_P}{505.015 - y_P}\right) + 180° \tag{6.53}$$

$$d_1 = \sqrt{(x_A - x_P)^2 + (y_A - y_P)^2} \tag{6.54}$$

$$d_2 = \sqrt{(x_B - x_P)^2 + (y_B - y_P)^2} \tag{6.55}$$

$$d_3 = \sqrt{(x_C - x_P)^2 + (y_C - y_P)^2} \tag{6.56}$$

The A-matrix in the order of the equations and with respect to x_P and y_P (and using their approximate values):

$$A = \begin{bmatrix} -0.006\ 43 & -0.004\ 79 \\ 0.000\ 42 & 0.009\ 17 \\ 0.009\ 31 & -0.003\ 37 \\ 0.006\ 85 & 0.013\ 96 \\ 0.597\ 40 & -0.801\ 94 \\ -0.999\ 0 & 0.045\ 79 \\ 0.340\ 14 & 0.940\ 38 \end{bmatrix}$$

Using approximate coordinates $x_P^0 = 391.165$ m and $y_P^0 = 405.029$ m, the following misclosure vector (w) is obtained:

$$w = f(x^0) - \ell = \begin{bmatrix} 5.642\ 934\ 1 \\ 1.616\ 602 \\ 3.488\ 657\ 8 \\ 2.256\ 853\ 4 \\ 124.679\ 25 \\ 108.954\ 28 \\ 101.020\ 336 \end{bmatrix} - \begin{bmatrix} 5.642\ 988\ 8 \\ 1.616\ 538\ 5 \\ 3.488\ 622\ 3 \\ 2.256\ 783\ 4 \\ 124.679 \\ 108.957 \\ 101.015 \end{bmatrix} = \begin{bmatrix} -0.000\ 054\ 7 \\ 0.000\ 063\ 7 \\ 0.000\ 035\ 5 \\ 0.000\ 069\ 9 \\ 0.000\ 258\ 4 \\ -0.002\ 717\ 3 \\ 0.005\ 336\ 2 \end{bmatrix}$$

Weight matrix (P) of observations using the inverses of the variances of the observations on the diagonal:

$$P = \begin{bmatrix} 2.6591E+9 & & & & & & \\ & 2.6591E+9 & & & & & \\ & & 2.6591E+9 & & & & \\ & & & 1.1818E+9 & & & \\ & & & & 2.5E+5 & & \\ & & & & & 2.5E+5 & \\ & & & & & & 2.5E+5 \end{bmatrix}$$

Matrix of coefficients of normal equations (N) and the vector of corrections (δ) to the approximate parameters (x^0):

$$N = A^T PA = \begin{bmatrix} 7.6401E + 5 & 7.0663E + 4 \\ 7.0663E + 4 & 9.2741E + 5 \end{bmatrix} \quad u = A^T Pw = \begin{bmatrix} 3.6234E + 3 \\ 4.2565E + 3 \end{bmatrix}$$

$$\delta = -N^{-1}u = \begin{bmatrix} -0.0043 \\ -0.0043 \end{bmatrix}$$

Adjusted parameters:

$$\hat{x} = x^0 + \delta = \begin{bmatrix} 391.165 - 0.0043 \\ 405.029 - 0.0043 \end{bmatrix} \rightarrow \hat{x} = \begin{bmatrix} 391.1607 \\ 405.0247 \end{bmatrix}$$

Residual vector:

$$v = A\delta + w = \begin{bmatrix} -6.3226E - 6 \\ 2.2823E - 5 \\ 9.3918E - 6 \\ -1.9336E - 5 \\ 1.0754E - 3 \\ 1.4320E - 3 \\ -1.4742E - 4 \end{bmatrix}$$

A posteriori variance factor of unit weight:

$$s_0^2 = \frac{v^T P v}{df} \rightarrow s_0^2 = \frac{2.974\,961}{5} = 0.594\,99$$

Cofactor matrix of the adjusted parameters:

$$Q_{\hat{x}} = N^{-1} = \begin{bmatrix} 1.318E - 6 & -1.00E - 7 \\ -1.00E - 7 & 1.086E - 6 \end{bmatrix}$$

Covariance matrix of the adjusted parameters:

$$C_{\hat{x}} = s_0^2 Q_{\hat{x}} = \begin{bmatrix} 7.843E - 7 & -5.98E - 8 \\ -5.98E - 8 & 6.461E - 7 \end{bmatrix}$$

The MATLAB code for the parametric adjustment of the resection problem is given in Table 6.22.

Table 6.22 MATLAB code for the parametric adjustment.

```
>> % Use syms to define the variables
>> syms xA yA xB yB xC yC xP yP
>> % Form the parametric Equations (6.50) - (6.56)
>> Apa=atan((xA-xP)/(yA-yP))+2*pi;
>> Apb=atan((xB-xP)/(yB-yP)) + pi;
>> Apc=atan((xC-xP)/(yC-yP))+pi;
>> theta=atan((xB-xP)/(yB-yP))-atan((xA-xP)/(yA-yP)) +pi;
>> d1=sqrt((xA-xP)^2+(yA-yP)^2);
>> d2=sqrt((xB-xP)^2+(yB-yP)^2);
>> d3=sqrt((xC-xP)^2+(yC-yP)^2);
>> % Determine Jacobian wrt unknown parameters xP and yP
>> J=jacobian([Apa;Apb;Apc;theta;d1;d2;d3],[xP yP]);
>> % input given values and evaluate Jacobian as A matrix
>>
   xA=316.682;yA=505.015;xB=500.005;yB=400.040;
   xC=356.804;yC=310.032;
>> xP=391.165;yP=405.029;
>> format  long
>> A=eval(J)
A =
  -0.006432070166759  -0.004791469628055
   0.000420266615083   0.009168534452934
   0.009308769385695  -0.003367039220837
   0.006852336781843   0.013960004080989
   0.597396880060841  -0.801945738621743
  -0.998951094827644   0.045789847593670
   0.340139434154846   0.940375013137217
>> % Determine computed observations L0 and given
   observations L and form the misclosure w
>> L0=eval([Apa;Apb;Apc;theta;d1; d2; d3]);
>> L=[(323+19/60+10/3600)*pi/180; (92+37/60+15/3600)
   *pi/180; (199+53/60)*pi/180; (129+18/60+15/3600)*pi/
   180; 124.679; 108.957; 101.015];
>> w=L0-L;
>> % Form weight matrix P and solution d
>> P=inv(diag([(4/206265)^2 (4/206265)^2 (4/206265)^2
   (6/206265)^2 (0.002)^2 (.002)^2 (0.002)^2]));
>> N=A'*P*A
N =
   1.0e+05 *
   7.640091030567186   0.706627045540446
   0.706627045540446   9.274142898733244
```

(*Continued*)

Table 6.22 (Continued)

```
>> u=A'*P*w
u =
   1.0e+03 *
   3.623420211039340
   4.256548302052698
>> d=-inv(N)*u
d =
  -0.004348788431742
  -0.004258346235394
>> % Determine the adjusted parameters xPa and yPa
>> xPa=xP+d(1)
xPa =
     3.911606512115683e+02
>> yPa=yP+d(2)
yPa =
     4.050247416537646e+02
>> % Determine residuals v, aposteriori variance S02
>> v=A*d+w;
>> S02=v'*P*v/5
S02 =
   0.5950
>> % Determine cofactor Qx and covariance Cx of parameters
>> Qx=inv(N)
Qx =
   1.0e-05 *
   0.1318 -0.0100
  -0.0100 0.1086
>> Cx=S02*Qx
Cx =
   1.0e-06 *
   0.7843 -0.0598
  -0.0598  0.6461
>>
```

6.6 Curve-fitting Example

In order to determine the accuracy of an EDM instrument, it was calibrated on a government calibration baseline. The measured distances corrected for meteorological condition and reduced to mark to mark and the corresponding published mark-to-mark distances provided for the baseline are as given in

Table 6.23 Published and measured mark-to-mark distances for an EDM instrument.

Line	Published mark-to-mark distance (m)	Measured mark-to-mark distance (m)	Std. (m)
1_2	70.0002	70.0020	0.001 01
1_3	219.9998	219.9982	0.001 03
1_4	670.0132	670.0139	0.001 21
1_5	970.0067	970.0070	0.001 40
1_6	1469.9896	1469.9906	0.001 79
2_3	149.9996	149.9985	0.001 02
2_4	600.013	600.0118	0.001 18
2_5	900.0066	900.0036	0.001 36
2_6	1399.9894	1399.9921	0.001 73
3_4	450.0134	450.0116	0.001 11
3_5	750.007	750.0046	0.001 26
3_6	1249.9898	1249.9868	0.001 61
4_5	299.9936	299.9925	0.001 05
4_6	799.9765	799.9759	0.001 29
5_6	499.9829	499.9839	0.001 13

Table 6.23. A linear regression line in form of $y = a + bx$ is to be fitted to the data set with y representing the corrected distance measurements, x the vector of published distances (considered as constant values), and a and b the unknown parameters to be determined. Given the standard deviations of the measurements in the last column of Table 6.23, determine the parametric least squares adjusted values of the parameters a and b, and give the plot of the data sets showing the fitted regression line. (Assume the approximate values of the parameters are $a^0 = 0$, $b^0 = 1$.)

Solution:

Formulate the parametric model equations ($\ell = f(x)$) (hats ignored) as follows:

The parameter vector: $x = \begin{bmatrix} a \\ b \end{bmatrix}$

$y_1 = a + b(70.0002)$

$y_2 = a + b(219.9998)$

$y_3 = a + b(670.0132)$

$$y_4 = a + b(970.0067)$$

$$y_5 = a + b(1469.9896)$$

$$y_6 = a + b(149.9996)$$

$$y_7 = a + b(600.013)$$

$$y_8 = a + b(900.0066)$$

$$y_9 = a + b(1399.9894)$$

$$y_{10} = a + b(450.0134)$$

$$y_{11} = a + b(750.0070)$$

$$y_{12} = a + b(1249.9898)$$

$$y_{13} = a + b(299.9936)$$

$$y_{14} = a + b(799.9765)$$

$$y_{15} = a + b(499.9829)$$

The first design matrix (A) and the misclosure vector (w) using the approximate values of the parameters ($a^0 = 0$, $b^0 = 1$):

$$
A = \begin{bmatrix}
1 & 70.0002 \\
1 & 219.9998 \\
1 & 670.0132 \\
1 & 970.0067 \\
1 & 1469.9896 \\
1 & 149.9996 \\
1 & 600.0130 \\
1 & 900.0066 \\
1 & 1399.9894 \\
1 & 450.0134 \\
1 & 750.0070 \\
1 & 1249.9898 \\
1 & 299.9936 \\
1 & 799.9765 \\
1 & 499.9829
\end{bmatrix}
\quad
w = f(x^0) - \ell =
\begin{bmatrix}
-0.0018 \\
0.0016 \\
-0.0007 \\
-0.0003 \\
-0.0010 \\
0.0011 \\
0.0012 \\
0.0030 \\
-0.0027 \\
0.0018 \\
0.0024 \\
0.0030 \\
0.0011 \\
0.0006 \\
-0.0010
\end{bmatrix}
$$

Weight matrix (P) of measurements as the inverses of the variances of the measurements:

$$P = \text{diag}[10.0 \ 10.0 \ 6.944 \ 5.102 \ 3.086 \ 10.0 \ 6.944 \ 5.102 \ 3.460 \ 8.264$$
$$5.917 \ 3.906 \ 8.264 \ 5.917 \ 8.264] \times 1E + 05$$

The matrix of the coefficients of normal equations (N) and the vector of corrections (δ) to the approximate parameters (x^0):

$$N = A^T P A = \begin{bmatrix} 1.011\ 735\ 6E + 7 & 5.652\ 653\ 32E + 9 \\ 5.652\ 653\ 32E + 9 & 4.646\ 217\ 3E + 12 \end{bmatrix}$$

$$u = A^T Pw = \begin{bmatrix} 5.899\ 146\ 07E + 3 \\ 5.375\ 650\ 18E + 6 \end{bmatrix}$$

$$\delta = -N^{-1}u = \begin{bmatrix} -5.531E - 3 \\ -5.359E - 8 \end{bmatrix}$$

Adjusted parameters: $\hat{x} = \begin{bmatrix} \hat{a} \\ \hat{b} \end{bmatrix}$

$$\hat{x} = x^0 + \delta = \begin{bmatrix} 0.0 \\ 1.0 \end{bmatrix} + \begin{bmatrix} -5.531E - 4 \\ -5.359E - 8 \end{bmatrix} \rightarrow \hat{x} = \begin{bmatrix} -0.55 \text{ mm} \\ 0.999\ 999\ 946\ 4 \end{bmatrix}$$

Residual vector (v): $v = A\delta + w$

$$v = [-2.4 \ 1.0 \ -1.3 \ -0.9 \ -1.6 \ 0.5 \ 0.6 \ 2.4 \ -3.3 \ 1.2 \ 1.8 \ 2.4 \ 0.5 \ 0.0 \ -1.6]^T$$
$$\times E - 3$$

A posteriori variance factor of unit weight:

$$s_0^2 = \frac{v^T Pv}{df} \quad s_0^2 = \frac{24.016\ 32}{13} = 1.847\ 4$$

Covariance matrix of parameters:

$$C_{\hat{x}} = s_0^2 N^{-1} = \begin{bmatrix} 5.7014E - 7 & -6.94E - 10 \\ -6.94E - 10 & 1.24E - 12 \end{bmatrix}$$

The fitted curve to the data set in Table 6.23 is shown in Figure 6.11.

The MATLAB code for the parametric adjustment of the regression problem is given in Table 6.24.

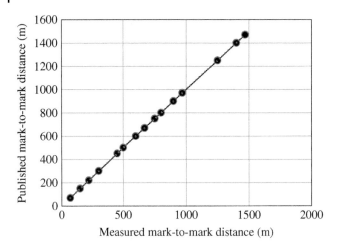

Figure 6.11 Regression line fitted to data set for the EDM calibration.

Table 6.24 MATLAB code for the least squares adjustment.

```
>> % Use syms to define unknown parameters a and b
>> syms a b
>> % Form parametric Equations
>> y=a+70.0002*b;
>> y1=a+70.0002*b;
>> y2=a+219.9998*b;
>> y3=a+670.0132*b;
>> y4=a+970.0067*b;
>> y5=a+1469.9896*b;
>> y6=a+149.9996*b;
>> y7=a+600.013*b;
>> y8=a+900.0066*b;
>> y9=a+1399.9894*b;
>> y10=a+450.0134*b;
>> y11=a+750.0070*b;
>> y12=a+1249.9898*b;
>> y13=a+299.9936*b;
>> y14=a+799.9765*b;
>> y15=a+499.9829*b;
>> % Determine and evaluate the Jacobian (A) wrt a and b
>> J=jacobian([y1;y2;y3;y4;y5;y6;y7;y8;y9;y10;y11;y12;
   y13;y14;y15],[a b]);
>> A=eval(J);
>> % Form the weight matrix P, observations L, computed
   observations L0 and misclosure w
```

Table 6.24 (Continued)

```
>> P=inv(diag([0.001^2 .001^2 .0012^2 .0014^2 .0018^2
   .001^2 .0012^2 .0014^2 .0017^2 .0011^2 .0013^2 .0016^2
   .0011^2 .0013^2 .0011^2]));
>>
L=[70.0020;219.9982;670.0139;970.0070;1469.9906;
   149.9985;600.0118;900.0036;1399.9921;450.0116;
   750.0046;1249.9868;299.9925;799.9759;499.9839];
>> a=0;b=1.0;
>> L0=eval([y1;y2;y3;y4;y5;y6;y7;y8;y9;y10;y11;y12;y13;
y14;y15]);
>> w=L0-L;
>> % determine the solution d
>> N=A'*P*A
N =
  1.0e+12 *
  0.000010117355584 0.005652653318765
  0.005652653318765 4.646217294087624
>> u=A'*P*w;
>> d=-inv(N)*u
d =
  1.0e-03 *
 -0.553129582912654
 -0.000053592071413
>> % compute the adjusted parameters aa and ba
>> aa=a+d(1)
aa =
  -5.531295829126544e-04
>> ba=b+d(2)
ba =
  0.999999946407929
>> % Determine residuals v, aposteriori variance s02,
covariance matrix Cx of parameters
>> v=A*d+w;
>> vtpv=v'*P*v
vtpv =
24.016324918065195
>> s02=vtpv/13
s02 =
  1.847409609081938
>> Cx=s02*inv(N)
Cx =
  1.0e-06 *
  0.570142743598938 -0.000693643682114
 -0.000693643682114 0.000001241512505
>>
```

6.7 Weight Constraint Adjustment Steps

The steps for weight-constrained parametric least squares adjustment method discussed in Chapter 5 are summarized in this section as follows:

1) Give the parametric model ($\hat{\ell} = f(\hat{x})$) with P as the weight matrix of the observations and C_x as the a priori covariance matrix of the vector of all the parameters or of a sub-vector of the parameters.

2) Form the first design matrix (A), which is the matrix containing the partial derivatives of the parametric equations with respect to the unknown parameters. The size of this matrix will be equal to the number of equations (or number of observations) forming the size of row by the size of the column representing the number of parameters including those parameters associated with the given covariance matrix C_x.

3) Choose approximate values for the parameters (x^0); the values of those parameters with given covariance matrix can be used as approximate values for the adjustment. Usually if the values chosen are not close enough to what the final values should be, the final values computed can be used again as approximate values in another round of adjustment in an iterative way until the final values are not too different from the approximate (or initial) values. In this case, the least squares adjustment may have to be repeated as many times as possible until the final solution and the previous one converge to a specific tolerance limit. Substitute these approximate values into the parametric equations formed in step 1 to obtain the derived or simulated values of the observations ($\ell^0 = f(x^0)$).

4) Find the misclosure vector ($w = f(x^0) - \ell$) as the difference between the derived observations ($\ell^0 = f(x^0)$) calculated in step 3 and the actual observations (ℓ).

5) Form the cofactor matrix (Q) of the observations (assuming the a priori variance factor σ_0^2 is 1) by using the variances of the observations along the diagonal of the matrix (in the order in which the parametric equations are formed in step 1). The inverse of this cofactor matrix is the weight matrix (P) of the observations.

6) Take the inverse (C_x^{-1}) of the covariance matrix of the parameters as the weight of parameters, $P_x = C_x^{-1}$; line them up along the diagonal of the matrix (in the order in which the parametric equations are formed in step 1); and assign zero weights to those parameters with unknown covariance matrix.

7) Determine the weight constraint solution of the corrections to the approximate parameters as follows:

$$\delta = -\left(P_x + A^T P A\right)^{-1} A^T P w \qquad (6.57)$$

8) Determine the adjusted parameters: $\hat{x} = x^0 + \delta$.
9) If required, determine the residuals (v) of observations and the adjusted observations ($\hat{\ell}$):$v = A\delta + w$ and $\hat{\ell} = \ell + v$.
10) Determine the cofactor matrix ($Q_{\hat{x}}$) and covariance matrix ($C_{\hat{x}}$) of the adjusted parameters, respectively, as follows:

$$Q_{\hat{x}} = \left(P_x + A^T PA\right)^{-1} \tag{6.58}$$

$$C_{\hat{x}} = s_0^2 Q_{\hat{x}} \tag{6.59}$$

where the APVF of unit weight s_0^2 can be given as follows:

$$s_0^2 = \frac{v^T P v + \delta^T P_x \delta}{n_d - u_d} \tag{6.60}$$

with n_d as the number of direct (or actual) observations and u_d as the number of direct parameters (whose values have no standard deviations associated with them).

11) Determine the cofactor matrix ($Q_{\hat{\ell}}$) and covariance matrix ($C_{\hat{\ell}}$) of the adjusted observations, respectively, as follows:

$$Q_{\hat{\ell}} = A\left(P_x + A^T PA\right)^{-1} A^T \tag{6.61}$$

$$C_{\hat{\ell}} = s_0^2 Q_{\hat{\ell}} \tag{6.62}$$

where the variance factor s_0^2 is determined from Equation (6.60).

6.7.1 Weight Constraint Examples

Generally, weight constraint problems have weights (or variances) already associated with the parameters to be determined in a subsequent adjustment. For example, continuing from the Example 6.5 in Section 6.2.4, assume the azimuth $x_1 = 90°00'00''$ with an associated standard deviation of $\pm 3''$. Perform weight constraint adjustment to determine the least squares adjusted azimuths x_2 and x_3 using the other information provided on this problem in Example 6.5.

Solution:

Given the following

$$\text{Parameters}: \quad x = \begin{bmatrix} x_1 \\ x_2 \\ x_3 \end{bmatrix} \quad x^0 = \begin{bmatrix} 90°00'00'' \\ 302°03'14'' \\ 46°42'10'' \end{bmatrix}$$

Form the parametric model equations ($\ell = f(x)$) (hats ignored):

$$\beta = x_1 - x_3 \tag{6.63}$$

$$\alpha - 180 + x_3 - (x_2 - 180)$$

or

$$\alpha = x_3 + x_2 + 360 \tag{6.64}$$

$$\gamma = x_2 - x_1 - 180 \tag{6.65}$$

From Equations (6.63)–(6.65), the first design matrix (A) is obtained with respect to the parameters as follows:

$$A = \frac{\partial f}{\partial x} = \begin{bmatrix} 1 & 0 & -1 \\ 0 & -1 & 1 \\ -1 & 1 & 0 \end{bmatrix}$$

Weight matrices of observations (P) and of parameters (P_x) are given as follows:

$$P = \begin{bmatrix} \dfrac{1}{44.89} & & \\ & \dfrac{1}{98.01} & \\ & & \dfrac{1}{18.49} \end{bmatrix} \quad P_x = \begin{bmatrix} \dfrac{1}{9} & & \\ & 0 & \\ & & 0 \end{bmatrix}$$

The matrix of the coefficients of the normal equations for the weight constraint problem is formulated as follows:

$$N = A^T P A = \begin{bmatrix} 7.635\ 996\ 5\text{E}-2 & -5.408\ 328\ 8\text{E}-2 & -2.227\ 667\ 6\text{E}-2 \\ -5.408\ 328\ 8\text{E}-2 & 0.064\ 286\ 329 & -1.020\ 304\ 1\text{E}-2 \\ -2.227\ 667\ 6\text{E}-2 & -1.020\ 304\ 1\text{E}-2 & 0.032\ 479\ 716\ 8 \end{bmatrix}$$

$$A^T P A + P_x = \begin{bmatrix} 18.747\ 107\ 6\text{E}-2 & -5.408\ 328\ 8\text{E}-2 & -2.227\ 667\ 6\text{E}-2 \\ -5.408\ 328\ 8\text{E}-2 & 0.064\ 286\ 329 & -1.020\ 304\ 1\text{E}-2 \\ -2.227\ 667\ 6\text{E}-2 & -1.020\ 304\ 1\text{E}-2 & 0.034\ 797\ 168 \end{bmatrix}$$

$$\left(A^T P A + P_x\right)^{-1} = \begin{bmatrix} 9.0 & 9.0 & 9.0 \\ 9.0 & 25.371\ 653 & 14.142\ 921\ 5 \\ 9.0 & 14.142\ 921\ 5 & 41.404\ 021\ 3 \end{bmatrix}$$

Substitute the approximate values (x^0) of the parameters into Equations (6.63)–(6.65) to obtain the following derived observations $(f(x^0))$ and the misclosure vector (w):

$$f(x^0) = \begin{bmatrix} 90° - 46°42'10'' \\ 46°42'10'' - 302°03'14'' + 360° \\ 302°03'14'' - 270° \end{bmatrix} = \begin{bmatrix} 43°17'50'' \\ 104°38'56'' \\ 32°03'14'' \end{bmatrix}$$

$$w = f(x^0) - \ell = \begin{bmatrix} 43°17'50'' \\ 104°38'56'' \\ 32°03'14'' \end{bmatrix} - \begin{bmatrix} 43°17'35'' \\ 104°38'56'' \\ 32°03'14'' \end{bmatrix} = \begin{bmatrix} 15'' \\ 0 \\ 0 \end{bmatrix}$$

$$A^T P w = \begin{bmatrix} 0.334\ 150 \\ 0 \\ -0.334\ 150 \end{bmatrix}$$

The corrections to the approximate parameters:

$$\delta = (A^T P A + P_x)^{-1} A^T P w = \begin{bmatrix} 0.0 \\ 1.719'' \\ 10.828'' \end{bmatrix}$$

The adjusted parameters (azimuths $\hat{x}_1$, $\hat{x}_2$, and $\hat{x}_3$):

$$\hat{x} = x^0 + \delta = \begin{bmatrix} 90°00'00'' \\ 302°03'14'' \\ 46°42'10'' \end{bmatrix} + \begin{bmatrix} 0.0 \\ 1.719'' \\ 10.828'' \end{bmatrix}$$

$$\hat{x} = \begin{bmatrix} \hat{x}_1 \\ \hat{x}_2 \\ \hat{x}_3 \end{bmatrix} = \begin{bmatrix} 90°00'00'' \\ 302°03'15.72'' \\ 46°42'20.83'' \end{bmatrix}$$

Example 6.11 Table 6.25 consists of the differential leveling field notes for the leveling network shown in Figure 6.12. The elevations of stations FH1 and FH2 are known as 100.000 m ± 0.002 m and 99.729 m ± 0.002 m, respectively.

Table 6.25 Leveling field notes.

Leg i	BS	FS	dh_i (m)	Standard dev. s_i (m)
1	FH1	A	−7.341	0.008
2	A	B	2.495	0.005
3	B	FH2	5.107	0.004

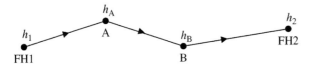

Figure 6.12 Leveling net.

a) Determine the adjusted elevations of points A and B using the parametric model adjustment method (with weight constraint).

Solution:

Since the benchmarks have standard deviations associated with them, they will be considered as both parameters (since they will change after adjustment) and observations (since they have standard deviations). The observations and parameters can be given (remembering that the benchmark values are also observations) as follows:

$$\hat{\ell} = \begin{bmatrix} d\hat{h}_1 \\ d\hat{h}_2 \\ d\hat{h}_3 \end{bmatrix} \quad \hat{x} = \begin{bmatrix} \hat{h}_1 \\ \hat{h}_A \\ \hat{h}_B \\ \hat{h}_2 \end{bmatrix}$$

The parametric model equations $(\hat{\ell} = f(\hat{x}))$ are formulated as follows:

$$d\hat{h}_1 = \hat{h}_A - \hat{h}_1$$

$$d\hat{h}_2 = \hat{h}_B - \hat{h}_A$$

$$d\hat{h}_3 = \hat{h}_2 - \hat{h}_B$$

Let the approximate heights of the unknown points be

$$x^0 = \begin{bmatrix} h_1^0 \\ h_A^0 \\ h_B^0 \\ h_2^0 \end{bmatrix} = \begin{bmatrix} 100.000 \\ 92.659 \\ 95.154 \\ 99.729 \end{bmatrix}$$

The first design matrix (A) with respect to all the parameters is as follows:

$$A = \begin{bmatrix} -1 & 1 & 0 & 0 \\ 0 & -1 & 1 & 0 \\ 0 & 0 & -1 & 1 \end{bmatrix}$$

Substitute the approximate parameters into the parametric equations to obtain the derived observations as follows:

$$f(x^0) = \begin{bmatrix} -7.341 \\ 2.495 \\ 4.575 \end{bmatrix}$$

Subtracting the corresponding observations from the derived observations gives the following misclosure vector:

$$w = f(x^0) - \ell$$

$$w = \begin{bmatrix} -7.341 \\ 2.495 \\ 4.575 \end{bmatrix} - \begin{bmatrix} -7.341 \\ 2.495 \\ 5.107 \end{bmatrix} \rightarrow \begin{bmatrix} 0 \\ 0 \\ -0.532 \end{bmatrix}$$

The weight matrix of observations (P) can be constructed from the covariance matrix (C_ℓ) of the observations as follows:

$$C_\ell = \begin{bmatrix} (0.008)^2 & & \\ & (0.005)^2 & \\ & & (0.004)^2 \end{bmatrix}$$

$$P = C_\ell^{-1} = \begin{bmatrix} 15\ 625 & 0 & 0 \\ 0 & 40\ 000 & 0 \\ 0 & 0 & 62\ 500 \end{bmatrix}$$

The weight matrix of parameters is constructed from the given standard deviations of the parameters (with zero weights for parameters with no standard deviations). Points A and B in the following weight matrix (P_x) have no standard deviations, so their weights are set to zero:

$$P_x = \begin{bmatrix} 1/(0.002)^2 & & & \\ & 0.0 & & \\ & & 0.0 & \\ & & & 1/(0.002)^2 \end{bmatrix} \quad \text{or} \quad P_x = \begin{bmatrix} 250\ 000 & & & \\ & 0.0 & & \\ & & 0.0 & \\ & & & 250\ 000 \end{bmatrix}$$

From Equation (6.57), the corrections to be applied to the initial coordinates can be computed as follows:

$$\delta = -\left(P_x + A^T PA\right)^{-1} A^T Pw$$

$$A^T Pw = \begin{bmatrix} 0 \\ 0 \\ 33\ 250 \\ -33\ 250 \end{bmatrix}$$

$$A^T PA = \begin{bmatrix} 15\ 625 & -15\ 625 & 0.0 & 0.0 \\ -15\ 625 & 55\ 625 & -40\ 000 & 0.0 \\ 0.0 & -40\ 000 & 102\ 500 & -62\ 500 \\ 0.0 & 0.0 & -62\ 500 & 62\ 500 \end{bmatrix}$$

$$\left(P_x + A^T PA\right)^{-1} = \begin{bmatrix} 3.8584E-6 & 1.5929E-6 & 7.0793E-7 & 1.4159E-7 \\ 1.5929E-6 & 2.7079E-5 & 1.2035E-5 & 2.4070E-6 \\ 7.0793E-7 & 1.2035E-5 & 1.6460E-5 & 3.2920E-6 \\ 1.4159E-7 & 2.4070E-6 & 3.2920E-6 & 3.8584E-6 \end{bmatrix}$$

$$\delta = -\left(P_x + A^T PA\right)^{-1} A^T Pw \quad \text{or} \quad \delta = \begin{bmatrix} -0.0188 \\ -0.3201 \\ -0.4378 \\ 0.0188 \end{bmatrix}$$

The adjusted parameters (with weight constraint) are

$$\hat{x} = x^0 + \delta \;\rightarrow\; \begin{bmatrix} 100.000 \\ 92.659 \\ 95.154 \\ 99.729 \end{bmatrix} + \begin{bmatrix} -0.0188 \\ -0.3201 \\ -0.4378 \\ 0.0188 \end{bmatrix} \;\rightarrow\; \begin{bmatrix} 99.981 \\ 92.339 \\ 95.716 \\ 99.748 \end{bmatrix} \text{m}$$

b) Calculate the adjusted observations.

Solution:

The residuals can be calculated in two ways:

$$v = A\delta + w = \begin{bmatrix} -0.3013 \\ -0.1177 \\ -0.0753 \end{bmatrix}$$

or using the vector of correlates (k) as follows:

$$k = -P(A\delta + w) = \begin{bmatrix} -4\ 707.759\ 625 \\ -4\ 708.079\ 753 \\ -4\ 708.517\ 583 \end{bmatrix}$$

$$v = -P^{-1}k = \begin{bmatrix} -0.3013 \\ -0.1177 \\ -0.0753 \end{bmatrix}$$

The adjusted observations:

$$\hat{\ell} = \ell + v = \begin{bmatrix} -7.341 \\ 2.495 \\ 5.107 \end{bmatrix} + \begin{bmatrix} -0.3013 \\ -0.1177 \\ -0.0753 \end{bmatrix} \rightarrow \begin{bmatrix} -7.642 \\ 2.377 \\ 5.032 \end{bmatrix}$$

c) Calculate the a posteriori variance factor of unit weight for this problem.

Solution:

From Equation (6.60)

$$s_0^2 = \frac{v^T P v + \delta^T P_x \delta}{n_d - u_d}$$

use the P, P_x, and δ given in part (a) above and then the observation residuals (v) in part (b), giving the following:

$$s_0^2 = \frac{2326.976 + 176.72}{3 - 2} = 2503.696$$

The following two important observations can be made from the adjustment of the network in Example 6.11:

1) The corrections (±0.0188 m) to the coordinates of the weighted points FH1 and FH2 are larger than the standard deviations (±0.002 m) associated with the points. This should not be the case if the measurements are only affected by random errors; the corrections are expected to be within the given standard deviations of the points if there are no systematic or gross errors in the measurements. This result, however, demonstrates that only random errors causing misclosures in a network will be adjusted (or distributed) properly by the least squares method.

2) The APVF of unit weight for the adjustment is extremely large (2503.696); a value close to one is expected. This is an indication that there is at least an outlier in the measurements. The outlier must therefore be removed before

the least squares method can properly adjust the remaining (outlier-free) measurements for correct solution. In Example 6.12, the measurement B-FH2 was remeasured and found to be in gross error. The adjustment of the measurements are then done all over again as follows.

Example 6.12 The leveling network in Figure 6.12 was remeasured, and the differential leveling field notes in Table 6.26 were obtained. The elevations of stations FH1 and FH2 are known as 100.000 m ± 0.002 m and 99.729 m ± 0.002 m, respectively.

Table 6.26 Leveling field notes.

Leg i	BS	FS	dh_i (m)	Standard dev. s_i (m)
1	FH1	A	−7.341	0.008
2	A	B	2.495	0.005
3	B	FH2	4.583	0.004

a) Determine the adjusted elevations of points A and B using the parametric model adjustment method (with weight constraint).

Solution:

Following Example 6.11,

$$\hat{\ell} = \begin{bmatrix} d\hat{h}_1 \\ d\hat{h}_2 \\ d\hat{h}_3 \end{bmatrix} \quad \hat{x} = \begin{bmatrix} \hat{h}_1 \\ \hat{h}_A \\ \hat{h}_B \\ \hat{h}_2 \end{bmatrix}$$

The parametric model equations ($\hat{\ell} = f(\hat{x})$) are formulated as follows:

$$d\hat{h}_1 = \hat{h}_A - \hat{h}_1$$

$$d\hat{h}_2 = \hat{h}_B - \hat{h}_A$$

$$d\hat{h}_3 = \hat{h}_2 - \hat{h}_B$$

Let the approximate heights of the unknown points and the first design matrix (A) be

$$x^0 = \begin{bmatrix} h_1^0 \\ h_A^0 \\ h_B^0 \\ h_2^0 \end{bmatrix} = \begin{bmatrix} 100.000 \\ 92.659 \\ 95.154 \\ 99.729 \end{bmatrix} \quad A = \begin{bmatrix} -1 & 1 & 0 & 0 \\ 0 & -1 & 1 & 0 \\ 0 & 0 & -1 & 1 \end{bmatrix}$$

Substitute the approximate parameters into the parametric equations to obtain the following vector of derived observations:

$$f\left(x^0\right) = \begin{bmatrix} -7.341 \\ 2.495 \\ 4.575 \end{bmatrix}$$

Subtracting the corresponding observations (ℓ) from the derived observations ($\ell^0 = f(x^0)$) gives the following misclosure vector:

$$w = f\left(x^0\right) - \ell \quad \rightarrow \quad w = \begin{bmatrix} -7.341 \\ 2.495 \\ 4.575 \end{bmatrix} - \begin{bmatrix} -7.341 \\ 2.495 \\ 4.583 \end{bmatrix} \quad \rightarrow \quad \begin{bmatrix} 0 \\ 0 \\ -0.008 \end{bmatrix}$$

Using the weight matrix of observations (P) and the weight matrix of parameters (P_x) given in Example 6.11, the corrections to be applied to the initial coordinates can be computed from Equation (6.57) as follows:

$$\delta = -\left(P_x + A^T P A\right)^{-1} A^T P w$$

$$A^T P w = \begin{bmatrix} 0 \\ 0 \\ 500 \\ -500 \end{bmatrix}$$

$$A^T P A = \begin{bmatrix} 15\,625 & -15\,625 & 0.0 & 0.0 \\ -15\,625 & 55\,625 & -40\,000 & 0.0 \\ 0.0 & -40\,000 & 102\,500 & -62\,500 \\ 0.0 & 0.0 & -62\,500 & 62\,500 \end{bmatrix}$$

$$\left(P_x + A^T P A\right)^{-1} = \begin{bmatrix} 3.8584\mathrm{E}{-6} & 1.5929\mathrm{E}{-6} & 7.0793\mathrm{E}{-7} & 1.4159\mathrm{E}{-7} \\ 1.5929\mathrm{E}{-6} & 2.7079\mathrm{E}{-5} & 1.2035\mathrm{E}{-5} & 2.4070\mathrm{E}{-6} \\ 7.0793\mathrm{E}{-7} & 1.2035\mathrm{E}{-5} & 1.6460\mathrm{E}{-5} & 3.2920\mathrm{E}{-6} \\ 1.4159\mathrm{E}{-7} & 2.4070\mathrm{E}{-6} & 3.2920\mathrm{E}{-6} & 3.8584\mathrm{E}{-6} \end{bmatrix}$$

$$\delta = -\left(P_x + A^T P A\right)^{-1} A^T P w \quad \text{or} \quad \delta = \begin{bmatrix} -0.0003 \\ -0.0048 \\ -0.0066 \\ 0.0003 \end{bmatrix}$$

The adjusted parameters (with weight constraint):

$$\hat{x} = x^0 + \delta \;\rightarrow\; \begin{bmatrix} 100.000 \\ 92.659 \\ 95.154 \\ 99.729 \end{bmatrix} + \begin{bmatrix} -0.0003 \\ -0.0048 \\ -0.0066 \\ 0.0003 \end{bmatrix} \;\rightarrow\; \begin{bmatrix} 100.000 \\ 92.654 \\ 95.147 \\ 99.729 \end{bmatrix} \; m$$

b) Calculate the adjusted observations.

Solution:

The residuals can be calculated as follows:

$$v = A\delta + w = \begin{bmatrix} -0.0045 \\ -0.0018 \\ -0.0011 \end{bmatrix}$$

The adjusted observations:

$$\hat{\ell} = \ell + v = \begin{bmatrix} -7.341 \\ 2.495 \\ 4.583 \end{bmatrix} + \begin{bmatrix} -0.0045 \\ -0.0018 \\ -0.0011 \end{bmatrix} \;\rightarrow\; \begin{bmatrix} -7.346 \\ 2.493 \\ 4.582 \end{bmatrix}$$

c) Calculate the a posteriori variance factor of unit weight for this problem.

Solution:

From Equation (6.60)

$$s_0^2 = \frac{0.5263 + 0.0401}{3 - 2} = 0.5664$$

d) Determine the standard deviations of the adjusted parameters.

Solution:

From Equation (6.58) the cofactor of the adjusted parameters can be determined as follows:

$$Q_{\hat{x}} = \left(P_x + A^T P A\right)^{-1} \;\rightarrow\; Q_{\hat{x}} = \begin{bmatrix} 0.0386 & 0.0159 & 0.0071 & 0.0014 \\ 0.0159 & 0.2708 & 0.1204 & 0.0241 \\ 0.0071 & 0.1204 & 0.1646 & 0.0329 \\ 0.0014 & 0.0241 & 0.0329 & 0.0386 \end{bmatrix} \times 1E-4 \; m^2$$

From Equation (6.59), the covariance matrix of the adjusted parameters is given as follows:

$$C_{\hat{x}} = s_0^2 Q_{\hat{x}} \quad \rightarrow \quad C_{\hat{x}} = \begin{bmatrix} 0.0219 & 0.0090 & 0.0040 & 0.0008 \\ 0.0090 & 0.1534 & 0.0682 & 0.0136 \\ 0.0040 & 0.0682 & 0.0932 & 0.0186 \\ 0.0008 & 0.0136 & 0.0186 & 0.0219 \end{bmatrix} \times 1E-4 \ m^2$$

The standard deviations of the adjusted parameters are $\sigma_{FH1} = \pm 0.0015$ m, $\sigma_A = \pm 0.0039$ m, $\sigma_B = \pm 0.0031$ m, and $\sigma_{FH2} = \pm 0.0015$ m. It can be seen from these results that the measurements improved the precisions of FH1 and FH2 from the given standard deviation of ± 0.002 m.

e) Determine the standard deviations of the adjusted observations.

Solution:

From Equation (6.61) the cofactor of the adjusted observations can be determined as follows:

$$Q_{\hat{\ell}} = A\left(P_x + A^T P A\right)^{-1} A^T \quad \rightarrow \quad Q_{\hat{\ell}} = \begin{bmatrix} 0.2775 & -0.1416 & -0.0906 \\ -0.1416 & 0.1947 & -0.0354 \\ -0.0906 & -0.0354 & 0.1373 \end{bmatrix} \times 1E-4 \ m^2$$

From Equation (6.62), the covariance matrix of the adjusted observations is given as follows:

$$C_{\hat{\ell}} = s_0^2 Q_{\hat{\ell}} \quad \rightarrow \quad C_{\hat{\ell}} = \begin{bmatrix} 0.1572 & -0.0802 & -0.0513 \\ -0.0802 & 0.1103 & -0.0200 \\ -0.0513 & -0.0200 & 0.0778 \end{bmatrix} \times 1E-4 \ m^2$$

Standard deviations of the adjusted observations are $\sigma_{d\hat{h}_1} = \pm 0.0040$ m, $\sigma_{d\hat{h}_2} = \pm 0.0033$ m, and $\sigma_{d\hat{h}_3} = \pm 0.0028$ m. It can be seen that the adjusted measurements are improved after the adjustment.

The following two important observations can be made from the adjustment of the network in Example 6.12:

1) The corrections (± 0.0003 m) to the coordinates of the weighted points FH1 and FH2 are within the given standard deviations (± 0.002 m) associated with the points, as expected.

2) The APVF of unit weight for the adjustment is 0.5664, which is closer to one, as expected. The measurements are considered free from outliers but are only affected by the unavoidable random errors.

Problems

6.1 Figure P6.1 contains a diagram of a leveling network to be adjusted with the table containing all of the height difference measurements Δh_{ij} and their corresponding standard deviations (Std.) necessary to carry out the adjustment. P1 is a fixed point with known elevation of $H_1 = 107.500$ m. Answer the following questions.

No.	i	j	Δh_{ij} (m)	Std. (m)
1	P2	P1	1.340	0.010
2	P1	P3	−5.000	0.015
3	P2	P3	−3.680	0.015

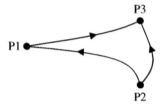

Figure P6.1

Let the approximate elevations of point P2 and P3 be 106.000 and 102.500 m, respectively.

a) Calculate the adjusted heights of points P2 and P3 and the adjusted height differences.

b) Calculate the standard deviations of the adjusted elevations.

c) Given the covariance matrix of the adjusted observations as

$$C_{\hat{\ell}} = \begin{bmatrix} 6.0 & -3.0 & 3.0 \\ -3.0 & 9.7 & 6.7 \\ 3.0 & 6.7 & 9.7 \end{bmatrix} \times \text{E}-5 \text{ m}^2$$

compute the covariance matrix of the residual vectors.

d) Let the standard deviations of the elevations of points P1 and P2 be 0.015 m with their elevations as 107.500 and 106.000 m, respectively. Recalculate the adjusted elevations of all the network points and their standard deviations by readjusting the network and the figure using the least squares method with weight constraint on points P1 and P2 (showing every intermediate steps and calculations).

6.2 Use the leveling notes in the following table and the sketch of the leveling network in Figure P6.2 to answer the following questions, given that the elevation of the benchmark BMA is 100.000 m.

Line	Observed elevation difference (m)	Weight
1	−21.270	0.4
2	−21.200	0.5
3	−21.290	0.444 44

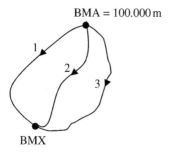

BMA = 100.000 m

BMX

Figure P6.2

a) List the measurements and the parameter and formulate the functional model ($\hat{\ell} = f(\hat{x})$) relating the measurements with the unknown elevation **h** of BMX.

b) Form the first design matrix, A.

c) Calculate the approximate elevation of BMX (x^0) based on leveling line (1) and use it to determine the values of the vector $f(x^0)$.

d) Using the results obtained in (c), calculate the misclosure vector w.

e) Determine the degrees of freedom (or the redundancy) of measurements.

f) Determine the least squares adjusted elevation of BMX.

g) Calculate the adjusted elevation differences.

6.3 The horizontal coordinates of points B and C in the network in Figure P6.3 are to be determined. In order to do this, a traverse survey was run from a known point A and closed on C. The measurements include two bearings (α_1, α_2) and three distances (d_1, d_2, d_3) as shown in the diagram below.

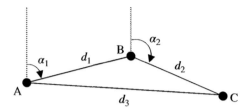

Figure P6.3 (Not to scale.)

The measurements are stochastically independent, and their values together with the corresponding standard deviations are as follows:

$\alpha_1 = 71°29'10''$, $s_{\alpha_1} = 3''$

$\alpha_2 = 121°14'40''$, $s_{\alpha_2} = 3''$

$d_1 = 561.020$ m, $s_{d_1} = 0.002$ m

$d_2 = 428.360$ m, $s_{d_2} = 0.002$ m

$d_3 = 899.295$ m, $s_{d_3} = 0.003$ m

The known coordinates of point A and the approximate coordinates of points B and C are given in the following table.

Point	X (easting) (m)	Y (northing) (m)
A	1000.000	1000.000
B	1531.985	1178.140
C	1898.206	955.940

Answer the following:

a) If point A is known and fixed, perform the parametric least squares adjustment to determine the adjusted coordinates of the unknown points, given the following first design matrix (A) determined with respect to the coordinates of points B and C, $x = \begin{bmatrix} x_B & y_B & x_C & y_C \end{bmatrix}^T$, respectively:

$$A = \begin{bmatrix} 5.660\text{E}-4 & -1.690\text{E}-3 & 0 & 0 \\ 1.211\text{E}-3 & 1.996\text{E}-3 & -1.211\text{E}-3 & -1.996\text{E}-3 \\ 0.948 & 0.318 & 0 & 0 \\ -0.855 & 0.519 & 0.855 & -0.519 \\ 0 & 0 & 0.999 & -0.049 \end{bmatrix}$$

(Use the above design matrix A in your calculations.)

b) Compute the a posteriori variance factor of unit weight.

c) Compute the standard deviations of the adjusted parameters.

7

Confidence Region Estimation

CHAPTER MENU

OBJECTIVES

After studying this chapter, you should be able to:

1) Define some basic statistical terms relating to parameter estimation in geomatics, such as mean squared error, biased and unbiased estimators, mathematical expectation, and point and interval estimators.
2) Construct confidence intervals for population means, population variances, and ratio of two population variances.
3) Construct standard and confidence error ellipses for absolute and relative cases in geomatics.

Understanding Least Squares Estimation and Geomatics Data Analysis, First Edition.
John Olusegun Ogundare.
© 2019 John Wiley & Sons, Inc. Published 2019 by John Wiley & Sons, Inc.
Companion website: www.wiley.com/go/ogundare/Understanding-lse-and-gda

7.1 Introduction

Estimation in statistics is a procedure for making deductions about population using information derived from a sample. Typical pieces of information that can be derived from a sample include statistics, such as sample mean, sample standard deviation, z-score, t-score, chi-square statistic, and F-statistic. An *estimator* is a particular example of a statistic, i.e. a function of that data that is used to infer the value of an unknown parameter in a statistical model. It is a rule for calculating sample statistics (mean, standard deviation, etc.) based upon sample data; the rule is generally expressed as a formula. It can generally be defined as a function of observables or sample data for estimating the unknown population parameter. Anything that produces an estimate is an estimator, such as the algorithm for calculating averages. Some statistics are good estimators and some are not estimators, such as test statistics. It can be said that every estimator is a statistic, but not every statistic is an estimator. However, statistics and estimators are similar since both are functions of random observables and they both have sampling distributions. An output of the estimator when it is applied to a particular sample data is an *estimate*. When the mean of the *sampling distribution* of a statistic is equal to a population parameter, that statistic is said to be an *unbiased estimator* of the parameter. An unbiased estimator is therefore an estimator whose expected value is equal to the parameter it is estimating. It is said to be a consistent estimator if the difference between the estimator and the parameter grows smaller as the sample size increases.

The unbiasedness of an estimator ensures that an average value of the estimator will tend to a value that is equal to the unknown parameter. Many different estimators are possible for any given parameter. The two types that will be discussed in this chapter are *point estimators* and *interval estimators*. While point estimators produce single estimates, interval estimators produce *confidence intervals*. In practice, confidence intervals are preferred since they provide measures of precision and uncertainty of estimates. In this book, confidence interval is used as a special case of *confidence region*, which is a multidimensional generalization of confidence intervals. In this case, a confidence interval is considered as a one-dimensional *confidence region*. Details of how to construct confidence regions are discussed in this chapter.

7.2 Mean Squared Error and Mathematical Expectation

7.2.1 Mean Squared Error

Mean squared error (MSE) of an estimator is the average of the square of the deviation of the estimator from the quantity estimated. This is one of the many ways of quantifying the amount by which an estimator differs from the true value of the quantity being estimated. It is a measure of the average of the square of the error (the error being the amount by which the estimator differs from the quantity to be estimated). In regression analysis, for example, MSE is a measure of how close a fitted line is to data points or a measure of the difference between the actual observations and the response predicted by the model. This can be used to determine if the model fits the data well or if the model can be simplified by removing some terms. The smaller the MSE, the closer the fit is to the data. In this case, MSE is the square of the deviations of points from their true position, which can be considered as estimate of true variances (a function of observed quantity rather than a function of unknown parameter). The squaring is done so that negative values do not cancel positive values when they are summed. From the above discussion, it can be seen that MSE is used as a means of selecting estimators. For example, the MSE for the sample mean ($\bar{x}$) when the true value $\theta = \mu$ (the population mean) can be given as follows:

$$\text{MSE}(\bar{x}) = E\left((\bar{x}-\mu)^2\right) \tag{7.1}$$

Equation (7.1), which is the second moment of the random observable, explains generally how close some data points are to a fitted line (in the case of linear regression). If the estimator $\bar{x}$ is unbiased, the following will result:

$$\text{MSE}(\bar{x}) = \left(\frac{\sigma}{\sqrt{n}}\right)^2 \tag{7.2}$$

Otherwise, if it is biased, the following will result:

$$\text{MSE}(\bar{x}) = \left(\frac{\sigma}{\sqrt{n}}\right)^2 + \text{Bias}[\bar{x}]^2 \tag{7.3}$$

An unbiased estimator such as $\text{MSE}(\bar{x})$ that is a minimum value among all unbiased estimators of θ is called a *minimum variance unbiased estimator.* Similarly, when the sample variance (s^2) is an unbiased estimator of the population variance (σ^2), the MSE when the true value $\theta = \sigma^2$ (the population variance) can be given as follows:

$$\text{MSE}\left(s^2\right) = E\left(s^2 - \sigma^2\right) = \frac{2\sigma^4}{n-1} \tag{7.4}$$

The MSE value will be zero, which is impossible, if the observations of the parameter are predicted perfectly by the estimator. The square root of MSE is known as the *root mean squared error (RMSE),* which is a measure of how close a curve fits some data set. RMSE is a measure of the differences between values predicted by a model or an estimator and the values observed in relation to the quantity being estimated. In another words, RMSE is the distance, on average, of a data point from the fitted line, measured along a vertical line.

7.2.2 Mathematical Expectation

Mathematical expectation, expected value, or simply *expectation* of a random observable is an important concept in probability and statistics. It is the predicted arithmetic (or weighted) average or the weighted sum of all possible values of the given observable, where the weights are the corresponding probabilities of these values. An expectation is an "average" taken over all possible samples of size n (e.g. expectation $(\bar{x}_i) = \mu$ for the average (μ) of the possible averages $(\bar{x}_i)$). As an example, consider an idealized *m-faced* die whose faces are marked with the values $x_1, x_2, x_3, ..., x_m$. Let a discrete observable X represent the value shown on the uppermost face of such a die after it has been rolled, with each of the values $x_1, x_2, x_3, ..., x_m$ having equal chance of being taken. Its *mean* or *expected value* is given for the discrete random observable by

$$\mu = E(X) = \sum_{i=1}^{m} x_i p(x_i) \tag{7.5}$$

and its variance (σ^2) by

$$\mathrm{var}(X) = E\left[(x_i - \mu)^2\right] = \sigma^2 \tag{7.6}$$

or

$$\sigma^2 = \sum_{i=1}^{m} (x_i - \mu)^2 p(x_i) \tag{7.7}$$

with $p(x_i)$ (for $i = 1, 2, ..., m$) as the corresponding probability distribution, and Equations (7.5) and (7.7) are formulas for the population mean and variance of the random variable X, respectively. An alternative way of illustrating Equations (7.5) and (7.7) is as follows:

$$\bar{x} = \frac{\bar{x}_1 + \bar{x}_2 + \bar{x}_3 + \bar{x}_4 + \cdots + \bar{x}_{n_s}}{n_s} \tag{7.8}$$

where $E(\bar{x}) = \mu$, n_s is the number of possible samples, and $1/n_s$ is the probability of each sample (assuming equal probability). In Equation (7.8), the average of all sample means is equal to the population mean (μ); individual sample means may either overstate or understate the population mean, but their arithmetic mean $(\bar{x})$ is always equal to the population mean. The standard deviation of the mean, which is called the standard error of the mean $(\sigma_{\bar{x}})$, can be determined as

$$\sigma_{\bar{x}} = \sqrt{\frac{\sum_{i}^{n} (\bar{x}_i - \mu)^2}{n_s}} \tag{7.9}$$

where n_s is the number of samples. This standard error of the mean indicates the "average" difference between the various means and the population mean. On the average, each sample mean differs from the population mean by $\sigma_{\bar{x}}$. If $\sigma_{\bar{x}}$ is small, any one sample mean will be a good estimate of the population mean. The approximate value of the standard error of the mean can be given as

$$\sigma_{\bar{x}} = \frac{\sigma}{\sqrt{n}} \tag{7.10}$$

where n is the sample size. According to Equation (7.10), the standard deviation of the mean becomes smaller as the sample size increases, meaning that large sample is more reliable than a small one. From Equations (7.2), (7.3), and (7.10), it can be seen that RMSE and the standard error of the mean are equal only for a zero mean random observable (where there is no bias). The sample means are approximately normally distributed whenever the sample size is 30 or more. If all possible samples of size n ($n \geq 30$) are selected from a given population, then the various sample means are approximately normally distributed, having an average that is equal to the population mean with a standard error that is equal to the population standard deviation divided by the square root of the sample size.

The mean or expected value for a case of continuous random observable can be given by

$$\mu = E(X) = \int_{-\infty}^{\infty} xf(x)\mathrm{d}x \qquad (7.11)$$

where $f(x)$ from negative infinity to positive infinity provides the probability distribution for the given random observable. In Equations (7.5) and (7.7), one needs to know the possible values of the given random observable (X) and the probabilities that correspond to each of these values in order to determine the mean. As it can be seen, expectations are interested in the possible outcomes of an event that has not yet happened. An arithmetic mean or a sample mean is a special case of expectation in which the associated probabilities of the values of the random observable are the same. Essentially, the same formulas are used in determining sample mean and sample variance, except that one is interested in actual outcomes of m events that had already occurred in the case of sample mean and sample variance.

Example 7.1   A discrete random observable Y has a probability distribution $p(y)$ and an expected value $E(Y)$, which can be represented as

$$E(Y) = \sum_{i=1}^{m} yp(y) \qquad (7.12)$$

with the elements summed over all the possible values (y) of the random observable Y. Let the random observable (Y) be the number of heads when tossing two coins simultaneously; coin number 1 has faces as (H1, T1) and coin number 2 has faces as (H2, T2). Tossing two coins at the same time will likely produce one of the following four different combinations: H1H2, H1T2, T1H2, or T1T2. If y is the number of heads observed in each case, the probability distribution of tossing two coins simultaneously can be given as shown in Table 7.1. In the table, it can be seen that no head in a toss ($y = 0$) will likely occur once out of four times, one head in a toss ($y = 1$) will occur twice out four times, and two heads in a toss ($y = 2$) will occur once in four times.

Table 7.1 Probability distribution of tossing two coins.

y	p(y)	Comments
0	1/4	When no head (H1 or H2) is expected
1	1/2	When only one head (H1 or H2) is expected
2	1/4	When two heads (H1 and H2) are expected

If the tossing of two coins simultaneously were to be performed 4 000 000 times, what would be the average value of y?

Solution:

From Table 7.1 and Equation (7.9),

$$E(y) = \frac{1}{4}(0) + \frac{1}{2}(1) + \frac{1}{4}(2) \rightarrow 1$$

This can be interpreted to mean that one head will be expected in average. As it can be seen, the actual tossing of the two coins is not involved in the calculation done.

7.3 Population Parameter Estimation

The common types of population parameters that are of interest in this chapter are the population mean (μ) and the population standard deviation (σ), which are estimated from sample data as sample mean ($\bar{x}$) and sample standard deviation (s), respectively. The sample standard deviation is a sufficiently good approximation of the population standard deviation (σ) if the sample size is large enough. The often quoted criterion for the required size of a sample, in statistical inferences, is that a size larger than 30 constitutes a large sample. The population mean (μ) can be estimated in two ways: as *point estimate* and as *interval estimate*.

7.3.1 Point Estimation of Population Mean

A *point estimate* is an estimate of a population parameter expressed as a single value with a precision of estimate stated at a certain probability, e.g. an average angle given as $60°30'27''$ is a point estimate. The point estimate of population mean provides a single estimate of the population mean (μ) as a specific value ($\bar{x}$) known as sample mean. This value, however, does not immediately reveal the uncertainty of the estimation until the probability of maximum error of estimate is provided. The probability is involved because uncertainty (precision or standard deviation) of estimation is involved. The error of point estimate ($\bar{x}$) can be given as

$$\varepsilon = \bar{x} - \mu \tag{7.13}$$

where ε can be considered as a measure of the precision of $\bar{x}$ as an estimate of μ. From the sampling distribution, the *z-score* for the mean statistic can be given as

$$z = \frac{\bar{x} - \mu}{\text{SE}} \tag{7.14}$$

where the standard error is ($\text{SE} = \sigma/\sqrt{n}$) for a case where the sample size (n) is greater than 30 with a known population standard deviation (σ) or $\text{SE} = s/\sqrt{n}$ in the case where the standard deviation of the population is unknown but the sample standard deviation (s) is determined. The t-statistic (for a case where the sample size is less than or equal to 30) can be given as

$$t = \frac{\bar{x} - \mu}{\text{SE}} \tag{7.15}$$

where SE is determined based on whether σ or s is known as discussed above. At a probability of $1 - \alpha$, the *critical values* of z and t (for two-tailed case) will be $z_{1 - \alpha/2}$ and $t_{1 - \alpha/2,\ df}$, respectively, and df is the number of degrees of freedom. From Equations (7.13) to (7.15), the sampling error ($e = |\bar{x} - \mu|_{1 - \alpha}$) of estimate (or what is sometimes known as *margin of error*) at probability $1 - \alpha$ can be expressed as

$$|\bar{x} - \mu|_{1 - \alpha} = (\text{SE})k_p \tag{7.16}$$

where $k_p = z_{p\ =\ 1 - \alpha/2}$ or $k_p = t_{p\ =\ 1 - \alpha/2,\ df}$ depending on whether the condition in Equation (7.14) or Equation (7.15) is satisfied; k_p can be considered as a factor for scaling the SE to an appropriate probability (p) level. In Equation (7.16), the probability of $|\bar{x} - \mu|_{1 - \alpha}$ being less than $(\text{SE})k_p$ is $1 - \alpha$. Equation (7.16) can be used to estimate the maximum error of point estimate. In this case, $\varepsilon = (\text{SE}) k_p$ at the given probability of $p = 1 - \alpha$.

7.3.2 Interval Estimation of Population Mean

An interval estimate of a population parameter is expressed by two numbers between which the parameter may be considered to lie, e.g. the average angle is $60°30'27'' \pm 2''$ represents interval estimate with $2''$ indicating the precision of the estimate. It should be mentioned that the use of standard deviation in expressing uncertainty works well when large sample sizes (more than about 30) are involved; intervals are more valid when small samples are involved. The *interval estimate* places the population mean (μ) within an interval and stipulates a degree of confidence as a measure of uncertainty of the estimated statistic (the mean). The interval estimate immediately reveals the uncertainty associated with the estimation of the population mean (μ). *Confidence interval* is used to describe the amount of uncertainty (or the region of doubt) associated with a sample estimate of a population parameter.

Confidence interval estimation is used in this section to mean the same thing as *one-dimensional confidence region estimation*. A confidence interval is, therefore, a random region containing the statistic with some confidence level $(1 - \alpha)$ or $(1 - \alpha)$ 100% associated with it so that the true value of the parameter can be claimed to fall within the two points (*confidence limits*) of the interval. For example, if $\alpha = 0.05$, the 95% confidence interval is the interval in which one is 95% confident that the true value of the mean or difference between two means will fall. Strictly speaking, a confidence interval should not be interpreted to mean that the population mean (μ) would lie within the limits with a 95% probability; rather, the interval is to be interpreted to mean that if one sets up many 95% confidence intervals for many different samples, then 95% of the time μ would fall within the limits of the confidence interval. In this case, the probability statement is not about the validation of the confidence interval, but about the validation of the process leading to the construction of the confidence interval. Remembering that the population parameters (μ, σ) are quantities with constant (or fixed) values, one cannot again treat the quantities as variables or statistics. Three pieces of information are needed in order to construct a confidence interval, such as the following:

- A confidence level $(1 - \alpha)$ or a significance level (α).
- Sample statistic ($\bar{x}$).
- Sampling error (or margin of error) of the statistic given in Equation (7.16).

The limits of confidence interval can then be defined as follows:

$$\text{Limits of confidence interval} = \text{sample statistic} \pm \text{margin of error} \quad (7.17)$$

$$\text{Margin of error} = \text{critical value} \times \text{sample standard deviation} \quad (7.18)$$

where the margin of error is considered as the uncertainty in the confidence interval. This confidence interval can be constructed as follows:

$$\mu = \bar{x} \pm b \quad (7.19)$$

or

$$\bar{x} - b < \mu < \bar{x} + b \quad (7.20)$$

where b (the margin of error of estimate) is determined based on the selected confidence level $(1 - \alpha)$ and on whether the population standard deviation σ is known as discussed in Section 7.3.1. It should be noted that whenever a problem of interval construction is involved, one is dealing with two-tailed problem since interval only makes sense when two points are involved – hence the two-tailed test. Concerning the interval, it can be said that one is certain with probability $(1 - \alpha)$ that the true value μ will fall within the confidence limits; some books use $\leq$ and $\geq$, but $<$ and $>$ are used in this book since rounding off errors in calculations, distribution tables, and choosing significance level will make the equal sign meaningless. With regard to a confidence interval, it can be stated that the probability of a true value μ falling within the confidence region is $(1 - \alpha)$, which can be expressed mathematically as

$$P(\bar{x} - b < \mu < \bar{x} + b) = 1 - \alpha \tag{7.21}$$

Since the population mean μ is not known, the interval created based on the sample mean as shown in Equation (7.21) is not a guarantee that one would know if μ is really included within the calculated interval. One can say, however, that if the intervals are based on a $1 - \alpha$ probability, and if one obtains a great number of samples, one would expect μ to fall within the intervals $(1 - \alpha)$ % of the time. It should be noted that one cannot say that the probability is $1 - \alpha$ and that μ will fall within any particular interval since the probability is one if μ indeed falls within the interval and the probability is zero if it does not. Although one cannot make a probability statement about μ (which is a constant) being within a particular interval, the statisticians have devised a related concept, called the *degree of confidence*, that allows one to make statements about μ being within particular intervals. The degree of confidence is a measure of how confident one is that the stated interval will contain μ. Note that the end points of confidence interval are called *confidence limits* and the difference between upper and lower confidence limits is the *confidence interval width*. Just as the precision of estimate for a point estimate increases with an increase in the probability, so also the confidence interval width increases with an increase in the degree of confidence; the broader the interval, the more certain one can be that the interval will contain μ.

Confidence intervals can be constructed generally from Equations (7.19) and (7.20) as follows:

$$\mu = \bar{x} \pm (SE)k_p \tag{7.22}$$

or

$$\bar{x} - (SE)k_p < \mu < \bar{x} + (SE)k_p \tag{7.23}$$

where $b = (SE)k_p$, $k_p = z_{p = 1 - \alpha/2}$ when the sample size (n) is greater than 30 or the population standard deviation σ is known or both conditions are satisfied and otherwise, and $k_p = t_{p = 1 - \alpha/2, \, df}$ with df as the number of degrees of freedom. In this case, the SE in Equations (7.22) and (7.23) is the propagated standard deviation or standard error, which can be given as $SE = \sigma/\sqrt{n}$ for a case where the population standard deviation (σ) is known or $SE = s/\sqrt{n}$ in the case where

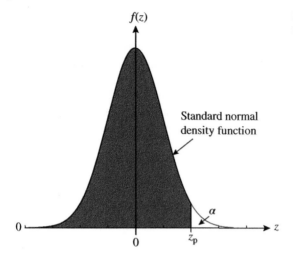

Figure 7.1 Standard normal curve.

the standard deviation of the population is unknown but the sample standard deviation (s) is determined. It can be seen from Equations (7.22) and (7.23) that confidence intervals get narrower as the number (n) of observations (or degrees of freedom) increases. The $z_{1-\alpha/2}$ value is obtained from the standard normal distribution table (where the standard normal curve areas are provided), and n is the number of observations. A standard normal curve is shown in Figure 7.1, where the standard normal curve areas (shaded) are related to their z_p values.

The area to the left of z_p (where $p = 1 - \alpha$) is the probability that a value less than z_p will be attained, which is the same as the confidence level $1 - \alpha$. This confidence level ($1 - \alpha$) can be expressed mathematically by integrating (or finding the area) from negative infinity ($-\infty$) to point z_p using the usual standard normal density function given as follows:

$$1 - \alpha = \int_{-\infty}^{z_p} \frac{1}{\sqrt{2\pi}} e^{-\frac{z^2}{2}} dx \tag{7.24}$$

Similarly, the area (α) to the right of z_p, which is the probability that the value of z_p will be attained or exceeded, can be obtained mathematically by integrating from point z_p to the positive infinity (∞) using the following:

$$\alpha = \int_{z_p}^{\infty} \frac{1}{\sqrt{2\pi}} e^{-\frac{x^2}{2}} dx \tag{7.25}$$

The integral in Equation (7.24), for example, can be determined using standard normal cumulative probability distribution tables or using some software package, such as Microsoft Excel 2013, that provides cumulative probability distribution routine NORM.S.DIST (z_p, TRUE)). If $z_p = 1.645$ is input to the routine, the value of $1 - \alpha = 0.95$ will be obtained. If $1 - \alpha$ is known, the inverse of Equation (7.24) will give the value of z_p. Instead of doing this mathematically, one can consult the standard normal distribution tables for z_p or the Microsoft Excel 2013, which provides NORM.S.INV (p) for the inverse calculation. For example, if $p = 0.95$ (or $\alpha = 0.05$), NORM.S.INV (0.95) gives $z_p = 1.645$. Remembering that z is symmetrical about the center ($z = 0$), one will notice that the values of z to the left of the center will be negative, while those to the right of the center will be positive. For example, using $p = 0.05$ in NORM.S. INV (p) will give $z_p = -1.645$.

In a confidence interval estimation where the number of observations or sample size (n) is less than or equal to 30, the student's t-distribution value will be used so that $k_p = t_{p\,=\,1\,-\,\alpha/2,\,df}$ in Equations (7.22) and (7.23). In this case, instead of using standard normal distribution, the $t_{1\,-\,\alpha/2,\,df}$ value will be obtained from the t-distribution table or from the Microsoft Excel 2013 routine T.INV(p, df) with the degrees of freedom (or redundancy) df. The *t-distribution* is similar to normal distribution, except that the degrees of freedom are now involved. For example, if $p = 0.95$ and df $= 5$ are input to T.INV (p, df), $t_{p,\,df} = 2.015$ will be obtained. The summary of interval estimation procedure for the population mean is given in Figure 7.2.

7.3.3 Relative Precision Estimation

Relative precision (RP) of a given measurement is the ratio of the precision of the measurement and the value of the measurement itself. Usually, the RP is determined for a given confidence level, such as 95% confidence level, and it is commonly expressed in ratio, such as $1 : p$ (e.g. $1 : 5000$). The RP of measurement at 95% confidence level can be given as follows:

$$\text{R.P.} = 1 : \frac{\text{Mean measurement}}{\text{Margin of error}} \tag{7.26}$$

In the case where the sample size is greater than 30, the RP at $(1 - \alpha)$ confidence level can be given as

$$\text{R.P.} = 1 : \frac{\bar{x}}{(\text{SE})k_p} \tag{7.27}$$

where $k_p = z_{p\,=\,1\,-\,\alpha/2}$. The $z_{1\,-\,\alpha/2}$ value is obtained from the normal distribution curve: $\text{SE} = \sigma/\sqrt{n}$ for a case where the population standard deviation (σ) is known or $\text{SE} = s/\sqrt{n}$ in the case where the standard deviation of the population is unknown but the sample standard deviation (s) is determined. In the case where the sample size is less than or equal to 30, the RP at $(1 - \alpha)$ confidence level will be calculated using Equation (7.27) with $k_p = t_{p\,=\,1\,-\,\alpha/2,\,df}$ and $t_{1\,-\,\alpha/2,\,df}$ value is obtained from the t-distribution curve with the degrees of freedom (or redundancy) df.

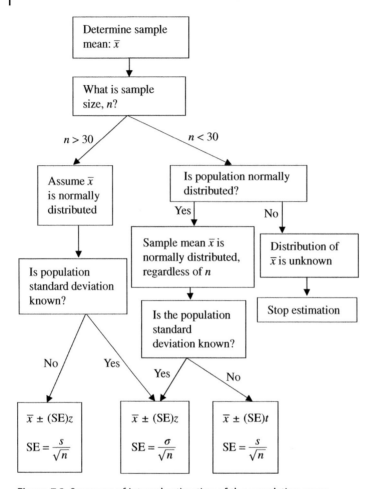

Figure 7.2 Summary of interval estimation of the population mean.

Example 7.2 The mean of 75 angle measurements is 55°27′23.9″ with its standard deviation (computed from the measurements) as 3.1″. Find the 95% confidence interval for the true mean.

Solution:

Since $n > 30$ use Equation (7.22) or Equation (7.23) with $k_p = z_{p \,=\, 1 \,-\, \alpha/2}$ as follows:

$$\mu = \bar{x} \pm (\text{SE})z_{1-\alpha/2}$$

or

$$\bar{x} - (\text{SE})z_{1-\alpha/2} < \mu < \bar{x} + (\text{SE})z_{1-\alpha/2}$$

For $\alpha = 0.05$, $z_{1-0.025} = 1.96$ (from the standard normal table for two-tailed test; refer to Table B.1). In order to extract the z-value, you can use the standard normal distribution program (NORM.S.INV) in Microsoft Excel 2013. This program gives the $z_{1-\alpha/2}$ value directly by inputting $1 - \alpha/2$. The z-values can also be extracted from Table B.1:

$$\bar{x} = 55°27'23.9'' \quad s = 3.1'' \quad SE = \frac{s}{\sqrt{n}} \quad z_{1-0.025} = 1.96$$

$$55°27'23.9'' - \left(\frac{3.1}{\sqrt{75}}\right)1.96 < \mu < 55°27'23.9'' + \left(\frac{3.1}{\sqrt{75}}\right)1.96$$

$$55°27'23.9'' - 0.7'' < \mu < 55°27'23.9'' + 0.7''$$

The 95% confidence interval: $55°27'23. 2'' < \mu < 55°27'24. 6''$

Example 7.3 The mean of 10 distance measurements is 16 183.677 m with its standard deviation (computed from the measurements) as 0.196 m. Find the 95% confidence interval for the true mean.

Solution:

Since $n < 30$ use Equation (7.22) or Equation (7.23) with $k_p = t_p = 1 - \alpha/2$, df as follows:

$$\mu = \bar{x} \pm (SE)t_{1-\alpha/2,\,df}$$

or

$$\bar{x} - (SE)t_{1-\alpha/2} < \mu < \bar{x} + (SE)t_{1-\alpha/2,\,df}$$

For $\alpha = 0.05$ at degrees of freedom of 9, $t_{1-0.05/2,\,df\,=\,10-1} = 2.26$ (for two-tailed test, extracted from Table B.2). The t-value at $\alpha = 0.025$ or at $\alpha = 0.975$ (the same absolute value is obtained for both $\alpha = 0.025$ and $\alpha = 0.975$ since the t-distribution gives symmetric values about 0.0 value) can also be obtained from the Student's t-distribution program (T.INV) in Microsoft Excel 2013 using the significance level α and the number of degrees of freedom. This example is based on the use of Table B.2:

$$\text{Given}: \bar{x} = 16\ 183.677\ \text{m} \quad s = 0.196\ \text{m} \quad t_{1-0.05/2,\,df\,=\,10-1} = 2.26 \quad SE = \frac{s}{\sqrt{n}}$$

The confidence interval can be given as

$$16\ 183.677 - 2.26 \times \frac{0.196}{\sqrt{10}} < \mu < 16\ 183.677 + 2.26 \times \frac{0.196}{\sqrt{10}}$$

$$16\ 183.677 - 0.140 < \mu < 16\ 183.677 + 0.140$$

The 95% confidence interval: $16\ 183.537 < \mu < 16\ 183.817$

Example 7.4   Continue from Example 7.3, if the mean of 10 distance measurements is 16 183.677 m with its standard deviation (computed from the measurements) as 0.196 m, find the 95% RP of the measurement.

Solution:

Given : $\bar{x} = 16\ 183.677$ m $s = 0.196$ m $t_{1-0.05/2,\ df\ =\ 10-1} = 2.26$ $SE = \dfrac{s}{\sqrt{n}} = 0.062\ 0$

Margin of error at 95% level $= (SE)t_{1-\alpha/2,\ df}$ or 0.140 m

$$R.P. = 1 : \frac{\bar{x}}{(SE)t_{1-\alpha/2,\ df}} \rightarrow 1 : \frac{16\ 183.677}{0.140} \rightarrow 1\ :\ 115\ 600$$

7.3.4   Interval Estimation for Population Variance

From the sampling distribution, the χ^2 statistic *(or chi-square statistic)* for the sample variance (s^2) can be given as

$$\chi^2 = \frac{(\,df)s^2}{\sigma^2} \tag{7.28}$$

The confidence interval for the population variance (σ) in the case of two-tailed problem with one observable involved can be given as follows (refer to Figure 7.3 in order to use this equation):

$$\chi^2_{\frac{\alpha}{2},\ df} < \chi^2 < \chi^2_{1-\frac{\alpha}{2},\ df} \tag{7.29}$$

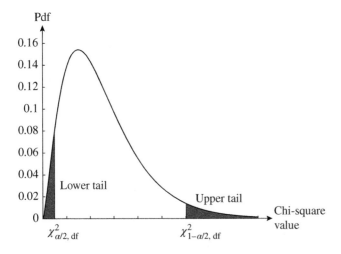

Figure 7.3 Typical two-tailed chi-square distribution (lower area case).

Or by substituting Equation (7.28) into Equation (7.29), the following can be deduced:

$$\frac{(df)s^2}{\chi^2_{1-\frac{\alpha}{2}, df}} < \sigma^2 < \frac{(df)s^2}{\chi^2_{\frac{\alpha}{2}, df}} \tag{7.30}$$

where n is the number of observations

α = significance level
df = degrees of freedom (e.g. $n - 1$ for single observable)
σ = population standard deviation
χ^2 = chi-square value
s = computed standard deviation

The chi-square in Table B.3 is based on p being an area to the left of the distribution curve. In this standard chi-square table, the areas (i.e. p) given across the top are the areas to the left of the critical value. To look up, for example, an area $1 - \alpha/2$ to the left of a critical value, take $p = 1 - \alpha/2$ and then look it up in the table. Refer to the Figure 7.3 and Table B.3 for chi-square values with p referring to the area to the left of a critical value.

For an example, the chi-square values corresponding to $\alpha = 0.010$ at the degrees of freedom, df = 4 (in the Figure 7.3) for two-tailed case, will be $\chi^2_{0.01/2, df = 4} = 0.207$ and $\chi^2_{1-0.01/2, df = 4} = 14.860$ for $p = 0.005$ and 0.995, respectively. Refer to Table B.3 and also to the Microsoft Excel 2013 routine CHISQ.INV (p, df) to obtain the chi-square values at the given p-values and the degrees of freedom df. The relationship between the chi-squared value (e.g. $\chi^2_{0.01/2, df = 4}$) and the level of significance α (which is the area from negative infinity ∞ to $\chi^2_{0.01/2, df = 4}$) is given by the equation

$$p = 0.01/2 = \int_{-\infty}^{\chi^2_{, p = 0.005, df = 4}} \chi^2(x)dx \tag{7.31}$$

If chi-square distribution with areas to the right of critical values is considered, the integral in Equation (7.31) will be done from the given critical value to positive infinity (∞) in order to obtain $p = \alpha/2$. If a nonstandard chi-square distribution (the one that gives areas to the right of critical values) is used, the following confidence interval (modified from Equation (7.30)) should be used:

$$\frac{(df)s^2}{\chi^2_{\frac{\alpha}{2}, df}(\text{upper area})} < \sigma^2 < \frac{(df)s^2}{\chi^2_{1-\frac{\alpha}{2}, df}(\text{upper area})} \tag{7.32}$$

The Microsoft Excel 2013 routine CHISQ.INV.RT (p, df) gives the upper area chi-square values, which are based on nonstandard chi-square statistical distribution (using upper tail area value of α). In order to obtain the chi-square value given in Figure 7.3, however, the Microsoft Excel 2013 program CHISQ. INV(p, df) is to be used.

7.3.5 Interval Estimation for Ratio of Two Population Variances

Confidence interval can be constructed for ratio of two variances from two normally distributed populations. Consider two experimental standard deviations s_1 and s_2 computed from two independent samples (Sample I and Sample II) of sizes n_1 and n_2, respectively, and let the two samples (Sample I and Sample II) be taken from two normally distributed populations with population standard deviations of σ_1 and σ_2, respectively. From the sampling distribution, the *F-statistic* for the ratio of sample variances can be given as

$$F = \frac{s_1^2 / s_2^2}{\sigma_1^2 / \sigma_2^2} \tag{7.33}$$

where F is distributed as Fisher distribution. A typical F (Fisher)-distribution for two-tailed case is given in Figure 7.4. The confidence interval for the ratio of the population variances (σ_1^2/σ_2^2) can be given as follows:

$$F_{\frac{\alpha}{2}, df_1, df_2} < F < F_{1-\frac{\alpha}{2}, df_1, df_2} \tag{7.34}$$

or by substituting Equation (7.33) into Equation (7.34), the following can be deduced:

$$\frac{s_1^2 / s_2^2}{F_{1-\alpha/2, df_1, df_2}} < \frac{\sigma_1^2}{\sigma_2^2} < \frac{s_1^2 / s_2^2}{F_{\alpha/2, df_1, df_2}} \tag{7.35}$$

where df_1 and df_2 are the degrees of freedom for computing the sample variances s_1^2 and s_2^2, respectively; α is the significance level; and $F_{\alpha/2, df_1, df_2}$ and $F_{1-\alpha/2, df_1, df_2}$ are the lower tailed and upper tailed F-distribution critical values, as shown in Figure 7.4.

The critical values for the F (Fisher)-distribution can be extracted using Microsoft Excel 2013 program F.INV(p, df_1, df_2) for two-tailed case shown in Figure 7.4. In addition, the F-values for lower and upper tails can be related as follows:

$$F_{\alpha, df_1, df_2} = \frac{1}{F_{1-\alpha, df_2, df_1}} \tag{7.36}$$

The Table B.4 gives F critical values based on areas to the left of critical values. From Table B.4, for example, to extract F-value at $df_1 = 4$, $df_2 = 8$, and $\alpha = 0.05$,

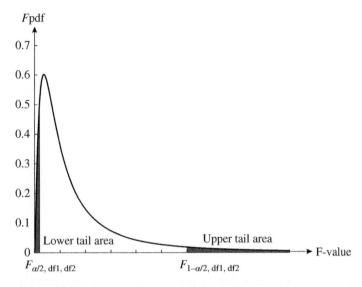

Figure 7.4 Typical two-tailed F-distribution (lower area case).

locate "4" under "df_1" column and 0.05 under "α" column; trace further on a straight line, and locate the F-value under "8" column under the "df_2" column; F-value will be 0.166. The Microsoft Excel 2013 program F.INV(p, df_1, df_2) can also be used to extract the F-value as follows: enter the p value (e.g. $p = 0.05$) as the probability and then the first and second degrees of freedom and obtain 0.166. The relationship expressed in Equation (7.36) can be tested by switching around the given two values of the degrees of freedom and using $p = 1 - \alpha$ instead of $p = \alpha$. For example, using Table B.4, the F-value at $df_1 = 8$, $df_2 = 4$, and $p = 0.95$ (switching the values of df_1 and df_2 and using $1 - \alpha$ instead of α) will be 6.041; the reciprocal of this F-value is 0.166 that is the same as that obtained above for F-value at $df_1 = 4$, $df_2 = 8$, and $\alpha = 0.05$.

Other useful statistical distribution tables are provided in the appendices, such as Extracts from Baarda's Nomogram (Baarda 1968) in Appendix A, Basic Statistical Tables (computed by the author) in Appendix B, and the Pope's Tau Distribution Table (1976), which was computed using Visual Basic computer codes by the author, in Appendix C.

Example 7.5 A baseline of calibrated length (μ) 1153.00 m is measured 5 times. Each measurement is independent and made with the same precision. The sample mean ($\bar{x}$) and sample standard deviation (s) are calculated from the measurements and given as follows:

$$\bar{x} = 1153.39\,\text{m}; \quad s = 0.06\,\text{m}.$$

Construct the confidence interval for the population variance at 10% significance level.

Solution:

Using the chi-square distribution in Table B.3 with areas to the left of critical values (shaded area in Figure 7.3), Equation (7.30) can be used as follows:

$$\frac{(df)s^2}{\chi^2_{1-\frac{\alpha}{2}, df}} < \sigma^2 < \frac{(df)s^2}{\chi^2_{\frac{\alpha}{2}, df}}$$

$$s = 0.06 \text{ m}; \quad \chi^2_{0.95, df = 4} = 9.488; \quad \chi^2_{0.05, df = 4} = 0.711$$

$$\frac{(4) \times 0.06^2}{9.488} < \sigma^2 < \frac{(4) \times 0.06^2}{0.711} \rightarrow 0.0015 < \sigma^2 < 0.020$$

Example 7.6 In order to determine that the precision of leveling equipment is appropriate for the intended measuring task, two calibration tests were carried out with two samples of measurements by the same instrument but different observers. The results of the tests are as follows:

Test 1: Computed standard deviation of instrument (s_1) = 2.5 mm, number of degrees of freedom (df_1) = 25.

Test 2: Computed standard deviation of instrument (s_2) = 2.0 mm, number of degrees of freedom (df_2) = 18.

Construct 95% confidence interval for the ratio of the two population variances associated with the two tests.

Solution:

Given s_1 = 2.5 mm, s_2 = 2.0 mm, df_1 = 25, and df_2 = 18.
  Using the confidence interval given in Equation (7.35)

$$\frac{s_1^2 / s_2^2}{F_{1-\alpha/2, \, df_1, \, df_2}} < \frac{\sigma_1^2}{\sigma_2^2} < \frac{s_1^2 / s_2^2}{F_{\alpha/2, \, df_1, \, df_2}} \tag{7.37}$$

evaluate the confidence interval at α = 0.05. Using Microsoft Excel 2013 software routine for $\alpha/2$ in the lower tail area of F-distribution, F.INV(0.025, 25, 18) gives 0.428, and for $\alpha/2$ in the upper tail area, F.INV(0.975, 25, 18) gives 2.491. The 95% confidence interval for the ratio of the two population variances can be given as follows:

$$\frac{(2.5)^2 \Big/ (2.0)^2}{2.491} < \frac{\sigma_1^2}{\sigma_2^2} < \frac{(2.5)^2 \Big/ (2.0)^2}{0.428}$$

$$0.627 < \frac{\sigma_1^2}{\sigma_2^2} < 3.651$$

7.4 General Comments on Confidence Interval Estimation

Confidence intervals constructed for population parameters of survey observables can be interpreted better if the concepts of confidence intervals are understood. The following comments are to aid in this direction:

1) Confidence intervals are created for population parameters and not for sample statistics. It should be noted that the population parameter of a variable or observable is a constant value, while the sample statistics are random variables (that can have many values depending on the conditions of their determination). Each time a particular method of constructing an interval is applied to a new sample, a new interval is generated that may be different from the other previous intervals created for similar samples. In this case, it will be wrong, for example, to say that the probability of a true average being in a given interval is 95% since the true value is fixed and will always have 50% probability of being within the given interval or not being within.

2) Considering the case of 95% confidence interval, one can say that if a surveyor were to take 100 samples from the pool of measurements of the same observable, 95% times out of 100, the true value of the observable will fall between the intervals created. This can also be stated differently that the surveyor is 95% confident that the true value will fall between the intervals created. This is an objective statement about how often, in the long run, the random interval will contain the fixed but the unknown value of the observable, according to the randomness in the sampling. In this case, the subject of the probability statement is actually on the interval and not on the unknown parameters being investigated. In reality, the idea of probability will not make sense in this case since the true value will either fall within the intervals or not (with 50% probability). This explains the concepts that a confidence interval refers only to the randomness that results in the creation of the interval and not to the actual interval itself; this is to say that it is not that the particular interval contains the true value with 95% probability, but rather that the interval constructed in that particular way will contain the true

value 95% of the time. In practice, the surveyor will draw a single sample and construct a single confidence interval. That single confidence interval may or may not contain the true population parameter, but the surveyor can only say that they are 95% confident that their single interval captures the population parameter.

7.5   Error Ellipse and Bivariate Normal Distribution

The density function for one-dimensional case is expressed in Equation (3.1), where the standard deviation (σ) can be seen as error bar around the mean. The density function of n-dimensional multivariate (for multiple observables) normal distribution is given from Equation (3.19) as

$$f(X) = \frac{1}{(2\pi)^{n/2}\sqrt{\det(\Sigma)}} e^{-\frac{1}{2}(X-\mu)^T \Sigma^{-1}(X-\mu)} \tag{7.38}$$

where X is a vector of the measured values of the observables, Σ is n by n true covariance matrix of the vector of measured values, μ is the vector of the true values of the observables, n is the number of observables (not the number of observations) considered in the multivariate case, $(2\pi)^{n/2}\sqrt{\det(\Sigma)}$ is the normalizing constant that makes the volume under the multivariate normal distribution probability density equal to one, and $\det(\Sigma)$ is the determinant of the covariance matrix Σ. The covariance matrix Σ can be considered as an error ellipsoid around the means of observables involved. The surface plot of a two-dimensional (x–y) normal probability density function expressed by Equation (7.38) is given in Figure 7.5. In this figure, when a plane parallel to the x, y coordinate plane intersects the density surface at a given height, an ellipse of intersection is formed. The ellipse formed is the contour plot of the line of constant probability density; a series of ellipses (or contours) are formed corresponding to different heights of the probability function. The contours so plotted will be lines of equal altitude (equal probability) on a topographic map. The equation for the intersecting ellipse in the two-dimensional (bivariate) case is determined by requiring that the exponent in the density function in Equation (7.38) be a constant (k^2). This exponent, which represents a family of error ellipses centered on the means of the variables, is a quadratic form

$$k^2 = (X-\mu)^T \Sigma^{-1}(X-\mu) \tag{7.39}$$

which is distributed as chi-square (χ^2) with two degrees of freedom (for the two-dimensional case). Equation (7.39) is simply the sum of squares of independent data samples that are normally distributed, which produces another statistic known as chi-square statistic χ^2. It can be shown that the probability is $1 - \alpha$

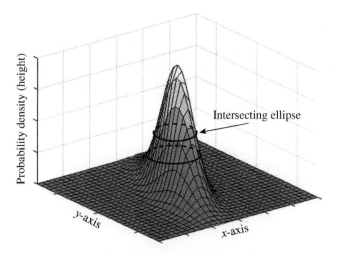

Figure 7.5 Bivariate normal probability density function constituting a series of error ellipses.

and that the value of a random vector will fall inside the ellipsoid defined by Equation (7.38); this probability can be expressed as

$$P\left[k^2 \leq \chi^2_{\mathrm{df}=n,\,1-\alpha}\right] = 1-\alpha \tag{7.40}$$

or

$$P\left\{(X-\mu)^T \Sigma^{-1}(X-\mu) < \chi^2_{\mathrm{df}=n,\,1-\alpha}\right\} = 1-\alpha \tag{7.41}$$

where α is the significance level (representing the lower tail area of the chi-square distribution), df is the number of degrees of freedom, n is the number of variable, and $\chi^2_{\mathrm{df}=n,1-\alpha}$ is the chi-square value at significance level of α. The quadratic form in Equation (7.39) is a squared statistical distance of X from μ accounting for the fact that the variances of the n variables may be different and they may be correlated. The X-values that yield a constant height for the probability density form ellipsoids centered at μ. For two-dimensional case, Equation (7.39) produces an ellipse that is centered at the point corresponding to the means (μ_x, μ_y) for the observables X. For a bivariate distribution, for example, $f(X)$ represents probability surface with the total volume bounded by the surface and the x–y plane equal to one.

The following can be considered as the typical elements involved in the function $f(X)$ in Equation (7.38):

$$X = \begin{bmatrix} x \\ y \end{bmatrix} \quad \mu = \begin{bmatrix} \mu_x \\ \mu_y \end{bmatrix} \quad \text{and} \quad \Sigma = \begin{bmatrix} \sigma_1^2 & \sigma_{12} \\ \sigma_{21} & \sigma_2^2 \end{bmatrix} \tag{7.42}$$

The 2×2 covariance matrix Σ is first diagonalized to produce the maximum and minimum eigenvalues λ_1 and λ_2, respectively, with their corresponding eigenvectors. The eigenvectors, which define the axes aligned with the semi-major and semi-minor axes of the error ellipse, produce the rotated coordinate system (u, v) so that the random variables in the directions of these axes are uncorrelated. Equation (7.39), for bivariate probability distribution centered at (μ_x, μ_y), can be given as

$$\left(\frac{x-\mu_x}{\sqrt{\lambda_1}}\right)^2 + \left(\frac{y-\mu_y}{\sqrt{\lambda_2}}\right)^2 = k^2 \tag{7.43}$$

which can be related to the usual equation of an ellipse centered at an origin $(u = 0, v = 0)$, given as

$$\frac{u^2}{a^2} + \frac{v^2}{b^2} = 1 \tag{7.44}$$

with u and v representing the directions of the eigenvectors with the semi-major axis value (a) and semi-minor axis value (b) of the ellipse, given, respectively, as follows:

$$a = k\sqrt{\lambda_1} \tag{7.45}$$
$$b = k\sqrt{\lambda_2} \tag{7.46}$$

where k is the scale of the ellipse, representing a chosen confidence level. From Equation (7.40), the marginal value of k can be given from the following:

$$k = \sqrt{\chi^2_{df = n, \, 1-\alpha}} \tag{7.47}$$

The error ellipse represented by Equations (7.43) and (7.44) is shown in Figure 7.6 with u and v representing the directions defined by corresponding eigenvectors.

In the case where the vector of true values (μ) of the observables and their corresponding covariance matrix (Σ) are unavailable, the estimated values of the observables $(\bar{X})$ and estimated covariance matrix (C) can be used instead of Equation (7.39) as follows:

$$k^2 = (X - \bar{X})^T C^{-1} (X - \bar{X}) \tag{7.48}$$

where all values of the vector X that satisfy Equation (7.48) define an ellipse with the axes in the directions of the eigenvectors of C and the lengths of the axes defined by the square root of the corresponding eigenvalues. Given the values of the quantities in Equation (7.48) as follows:

$$X = \begin{bmatrix} x \\ y \end{bmatrix}, \quad \bar{X} = \begin{bmatrix} \bar{x} \\ \bar{y} \end{bmatrix} \quad \text{and} \quad C = \begin{bmatrix} \lambda_1 & \\ & \lambda_2 \end{bmatrix} \tag{7.49}$$

Figure 7.6 Error ellipse representing bivariate contours.

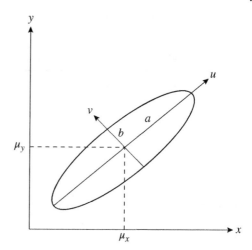

where λ_1 and λ_2 are the eigenvalues of the covariance matrix C (when diagonalized), the Equation (7.48) can be expressed from Equation (7.43) as follows:

$$\frac{(x-\bar{x})^2}{\left(\sqrt{\lambda_1}\right)^2} + \frac{(y-\bar{y})^2}{\left(\sqrt{\lambda_2}\right)^2} = k^2 \tag{7.50}$$

which represents the equation of two-dimensional ellipse centered at point $(\bar{x}, \bar{y})$ with the semi-major axis and semi-minor axis values expressed by Equations (7.45) and (7.46), respectively, where α is the significance level and df = 2 is for the number of degrees of freedom, which is 2 for two dimensions. Figure 7.6 represents a line of contour plot (iso-density contour) that denotes the set of points for which the values of x and y give the same value for the density function $f(X)$ as defined by the constant k. This contour is given for a fixed value of k, which defines a constant probability and centered at $(\bar{x}, \bar{y})$. Various values of k^2 in Equation (7.43) will produce a family of concentric ellipses (at different cross sections of the probability density surface with the intersecting planes at various heights of the surface). Equations (7.43) and (7.50) will become the equations of standard error ellipses if the value of k is one. Given the probability (p) and the dimension or degrees of freedom (df), the Microsoft Excel 2013 software routine *CHISQ.INV(p, df)* can be used to determine the equivalent chi-square $(\chi^2_{p, df})$ and k-values as shown in Table 7.2 (for significance level α being the lower tail area).

Given the dimension or degrees of freedom (df) and the constant (k), the Microsoft Excel 2013 software routine *CHISQ.DIST(χ^2, df, TRUE)* can be used to determine the equivalent probability value (p) as shown in Table 7.3 (for significance level α being the lower tail area).

Table 7.2 Chi-square and k-values at various probability values and dimensions.

p	df	$\chi^2_{p,\,df}$	$k = \sqrt{\chi^2_{p,\,df}}$	df	$\chi^2_{p,\,df}$	$k = \sqrt{\chi^2_{p,\,df}}$
0.3935	2	1.0000	1.0000	3	1.8389	1.3561
0.5000	2	1.3863	1.1774	3	2.3660	1.5382
0.9000	2	4.6052	2.1460	3	6.2514	2.5003
0.9500	2	5.9915	2.4477	3	7.8147	2.7955
0.9900	2	9.2103	3.0349	3	11.3449	3.3682

Table 7.3 Probability values at various k-values and dimensions.

k	$\chi^2_{p,\,df} = k^2$	df	p	df	p
0.5000	0.2500	2	0.1175	3	0.0309
1.0000	1.0000	2	0.3935	3	0.1987
2.0000	4.0000	2	0.8647	3	0.7385
2.4477	5.9915	2	0.9500	3	0.8880
3.0000	9.0000	2	0.9889	3	0.9707

7.6 Error Ellipses for Bivariate Parameters

Bivariate parameters are quantities that are associated with two dimensions, such as locations of points (involving x, y coordinates). The immediate local measure of accuracy of such quantities, such as the adjusted coordinates of a point, is usually related to the covariance matrix of the adjusted coordinates of that point. Assume the covariance matrix of adjusted coordinates $\hat{p}$ of a point (using 2×2-block covariance matrix corresponding to the point) is given as follows:

$$C_{\hat{p}} = \begin{bmatrix} s_{xx}^2 & s_{xy} \\ s_{yx} & s_{yy}^2 \end{bmatrix} \tag{7.51}$$

where the standard deviations (or precisions) of the adjusted coordinates $\hat{x}$ and $\hat{y}$ are s_{xx} and s_{yy}, respectively, and s_{yx} and s_{xy} are the covariances of the adjusted coordinates, which are equal to each other in the case of a symmetric covariance matrix. If the a priori variance factor of unit weight (σ_0^2) is unknown but set equal to 1, a new value (s_0^2) called a posteriori variance factor of unit weight must be calculated and used to scale the cofactor of the adjusted coordinates in order to

Figure 7.7 A typical error ellipse.

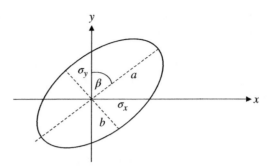

obtain a more realistic covariance matrix of the adjusted coordinates. In practice, it is popular to display precision of a network by its two-dimensional point
standard error ellipses or *confidence error ellipses*. The *standard error ellipses*
are the generalizations of the standard deviations used in one-dimensional
cases, while *confidence error ellipses* are the two-dimensional equivalent of
the confidence intervals used in one-dimensional cases.

Three quantities (parameters) are required to define an error ellipse: the semi-
major axis (a), semi-minor axis (b), and the bearing of the semi-major axis (β).
A typical error ellipse is illustrated in Figure 7.7.

There are two types of error ellipses depending on where the error ellipses are
situated: *absolute error ellipses*, which refer to given station points, are usually
situated at those station points, and *relative error ellipses*, which refer to
position differences of pairs of points, are situated in between the pairs of station
points that are connected by observations. In general, sizes of error ellipses
depend on geometry of network points, types of quantities observed, and
standard deviations of observations; they do not depend on the corrections
(or residuals) to be applied to the observations except the error ellipses are
scaled by the a posteriori variance factor of unit weight.

7.6.1 Absolute Error Ellipses

Absolute error ellipses are commonly used as precision criteria especially in
networks where only a few points in the network are of interest or have to be
positioned to a specified precision, e.g. in defining the center point of a
vertical shaft, the other control points only exist to allow that point to be
precisely positioned. In parametric least squares adjustment where control
points are fixed to define the datum for the adjustment, the associated
absolute error ellipses increase proportionally to distance from those fixed
points (meaning that the sizes of the absolute error ellipses are datum
dependent).

An absolute error ellipse can be constructed for a given point by using the
covariance matrix ($C_{\hat{p}}$) of the adjusted coordinates of the point. If the cofactor

matrix of the point is estimated, it must be multiplied by the a posteriori variance factor of unit weight (s_0^2) if the a priori variance factor of unit weight ($\sigma_0^2 = 1$) is not well known or the standard deviations of the observations are not well estimated. The parameters of an absolute error ellipse, for example, can be computed from the covariance matrix in Equation (7.51) depending on whether σ_0^2 is well known or not. The steps for the computations are as follows:

i) Compute the eigenvalues λ_1 (maximum value) and λ_2 (minimum value) from the covariance matrix $C_{\hat{p}}$ of the adjusted coordinates of the point by solving the determinant of the following composed from Equation (7.51):

$$\det\left(C_{\hat{p}} - \lambda I\right) = 0 \tag{7.52}$$

where I is an identity matrix and $\lambda = [\lambda_1 \ \lambda_2]^T$ is a vector of the eigenvalues. The solution of Equation (7.52) can be given as follows:

$$\lambda_1 = \frac{1}{2}\left(s_{xx}^2 + s_{yy}^2 + z\right) \tag{7.53}$$

$$\lambda_2 = \frac{1}{2}\left(s_{xx}^2 + s_{yy}^2 - z\right) \tag{7.54}$$

where

$$z = \left[\left(s_{xx}^2 - s_{yy}^2\right)^2 + 4s_{xy}^2\right]^{1/2} \tag{7.55}$$

and $s_{xy} = s_{yx}$, which is usually the case in adjustment computations. It can be seen from Equations (7.53) and (7.54) that the eigenvalue λ_1 is always numerically greater than the eigenvalue λ_2; take note also that it is not a typo that the covariance s_{xy} is squared as s_{xy}^2 in Equation (7.55) but not squared in the original covariance matrix in Equation (7.51).

ii) The parameters of the standard error ellipse can be given as follows:

$$a_{st} = \sqrt{\lambda_1} \tag{7.56}$$

$$b_{st} = \sqrt{\lambda_2} \tag{7.57}$$

$$\beta = \operatorname{atan}\left(\frac{s_{xy}}{\lambda_1 - s_{xx}^2}\right) \tag{7.58}$$

where a_{st}, b_{st}, and β are the values of the semi-major axis, semi-minor axis, and the bearing of the semi-major axis of the standard error ellipse and λ_1 and λ_2 are the eigenvalues defined above (with λ_1 always greater than λ_2).

iii) In the case where σ_0^2 is known, the parameters of the confidence error ellipse (at $1 - \alpha$ confidence level) can be given as follows (using $\chi^2_{1-\alpha,\,df=2}$

distribution with the degrees of freedom df = 2 and the 2 representing two coordinates associated with the point):

$$a_{(1-\alpha)100\%} = a_{st}k \qquad (7.59)$$

$$b_{(1-\alpha)100\%} = b_{st}k \qquad (7.60)$$

$$\beta = \mathrm{atan}\left(\frac{s_{xy}}{\lambda_1 - s_{xx}^2}\right) \qquad (7.61)$$

$$k = \sqrt{\chi^2_{1-\alpha,\,df=2}} \qquad (7.62)$$

where $a_{(1-\alpha)100\%}$, $b_{(1-\alpha)100\%}$, and β are the values of the semi-major axis, semi-minor axis, and the bearing of the semi-major axis of the $a_{(1-\alpha)100\%}$ confidence error ellipse; k is the *scaling factor* for transforming standard ellipses into confidence ellipses; df = 2 is the number of degrees of freedom, which represents the dimensions of the coordinate system involved (e.g. 2 for x-y coordinate system and 3 for x-y-z coordinate system); and λ_1 and λ_2 are the eigenvalues defined above (with λ_1 always greater than λ_2). For example, $\chi^2_{1-\alpha,\,df=2}$ = 5.99 and 9.21 for α = 0.05 and 0.01, respectively.

iv) In the case where σ_0^2 is unknown and s_0^2 was used in scaling the cofactor of the adjusted coordinates, the parameters of the confidence error ellipse (at $1 - \alpha$ confidence level) can be given as follows. The $F_{1-\alpha,\,df_1=2,\,df_2=n-u}$ distribution is used with the degrees of freedom df_1 = 2, $df_2 = n - u$, where the 2 represents the two coordinates associated with the point, n is the number of observations, and u is the number of unknown parameters (coordinates) determined in the original adjustment:

$$a_{(1-\alpha)100\%} = a_{st}k \qquad (7.63)$$

$$b_{(1-\alpha)100\%} = b_{st}k \qquad (7.64)$$

$$\beta = \arctan\left(\frac{s_{xy}}{\lambda_1 - s_{xx}^2}\right) \qquad (7.65)$$

$$k = \sqrt{2F_{1-\alpha,\,df_1=2,\,df_2=n-u}} \qquad (7.66)$$

where $a_{(1-\alpha)100\%}$, $b_{(1-\alpha)100\%}$, and β are the values of the semi-major axis, semi-minor axis, and the bearing of the semi-major axis of the $(1-\alpha)$ 100% confidence error ellipse and λ_1 and λ_2 are the eigenvalues defined above (with λ_1 always greater than λ_2).

Example 7.7   Given the covariance matrix of the adjusted coordinates (x, y) of a point P as

$$C_{\hat{p}} = \begin{bmatrix} 8.49\text{E}-5 & 2.27\text{E}-5 \\ 2.27\text{E}-5 & 6.56\text{E}-5 \end{bmatrix}$$

Calculate the following:

a) The parameters of the absolute standard error ellipse assuming that σ_0 is known.
b) The parameters of the absolute 95% confidence error ellipse assuming that σ_0 is known.
c) The parameters of the absolute 95% confidence error ellipse assuming that σ_0 is unknown and the original adjustment was done with a redundancy of 10.

Solution:

a) For the parameters of the absolute standard error ellipse assuming that σ_0 is known, for the eigenvalues, use Equations (7.53)–(7.55):

$$z = \left[(8.49\text{E}-5-6.56\text{E}-5)^2 + 4(2.27\text{E}-5)^2 \right]^{1/2} = 4.933\text{E}-5$$

$$\lambda_1 = \frac{1}{2}(8.49\text{E}-5+6.56\text{E}-5+4.933\text{E}-5) = 9.992\text{E}-5$$

$$\lambda_2 = \frac{1}{2}(8.49\text{E}-5+6.56\text{E}-5-4.933\text{E}-5) = 5.058\text{E}-5$$

where λ_1 and λ_2 are the eigenvalues; it can be seen that λ_1 has a value that is greater than that of λ_2. The MATLAB code snippets for computing the above eigenvalues are given in Tables 7.4 and 7.5. In Table 7.5, the eigenvalues D and the corresponding eigenvectors V are computed from the given covariance matrix C. For the standard error ellipse, use Equations (7.56)–(7.58) as follows:

$$a_{st} = \sqrt{9.992\text{E}-5} = 0.010\,\text{m}$$

$$b_{st} = \sqrt{5.058\text{E}-5} = 0.007\,\text{m}$$

$$\beta = \arctan\left(\frac{2.27\text{E}-5}{9.992\text{E}-5-8.49\text{E}-5} \right) = \arctan(1.5113)$$

$$= 56°30'$$

Note that the greater of the two eigenvalues calculated above is used in obtaining the semi-major axis value (a_{st}) of the standard error ellipse and the smaller eigenvalue is used to obtain the semi-minor axis value (b_{st}) of the standard error ellipse. Another approach (Equation (7.67)), which is different from the one given in Equation (7.65), is used to compute the bearing

(β) of the semi-major axis; the MATLAB code uses the arctangent of the ratio of the elements corresponding to the larger eigenvalue, in the eigenvectors V, as follows:

$$\beta = \operatorname{atan}\left(\frac{V(1,2)}{V(2,2)}\right) \tag{7.67}$$

where $V(1, 2)$ and $V(2, 2)$ are the elements in column two (corresponding to the bigger eigenvalue) of the eigenvectors V.

Table 7.4 MATLAB code for calculating eigenvalues.

```
>> % Form the  covariance matrix C
>> C= [8.49E-5 2.27E-5;2.27E-5 6.56E-5]
C =
   1.0e-04 *
   0.849000000000000   0.227000000000000
   0.227000000000000   0.656000000000000
>> % determine the eigenvalues
>> e=eig(C)
e =
   1.0e-04 *
   0.505839808643551
   0.999160191356449
```

Table 7.5 MATLAB code for calculating eigenvalues D, eigenvectors V, and corresponding bearing of the semi-major axis.

```
>> % Continuing from Table 7.4, determine
eigenvalues (D) and eigenvectors (V)
>> [V,D]=eig(C)
V =
   0.551712567192598   -0.834034317760218
  -0.834034317760218   -0.551712567192598
D =
   1.0e-04 *
   0.505839808643551                    0
                   0   0.999160191356449
>> % determine the bearing of the semi-major axis
>> B=(180/pi)*atan(V(1,2)/V(2,2))
B =
  56.515418257393392
>>
```

b) For the parameters of the absolute 95% confidence error ellipse assuming that σ_0 is known, use Equations (7.59)–(7.61) and $\alpha = 0.05$, $\chi^2_{1-\alpha, \, df = 2} = 5.99$:

$$k = \sqrt{5.99} = 2.4474$$

$$a_{95\%} = 0.010 \times 2.4474 \quad \text{or} \quad 0.024 \, \text{m}$$

$$b_{95\%} = 0.007 \times 2.4474 \quad \text{or} \quad 0.017 \, \text{m}$$

$$\beta = 56°30'$$

c) For the parameters of the absolute 95% confidence error ellipse assuming that σ_0 is unknown and the original adjustment was done with a redundancy of 10, use Equations (7.63)–(7.65) and $\alpha = 0.05$, $F_{1-\alpha, \, df_1 = 2, \, df_2 = 10} = 4.10$.

$$k = \sqrt{2 \times 4.10} = 2.8636$$

$$a_{95\%} = 0.010 \times 2.8636 \quad \text{or} \quad 0.029 \, \text{m}$$

$$b_{95\%} = 0.007 \times 2.8636 \quad \text{or} \quad 0.020 \, \text{m}$$

$$\beta = 56°30'$$

Example 7.8 In a least squares adjustment, the covariance matrix of the unknown point C is given as follows:

$$C_{\hat{x}} = \begin{bmatrix} 2.485\text{E}-5 & -1.331\text{E}-5 \\ -1.331\text{E}-5 & 5.942\text{E}-5 \end{bmatrix}$$

The estimated eigenvalues are 2.031E-5 m^2 and 6.395E-5 m^2, and the number of degrees of freedom for the least squares adjustment is 3. Answer the following.

a) Confirm the estimated eigenvalues using MATLAB code, and use the code to determine the orientation of the semi-major axis of the standard error ellipse.

b) Calculate the semi-minor axis and semi-major axis values of the 99% confidence error ellipse for point C (assuming $\sigma_0^2 = 1$ is *unknown*).

Solution (a):

The eigenvalues (D) and eigenvectors (V) for the given covariance matrix are calculated using MATLAB code given in Table 7.6. The eigenvectors (V) are used to determine the orientation (β) of the semi-major axis using Equation (7.67). The eigenvalues are confirmed to be $\lambda_2 = 2.032\text{E-5} \, \text{m}^2$ and $\lambda_1 = 6.395\text{E-5} \, \text{m}^2$; the calculated orientation (β) of the semi-major axis is $-18.80°$.

Table 7.6 MATLAB code for eigenvalues and orientation calculation.

```
>> % form covariance matrix C
>> C=[2.485E-5 -1.331E-5;-1.331E-5 5.942E-5]
C =
   1.0e-04 *
   0.248500000000000  -0.133100000000000
  -0.133100000000000   0.594200000000000
>> % determine eigenvalues (D) and eigenvectors D
>> [V, D]=eig(C)
V =
  -0.946656527810325  -0.322244345728672
  -0.322244345728672   0.946656527810325
D =
   1.0e-04 *
   0.203192413608878                    0
                   0   0.639507586391122
>> % Determine the bearing of semi-major axis
>> B=(180/pi)*atan(V(1,2)/V(2,2))
B =
 -18.798707830126695
>>
```

Solution (b):

The assumption is that σ_0^2 is *unknown*; note also from the question that the eigenvalues are given with the smaller one given first; this means that $\lambda_2 = $ 2.031E-5 m^2 and $\lambda_1 = $ 6.395E-5 m^2. On the basis that σ_0^2 is *unknown*, Equations (7.63)–(7.65) should be used as follows:

$$F_{(0.99,\text{df}_1 = 2,\text{df}_2 = 3)} = 30.8 \quad (\text{From the Fishers } F\text{-distribution table})$$

$$k = \sqrt{2F_{1-\alpha,\, \text{df}_1 = 2,\, \text{df}_2 = n-u}} \rightarrow k = 7.848\ 566\ 7$$

$$a_{st} = \sqrt{\lambda_1} = 7.996\ 874\text{E-}3 \quad b_{st} = \sqrt{\lambda_2} = 4.506\ 66\text{E-}3$$

$$a = a_{st}k \rightarrow 0.0628\ \text{m}$$

$$b = b_{st}k \rightarrow 0.0354\ \text{m}$$

7.6.2 Relative Error Ellipses

Relative error ellipses are traditionally centered at midpoint of two stations involved. In this case, the variance–covariance matrix of the coordinate differences between the two points will be used to construct the relative error ellipses.

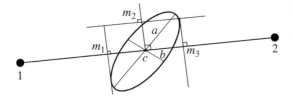

Figure 7.8 Relative error ellipse between stations 1 and 2.

For example, the relative error ellipse between two stations (1 and 2) can be constructed as given in Figure 7.8. In the figure, the relative error ellipse is centered at point c, and there are two specific points at both ends of the ellipse with tangent lines to those points intersecting perpendicularly the line 1–2 (at points m_1 and m_3) to which the relative error ellipse relates; another perpendicular line from the center of the relative error ellipse will intersect the error ellipse at point m_2. Distance cm_1 or cm_3 is the standard deviation σ_D of the distance 1–2 (of length D), and the distance cm_2 is $D\sigma_\beta$ for the bearing β with σ_β (in radians) as the standard deviation of the bearing, which can be computed from the adjusted coordinates of points 1 and 2.

To illustrate the computation of the parameters of a relative error ellipse, let the variance–covariance matrix (from the least squares adjustment) for two stations (1 and 2) be given as follows:

$$C_{\hat{x}} = \begin{bmatrix} s_{x_1}^2 & s_{x_1 y_1} & s_{x_1 x_2} & s_{x_1 y_2} \\ s_{y_1 x_1} & s_{y_1}^2 & s_{y_1 x_2} & s_{y_1 y_2} \\ s_{x_2 x_1} & s_{x_2 y_1} & s_{x_2}^2 & s_{x_2 y_2} \\ s_{y_2 x_1} & s_{y_2 y_1} & s_{y_2 x_2} & s_{y_2}^2 \end{bmatrix} \tag{7.68}$$

The coordinate differences between the two points 1 (x_1, y_1) and 2 (x_2, y_2) can be given as follows:

$$\Delta x = x_2 - x_1 \tag{7.69}$$

$$\Delta y = y_2 - y_1 \tag{7.70}$$

By variance–covariance propagation law on Equations (7.69) and (7.70), the relative covariance matrix $(C_{\Delta 12})$ for the two points can be given as

$$C_{\Delta 12} = B C_{\hat{x}} B^T \tag{7.71}$$

where B is the Jacobian of the Equations (7.69) and (7.70) with respect to the coordinates of points 1 (x_1, y_1) and 2 (x_2, y_2):

$$B = \begin{bmatrix} \dfrac{\partial \Delta x}{\partial x_1} & \dfrac{\partial \Delta x}{\partial y_1} & \dfrac{\partial \Delta x}{\partial x_2} & \dfrac{\partial \Delta x}{\partial y_2} \\ \dfrac{\partial \Delta y}{\partial x_1} & \dfrac{\partial \Delta y}{\partial y_1} & \dfrac{\partial \Delta y}{\partial x_2} & \dfrac{\partial \Delta y}{\partial y_2} \end{bmatrix} = \begin{bmatrix} -1 & 0 & 1 & 0 \\ 0 & -1 & 0 & 1 \end{bmatrix} \tag{7.72}$$

Using Equations (7.68) and (7.72) in Equation (7.71) gives

$$C_{\Delta 12} = \begin{bmatrix} s_{\Delta x}^2 & s_{\Delta x \Delta y} \\ s_{\Delta y \Delta x} & s_{\Delta y}^2 \end{bmatrix} \tag{7.73}$$

where

$$s_{\Delta x}^2 = s_{x_1}^2 + s_{x_2}^2 - 2s_{x_1 x_2} \tag{7.74}$$

$$s_{\Delta y}^2 = s_{y_1}^2 + s_{y_2}^2 - 2s_{y_1 y_2} \tag{7.75}$$

$$s_{\Delta x \Delta y} = s_{x_1 y_1}^2 + s_{x_2 y_2}^2 - s_{x_1 y_2} - s_{y_1 x_2} \tag{7.76}$$

To compute the parameters of the relative error ellipse between points 1 and 2, use the relative covariance matrix $(C_{\Delta 12})$ in Equation (7.73) as follows:

$$\lambda_1 = \frac{1}{2}\left(s_{\Delta x}^2 + s_{\Delta y}^2 + R\right) \tag{7.77}$$

$$\lambda_2 = \frac{1}{2}\left(s_{\Delta x}^2 + s_{\Delta y}^2 - R\right) \tag{7.78}$$

$$\theta = \text{atan}\left(\frac{s_{\Delta x \Delta y}}{\lambda_1 - s_{\Delta x}^2}\right) \tag{7.79}$$

$$R = \left[\left(s_{\Delta x}^2 - s_{\Delta y}^2\right)^2 + 4s_{\Delta x \Delta y}^2\right]^{1/2} \tag{7.80}$$

where λ_1 and λ_2 are the maximum and the minimum eigenvalues of the relative covariance matrix $C_{\Delta 12}$ (note also here that λ_1 is always greater than λ_2) and θ is the bearing of the semi-major axis of the relative error ellipse. The confidence and standard relative error ellipses can be obtained similarly as in the case of absolute error ellipses, by substituting the eigenvalues in the appropriate equations in steps (ii)–(iv) in Section 7.6.1. In comparison with absolute error ellipses, it should be pointed out that relative error ellipses are invariant with respect to translation of datum, while absolute error ellipses depend on the underlying datum.

Example 7.9   Given the vector of adjusted parameters, $\hat{x} = \hat{x} = [\hat{x}_B \ \hat{y}_B \ \hat{x}_C \ \hat{y}_C]^T$, by least squares method with the degrees of freedom of 10, the covariance matrix of the vector of the adjusted parameters, $\hat{x}$, is as follows:

$$C_{\hat{x}} = \begin{bmatrix} 8.04E-5 & -4.98E-5 & 8.04E-5 & -4.98E-5 \\ -4.98E-5 & 8.07E-5 & -4.98E-5 & 8.07E-5 \\ 8.04E-5 & -4.98E-5 & 2.90E-4 & 1.18E-4 \\ -4.98E-5 & 8.07E-5 & 1.18E-4 & 1.41E-3 \end{bmatrix} m^2$$

Compute the elements (axes and orientation) of the relative error ellipse for a pair of network points B and C at 5% significance level.

Solution:

Let the adjusted coordinate differences be given as

$$\Delta\hat{x} = \hat{x}_C - \hat{x}_B \tag{7.81}$$

$$\Delta\hat{y} = \hat{y}_C - \hat{y}_B \tag{7.82}$$

The Jacobian matrix (G) of Equations (7.81) and (7.82) with respect to the adjusted coordinates, $\hat{x} = [\hat{x}_B \ \hat{y}_B \ \hat{x}_C \ \hat{y}_C]^T$:

$$G = \frac{\partial R}{\partial x} = \begin{bmatrix} -1 & 0 & 1 & 0 \\ 0 & -1 & 0 & 1 \end{bmatrix}$$

By the laws of variance–covariance propagation, covariance matrix of the coordinate differences (referring to Equation (7.73)) can be given as

$$C_R = GC_{\hat{x}}G^T = \begin{bmatrix} 0.000\ 209\ 6 & 0.000\ 167\ 8 \\ 0.000\ 167\ 8 & 0.001\ 329\ 3 \end{bmatrix}$$

Find the eigenvalues using Equations (7.77)–(7.80):

$$R = \left(\left(\sigma^2_{\Delta x} - \sigma^2_{\Delta y} \right)^2 + 4\sigma^2_{\Delta x \Delta y} \right)^{1/2} = 1.168\ 912\ 1E - 3$$

$$\lambda_1 = \frac{\left(\sigma^2_{\Delta x} + \sigma^2_{\Delta y} + R \right)}{2} = 1.353\ 906E - 3$$

$$\lambda_2 = \frac{\left(\sigma^2_{\Delta x} + \sigma^2_{\Delta y} - R \right)}{2} = 1.849\ 940E - 4$$

Orientation of the semi-major axis of the error ellipse:

$$\theta = \arctan\left(\frac{\sigma_{\Delta x \Delta y}}{\lambda_1 - \sigma^2_{\Delta x}} \right) \rightarrow \arctan(0.146\ 639\ 1)$$

$$\theta = 8.342\ 35° \quad (\text{or } 8°20'32'')$$

Compute the 95% confidence relative ellipse for B–C using Equations (7.63) and (7.64) since there is no given information about the estimation of the a priori variance factor of unit weight and of the standard deviations of the observations used in the adjustment:

$$F_{2,10,0.95} = 4.10 \quad (\text{from } F-\text{distribution})$$

$$k = \sqrt{2F_{1-a,\ df_1\ =2,\ df_2\ =n-u}} \rightarrow k = 2.863\ 56$$

$a_{st} = \sqrt{\lambda_1} = 3.679\ 55\mathrm{E}-2 \quad b_{st} = \sqrt{\lambda_2} = 1.360\ 12\mathrm{E}-2$

$a = a_{st}k \rightarrow 0.1054\,\mathrm{m}$

$b = b_{st}k \rightarrow 0.0389\,\mathrm{m}$

Problems

7.1 In order to check the precision of a theodolite measurement, 32 direction readings of a single line were measured. The mean (seconds' portion only) of the readings was 25.2″, and the standard deviation of the mean measurement is ±0.4″. Answer the following.
a) Determine the 95% confidence interval for the population mean.
b) Compute the 99% confidence range for the population variance.

7.2 A baseline of calibrated length (μ) 100.0 m is measured 5 times. Each measurement is independent and made with the same precision. The sample mean ($\bar{x}$) and sample standard deviation (s) are calculated from the measurements: $\bar{x} = 100.5$ m; $s = 0.05$m.
a) Determine the 95% confidence interval for the population mean.
b) Compute the relative precision at 95% confidence level for the above baseline measurement.

7.3 The horizontal coordinates of point B (x_B, y_B) in the network shown in Figure P7.3 are to be determined in a traverse survey. In order to determine those coordinates, two bearings (α_1, α_2) and two distances (d_1, d_2) as shown in the figure were measured:

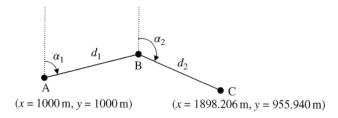

A
(x = 1000 m, y = 1000 m)   C
                                   (x = 1898.206 m, y = 955.940 m)

Figure P7.3

If the semi-major and the semi-minor axis values of the standard error ellipse at point B are 0.002 and 0.001 m, respectively, determine the semi-major and the semi-minor axis values of the 95% confidence ellipse for the point.

7.4 The variance–covariance matrix ($C_{\hat{x}}$) of the adjusted horizontal coordinates (x, y) of a survey station was determined from the method of least squares adjustment. The eigenvalues computed from the variance–covariance matrix are 0.0055 and 0.0625 m^2.

a) Calculate the semi-minor and semi-major axes of the standard error ellipse associated with this position error.

b) Calculate the semi-minor and semi-major axes of the 99% confidence error ellipse for this position error given the degrees of freedom for the adjustment as 13 (but σ_0^2 is unknown in estimating the covariance matrix of the observations).

8

Introduction to Network Design and Preanalysis

Objectives
After studying this chapter, you should be able to:

1) Discuss the problems of network design.
2) Explain different design variables and how they relate to each other, including their uses and importance.
3) Perform simple preanalysis of survey observations.
4) Perform network design and simulation involving one-dimensional, two-dimensional, and three-dimensional cases.

8.1 Introduction

Network design is about selecting observables to measure, measurement procedures, instrumentation, etc. for a project in order to achieve the specific goals of the project. Network design methods have evolved over time to solve

Understanding Least Squares Estimation and Geomatics Data Analysis, First Edition.
John Olusegun Ogundare.
© 2019 John Wiley & Sons, Inc. Published 2019 by John Wiley & Sons, Inc.
Companion website: www.wiley.com/go/ogundare/Understanding-lse-and-gda

different cases of network design problems, which can be summarized as follows (Grafarend 1974; Vanicek and Krakiwsky 1986; Cooper 1987):

1) *Zero-order design (ZOD) problem* or *datum problem*. In this problem, a suitable reference frame or a coordinate system is selected (or fixed) for the yet to be determined unknown coordinate parameters and their covariance matrices; this is about what points or lines to fix for the network adjustment. This is important in parametric model and general model least squares adjustment, where the network points fixed for datum definition also constitute the *zero-variance reference* base for adjustment.

2) *First-order design (FOD) problem* or *configuration problem*. In this problem, a suitable geometric layout or configuration (first design matrix or A-matrix) of network observables to be measured and the locations of network points must be determined based on the given covariance matrices of the unknown parameters and the measurements. This configuration must be selected with a consideration for intervisibility between network points and the nature of the topography of the project site.

3) *Second-order design (SOD) problem* or *generalized weight problem*. This problem requires determining the covariance matrix (C_ℓ) of measurements of the observables selected in the FOD solution (based on the given configuration or first design matrix (A-matrix) and the covariance matrix (C_x) of the unknown parameters). Since precisions of measurements are related to instruments and observation procedures (including the number of repetitions of a measurement), the SOD problem can be seen as a problem of selecting suitable instrument and observation procedures.

4) *Third-order design (ThOD) problem* or *densification problem*. This problem involves selecting observables, measurements, and weights for the purpose of improving an existing network. It can be seen as a problem of selecting how to best connect or integrate a new survey to an existing one (e.g. a national survey), which may involve considering the estimated coordinates of the existing survey and their covariances as a priori values in an adjustment.

Another case of design problem is described by Grafarend et al. (1979) as *combined design* (COMD) problem, where optimal solution to FOD and SOD problems is determined simultaneously, i.e. the network configuration (A-matrix) and the covariance matrix (C_ℓ) of measurements are determined, given the covariance matrix (C_x) of the unknown parameters. When a design is performed with consideration for the available instruments, most economic field survey, intervisibility of survey points, ability of network to allow identification and elimination of gross errors in observations, and the effects of undetected gross errors in observations on the network, in achieving a minimum value of a set objective function, the design is said to be *optimized*. An objective function to be maximized or minimized within some constraints could be reliability and sensitivity; in this case, the optimized design will produce the most reliable and sensitive network possible under the constraints of instrumentation, time, cost, etc. The *optimization of a design* is to achieve the best design in an analytical way (e.g. Grafarend 1974; Cross 1985; Kuang 1991). In his analytical approach, Kuang (1991) used the so-called multiobjective optimization method to solve all the cases of network design problems in a single mathematical way. Since the analytical approach is difficult to understand and implement, it is common to design networks that are just acceptable in terms of precision, reliability, cost, etc., but not necessarily the best.

As discussed above, the main variables of a design are reference frame, configuration, and observation precisions (or weights). Design and preanalysis allow one to experiment with these variables in the process of trying to meet a given accuracy specification for a project. According to Vanicek and Krakiwsky (1986), the main objective of preanalysis is to come up with a set of guidelines on what observables to measure and the acceptable accuracy of those measurements, given the expected tolerance limits of the unknown parameters. In this case, preanalysis will determine optimal accuracy (or precisions) of measurements by solving FOD, SOD, or COMD problems.

The variables selected in a survey network design usually depend on the type of network involved: *simple* or *complex*. Simple networks may have observation precisions as the only variables, while complex networks, which usually require least squares adjustment, may have variables that include reference frame, configuration, and observation precisions. After the initial selection of values for these variables (initial design), these variables are changed in the process of achieving an optimal results by the procedure of *preanalysis* or *simulation*. Simple network (survey) design will require simple preanalysis, while complex network design will require complex preanalysis. In this chapter, simple survey design and preanalysis will be referred to as simply *preanalysis of survey observations*; complex network design (or network design) and preanalysis will be treated under *network design* with the preanalysis aspect treated under *network simulation*.

8.2 Preanalysis of Survey Observations

Preanalysis of survey observations discussed in this section is about analyzing simple survey observations for a project before the project is actually started. A simple network design involved in this section is the SOD type with the variables being the precisions of observations. This means that preanalysis of survey is done in order to determine precisions of observations that satisfy a specified tolerance limit for the unknown quantities to be determined. At the end of a preanalysis, it may be concluded by the surveyor that the requirements for the accuracies of measurements are within or beyond what can be achieved. If the requirements are beyond the capabilities of the surveyor, the client must be told that the survey tolerance limits specified are beyond what can be satisfied.

8.2.1 Survey Tolerance Limits

Survey tolerance limits are the intervals within which the maximum allowable error of observation must fall. Based on the concepts of interval estimation discussed in Section 7.3.2, Equation (7.17) or Equation (7.18) can be used to determine the survey tolerance. Usually, at the preanalysis stage, Equation (7.20) should be used, such that the survey tolerance will be given as $z_{1-\alpha/2}(SE)$. The most commonly used uncertainty for survey tolerance is at a probability of 99.7% or $\alpha = 0.003$. Using Equation (7.20) with $z_{1-0.003/2} = 3.0$, the survey tolerance or maximum error acceptable can be given as 3(SE) or three times the standard error (SE) of measurement. Note that if the error of a single measurement is of interest, the SE will be taken as the standard deviation of the single measurement. For example, a measurement is said to meet its specified tolerance if its standard deviation times three (3σ) is less than the given tolerance. For example, if the tolerance of ±15 mm is allowed in a measurement, the allowable standard deviation should be ±5 mm.

8.2.2 Models for Preanalysis of Survey Observations

The mathematical models for performing preanalysis of survey observations relate with the laws of random error (variance–covariance) propagation in reversed form. In this case the mathematical model relating the unknown parameter (x) is first formulated as a function of the expected measurements (ℓ) in the form of $x = f(\ell)$. The usual variance–covariance propagation laws discussed in Chapter 2 are then applied to the functional model with the covariance matrix of observations considered unknown to be solved for. A simple example can be used to illustrate the model with reference to variance–covariance propagation laws in Chapter 2. For example, consider a simple case where the total random error expected in the measurement of a 500 m distance is to be 16 mm. The expected random error in each of the 50 m tape measurement can be determined when the tape is used to measure the 500 m distance as follows. This is simply an error propagation problem in reversed form. By using 50 m tape, the total distance $D = 500$ m will have to be measured in 10 segments (with each segment $d_i = 50$ m). This can be expressed mathematically as follows:

$$D = d_1 + d_2 + \cdots + d_{10} \tag{8.1}$$

Equation (8.1) is a form of a model $x = f(\ell)$ given in Equation (2.1), where in this problem, $x = D$ is the parameter whose error propagation is to be made and the vector of observations is $\ell = [d_1 \; d_2 \; \cdots \; d_{10}]^T$. Applying the variance–covariance propagation rules (referring to Equation (2.43) or (2.44)) to Equation (8.1) and assuming zero correlation between them,

$$\sigma_D^2 = \left(\frac{\partial D}{\partial d_1}\right)^2 \sigma_{d_1}^2 + \left(\frac{\partial D}{\partial d_2}\right)^2 \sigma_{d_2}^2 + \cdots + \left(\frac{\partial D}{\partial d_{10}}\right)^2 \sigma_{d_{10}}^2 \qquad (8.2)$$

where, for example, $\dfrac{\partial D}{\partial d_1}$ is the partial derivative of the parameter D with respect to the observation d_1. There is an additional assumption necessary in order to evaluate Equation (8.2), which is to assume that all observations will contribute the same amount of error to the overall error for the parameter (assuming *balanced accuracy of measurements*). Note that balanced accuracy of measurements is used to mean each term in the variance–covariance propagation for the unknown measurements will have equal contribution to the given variance of the unknown parameter. This can be expressed mathematically as

$$\left(\frac{\partial D}{\partial d_1}\right)^2 \sigma_{d_1}^2 = \left(\frac{\partial D}{\partial d_2}\right)^2 \sigma_{d_2}^2 = \cdots = \left(\frac{\partial D}{\partial d_{10}}\right)^2 \sigma_{d_{10}}^2 = \frac{\sigma_D^2}{10} \qquad (8.3)$$

As can be seen in Equation (8.3), each component in the variance–covariance propagation in Equation (8.2) is equated to the square of overall error divided by the number of observations (10) involved. All the partial derivatives of the Equation (8.1) are ones; substituting the partial derivatives and $\sigma_D = 16$ mm into Equation (3.37b) gives

$$(1)^2 \sigma_{d_1}^2 = \frac{\sigma_D^2}{10} \rightarrow \sigma_{d_1} = \sqrt{\frac{256}{10}} \text{ mm} \quad \text{or} \quad \pm 5 \text{ mm} \qquad (8.4)$$

which is the same for the remaining observations. The standard deviation of each 50 m tape measurement must be ±5 mm for the total random error of 16 mm to be achieved in the measurement of the 500 m distance. The following example is given to further illustrate how simple preanalysis of survey observations can be done.

Example 8.1   For visible and near-infrared radiation and neglecting the effects of water vapor pressure, the formula for computing the refractive index, n, in an EDM can be determined by

$$n - 1 = \frac{0.269\,578[n_0 - 1]}{273.15 + t} p$$

where n_0 is the constant refractive index set in the EDM, t is the temperature in °C, p is the pressure in mbar, and n is the realistic refractive index to be determined. The EDM has a set constant value, $n_0 = 1.000\,294\,497$; and the average temperature and pressure during the measurements are expected to be $+30\,°C$ and 950 mb, respectively. Assuming the standard deviation of measuring temperature, $\sigma_t = 1.0\,°C$, what would be the largest value of σ_p so that the error in σ_n will not be more than 2 ppm?

Solution:

By variance–covariance propagation laws in Chapter 2,

$$J = \begin{bmatrix} \dfrac{\partial n}{\partial p} & \dfrac{\partial n}{\partial t} \end{bmatrix} \quad J = \begin{bmatrix} \dfrac{0.269578[n_0 - 1]}{273.15 + t} & -\dfrac{0.269578[n_0 - 1]p}{(273.15 + t)^2} \end{bmatrix}$$

$\sigma_t = 1.0°C, \; n_0 = 1.000\,294\,497; \; t = +30°C; \; p = 950 \text{ mb},$

or

$$J = \begin{bmatrix} 2.618\,833\text{E}-7 & -8.206\,798\,7\text{E}-7 \end{bmatrix}$$

Covariance matrix (C) of measurements and variance–covariance propagation:

$$C = \begin{bmatrix} (1)^2 & \\ & \sigma_t^2 \end{bmatrix} \quad \sigma_n^2 = JCJ^T$$

or

$$\sigma_n^2 = \left(\frac{0.269\,578[n_0 - 1]}{273.15 + t} \right)^2 \sigma_p^2 + \left(\frac{0.269\,578[n_0 - 1]}{(273.15 + t)^2} p \right)^2 \sigma_t^2$$

$$(2.0\text{E}-6)^2 = (2.618\,832\,665\,87\text{E}-7\sigma_p)^2 + (8.206\,798\,722\,02\text{E}-7\sigma_t)^2$$

or

$$(2.0\text{E}-6)^2 = (8.206\,798\,722\,02\text{E}-7(1))^2 + (2.618\,832\,665\,87\text{E}-7\sigma_p)^2$$

$$(2.0\text{E}-6)^2 - (8.206\,798\,722\,02\text{E}-7(1))^2 = (2.618\,832\,665\,87\text{E}-7\sigma_p)^2$$

$$\sigma_p = \sqrt{\frac{3.326\,484\,547\,36\text{E}-12}{6.858\,284\,531\,83\text{E}-14}} = \sigma_p = \sqrt{48.503\,157\,5} \text{ or } 6.96 \text{ mb}$$

The largest value of the pressure so that the error in refractive index will not be more than 2 ppm is 6.96 mb.

8.2.3 Trigonometric Leveling Problems

The elevation difference when leveling between backsight point A and foresight point B with the total station instrument set at the midpoint based on leapfrog trigonometric leveling procedure (assuming the effect of earth curvature is negligible) can be given as

$$\Delta h = \left(S_B \cos Z_B - S_A \cos Z_A\right) - \frac{1}{2R}\left(k_B S_B^2 \sin^2 Z_B - k_A S_A^2 \sin^2 Z_A\right) \tag{8.5}$$

where S_A and S_B are the backsight and foresight slope distances to points A and B, respectively; Z_A and Z_B are the zenith angle measurements to points A and B, respectively; k_A and k_B are the coefficients of refraction to A and B, respectively; and $R = 6371$ km is the radius of the earth. Assume the average slope of the terrain (which is covered with the same material) is $10°$ (or zenith angle of $80°$), $\sin Z_A = \sin Z_B$, $\cos Z_A = -\cos Z_B$, $\Delta k = k_B - k_A$, or $\Delta k = 0.3$; the average sight lengths (with $S_A = S_B$) are to be 250 m, and the deflection of the vertical at the two stations are assumed negligible. Simplify Equation (8.5) based on some of the assumptions given above, and determine the standard deviations of the zenith angle and the slope distance measurements and the standard deviation of the difference in the coefficient of refraction so that the standard deviation of the elevation difference at this setup will be less than 2 mm.

Given $S_A = S_B$, $\sin Z_A = \sin Z_B$, $\cos Z_A = -\cos Z_B$, $k_A - k_B = 0.3$, substitute them into Equation (8.5):

$$\Delta h = 2S \cos Z - \frac{(\Delta k)S^2 \sin^2 Z}{2R} \tag{8.6}$$

By error propagation law,

$$\sigma_{\Delta h}^2 = \left(\frac{\partial \Delta h}{\partial S}\right)^2 \sigma_S^2 + \left(\frac{\partial \Delta h}{\partial Z}\right)^2 \sigma_Z^2 + \left(\frac{\partial \Delta h}{\partial \Delta k}\right)^2 \sigma_{\Delta k}^2 \tag{8.7}$$

For $S = 250$ m, $Z = 80°$, $\sigma_S = ?$, $\sigma_Z = ?$

$$\frac{\partial \Delta h}{\partial S} = 2\cos Z - \frac{2(\Delta k)S\sin^2 Z}{2R}$$

$$= 0.347\,296\,355\,334 - 1.902\,853\,379\,36\mathrm{E}-5 \quad (\text{or } 0.347\,277\,326\,8)$$

$$\frac{\partial \Delta h}{\partial Z} = -2S \times \sin Z - \frac{(\Delta k)S^2 \sin 2Z}{2R} -$$

$$= -492.403\,876\,506 - 8.388\,109\,777\,9\mathrm{E}-4 \quad (\text{or} -492.404\,715\,317)$$

$$\frac{\partial \Delta h}{\partial \Delta k} = \frac{S^2 \sin^2 Z}{2R} = 4.757\,133\,448\,4\mathrm{E}-3$$

If it is assumed that the measurements contribute equally to the overall accuracy, then

$$\left(\frac{\partial \Delta h}{\partial S}\right)^2 \sigma_S^2 = \left(\frac{\partial \Delta h}{\partial Z}\right)^2 \sigma_Z^2 = \left(\frac{\partial \Delta h}{\partial \Delta k}\right)^2 \sigma_{\Delta k}^2 = \frac{\sigma_{\Delta h}^2}{3}$$

$\sigma_{\Delta h} = 2$ mm

$$(0.347\,277\,326\,8)^2 \times \sigma_S^2 = \frac{2^2}{3} \quad \rightarrow \quad \sigma_S = \pm\,3.3\,\text{mm}$$

$$(-492.404\,715\,317)^2 \times \sigma_Z^2 = \frac{2^2}{3} \quad \rightarrow \quad \sigma_Z = 5.499\,134\,4\text{E}{-}6 \text{ rad (or } 1.1''\text{)}$$

$$(4.757\,133\,448\,4\text{E}{-}3)^2 \times \sigma_{\Delta k}^2 = \frac{2^2}{3} \quad \rightarrow \quad \sigma_{\Delta k} = 242.730$$

Example 8.2   A slope distance D and a zenith angle Z must be measured in order to calculate an elevation difference h. What should be the accuracy of D and Z in order to obtain h with a standard deviation $\sigma_h < 5$ mm? Assume that $D = 500$ m and $Z = 60°$.

Solution:

Equation for calculating the elevation difference h:

$$h = D \cdot \cos Z \tag{8.8}$$

Applying the variance–covariance propagation rules to Equation (8.8) with h as a function of D and Z and assuming zero correlation between D and Z,

$$\sigma_h^2 = \left(\frac{\partial h}{\partial D}\right)^2 \sigma_D^2 + \left(\frac{\partial h}{\partial Z}\right)^2 \sigma_Z^2 \tag{8.9}$$

$$\sigma_h^2 = \left(\cos^2 Z\right)\sigma_D^2 + \left(D^2 \sin^2 Z\right)\sigma_Z^2 \tag{8.10}$$

Substituting $Z = 60°$ and $D = 500\,000$ mm so that the right-hand side result will be in the same unit (mm^2) as the left-hand side of the equal sign and assuming σ_z will be in radians, the following is obtained:

$$\sigma_h^2 = 0.25\sigma_D^2 + 1.875\text{E}^{11}\sigma_z^2 \ \ (\text{mm}^2) \tag{8.11}$$

where σ_D and σ_z are the standard deviations of the slope distance and zenith angle measurements, respectively (which are unknown to be determined). Note that the distance D is converted to millimeters and the random error in zenith is expressed in radians for the purpose of making the unit in the equation consistent. Assuming that the distance and the angular measurements contribute the same amount of error into the elevation difference calculation, each of the terms of the Equation (8.11) will contribute $\dfrac{\sigma_h^2}{2} = 12.5$ mm^2, giving the following:

$$0.25\sigma_D^2 = 12.5 \quad \rightarrow \quad \sigma_D = \sqrt{\frac{12.5}{0.25}} < 7\,\text{mm}$$

$$1.875\text{E}^{11}\sigma_z^2 = 12.5 \quad \rightarrow \quad \sigma_z = \sqrt{\dfrac{12.5}{1.875\text{E}^{11}}} < 8.165\text{E}^{-6} \text{ rad } (\text{or } 1.7'')$$

The elevation difference h can be measured to a standard deviation of less than 5 mm if the distance and zenith angle can be measured with standard deviations of less than 7 mm and $1.7''$, respectively.

Example 8.3 A total station is to be used to measure the elevation difference between a setup point and another point Q as shown in Figure 8.1. In the process, the slope distance d_S, the zenith angle z, the height of instrument (HI), and the height of reflector (HR) will be measured in order to determine V (the simple height difference from the total station to the reflector). Answer the following.

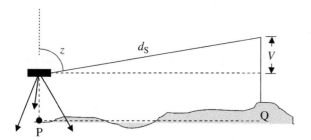

Figure 8.1 Illustration of the total station setup.

a) If the error in V (at 99.7% confidence level) is not to exceed ±15 mm, determine expected standard deviations in measuring the zenith angle (z), d_S, HI, and HR, assuming balanced accuracies with approximate values of these quantities as $z = 100°$, HI = 1.6 m, and $d_S = 200.0$ m.

Solution:

For the total station setup:
In one direction:

$$z = 100° \pm 5''; \ \text{HI} = 1.6\,\text{m} \pm 0.003; \ d_S = 200.0\,\text{m} \pm 0.003.$$

Elevation difference:

$$V = \text{HI} + d_S \cos z - \text{HR}$$

$$\sigma_V^2 = \sigma_{\text{HI}}^2 + (\cos z)^2 \sigma_{d_S}^2 + (d_S \sin z)^2 \sigma_z^2 + \sigma_{\text{HR}}^2 \tag{8.12}$$

The error in V at 99.7% confidence level is ±15 mm; the standard deviation can be derived from this using the relation that (SE) $z_{1-\alpha/2} = 15$ mm; for $z_{1-0.003/2} = 3$, the standard deviation (SE) can be given as

$$\sigma_V = \frac{0.015}{3} = 0.005 \text{ m}$$

Balancing the accuracies in Equation (8.12) gives

$$\sigma_{HI}^2 = \frac{\sigma_V^2}{4} \quad \rightarrow \quad \sigma_{HI} = \frac{\sigma_V}{2} = \frac{0.005}{2}\text{m} \text{ (or 2.5 mm)}$$

Similarly, $\sigma_{HRI} = \dfrac{\sigma_V}{2} = 2.5\,\text{mm}$

$$\sigma_z = \frac{\sigma_V}{2(d_S \sin Z)} = \frac{0.005}{393.923\,1} = 1.269\,283\,3\text{E}-5 \text{ rad (or 2.62'')}$$

$$\sigma_{d_S} = \frac{\sigma_v}{2(\cos Z)} = \frac{0.005}{2(0.173\,65)}\text{m} = 0.014\,\text{m}$$

The standard deviations of measuring the zenith, slope distance, HI, and HR are 2.6″, 0.014, 0.0025, and 0.0025 m, respectively.

b) The technical specifications for Leica TCRA 702 total station instrument are as follows:

- Standard deviation for horizontal (HZ) and vertical (Z) angles (ISO 17123-3) is 2″.
- Standard deviation for distance measurement (ISO 17123-4) (IR fine mode) is 2 mm ± 3 ppm.

If the TCRA 702 instrument is used in (a) above, determine the expected standard deviations in measuring HI and HR, assuming balanced accuracies and the centering error of 2 mm each for the instrument and target centering procedures.

Solution:

The standard deviations of measuring the zenith and the distance are known from the given specifications; the only unknowns are the errors in HI and HR.

Standard deviation of slope distance (given):

$$\sigma_{d_S} = \sqrt{2^2 + (3 \times 0.2)^2 + 2(2)^2} = 0.0035\,\text{m} \text{ (or 3.5 mm)}$$

For zenith angle measurement (in one set), $\sigma_z = 2''$.
From error propagation in Equation (8.12),

$$\sigma^2 = (\cos 100)^2 (0.003\,5)^2 + (200\sin 100)^2 \left(\frac{2}{206\,265}\right)^2 + \sigma_{HR}^2 + \sigma_{HI}^2$$

$$= 3.693\,826\,98E - 7 + 3.647\,302\,79E - 6 + \sigma_{HR}^2 + \sigma_{HI}^2$$

$$(8.13)$$

Again using the standard deviation of V from part (a) as $\sigma_V = 0.005$ m in Equation (8.13) and rearranging the terms give

$$(0.005)^2 - 4.016\,685\,483\,35E - 6 = \sigma_{HI}^2 + \sigma_{HR}^2$$

$$\sigma_{HI}^2 + \sigma_{HR}^2 = 2.098\,331\,4E - 5$$

For equal distribution from the remaining two components (HI and HR), the standard deviation of each component can be given as

$$\sigma_{HI}^2 = \frac{2.098\,331\,452E - 5}{2} \quad \rightarrow \quad \sigma_{HR} = \sigma_{HI} = \sqrt{1.049\,165\,726} = \underline{\mathbf{0.003\,2\,m}}$$

The HI and HR must be measured to an accuracy of 0.003 m in order to achieve an overall standard deviation of 0.005 m in height difference with TCRA 702 instrument used to measure the zenith and slope distance.

c) Continuing from (a) above, if the HI and HR will be measured to an accuracy of ± 3 mm, and the Leica TCRA in (b) will be used to measure the slope distance (with centering error of 2 mm each in the instrument and target centering procedures), determine new expected standard deviation of measuring the zenith angle (z).

Solution:

Substitute $\sigma_{d_S} = 0.0035$ m, $\sigma_{HR} = \sigma_{HI} = \pm 3$ mm, and $\sigma_V = 0.005$ m into Equation (8.14) rearranged from Equation (8.12), and solve for σ_z directly:

$$\sigma_V^2 - \sigma_{HR}^2 - \sigma_{HI}^2 - (\cos z)^2 \sigma_{d_S}^2 = (d_S \sin z)^2 \sigma_z^2 \qquad (8.14)$$

$$0.005^2 - 2(0.003)^2 - (0.173\,648)^2 (0.003\,5)^2 = (d_S \sin z)^2 \sigma_z^2$$

$$6.630\,617E - 6 = (d_S \sin z)^2 \sigma_z^2$$

$$\sigma_z = \frac{\sqrt{6.630\,617E - 6}}{(d_S \sin z)} \quad \rightarrow \quad \sigma_z = \frac{2.575E - 3}{(200 \sin 100)}$$

$$\sigma_z = \frac{2.575E - 3}{196.961\,55} = 1.307\,361E - 5\,\text{rad}\,(\text{or}\,2.7'')$$

The new expected standard deviation of measuring the zenith angle (z) is $2.7''$.

8.3 Network Design Model

In the design of a complex network to be adjusted by the method of least squares, the procedure for simple preanalysis given in Section 8.3 cannot be applied. A more rigorous design model (variance–covariance matrix of adjusted parameters) from the parametric least squares adjustment must be used as discussed in this section. The variance–covariance matrix of adjusted parameters can be expressed as follows (referring to Section 5.8):

$$C_{\hat{x}} = \sigma_0^2 \left(A^T P A \right)^{-1} \tag{8.15}$$

where $C_{\hat{x}}$ is the given variance–covariance matrix desired for the adjusted parameters, A is the first design matrix (or the network configuration discussed in Section 8.2) that can be determined from the approximate coordinates of the network, P is the weight matrix of measurements containing the standard deviations of individual measurements, and σ_0^2 is the a priori variance factor of unit weight (if unknown, the a posteriori variance factor of unit weight s_0^2 should be used). For network design, $\sigma_0^2 = 1$ is to be used; and weight matrix of observations P is the inverse of the covariance matrix (C_ℓ^{-1}) of the observations. Equation (8.15) can then be rewritten as

$$C_{\hat{x}} = \left(A^T C_\ell^{-1} A \right)^{-1} \tag{8.16}$$

Equation (8.16) expresses generally the network design problem. If inverted, Equation (8.16) will provide the covariance matrix of observations (C_ℓ), from which the standard deviations of the corresponding measurements can be extracted from its principal diagonal. Direct inversion of this type of covariance matrix, which is usually done by mathematical programming approach, is beyond the scope of this book. The approach adopted in this book is by using indirect method of inversion with computer software based on trial-and-error procedure known as simulation. If the network involved is simple enough, as in the case of one-dimensional network (e.g. leveling network), a direct inversion to determine the covariance matrix of the observations may be less complicated. Examples of simple network design are provided in the following sections.

8.4 Simple One-dimensional Network Design

With reference to Equations (7.13) and (7.14) in Section 7.3.1, the value $(a_{1-\alpha/2})$ of the *margin of error* at $(1 - \alpha)\%$ confidence level can be derived from the variance–covariance matrix of the adjusted parameters. Assuming s is the unknown standard deviation of a measurement (which is part of the covariance matrix of the adjusted parameters, through the covariance matrix of observations, C_ℓ), one can work forward from Equation (8.16) to obtain the formulas for estimating the margins of errors at $(1 - \alpha)\%$ confidence level as follows. After least squares adjustment of a leveling network, for an example, the value of the margin of error at $(1 - \alpha)\%$ can be given as follows (refer to Section 7.3.1 for further discussion on this):

$$a_{1-\alpha} = \pm (\text{SE})z_{1-\alpha/2} \tag{8.17}$$

or

$$a_{1-\alpha} = \pm (\text{SE})t_{1-\alpha/2,\text{df}} \tag{8.18}$$

where the SE in this case is the standard deviation of the adjusted parameter, $z_{1-\alpha/2}$ is the value from the standard normal distribution curve, and $t_{1-\alpha/2,\,\text{df}}$ is from the Student's t-distribution curve. Equation (8.17) should be used when the degrees of freedom of the adjustment is greater than 30; otherwise Equation (8.18) is used. In this case, the square of the calculated SE will be used in the covariance matrix $(C_{\hat{x}})$ of the unknown parameter in Equation (8.16). For example, consider a leveling network in Figure 8.2 where A, B, and C are control points with known heights and Δh_1, Δh_2, and Δh_3 are the three height difference measurements with standard deviations of σ_1, σ_2, and σ_3, respectively. If the relationship among the standard deviations is such that $\sigma_1 = \sigma_2$ and $\sigma_1 = 3\sigma_3$, determine the values of σ_1, σ_2, and σ_3 so that the margin of error at 95% confidence level for the height solution for point P using least adjustment is equal to 8.6 mm.

Figure 8.2 Leveling network.

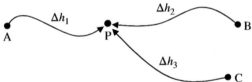

Since the degrees of freedom in this problem is df < 30, and $s_{\hat{h}_P}$ was computed with degrees of freedom df $= 2$, the Student's t-value should be used in Equation (8.18):

Given $t_{0.975,\,\text{df}\,=\,2} = 4.303$, SE $= s_{\hat{h}_P}$, $\alpha = 0.05$, $a_{95\%} = 8.6$ mm, substituting the given quantities into Equation (8.18) gives the following:

$$8.6 = s_{\hat{h}_P} \times 4.303, \text{then}$$

$$s_{\hat{h}_P} = \frac{8.6}{4.303} \rightarrow 2\,\text{mm} \tag{8.19}$$

Unknown parameter $= [\text{Elevation of point P}]$ or $x = [h_P]$

Vector of observations: $\ell = \begin{bmatrix} \Delta h_1 \\ \Delta h_2 \\ \Delta h_3 \end{bmatrix}$

Since the problem is based on parametric least squares adjustment, formulate the parametric equations: $\hat{\ell} = f(\hat{x})$

$$\Delta h_1 = h_P - h_A$$
$$\Delta h_2 = h_P - h_B \qquad\qquad (8.20)$$
$$\Delta h_3 = h_P - h_C$$

Form the first design matrix (A) from Equation (8.20):

$$A = \frac{\partial f}{\partial x} = \begin{bmatrix} \dfrac{\partial \Delta h_1}{\partial h_P} \\[2mm] \dfrac{\partial \Delta h_2}{\partial h_P} \\[2mm] \dfrac{\partial \Delta h_3}{\partial h_P} \end{bmatrix} \qquad A = \begin{bmatrix} 1 \\ 1 \\ 1 \end{bmatrix} \qquad\qquad (8.21)$$

Form the weight matrix, P.
Actual covariance matrix will be

$$C_\ell = \begin{bmatrix} \sigma_1^2 & & \\ & \sigma_2^2 & \\ & & \sigma_2^2 \end{bmatrix} \qquad\qquad (8.22)$$

Unify the covariance matrix by substituting $\sigma_1 = \sigma_2$ and $\sigma_1 = 3\sigma_3$:

$$C_\ell = \begin{bmatrix} \sigma_1^2 & & \\ & \sigma_1^2 & \\ & & \dfrac{\sigma_1^2}{9} \end{bmatrix} \rightarrow C_\ell^{-1} = \frac{1}{\sigma_1^2} \begin{bmatrix} 1 & & \\ & 1 & \\ & & 9 \end{bmatrix}$$

Form the $A^T C_\ell^{-1} A$ – Matrix:

$$A^T C_\ell^{-1} A = \frac{1}{\sigma_1^2}[11] \qquad\qquad (8.23)$$

Form the covariance matrix of the adjusted parameters $C_{\hat{x}} = \left(A^T C_\ell^{-1} A\right)^{-1}$:

$$C_{\hat{h}_P} = \left(A^T C_\ell^{-1} A\right)^{-1} \quad \rightarrow \quad C_{\hat{h}_P} = \left(\frac{1}{\sigma_1^2}[11]\right)^{-1}$$
$$C_{\hat{h}_P} = \sigma_1^2 [0.091] \qquad\qquad (8.24)$$

Note that $C_{\hat{h}_P} = s_{\hat{h}_P}^2$. Taking the value of $s_{\hat{h}_P}^2$ from Equation (8.19) and substituting into Equation (8.24) gives

$$4\,\text{mm}^2 = \sigma_1^2[0.091] \quad \rightarrow \quad \sigma_1 = \sqrt{44} = 6.63 \text{ mm}$$

Since $\sigma_1 = \sigma_2$ and $\sigma_1 = 3\sigma_3$,

$$\sigma_2 = 6.63 \text{ mm and } \sigma_3 = \frac{1}{3} \times 6.63 = 2.21 \text{ mm}$$

8.5 Simple Two-dimensional Network Design

Consider a case where the coordinates of control points P and Q are known as P (x_P = 1000.000 m, y_P = 1000.000 m) and Q (x_Q = 1500.000 m, y_Q = 1000.000 m), respectively, and the surveyor is to design a measurement scheme by preanalysis to determine the coordinates of point R. The following four options are to be considered in the preanalysis and the best network chosen based on the one that produces an SE ellipse with the smallest semi-major (assuming the approximate coordinates of point R are x_R = 1250 m, y_R = 1200 m; and an angle can be measured to a precision of 5″, an azimuth to a precision of 3″, and a distance to a precision of 0.003 m):

a) Measure angle θ and distance d as shown in Figure 8.3 to establish point R.
b) Measure all the angles θ_1, θ_2, and θ_3, as shown in Figure 8.4 to establish point R.
c) Measure the azimuths Az_1 and Az_2 from points P and Q, respectively, to point R, as shown in Figure 8.5.
d) Measure distances d_1 and d_2 from points P and Q, respectively, to point R, as shown in Figure 8.6.

Option a: Angle θ and distance d are measured according to Figure 8.3.
The parametric model equations can be formulated for the two measurements as follows:

$$d = \sqrt{(x_R - x_P)^2 + (y_R - y_P)^2} \tag{8.25}$$

Figure 8.3 Angle and distance measurements in design option a.

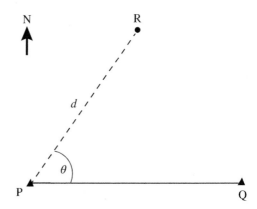

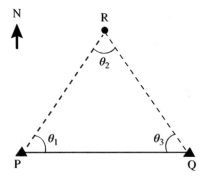

Figure 8.4 Angle measurements only in design option b.

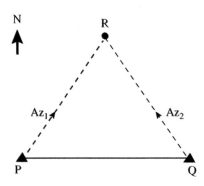

Figure 8.5 Azimuth measurements only in design option c.

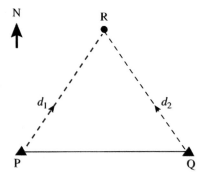

Figure 8.6 Distance measurements only in design option d.

$$\theta = \frac{\pi}{2} - \text{atan}\left(\frac{x_R - x_P}{y_R - y_P}\right) \tag{8.26}$$

The partial derivatives of Equations (8.25) and (8.26) with respect to coordinates (x_R, y_R) can be given as

$$\frac{\partial d}{\partial x_R} = \frac{-(x_P - x_R)}{d_{PR}^0} = 0.7809; \quad \frac{\partial d}{\partial y_R} = \frac{-(y_P - y_R)}{d_{PR}^0} = 0.6247$$

$$\frac{\partial \theta}{\partial x_R} = \frac{(y_P - y_R)}{\left(d_{PR}^0\right)^2} = -0.0020; \quad \frac{\partial \theta}{\partial y_R} = \frac{(x_P - x_R)}{\left(d_{PR}^0\right)^2} = 0.0024$$

where $d_{PR}^0 = 320.1562$ m is the approximate distance from point P to point R based on the given approximate coordinates of point R and the known coordinates of point P.

The first design matrix A can be given from the above partial derivatives as

$$A = \begin{bmatrix} 0.7809 & 0.6247 \\ -0.0020 & 0.0024 \end{bmatrix} \tag{8.27}$$

The weight matrix (P) based on the standard deviations of the distance and angle (in radians) measurements can be given as

$$P = \begin{bmatrix} 1.000\text{E}5 & 0.0 \\ 0.0 & 1.702\text{E}9 \end{bmatrix} \tag{8.28}$$

The covariance matrix (C_x) of the parameters (x_R, y_R), assuming the a priori variance factor is one, can be given as

$$C_x = \left(A^T P A\right)^{-1} = \begin{bmatrix} 0.2899 & -0.2499 \\ -0.2499 & 0.4024 \end{bmatrix} \times \text{E} - 4\,\text{m}^2 \tag{8.29}$$

The maximum and minimum eigenvalues of C_x are $\lambda_1 = 6.023\text{E}{-5}\,\text{m}^2$ and $\lambda_2 = 9.00\text{E}{-6}\,\text{m}^2$, respectively; the parameters of the SE ellipse are semi-major axis value, $a = 0.0078$ m; semi-minor axis value, $b = 0.003$ m; and the orientation of the semi-major axis, $\beta = 321°20'$.

Option b: Measure all the angles θ_1, θ_2, and θ_3, as shown in Figure 8.4 to establish point R.

The parametric model equations can be formulated for the three measurements as follows:

$$\theta_1 = \frac{\pi}{2} - \text{atan}\left(\frac{x_R - x_P}{y_R - y_P}\right) \tag{8.30}$$

$$\theta_2 = \text{atan}\left(\frac{x_P - x_R}{y_P - y_R}\right) - \text{atan}\left(\frac{x_Q - x_R}{y_Q - y_R}\right) \tag{8.31}$$

$$\theta_3 = \frac{\pi}{2} + \text{atan}\left(\frac{x_R - x_Q}{y_R - y_Q}\right) \tag{8.32}$$

The partial derivatives of Equations (8.30)–(8.32) with respect to coordinates (x_R, y_R) can be given as

$$\frac{\partial \theta_1}{\partial x_R} = \frac{(y_P - y_R)}{\left(d_{PR}^0\right)^2} = -0.0020; \quad \frac{\partial \theta_1}{\partial y_R} = \frac{(x_P - x_R)}{\left(d_{PR}^0\right)^2} = 0.0024$$

$$\frac{\partial \theta_2}{\partial x_R} = \frac{(y_Q - y_R)}{\left(d_{RQ}^0\right)^2} - \frac{(y_P - y_R)}{\left(d_{PR}^0\right)^2} = 0.000;$$

$$\frac{\partial \theta_2}{\partial y_R} = \frac{(x_P - x_R)}{\left(d_{PR}^0\right)^2} - \frac{(x_Q - x_R)}{\left(d_{RQ}^0\right)^2} = -0.0049$$

$$\frac{\partial \theta_3}{\partial x_R} = \frac{-(y_Q - y_R)}{\left(d_{RQ}^0\right)^2} = 0.0020; \quad \frac{\partial \theta_3}{\partial y_R} = \frac{(x_Q - x_R)}{\left(d_{RQ}^0\right)^2} = 0.0024$$

where $d_{RQ}^0 = 320.1562$ m is the approximate distance from point R to point Q based on the given approximate coordinates of point R and the known coordinates of point Q. The first design matrix A can be given from the above partial derivatives as

$$A = \begin{bmatrix} -0.0020 & 0.0024 \\ 0.0 & -0.0049 \\ 0.0020 & 0.0024 \end{bmatrix} \tag{8.33}$$

The weight matrix (P) based on the standard deviations of the angle (in radians) measurements can be given as

$$P = \begin{bmatrix} 1.7018E9 & & \\ & 1.7018E9 & \\ & & 1.7018E9 \end{bmatrix} \tag{8.34}$$

The covariance matrix (C_x) of the parameters (x_R, y_R), assuming the a priori variance factor is one, can be given as

$$C_x = \left(A^T P A\right)^{-1} = \begin{bmatrix} 0.7717 & 0.0 \\ 0.0 & 0.1646 \end{bmatrix} \times E{-}4 \ \text{m}^2 \tag{8.35}$$

The maximum and minimum eigenvalues of C_x are $\lambda_1 = 7.717E{-}5$ m^2 and $\lambda_2 = 1.646E{-}5$ m^2, respectively; the parameters of the SE ellipse are semi-major axis value, $a = 0.0088$ m; semi-minor axis value, $b = 0.0041$ m; and the orientation of the semi-major axis, $\beta = 90°00'$.

Option c: Measure the azimuths Az_1 and Az_2 from points P and Q, respectively, to point R, as shown in Figure 8.5.

The parametric model equations can be formulated for the two measurements as follows:

$$Az_1 = \operatorname{atan}\left(\frac{x_R - x_P}{y_R - y_P}\right) \tag{8.36}$$

$$Az_2 = 2\pi + \operatorname{atan}\left(\frac{x_R - x_Q}{y_R - y_Q}\right) \tag{8.37}$$

The partial derivatives of Equations (8.36)–(8.37) with respect to coordinates (x_R, y_R) can be given as

$$\frac{\partial Az_1}{\partial x_R} = \frac{-(y_P - y_R)}{\left(d_{PR}^0\right)^2} = 0.001\,951\,2; \quad \frac{\partial Az_1}{\partial y_R} = \frac{(x_P - x_R)}{\left(d_{PR}^0\right)^2} = -0.002\,439\,0$$

$$\frac{\partial Az_2}{\partial x_R} = \frac{-(y_Q - y_R)}{\left(d_{RQ}^0\right)^2} = 0.001\,951\,22; \quad \frac{\partial Az_2}{\partial y_R} = \frac{(x_Q - x_R)}{\left(d_{RQ}^0\right)^2} = 0.002\,439\,02$$

The first design matrix A can be given from the above partial derivatives as

$$A = \begin{bmatrix} 0.001\,951\,2 & -0.002\,439\,0 \\ 0.001\,951\,22 & 0.002\,439\,02 \end{bmatrix} \tag{8.38}$$

The weight matrix (P) based on the standard deviations of the angle (in radians) measurements can be given as

$$P = \begin{bmatrix} 4.727\,25E9 & 0.0 \\ 0.0 & 4.727\,25E9 \end{bmatrix} \tag{8.39}$$

The covariance matrix (C_x) of the parameters (x_R, y_R), assuming the a priori variance factor is one, can be given as

$$C_x = \left(A^T P A\right)^{-1} = \begin{bmatrix} 0.277\,811 & 0.0 \\ 0.0 & 0.177\,799 \end{bmatrix} \times E - 4 \text{ m}^2 \tag{8.40}$$

The maximum and minimum eigenvalues of C_x are $\lambda_1 = 2.7781E\text{-}5 \text{ m}^2$ and $\lambda_2 = 1.777\,99E\text{-}5 \text{ m}^2$, respectively; the parameters of the SE ellipse are semi-major axis value, $a = 0.0053$ m; semi-minor axis value, $b = 0.0042$ m; and the orientation of the semi-major axis, $\beta = 90°00'$.

Option d: Measure distances d_1 and d_2 from points P and Q, respectively, to point R, as shown in Figure 8.6.

The parametric model equations can be formulated for the two measurements as follows:

$$d_1 = \sqrt{(x_R - x_P)^2 + (y_R - y_P)^2} \tag{8.41}$$

$$d_2 = \sqrt{(x_R - x_Q)^2 + (y_R - y_Q)^2} \tag{8.42}$$

The partial derivatives of Equations (8.41)–(8.42) with respect to coordinates (x_R, y_R) can be given as

$$\frac{\partial d_1}{\partial x_R} = \frac{-(x_P - x_R)}{d_{PR}^0} = 0.7809; \quad \frac{\partial d_1}{\partial y_R} = \frac{-(y_P - y_R)}{d_{PR}^0} = 0.6247$$

$$\frac{\partial d_2}{\partial x_R} = \frac{-(x_Q - x_R)}{d_{RQ}^0} = -0.7809; \quad \frac{\partial d_2}{\partial y_R} = \frac{-(y_Q - y_R)}{d_{RQ}^0} = 0.6247$$

The first design matrix A can be given from the above partial derivatives as

$$A = \begin{bmatrix} 0.7809 & 0.6247 \\ -0.7809 & 0.6247 \end{bmatrix} \tag{8.43}$$

The weight matrix (P) based on the standard deviations of the distance measurements can be given as

$$P = \begin{bmatrix} 1.111E5 & 0.0 \\ 0.0 & 1.111E5 \end{bmatrix} \tag{8.44}$$

The covariance matrix (C_x) of the parameters (x_R, y_R), assuming the a priori variance factor is one, can be given as

$$C_x = \left(A^T P A\right)^{-1} = \begin{bmatrix} 0.073\,80 & 0.0 \\ 0.0 & 0.115\,3\,13 \end{bmatrix} \times E - 4\,m^2 \tag{8.45}$$

The maximum and minimum eigenvalues of C_x are $\lambda_1 = 1.153\,13E\text{-}5\ m^2$ and $\lambda_2 = 7.38E\text{-}6\ m^2$, respectively; the parameters of the SE ellipse are semi-major axis value, $a = 0.0034$ m; semi-minor axis value, $b = 0.0027$ m; and the orientation of the semi-major axis, $\beta = 00°00'$. The summary of all of the options is given in Table 8.1.

As it can be seen in Table 8.1, option d (measurement of two distances) seems to be the best design; it produces an error circle and has the least standard semi-major axis value of 0.003 m, and only two measurements are required. The worst design is associated with measuring the three angles in the triangular network (option b).

Table 8.1 Summary of results of preanalysis of simple two-dimensional network.

Option	Semi-major axis (a) (m)	Semi-minor axis (b) (m)	Orientation of semi-major axis (β)
a	0.008	0.003	321°20′
b	0.009	0.004	90°00′
c	0.005	0.004	90°00′
d	0.003	0.003	00°00′

Example 8.4 Continuing from Section 8.6, assume that after determining the coordinates of point R based on option c, distance P–R and angle θ (in option a) were measured to improve the precisions of the coordinates of point R. Recalculate the covariance matrix of the coordinates of point R using the concept of weighted station approach, and determine from the calculated SE ellipses if there is any improvement on either of the two combined options.

Solution:

Using the cofactor matrix of the weight constraint adjusted parameters from Equation (6.58),

$$C_{\hat{x}} = \left(P_x + A^T PA \right)^{-1} \tag{8.46}$$

where P_x is the a priori weight matrix (from option "c") for the coordinates of point R, which can be given from Equation (8.40) as

$$P_x = \left(A^T PA \right) = \begin{bmatrix} 3.5996 & 0.0 \\ 0.0 & 5.6243 \end{bmatrix} \times E + 4 \ m^2 \tag{8.47}$$

From option "a" (Equation (8.29), the following is obtained:

$$\left(A^T PA \right) = \begin{bmatrix} 7.4236 & 4.6102 \\ 4.6102 & 5.3481 \end{bmatrix} \times E + 4 \, m^2 \tag{8.48}$$

Substituting Equations (8.47) and (8.48) into Equation (8.46) gives

$$C_x = \begin{bmatrix} 0.1101 & -0.0462 \\ -0.0462 & 0.1106 \end{bmatrix} \times E - 4 \, m^2 \tag{8.49}$$

The maximum and minimum eigenvalues of C_x are $\lambda_1 = 1.566E\text{-}5 \ m^2$ and $\lambda_2 = 6.41E\text{-}6 \ m^2$, respectively; the parameters of the SE ellipse are semi-major axis value, $a = 0.0040 \ m$; semi-minor axis value, $b = 0.0025 \ m$; and the orientation of the semi-major axis, $\beta = 315°09'$. As it can be seen in Table 8.1, the combined adjustment done in this example has $a = 0.004$ m, which is an improvement on either of the two options combined in the example.

Example 8.5 Consider two-dimensional survey network in Figure 8.7, in which the coordinates of point P are to be determined from three fixed points 1, 2, and 3. The planned measurements are distances s_1, s_2, and s_3. The observations will be uncorrelated. The approximate coordinates of the fixed and new points taken from a large-scale map are given in Table 8.2.

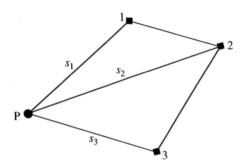

Figure 8.7 Two-dimensional network.

Table 8.2 Approximate coordinates.

Point	X (m)	Y (m)
1	600	800
2	900	700
3	600	100
P	200	400

What should be the accuracy of the three distances in order to obtain the semi-major axis of the absolute error ellipse at 95% confidence ($a_{95\%}$) of less than 10 mm? In solving this problem, the parametric equations, $\ell = f(\hat{x})$, with the parameters as $x = \begin{bmatrix} x \\ y \end{bmatrix}$ is formulated as follows:

$$s_1 = \left[(x-600)^2 + (y-800)^2 \right]^{\frac{1}{2}} \tag{8.50}$$

$$s_2 = \left[(x-900)^2 + (y-700)^2\right]^{\frac{1}{2}} \tag{8.51}$$

$$s_3 = \left[(x-600)^2 + (y-100)^2\right]^{\frac{1}{2}} \tag{8.52}$$

Solution:

Form the first design matrix, A, from Equations (8.50)–(8.52):

$$A = \frac{\partial f}{\partial x} = \begin{bmatrix} \dfrac{x-600}{s_1} & \dfrac{y-800}{s_1} \\[2mm] \dfrac{x-900}{s_2} & \dfrac{y-700}{s_2} \\[2mm] \dfrac{x-600}{s_3} & \dfrac{y-100}{s_3} \end{bmatrix} \qquad A = \begin{bmatrix} -0.707 & -0.707 \\ -0.919 & -0.394 \\ -0.800 & 0.600 \end{bmatrix}$$

Form the weight matrix (P) of the observations, assuming the same standard deviation (s) for all the measurements:

$$P = \begin{bmatrix} \left(\dfrac{1}{s}\right)^2 & 0 & 0 \\[2mm] 0 & \left(\dfrac{1}{s}\right)^2 & 0 \\[2mm] 0 & 0 & \left(\dfrac{1}{s}\right)^2 \end{bmatrix} \quad \text{or } P = s^{-2}\begin{bmatrix} 1 & 0 & 0 \\ 0 & 1 & 0 \\ 0 & 0 & 1 \end{bmatrix}$$

Form the matrix of coefficients of normal equations (N):

$$N = s^{-2}A^T A = s^{-2}\begin{bmatrix} 1.984\,83 & 0.382\,07 \\ 0.382\,07 & 1.015\,17 \end{bmatrix}$$

$$Q = N^{-1} = s^2\begin{bmatrix} 0.543\,17 & -2.044\,28 \\ -2.044\,28 & 1.061\,993 \end{bmatrix}$$

Use the cofactor Q to determine the eigenvalues as follows:

$$\lambda_1 = \frac{1}{2}\left(\sigma_x^2 + \sigma_y^2 + z\right) \quad z = \left[\left(\sigma_x^2 - \sigma_y^2\right)^2 + 4\sigma_{xy}^2\right]^{1/2}$$

$$z = 0.660\,558s^2 \qquad \lambda_1 = 1.132\,862s^2$$

Determine the semi-major axis value of the 95% error ellipse using Equation (7.52):

$$\chi^2_{0.95,\,df=2} = 5.99$$

$$a_{95\%} = \sqrt{\lambda_1 \times 5.99} \quad \rightarrow \quad \underline{2.604\,97s} \tag{8.53}$$

Equate the calculated semi-major axis value in Equation (8.53) with the given value $a_{95\%} \leq 10$ mm, and solve for the unknown standard deviation (s): 2.604 97$s = 10$ mm

$$s = \frac{10}{2.604\,97}\,\text{mm} \;\rightarrow\; \underline{3.8\,\text{mm}}$$

Each distance in the survey network must be measured to an accuracy of 3.8 mm in order to achieve $a_{95\%} \leq 10$ mm absolute error ellipse at 95% confidence level for the unknown point P.

Example 8.6 Consider the survey network in Figure 8.8. The coordinates of point P are to be determined from three fixed points 1, 2, 3. The planned measurements are angles β_1, β_2, and β_3 with standard deviation σ_β. The observations will be uncorrelated. The approximate coordinates of the fixed and new points taken from a large-scale map are as given in Table 8.2. Answer the following.

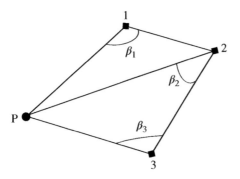

Figure 8.8 Two-dimensional network including angle measurements.

a) If the planned measurements are to have standard deviations of $\sigma_\beta = 2''$, calculate the expected positional error of point P at 95% confidence level.

Solution:

Formulate the parametric equations, $\ell = f(x)$, with the parameters as $x = \begin{bmatrix} x \\ y \end{bmatrix}$:

$$\beta_1 = \arctan\left(\frac{x-600}{y-800}\right) - \arctan\left(\frac{900-600}{700-800}\right) \tag{8.54}$$

$$\beta_2 = \arctan\left(\frac{x-900}{y-700}\right) - \arctan\left(\frac{600-900}{100-700}\right) \tag{8.55}$$

$$\beta_3 = \arctan\left(\frac{900-600}{700-100}\right) - \arctan\left(\frac{x-600}{y-100}\right) \qquad (8.56)$$

The approximate distances calculated using the approximate coordinates of point P from Table 8.2 are $s_1 = 565.685$, $s_2 = 761.577$, and $s_3 = 500.000$. Form the first design matrix, A, with respect to Equations (8.54)–(8.56):

$$A = \frac{\partial f}{\partial x} = \begin{bmatrix} \dfrac{y-800}{s_1^2} & -\left(\dfrac{x-800}{s_1^2}\right) \\[2ex] \dfrac{y-700}{s_2^2} & -\left(\dfrac{x-900}{s_2^2}\right) \\[2ex] -\left(\dfrac{y-100}{s_3^2}\right) & \dfrac{x-600}{s_3^2} \end{bmatrix} \quad A = \begin{bmatrix} -0.0125 & 0.001\,25 \\ -0.000\,517\,2 & 0.001\,206\,9 \\ -0.001\,20 & -0.001\,60 \end{bmatrix}$$

Form the weight matrix of the observations, P:

$$P = \begin{bmatrix} \left(\dfrac{206\,265}{2}\right)^2 & 0 & 0 \\[2ex] 0 & \left(\dfrac{206\,265}{2}\right)^2 & 0 \\[2ex] 0 & 0 & \left(\dfrac{206\,265}{2}\right)^2 \end{bmatrix}$$

or

$$P = \begin{bmatrix} 1.063\,631\,256\mathrm{E}10 & 0 & 0 \\ 0 & 1.063\,631\,256\mathrm{E}10 & 0 \\ 0 & 0 & 1.063\,631\,256\mathrm{E}10 \end{bmatrix}$$

Form the matrix of the coefficients of normal equations, N, and the cofactor matrix, Q, of the adjusted coordinates of point P:

$$N = A^T P A = \begin{bmatrix} 347\,81.153\,1 & -283\,7.309\,11 \\ -283\,7.309\,1 & 593\,41.643\,82 \end{bmatrix}$$

$$Q = N^{-1} = \begin{bmatrix} 2.886\,4\mathrm{E}-5 & 1.380\,082\mathrm{E}-6 \\ 1.380\,082\mathrm{E}-6 & 1.691\,773\mathrm{E}-5 \end{bmatrix}$$

Standard deviations of the adjusted coordinates of point P:

$$\sigma_x = \sqrt{2.8864\mathrm{E}-5} \rightarrow 0.0054\,\text{m}$$

$$\sigma_y = \sqrt{1.691\,773\mathrm{E}-5} \rightarrow 0.0041\,\text{m}$$

Determine the eigenvalues and the 95% confidence error ellipse using chi-square approach from Equations (7.52) and (7.53):

$$\lambda_1 = \frac{1}{2}\left(\sigma_x^2 + \sigma_y^2 + z\right) \quad \lambda_2 = \frac{1}{2}\left(\sigma_x^2 + \sigma_y^2 - z\right) \quad z = \left[\left(\sigma_x^2 - \sigma_y^2\right)^2 + 4\sigma_{xy}^2\right]^{1/2}$$

$z = 1.226\,078\,1\text{E}-5 \quad \lambda_1 = 2.902\,115\text{E}-5 \quad \lambda_2 = 1.676\,036\,7\text{E}-5$

$\chi^2_{0.95,df\,=\,2} = 5.99 \quad k = \sqrt{5.99} = 2.4474$

$a_{st} = \sqrt{\lambda_1} = 5.387\text{E}-3 \quad b_{st} = \sqrt{\lambda_2} = 4.094\text{E}-3$

$a_{95\%} = a_{st} \times k \quad \rightarrow \quad \underline{0.0132\text{ m}}$

$b_{95\%} = b_{st} \times k \rightarrow \underline{0.0100\text{ m}}$

$\theta = \text{atan}\left(\dfrac{\lambda_1 - \sigma_y^2}{\sigma_{xy}}\right) \quad \theta = 83.495°$

b) If the planned measurements are to be made with the same accuracy, $\sigma_\beta = s$, what should be the numerical value of the accuracy of the three angles in order to obtain the semi-major axis ($a_{95\%}$) of the 95% confidence absolute error ellipse of less than or equal to 10 mm at point P?

Solution:

The parametric equations and the A-matrix in question (a) are still applicable in this problem. Assume all the three distances would be measured with the same standard deviation (s) (with s in seconds); the following weight matrix is formed:

$$P = \begin{bmatrix} \left(\dfrac{206\,265}{s}\right)^2 & 0 & 0 \\ 0 & \left(\dfrac{206\,265}{s}\right)^2 & 0 \\ 0 & 0 & \left(\dfrac{206\,265}{s}\right)^2 \end{bmatrix}$$

or

$$P = s^{-2}\begin{bmatrix} 4.254\,525\text{E}10 & 0 & 0 \\ 0 & 4.254\,525\text{E}10 & 0 \\ 0 & 0 & 4.254\,525\text{E}10 \end{bmatrix}$$

Form the matrix of coefficients of normal equations, N, and the cofactor matrix of the adjusted coordinates, Q:

$$N = s^{-2}A^T A = s^{-2} \begin{bmatrix} 1.391\,246\,123\text{E}10 & -11\,349.236\,38 \\ -11\,349.236\,37 & 2.373\,641\,75\text{E}-5 \end{bmatrix}$$

$$Q = N^{-1} = s^2 \begin{bmatrix} 7.215\,95\text{E}-6 & 3.450\,204\text{E}-7 \\ 3.450\,204\text{E}-7 & 4.229\,43\text{E}-6 \end{bmatrix}$$

Compute the semi-major axis value of the 95% confidence error ellipse from the cofactor matrix, Q:

$$\lambda_1 = \frac{1}{2}\left(\sigma_x^2 + \sigma_y^2 + z\right) \qquad z = \left[\left(\sigma_x^2 - \sigma_y^2\right)^2 + 4\sigma_{xy}^2\right]^{1/2}$$

$$z = 3.065\,195\text{E}-6s^2 \qquad \lambda_1 = 7.255\,29\text{E}-6s^2$$

Using 95% error ellipse formula in Equation (7.52),

$$\chi_{0.95,\,\text{df}=2}^2 = 5.99$$

$$a_{95\%} = \sqrt{\lambda_1 \times 5.99} \quad \rightarrow \quad 0.006\,592\,36s \tag{8.57}$$

Equate the given $a_{95\%} \leq 10$ mm to the computed value in Equation (8.57), and solve for the value of s:

$$0.006\,592\,36s = 0.010$$

$$s = \frac{0.010}{0.006\,592\,36} \overset{''}{\quad} \rightarrow \quad 1.5''$$

The standard deviation for each angle measurement should be 1.5″.

Example 8.7 A detailed survey point P shown in Figure 8.9 was to be laid out by independent relocation traverse surveys from different control points within a simultaneously adjusted network. The point P with known coordinates

Figure 8.9 Relocation traverse surveys.

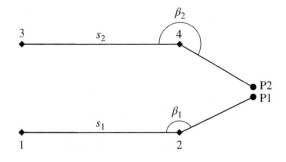

(Y_{P1}, X_{P1}) had been marked earlier on the ground as point P1 by the survey from control points 1 and 2. The second survey to the same point was carried from control points 3 and 4, and different coordinates P2 (Y_{P2}, X_{P2}) were obtained. The following variance–covariance matrix for points P1 and P2 are obtained and given in the order $x = [Y_{P1} \ X_{P1} \ Y_{P2} \ X_{P2}]^T$ as follows:

$$C_{xp} = 10^{-5} \begin{bmatrix} 5.134 & -0.784 & 0.928 & 0.166 \\ & 12.179 & 0.121 & 0.959 \\ & & 11.096 & -1.603 \\ & & & 8.220 \end{bmatrix}$$

Answer the following.

a) What are the maximum differences ΔX and ΔY between points P1 and P2 that could be allowed at the 95% confidence level for the given accuracy of the control and of the measurements of $\beta_1, \beta_2, s_1, s_2$?

Solution:

Equation for the coordinate differences in form of $p = f(x)$ is formulated as follows, where $x = [Y_{P1} \ X_{P1} \ Y_{P2} \ X_{P2}]^T$ is a vector of the adjusted coordinates of points P1 and P2 and $p = [\Delta x \ \Delta y]^T$ is a vector of corresponding coordinate differences:

$$\Delta x = X_{P2} - X_{P1} \tag{8.58}$$

$$\Delta y = Y_{P2} - Y_{P1} \tag{8.59}$$

The Jacobian matrix of Equations (8.58) and (8.59) with respect to the adjusted coordinates is given as

$$B = \frac{\partial p}{\partial x} = \begin{bmatrix} 0 & -1 & 0 & 1 \\ -1 & 0 & 1 & 0 \end{bmatrix}$$

By variance–covariance propagation laws,

$$C_p = B C_{x_p} B^T$$

$$C_p = \begin{bmatrix} 1.8481\text{E}-4 & -2.674\text{E}-5 \\ -2.674\text{E}-5 & 1.4374\text{E}-4 \end{bmatrix}$$

$$\sigma_{\Delta X} = \sqrt{1.8481\text{E}-4} \ \rightarrow \ 0.014\,\text{m}$$

$$\sigma_{\Delta Y} = \sqrt{1.4374E-4} \ \rightarrow \ 0.012\,\text{m}$$

At 95% confidence, the allowable values of ΔX and ΔY are determined as their error margins at 95% confidence level, using Equation (7.13): $\Delta = (SE)z_{1-\alpha/2}$ (for $z_{1-0.05/2} = 1.96$)

$$\Delta X_{max} = 1.96 \times 0.014\,\text{m} \rightarrow 0.0274\,\text{m}$$

$$\Delta Y_{max} = 1.96 \times 0.012\,\text{m} \rightarrow 0.0235\,\text{m}$$

b) What is the expected maximum distance between the points P1 and P2 to be marked on the ground from the two layouts?

Solution:

The expected distance is the semi-major axis of the relative error ellipse connecting P1 and P2 at 95% confidence level. By error propagation done in (a), the variance–covariance matrix for the differences in coordinates of the two points P1 and P2 is

$$C_p = \begin{bmatrix} 1.8481\text{E}-4 & -2.674\text{E}-5 \\ -2.674\text{E}-5 & 1.4374\text{E}-4 \end{bmatrix}$$

Determine the eigenvalues and the 95% confidence relative error ellipse from C_p:

$$\lambda_1 = \frac{1}{2}\left(\sigma_x^2 + \sigma_y^2 + z\right)$$

$$z = \left[\left(\sigma_x^2 - \sigma_y^2\right) + 4\sigma_{xy}^2\right]^{\frac{1}{2}}$$

$$z = 6.743\,04\text{E}-5; \quad \lambda_1 = 1.979\,9\text{E}-4; \quad \lambda_2 = 1.305\,60\text{E}-4$$

Using the 95% error ellipse formulas in Equations (7.52) and (7.53) with the value of $\chi_{1-\alpha,\text{df}=2}^2 = 5.991$,

$$a = \sqrt{\lambda_1 \times 5.991} \rightarrow 0.0344\,\text{m}$$

$$b = \sqrt{\lambda_2 \times 5.991} \rightarrow 0.0280\,\text{m}$$

$$\theta = \text{atan}\left(\frac{\lambda_1 - \sigma_y^2}{\sigma_{xy}}\right) \quad \theta = -63.76°$$

The maximum expected distance at 95% confidence is 0.034 m along the bearing of $-63.76°$.

8.6   Simulation of Three-dimensional Survey Scheme

Designing a survey scheme for the purpose of deciding the best choice of equipment and procedures for three-dimensional positioning usually involves the process of preanalysis or what is called computer simulation. The computer simulation steps are well documented (cf. Nickerson 1979; Cross 1985):

1) The simulation process usually starts with the input into the computer software, the standard deviations of the potential observables (such as horizontal and zenith angles and distances). The standard deviations of the observables are usually derived from the available equipment to be used in the survey scheme.
2) Input the potential geometry or preliminary location of points (as first design matrix *A*) using the given as approximate coordinates, taken from large-scale maps, aerial photographs, or other sources and supported with field reconnaissance survey to ensure intervisibility of points.
3) Simulate the quality of the network (using appropriate computer software) based on the initial design given in steps 1–2, and check the simulated quality (standard deviations and absolute and relative error ellipses of unknown parameters to determined later) against the expected positioning tolerance, which are sometimes the limit on relative error ellipses or absolute station error ellipses at 95% confidence level.
4) If step 3 is not satisfied, consider modifying and repeating steps 1–2 until step 3 is satisfied. Otherwise, consider the design complete.

Steps 1–2 constitute the initial network design, which is preanalyzed in steps 3–4. The following example illustrates the simulation process based on the use of MicroSurvey STAR∗NET v8 as the preanalysis or simulation software.

8.6.1   Typical Three-dimensional Micro-network

Eleven wall targets on four walls (*A*, *B*, *C*, and *D*) shown in Figure 8.10 are to be coordinated three dimensionally as a part of a micro-network establishment in an industrial environment. The targets are to be positioned to a relative positioning tolerance of 10.0 mm relative to the fixed point 1 at 95% confidence level, assuming the azimuth of line 1–2 will be considered known and fixed for the network datum. The approximate coordinates of the wall targets, possible locations of the instrument, and two well-calibrated scale bars (to provide scale) are shown in Table 8.3 and Figure 8.11. The scale bars whose positions will remain fixed throughout the project are 2 m in length with the calibrated accuracy of 0.02 mm; they will be used to improve precision of distance measurements instead of measuring distances to the wall targets since the lengths involved are less than 20 m. For example, using a total station with standard

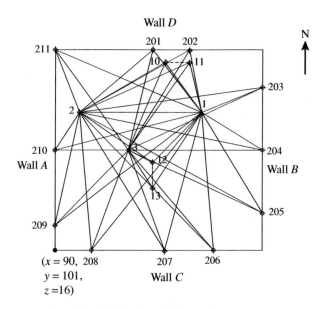

Figure 8.10 The plan locations of the wall targets, instrument, and scale bar locations in a micro-network survey.

Table 8.3 Approximate coordinates of wall targets.

Point no.	Northing (m)	Easting (m)	Elevation (m)	Description of point
201	117	98	20	Wall target
202	117	101	20	Wall target
203	114	107	20	Wall target
204	109	107	20	Wall target
205	104	107	20	Wall target
206	101	103	20	Wall target
207	101	99	20	Wall target
208	101	93	20	Wall target
209	103	90	20	Wall target
210	109	90	20	Wall target
211	117	90	20	Wall target
1	112	102	16	Instrument location (fixed)
2	112	92	16	Possible setup point (fixed)
3	109	96	16	Possible setup point (free)
10	116	99	18	First marker on scale bar 1
11	116	101	18.5	Second marker on scale bar 1
12	108	98	16	First marker on scale bar 2
13	106	98	16	Second marker on scale bar 2

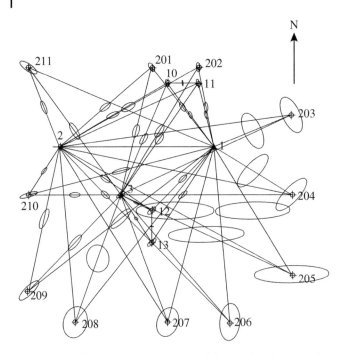

Figure 8.11 The two-dimensional view of the simulated survey scheme for a micro-network survey project.

deviation of distance measurement as 3 mm ± 2 ppm will produce constant precision of 3 mm for any distance measurement in this project, thereby reducing the precision of the whole project. The approximate coordinates provided in Table 8.3 for the wall targets, scale bar markers (10, 11, 12, 13), and the instrument locations (1, 2, 3) were extracted from the building drawings. Point 3 is given approximate coordinates to start with; these values are free to be changed when manipulating the location of the instrument in order to achieve the best geometry in relation to other target points to be measured to. The locations of the scale bars can also be changed during the preanalysis if that change will provide better results.

8.6.2 Simulation Results

The MicroSurvey STAR*NET v8 software was used in the simulation process. An instrument considered has a standard deviation for one set of direction measurements as 5″; point 1 and the azimuth of line 1–2 are fixed for datum definition, with the scale bars providing the scale. The total number of points whose coordinates are to be determined is 17 (with four of them representing the markers on the two scale bars); the total of 45 zenith angles and 45 horizontal directions are to be measured to the wall targets, markers on the scale bars, and targets on the other unoccupied setup stations, from three setup stations (1, 2, and 3) with the two scale bars providing additional two distance measurements and the azimuth of line 1–2 considered also as measurement. The total number of degrees of freedom for the adjustment is 76. The result of the simulation is given in Figure 8.11, showing the 95% confidence station and relative error ellipses and the survey scheme.

In Figure 8.11, it can be seen that the maximum station coordinate error ellipse at 95% confidence level is at station 205 with a semi-major axis value of 8 mm and its orientation along the azimuth 88°33′; the remaining 95% confidence station coordinate error ellipses have their semi-major axes values less than 5 mm. The maximum relative error ellipse at 95% confidence (the main consideration for this project) is on line 2–205 with a semi-major axis value of 8.5 mm (oriented along the azimuth 90°) with the remaining 95% relative error ellipses having their semi-major axes values less than 5 mm. This design is acceptable (but not necessarily the best) since the achieved 8.5 mm for the 95% confidence relative error ellipse is less than the required relative positioning tolerance of 10.0 mm. The MicroSurvey STAR∗NET code for the three-dimensional design is given in Table 8.4.

Table 8.4 The MicroSurvey STAR∗NET 8 code for the three-dimensional design.

```
# Three-dimensional design
.3D
.ORDER NE AtFromTo
.UNITS Meters DMS
C 10 116.0 99 18
C 11 116 101 18.5
C 201 117 98 20
C 202 117 101 20
C 203 114 107 20
C 204 109 107 20
C 205 104 107 20
C 206 101 103 20
C 207 101 99 20
C 208 101 93 20
C 209 103 90 20
C 210 109 90 20
C 211 117 90 20
```

(Continued)

Table 8.4 (Continued)

```
C 1 112 102 16 ! ! !
C 2 112 92 16
C 3 109 96 16
C 12 108 98 16
C 13 106 98 16
# Measurements- Fixed bearing
B 1-2 ? !
#Scale bar distance
D 10-11 ? 0.00002
D 12-13 ? 0.00002
# Zenith angle measurements
V 1-209 ? 5
V 1-210 ? 5
V 1-211 ? 5
V 1-201 ? 5
V 1-202 ? 5
V 1-10 ? 5
V 1-11 ? 5
V 1-12 ? 5
V 1-13 ? 5
V 2-209 ? 5
V 2-210 ? 5
V 2-211 ? 5
V 2-201 ? 5
V 2-202 ? 5
V 2-10 ? 5
V 2-11 ? 5
V 2-12 ? 5
V 2-13 ? 5
V 3-209 ? 5
V 3-210 ? 5
V 3-211 ? 5
V 3-201 ? 5
V 3-202 ? 5
V 3-10 ? 5
V 3-11 ? 5
V 3-12 ? 5
V 3-13 ? 5
#Wall A & D
# Horizontal direction measurements
DB 1
DN 209 ? 5
```

Table 8.4 (Continued)

```
DN 210 ? 5
DN 211 ? 5
DN 201 ? 5
DN 202 ? 5
DN 10 ? 5
DN 11 ? 5
DN 12 ? 5
DN 13 ? 5
DE
DB 2
DN 209 ? 5
DN 210 ? 5
DN 211 ? 5
DN 201 ? 5
DN 202 ? 5
DN 10 ? 5
DN 11 ? 5
DN 12 ? 5
DN 13 ? 5
DE
DB 3
DN 209 ? 5
DN 210 ? 5
DN 211 ? 5
DN 201 ? 5
DN 202 ? 5
DN 10 ? 5
DN 11 ? 5
DN 12 ? 5
DN 13 ? 5
DE
# To Wall B
# Zenith angles
V 1-205 ? 5
V 1-204 ? 5
V 1-203 ? 5
V 2-205 ? 5
V 2-204 ? 5
V 2-203 ? 5
V 3-205 ? 5
V 3-204 ? 5
V 3-203 ? 5
```

(*Continued*)

Table 8.4 (Continued)

```
# Horizontal directions
DB 1
DN 205 ? 5
DN 204 ? 5
DN 203 ? 5
DE
DB 2
DN 205 ? 5
DN 204 ? 5
DN 203 ? 5
DE
DB 3
DN 205 ? 5
DN 204 ? 5
DN 203 ? 5
DE
# To Wall C
# Zenith angles
V 1-208 ? 5
V 1-207 ? 5
V 1-206 ? 5
V 2-208 ? 5
V 2-207 ? 5
V 2-206 ? 5
V 3-208 ? 5
V 3-207 ? 5
V 3-206 ? 5
# Horizontal directions
DB 1
DN 208 ? 5
DN 207 ? 5
DN 206 ? 5
DE
DB 2
DN 208 ? 5
DN 207 ? 5
DN 206 ? 5
DE
DB 3
DN 208 ? 5
DN 207 ? 5
DN 206 ? 5
DE
```

Problems

8.1 Given the leveling network in Figure P8.1 where A and B are control points with known heights; Δh_1 and Δh_2 are two height difference measurements with standard deviations of σ_1 and σ_2, respectively; and $\sigma_1 = 0.25\sigma_2$. Determine the values of σ_1 and σ_2 so that the 90% confidence interval of the height solution for point P using least adjustment is equal to 10 mm.

Figure P8.1

8.2 You are to design two-dimensional FOD network for the monitoring of an object point OP (on a deformable body) to a relative positioning tolerance of 4 mm (at 95% confidence level) with the coordinates of point RBR1402 and y-coordinate of point RBR1408 fixed in a minimal constraint adjustment. The approximate coordinates (x^0) of the reference points for possible location of the instrument and the object point OP are as shown in the following table; they were extracted from a large scale topographic map of the region. In the design, you are to assume that the object point OP and the reference points RBR1402 and RBR1408 will not change; you are therefore left with the manipulation of points RBR1410 and BC1001 for the FOD design.

Approximate coordinates of network points

Point no.	Northing (m)	Easting (m)
RBR1402	4629.6	1306.8
RBR1408	4742.4	1191.0
OP	4695.0	1273.5
RBR1410	4768.8	1279.7
BC1001	4655.6	1314.3

Using appropriate measurement scheme with distances introduced in one direction only, perform the design work by completing the following. Based on the nature of the topography and the environment of the network location, you can assume positions of the moveable points (RBR1402 and RBR1408) can be shifted by ±25 m in the FOD design. Assume that you have only access to Leica TC703 total station. The specifications for TC703 are 3″ for direction measurements and 2 mm ± 2 ppm for distance measurements; and the centering error for the instrument and target is 0.2 mm (based on the use of forced centering devices). Perform the following tasks:

1) Input the accuracies of measuring all possible directions and distances with the Leica TC 703 total station into the simulation software.

2) Starting from the approximate coordinates given in the table, perform a SOD by considering using Leica TC 703. Give the summary of your best result (giving the worst relative 95% error ellipse and a plot of your design).

3) Perform a combined FOD (moving only points RBR1402 and RBR1408) and SOD of the network. Give the summary of your best result (giving the worst relative 95% error ellipse and a plot of your design).

4) Assuming the tolerance is achieved in step 4, delete distance or angular measurements that will have minimal effect (still barely satisfying 4.0 mm relative accuracy) on the design in step 3.

5) Conclude whether it is possible to achieve the relative positioning tolerance specified. In addition to the foregoing information, include the following:

   a) Maximum semiaxes at 95% level for your "best" network design.

   b) Provide for the best design (step 4), the final precisions of measurements, number, and types of observations.

   c) Coordinates of best location of reference points RBR1402 and RBR1408 in step 4.

   d) A plot showing your design with the appropriate 95% error ellipses from step 4.

9

Concepts of Three-dimensional Geodetic Network Adjustment

CHAPTER MENU

OBJECTIVES

After studying this chapter, you should be able to:

1) Formulate parametric model equations relating spatial observables, such as distances, zenith (vertical) angles, and azimuths (directions or angles), with the three-dimensional X, Y, Z Cartesian coordinates in conventional terrestrial (CT) system.
2) Formulate parametric model equations relating spatial observables, such as distances, zenith (vertical) angles, and azimuths (directions or angles), with the three-dimensional curvilinear geodetic coordinates (latitude, longitude, and ellipsoidal height) in geodetic system.
3) Formulate parametric model equations relating spatial observables, such as distances, zenith (vertical) angles, and azimuths (directions or angles), with the three-dimensional local Cartesian coordinates (n, e, u) in local astronomic (LA) system.

Understanding Least Squares Estimation and Geomatics Data Analysis, First Edition.
John Olusegun Ogundare.
© 2019 John Wiley & Sons, Inc. Published 2019 by John Wiley & Sons, Inc.
Companion website: www.wiley.com/go/ogundare/Understanding-lse-and-gda

9.1 Introduction

The adjustments of three-dimensional geodetic networks have been discussed in detail in a number of technical reports and books, such as Wolf (1963, 1975), Heiskanen and Moritz (1967), Vincenty and Bowring (1978), Vincenty (1979), Dragomir et al. (1982), Vanicek and Krakiwsky (1986), and Leick (2004). The discussion in this section is based on them. Further details can be found in those reports and books. There are three main classes of three-dimensional networks: those based on terrestrial measurements, such as spatial distance, horizontal and vertical angles, height differences, etc.; those based on photogrammetric and remote sensing measurements; and those based on the measurements made from tracking stations to orbiting satellites. This section is mainly interested in the networks based on terrestrial measurements.

9.2 Three-dimensional Coordinate Systems and Transformations

In the adjustment of three-dimensional geodetic networks, the measurements are not reduced to the ellipsoid as in two-dimensional cases, and computations are not done on the ellipsoid or on the conformal mapping planes; the computations are generally done in a three-dimensional Cartesian coordinate system. The differential shifts of coordinates are in linear units in a rectangular horizon system (local coordinate system) centered on the point where the measurements are made; this means that there will be as many local coordinate systems as there are measurement points. One of the important local coordinate systems commonly used is local geodetic coordinate system. Typical observables in modern surveying are horizontal angles (or directions), slope distances, zenith angles, GPS vectors, astronomic latitudes, longitudes, azimuths, and height differences. One of the limitations in classical three-dimensional adjustment is the uncertainty in the vertical refraction when measuring zenith angles.

The conventional terrestrial (CT) Cartesian coordinate system is a global system with the origin at the Earth's center of mass; the X, Y, Z Cartesian coordinates are in the equatorial system with X-axis passing though the Greenwich meridian, Z-axis is parallel to the mean rotation axis of the Earth, and Y-axis is perpendicular to the Z–X plane in a right-handed system. This is illustrated in Figure 9.1. An ellipsoid associating with latitude (ϕ), longitude (λ), and ellipsoidal height (h) can be positioned so as to be coaxial with the X, Y, Z of the CT system such that Z-axis coincides with the ellipsoid's rotation axis and the ellipsoid's center coinciding with the origin of the X, Y, Z system. The ellipsoid so-positioned and oriented with the CT system is a *reference ellipsoid* or global

Figure 9.1 Relationship between the conventional terrestrial (CT) system and the geodetic system.

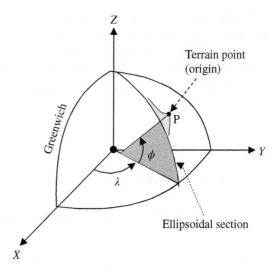

Terrain point (origin)

geodetic (G) system. The relationship between the Cartesian coordinates (X, Y, Z) and the G system coordinates (ϕ, λ, h) can be given as follows:

$$X = X_0 + (N + h)\cos\phi\cos\lambda \tag{9.1}$$

$$Y = Y_0 + (N + h)\cos\phi\sin\lambda \tag{9.2}$$

$$Z = Z_0 + \left(N\left(1 - e^2\right) + h\right)\sin\phi \tag{9.3}$$

where X_0, Y_0, and Z_0 are the coordinates of the center of the reference ellipsoid with respect to the CT system, N is the *radius of curvature in the prime vertical direction*, and M is the *radius of curvature in the meridian plane* given as follows:

$$N = \frac{a}{\left(1 - e^2\sin^2\phi\right)^{1/2}} \tag{9.4}$$

$$M = \frac{a\left(1 - e^2\right)}{\left(1 - e^2\sin^2\phi\right)^{3/2}} \tag{9.5}$$

with a and b as the semi-major axis and semi-minor axis values of the reference ellipsoid, respectively, and e as the first eccentricity of the ellipsoid. As an example, the parameters of the Geodetic Reference System of 1980 (GRS80 ellipsoid) are as follows:

$a = 6\ 378\ 137.0$ m

$b = 6\ 356\ 752.314\ 1$ m

$e^2 = 0.006\ 694\ 380\ 023$

By taking the partial derivatives of Equations (9.1)–(9.3) with respect to ϕ, λ, and h, the relationship between the coordinate differences (dX, dY, dZ) in Cartesian geodetic coordinate system and the coordinate differences in the curvilinear geodetic coordinate system (dϕ, dλ, dh) can be given as follows:

$$\begin{bmatrix} dX \\ dY \\ dZ \end{bmatrix} = \begin{bmatrix} -(M+h)\sin\phi\cos\lambda & -(N+h)\cos\phi\sin\lambda & \cos\phi\cos\lambda \\ -(M+h)\sin\phi\sin\lambda & (N+h)\cos\phi\cos\lambda & \cos\phi\sin\lambda \\ (M+h)\cos\phi & 0 & \sin\phi \end{bmatrix} \begin{bmatrix} d\phi \\ d\lambda \\ dh \end{bmatrix}$$

(9.6)

where dϕ and dλ are in radians.

9.2.1 Local Astronomic Coordinate Systems and Transformations

Measurements are usually made in the local astronomic (LA) system. The natural astronomic (physically meaningful) quantities usually measured between any two given points i and j are the spatial distance (s_{ij}), astronomic latitude (Φ_i), longitude (Λ_i), azimuth (Az_{ij}), vertical angle (v_{ij}) (or zenith angles z_{ij}), and orthometric height (H). The spatial orientation of this coordinate system is completely specified by the astronomic latitude and longitude with the Z-axis coinciding with the direction of the Conventional Terrestrial Pole (CTP). The LA coordinate system is a *topocentric coordinate system* or a local coordinate system as illustrated in Figure 9.2 and defined as follows:

- *Origin*: At the instrument setup station.
- *Z (or u)-axis*: Along the vertical (the direction of gravity) at the setup point.
- *X (or n)-axis*: A line tangent at the origin and aligned along the astronomical meridian, pointing toward the true north.

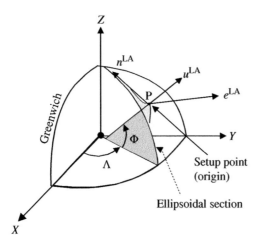

Figure 9.2 Relationship between the conventional terrestrial (CT) system and the local astronomic (LA) system.

Figure 9.3 Relationship between the azimuth (Az), vertical angle (v), zenith Angle (Z), and slope distance (s) in the local astronomic (LA) coordinate system.

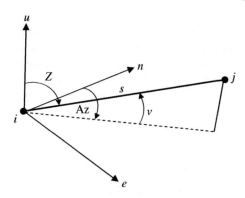

- *X–Y (n–e) plane*: Tangent to the geoid at the instrument setup point.
- *Y – (e) axis*: Defined by certain azimuth such that the coordinate system forms a right-handed system.
- Geocentric X_0, Y_0, Z_0 coordinates and the orthometric height H_0 are assigned to the origin.

In LA system, the north–east (*n–e*) plane shown in Figure 9.3 coincides with the physical horizontal plane. By using CT coordinate system, one can conveniently describe astro-geodetic networks extending over a large area such as states, provinces, continent, and the entire terrestrial globe.

The geoid in the region of measurement is defined as being tangent to the reference ellipsoid at the origin and the deflection of the vertical and the geoid undulations relative to the reference ellipsoid. With the instrument station *i* as the origin of the local coordinate system in Figure 9.3 and the target at point *j*, the coordinate differences between points *i* and *j* can be given as follows:

$$dn_{ij} = s_{ij} \cos v_{ij} \cos Az_{ij} \tag{9.7}$$

$$de_{ij} = s_{ij} \cos v_{ij} \sin Az_{ij} \tag{9.8}$$

$$du_{ij} = s_{ij} \sin v_{ij} \tag{9.9}$$

where s_{ij} is the slope distance, v_{ij} is the vertical angle, and Az_{ij} is the azimuth of line *i* to *j*. The inverses of Equations (9.7)–(9.9) can be given as

$$s_{ij} = \sqrt{dn_{ij}^2 + de_{ij}^2 + du_{ij}^2} \tag{9.10}$$

$$Az_{ij} = a \tan \left(\frac{de_{ij}}{dn_{ij}} \right) \tag{9.11}$$

$$v_{ij} = a \sin \left(\frac{du_{ij}}{s_{ij}} \right) \tag{9.12}$$

Total station instruments collect survey data in three dimensions at any given setup station, which is usually considered as the origin of that LA coordinate system; this system provides a natural system in which to perform the adjustment of the data. The relationship between the coordinate differences (dn, de, du) in the LA coordinate system and the coordinate differences (dX, dY, dZ) in the CT coordinate system can be given as

$$
\begin{bmatrix} dn_{ij} \\ de_{ij} \\ du_{ij} \end{bmatrix}^{LA} = \begin{bmatrix} -\sin\Phi_i\cos\Lambda_i & -\sin\Phi_i\sin\Lambda_i & \cos\Phi_i \\ -\sin\Lambda_i & \cos\Lambda_i & 0 \\ \cos\Phi_i\cos\Lambda_i & \cos\Phi_i\sin\Lambda_i & \sin\Phi_i \end{bmatrix} \begin{bmatrix} dX_{ij} \\ dY_{ij} \\ dZ_{ij} \end{bmatrix}^{CT}
\tag{9.13}
$$

where Φ_i and Λ_i are the astronomic latitude and astronomic longitude at point i corrected for the effect of polar motion so that they refer to the Conventional International Origin (CIO) of the CT system.

9.3 Parametric Model Equations in Conventional Terrestrial System

Equation (9.13) is exact, forming the basis of relating a measured quantity (e.g. a distance, an angle, a GPS vector, leveled height difference, etc.) to either the LG or LA coordinate differences between the stations involved in the measurement. By combining Equations (9.10)–(9.13), the following can be obtained (Vincenty and Bowring 1978):

$$
s_{ij} = \sqrt{dX_{ij}^2 + dY_{ij}^2 + dZ_{ij}^2}
\tag{9.14}
$$

$$
Az_{ij} = a\tan\left(\frac{-dX_{ij}\sin\Lambda_i + dY_{ij}\cos\Lambda_i}{-dX_{ij}\sin\Phi_i\cos\Lambda_i - dY_{ij}\sin\Phi_i\sin\Lambda_i + dZ_{ij}\cos\Phi_i}\right)
\tag{9.15}
$$

$$
v_{ij} = a\sin\left(\frac{dX_{ij}\cos\Phi_i\cos\Lambda_i + dY_{ij}\cos\Phi_i\sin\Lambda_i + dZ_{ij}\sin\Phi_i}{\sqrt{dX_{ij}^2 + dY_{ij}^2 + dZ_{ij}^2}}\right)
\tag{9.16}
$$

$$
z_{ij} = a\cos\left(\frac{dX_{ij}\cos\Phi_i\cos\Lambda_i + dY_{ij}\cos\Phi_i\sin\Lambda_i + dZ_{ij}\sin\Phi_i}{\sqrt{dX_{ij}^2 + dY_{ij}^2 + dZ_{ij}^2}}\right)
\tag{9.17}
$$

where z_{ij} is the zenith angle from point i to point j and Φ_i and Λ_i define the direction of gravity at the given point i and serve as reference direction in space to which Az_{ij} and v_{ij} (or z_{ij}) are referred. The Φ_i and Λ_i relate LA system to the CT system and are treated as additional unknown parameters in the adjustment. It should be mentioned that Φ_i and Λ_i may be replaced by the corresponding geodetic latitude (ϕ_i) and geodetic longitude (λ_i) in the coefficients of partial derivatives without losing accuracy. If observed latitudes and longitudes are available, they may be introduced to the parametric equations as observed parameters in the adjustment. Since horizontal angle is the difference between two azimuths, Equation (9.15) can be used to formulate horizontal angle equation. Equations (9.14)–(9.17) constitute the adjustment model of the usual parametric equations ($\hat{\ell} = f(\hat{x})$, for $\hat{\ell}$ as a vector of observations and $\hat{x}$ as vector of unknown parameters). The parameters to be estimated in the equations using the method of least squares adjustment are dX_{ij}, dY_{ij}, dZ_{ij}, where

$$\begin{bmatrix} dX_{ij} \\ dY_{ij} \\ dZ_{ij} \end{bmatrix} = \begin{bmatrix} X_j - X_i \\ Y_j - Y_i \\ Z_j - Z_i \end{bmatrix} \tag{9.18}$$

Equation (9.14) can be rewritten in symbolic forms as follows:

$$Az_{ij} = f\left(X_i, Y_i, Z_i, X_j, Y_j, Z_j\right) \tag{9.19}$$

$$v_{ij} = f\left(X_i, Y_i, Z_i, X_j, Y_j, Z_j\right) \tag{9.20}$$

$$s_{ij} = f\left(X_i, Y_i, Z_i, X_j, Y_j, Z_j\right) \tag{9.21}$$

The spatial distance s_{ij} given in Equation (9.14) relates to the CT system and can be rewritten as

$$\left[\left(X_j - X_i\right)^2 + \left(Y_j - Y_i\right)^2 + \left(Z_j - Z_i\right)^2\right]^{1/2} - s_{ij} = 0 \tag{9.22}$$

The linearized distance equation (required for the least squares adjustment) can be obtained by finding the partial derivatives of Equation (9.22) with respect to the unknown coordinates of points i and j given by

$$r_s = dX_{ij}\frac{\left(\delta X_j - \delta X_i\right)}{s_{ij}^0} + dY_{ij}\frac{\left(\delta Y_j - \delta Y_i\right)}{s_{ij}^0} + dZ_{ij}\frac{\left(\delta Z_j - \delta Z_i\right)}{s_{ij}^0} + s_{ij}^0 - s_{ij} \tag{9.23}$$

where dX_{ij}, dY_{ij}, dZ_{ij}, s_{ij}^0 are calculated values using approximate coordinates, s_{ij} is the measured distance and r_s is the residual, and δX_i, δY_i, δZ_i, δX_j, δY_j, and δZ_j are the unknown corrections to be determined and applied to the approximate Cartesian coordinates of points i and j in CT system. Equation (9.23) can also be given in matrix form as follows:

$$r_s = \begin{bmatrix} -\dfrac{dX_{ij}}{s_{ij}^0} & -\dfrac{dY_{ij}}{s_{ij}^0} & -\dfrac{dZ_{ij}}{s_{ij}^0} & \dfrac{dX_{ij}}{s_{ij}^0} & \dfrac{dY_{ij}}{s_{ij}^0} & \dfrac{dZ_{ij}}{s_{ij}^0} \end{bmatrix} \begin{bmatrix} \delta X_i \\ \delta Y_i \\ \delta Z_i \\ \delta X_j \\ \delta Y_j \\ \delta Z_j \end{bmatrix} + s_{ij}^0 - s_{ij} \tag{9.24}$$

or

$$r_s = a_{11}\delta X_i + a_{12}\delta Y_i + a_{13}\delta Z_i + a_{14}\delta X_j + a_{15}\delta Y_j + a_{16}\delta Z_j + s_{ij}^0 - s_{ij} \qquad (9.25)$$

Similarly, the linearized azimuth equation (9.15) and linearized vertical angle equation (9.16) can be given, respectively, as follows:

$$r_A = a_{21}\delta X_i + a_{22}\delta Y_i + a_{23}\delta Z_i + a_{24}\delta X_j + a_{25}\delta Y_j + a_{26}\delta Z_j + Az_{ij}^0 - Az_{ij}$$

$$(9.26)$$

$$r_v = a_{31}\delta X_i + a_{32}\delta Y_i + a_{33}\delta Z_i + a_{34}\delta X_j + a_{35}\delta Y_j + a_{36}\delta Z_j + v_{ij}^0 - v_{ij} \qquad (9.27)$$

where

$$a_{11} = \frac{\partial s_{ij}}{\partial X_i} = \frac{-dX_{ij}}{s_{ij}^0} = -a_{14} \qquad (9.28)$$

$$a_{12} = \frac{\partial s_{ij}}{\partial Y_i} = \frac{-dY_{ij}}{s_{ij}^0} = -a_{15} \qquad (9.29)$$

$$a_{13} = \frac{\partial s_{ij}}{\partial Z_i} = \frac{-dZ_{ij}}{s_{ij}^0} = -a_{16} \qquad (9.30)$$

$$a_{21} = \frac{\partial Az_{ij}}{\partial X_i} = \frac{-\sin\Phi_i \cos\Lambda_i \sin Az_{ij} + \sin\Lambda_i \cos Az_{ij}}{s_{ij}\cos v_{ij}} = -a_{24} \qquad (9.31)$$

$$a_{22} = \frac{\partial Az_{ij}}{\partial Y_i} = \frac{-\sin\Phi_i \sin\Lambda_i \sin Az_{ij} - \cos\Lambda_i \cos Az_{ij}}{s_{ij}\cos v_{ij}} = -a_{25} \qquad (9.32)$$

$$a_{23} = \frac{\partial Az_{ij}}{\partial Z_i} = \frac{\cos\Phi_i \sin Az_{ij}}{s_{ij}\cos v_{ij}} = -a_{26} \qquad (9.33)$$

$$a_{31} = \frac{\partial v_{ij}}{\partial X_i} = \frac{-s_{ij}\cos\Phi_i \cos\Lambda_i + \sin v_{ij}dX_{ij}}{s_{ij}^2 \cos v_{ij}} = -a_{34} \qquad (9.34)$$

$$a_{32} = \frac{\partial v_{ij}}{\partial Y_i} = \frac{-s_{ij}\cos\Phi_i \sin\Lambda_i + \sin v_{ij}dY_{ij}}{s_{ij}^2 \cos v_{ij}} = -a_{35} \qquad (9.35)$$

$$a_{33} = \frac{\partial v_{ij}}{\partial Z_i} = \frac{-s_{ij}\sin\Phi_i + \sin v_{ij}dZ_{ij}}{s_{ij}^2 \cos v_{ij}} = -a_{36} \qquad (9.36)$$

For a number of distance measurements, Equations (9.28)–(9.30) must be repeated for each measurement, making sure that the matrix elements a_{11}, a_{12}, etc. relate to appropriate parameters and columns in the overall design matrix A; similarly, for a number of azimuth (bearing) measurements, Equations (9.31)–(9.33) must be repeated for each measurement; and the same thing applies to Equations (9.34)–(9.36) for vertical angle measurements. Representing Equations (9.25)–(9.27) in matrix form will give the following:

$$
\begin{bmatrix} r_s \\ r_A \\ r_v \end{bmatrix} = \begin{bmatrix} a_{11} & a_{12} & a_{13} & a_{14} & a_{15} & a_{16} \\ a_{21} & a_{22} & a_{23} & a_{24} & a_{25} & a_{26} \\ a_{31} & a_{32} & a_{33} & a_{34} & a_{35} & a_{36} \end{bmatrix} \begin{bmatrix} \delta X_i \\ \delta Y_i \\ \delta Z_i \\ \delta X_j \\ \delta Y_j \\ \delta Z_j \end{bmatrix} + \begin{bmatrix} s_{ij}^0 - s_{ij} \\ Az_{ij}^0 - Az_{ij} \\ v_{ij}^0 - v_{ij} \end{bmatrix} \tag{9.37}
$$

or

$$
r = A\delta + w \tag{9.38}
$$

where A is the first design matrix, r is a vector of residuals, δ is a vector of unknown corrections to the approximate Cartesian coordinates in CT system, and w is a vector of misclosures.

9.4   Parametric Model Equations in Geodetic System

Sometimes it is preferred to work with the differences in geodetic latitude, longitude, and height ($d\phi_{ij}$, $d\lambda_{ij}$, dh_{ij}) as parameters instead of the Cartesian coordinate differences (dX_{ij}, dY_{ij}, dZ_{ij}); in this case, Equation (9.6) should be related to Equation (9.38). Equation (9.6) can be formulated for points i and then point j; for example, for point i the following will be obtained:

$$
\begin{bmatrix} \delta X_i \\ \delta Y_i \\ \delta Z_i \end{bmatrix} = \begin{bmatrix} -(M_i + h_i)\sin\phi_i\cos\lambda_i & -(N_i + h_i)\cos\phi_i\sin\lambda_i & \cos\phi_i\cos\lambda_i \\ -(M_i + h_i)\sin\phi_i\sin\lambda_i & (N_i + h_i)\cos\phi_i\cos\lambda_i & \cos\phi_i\sin\lambda_i \\ (M_i + h_i)\cos\phi_i & 0 & \sin\phi_i \end{bmatrix} \begin{bmatrix} \delta\phi_i \\ \delta\lambda_i \\ \delta h_i \end{bmatrix} \tag{9.39}
$$

or

$$
\delta_i = J_i \begin{bmatrix} \delta\phi_i \\ \delta\lambda_i \\ \delta h_i \end{bmatrix} \tag{9.40}
$$

Similarly, for point j, the following can be obtained:

$$
\delta_j = J_j \begin{bmatrix} \delta\phi_j \\ \delta\lambda_j \\ \delta h_j \end{bmatrix} \tag{9.41}
$$

Equations (9.40) and (9.41) can be combined to give

$$
\begin{bmatrix} \delta X_i \\ \delta Y_i \\ \delta Z_i \\ \delta X_j \\ \delta Y_j \\ \delta Z_j \end{bmatrix} = \begin{bmatrix} J_i & 0 \\ 0 & J_j \end{bmatrix} \begin{bmatrix} \delta\phi_i \\ \delta\lambda_i \\ \delta h_i \\ \delta\phi_j \\ \delta\lambda_j \\ \delta h_j \end{bmatrix}
\tag{9.42}
$$

If Equation (9.37) is partitioned according to points i and j, the following can be obtained:

$$
r = \begin{bmatrix} A_i & 0 \\ 0 & A_j \end{bmatrix} \begin{bmatrix} \delta X_i \\ \delta Y_i \\ \delta Z_i \\ \delta X_j \\ \delta Y_j \\ \delta Z_j \end{bmatrix} + w
\tag{9.43}
$$

By substituting Equation (9.42) into Equation (9.43), the following are obtained:

$$
r = \begin{bmatrix} A_i J_i & 0 \\ 0 & A_j J_j \end{bmatrix} \begin{bmatrix} \delta\phi_i \\ \delta\lambda_i \\ \delta h_i \\ \delta\phi_j \\ \delta\lambda_j \\ \delta h_j \end{bmatrix} + w
\tag{9.44}
$$

or

$$
\begin{bmatrix} r_s \\ r_A \\ r_v \end{bmatrix} = \begin{bmatrix} b_{11} & b_{12} & b_{13} & b_{14} & b_{15} & b_{16} \\ b_{21} & b_{22} & b_{23} & b_{24} & b_{25} & b_{26} \\ b_{31} & b_{32} & b_{33} & b_{34} & b_{35} & b_{36} \end{bmatrix} \begin{bmatrix} \delta\phi_i \\ \delta\lambda_i \\ \delta h_i \\ \delta\phi_j \\ \delta\lambda_j \\ \delta h_j \end{bmatrix} + \begin{bmatrix} s_{ij}^0 - s_{ij} \\ Az_{ij}^0 - Az_{ij} \\ v_{ij}^0 - v_{ij} \end{bmatrix}
\tag{9.45}
$$

Equation (9.45) can be given in matrix form in Equation (9.38) or in long form of linearized parametric equations based on the partial derivatives with respect to the unknown geodetic latitude, longitude, and height coordinates of points *i* and *j* as follows.

For *spatial distance measurement*, the parametric equation can be given as

$$r_s = b_{11}\delta\phi_i + b_{12}\delta\lambda_i + b_{13}\delta h_i + b_{14}\delta\phi_j + b_{15}\delta\lambda_j + b_{16}\delta h_j + s_{ij}^0 - s_{ij} \qquad (9.46)$$

For astronomical azimuth measurement, the parametric equation can be given as

$$r_A = b_{21}\delta\phi_i + b_{22}\delta\lambda_i + b_{23}\delta h_i + b_{24}\delta\phi_j + b_{25}\delta\lambda_j + b_{26}\delta h_j + b_{27}\delta\Phi_i + b_{28}\delta\Lambda_i + Az_{ij}^0 - Az_{ij}$$

$$(9.47)$$

where the astronomical coordinates (Φ_i, Λ_i) of the setup station are treated as unknown with the corrections as $\delta\Phi_i$ and $\delta\Lambda_i$ to be determined. For the vertical angle measurement, the parametric equation can be given as

$$r_v = b_{31}\delta\phi_i + b_{32}\delta\lambda_i + b_{33}\delta h_i + b_{34}\delta\phi_j + b_{35}\delta\lambda_j + b_{36}\delta h_j + b_{37}\delta\Phi_i + b_{38}\delta\Lambda_i - \delta v + v_{ij}^0 - v_{ij}$$

$$(9.48)$$

where δv is the unknown residual vertical angle refraction correction and the astronomical coordinates are considered unknown. The coefficients of Equations (9.46)–(9.48) are given in Vincenty and Bowring (1978), Vincenty (1979), and Vanicek and Krakiwsky (1986) as follows:

$$b_{11} = -(M_i + h_i)\cos Az_{ij} \cos v_{ij} \qquad (9.49)$$

$$b_{12} = -(N_i + h_i)\cos\phi_i \sin Az_{ij} \cos v_{ij} \qquad (9.50)$$

$$b_{13} = -\sin v_{ij} \qquad (9.51)$$

$$b_{14} = -(M_j + h_j)\cos Az_{ji} \cos v_{ji} \qquad (9.52)$$

$$b_{15} = -(N_j + h_j)\cos\phi_j \sin Az_{ji} \cos v_{ji} \qquad (9.53)$$

$$b_{16} = -\sin v_{ji} \qquad (9.54)$$

$$b_{21} = \frac{(M_i + h_i)\sin Az_{ij}}{s_{ij}\cos v_{ij}} \qquad (9.55)$$

$$b_{22} = -\frac{(N_i + h_i)\cos\phi_i \cos Az_{ij}}{s_{ij}\cos v_{ij}} \qquad (9.56)$$

$$b_{23} = 0 \qquad (9.57)$$

$$b_{24} = -\frac{\left[(M_j + h_j)\left(\sin\phi_i \sin\phi_j \cos\Delta\lambda \sin Az_{ij} + \sin\phi_j \sin\Delta\lambda \cos Az_{ij} + \cos\phi_i \cos\phi_j \sin Az_{ij}\right)\right]}{s_{ij}\cos v_{ji}}$$

$$(9.58)$$

$$b_{25} = \frac{\left[\left(N_j + h_j\right)\cos\phi_j\left(\cos\Delta\lambda\cos Az_{ij} - \sin\phi_i\sin\Delta\lambda\sin Az_{ij}\right)\right]}{s_{ij}\cos v_{ij}} \tag{9.59}$$

$$b_{26} = 0 \tag{9.60}$$

$$b_{27} = \sin Az_{ij}\tan v_{ij} \tag{9.61}$$

$$b_{28} = \sin\phi_i - \cos\phi_i\cos Az_{ij}\tan v_{ij} \tag{9.62}$$

$$b_{31} = \frac{\left(M_i + h_i\right)\cos Az_{ij}\sin v_{ij}}{s_{ij}} \tag{9.63}$$

$$b_{32} = \frac{\left(N_i + h_i\right)\cos\phi_i\sin Az_{ij}\sin v_{ij}}{s_{ij}} \tag{9.64}$$

$$b_{33} = \frac{-\cos v_{ij}}{s_{ij}} \tag{9.65}$$

$$b_{34} = \frac{-\left[\left(M_j + h_j\right)\left(\cos\phi_i\sin\phi_j\cos\Delta\lambda - \sin\phi_i\cos\phi_j - \cos Az_{ji}\sin v_{ij}\cos v_{ji}\right)\sec v_{ij}\right]}{s_{ij}}$$

$$\tag{9.66}$$

$$b_{35} = \frac{-\left[\left(N_j + h_j\right)\cos\phi_j\left(\cos\phi_i\sin\Delta\lambda - \sin Az_{ji}\sin v_{ij}\cos v_{ji}\right)\sec v_{ij}\right]}{s_{ij}}$$

$$\tag{9.67}$$

$$b_{36} = \frac{\left(\cos\phi_i\cos\phi_j\cos\Delta\lambda + \sin\phi_i\sin\phi_j + \sin v_{ij}\sin v_{ji}\right)\sec v_{ij}}{s_{ij}} \tag{9.68}$$

$$b_{37} = \cos Az_{ij} \tag{9.69}$$

$$b_{38} = \cos\phi_i\sin Az_{ij} \tag{9.70}$$

According to Vincenty and Bowring (1978), the following is acceptable:

$$\cos v_{ji} = \cos v_{ij}\left(\frac{a + h_i}{a + h_j}\right) \tag{9.71}$$

where a is the semi-major axis value of the reference ellipsoid and h_i and h_j are the ellipsoidal heights of points i and j, respectively.

Parametric equation for a *total station direction measurement* can be formulated from an azimuth equation by subtracting orientation parameter (γ) from the azimuth equation; in this case there will be an approximate value (γ^0) of the orientation parameter and an unknown correction ($\delta\gamma$) subtracted from Equation (9.47). Parametric equation for *horizontal angle measurement* will be obtained by subtracting parametric equations for two corresponding azimuth measurements. If astronomical latitude (Φ_i) and longitude (Λ_i) have also been measured, two more parametric equations can be added to the linearized model as

$$r_\Phi = \delta\Phi_i + \Phi_i^0 - \Phi_i \tag{9.72}$$

$$r_\Lambda = \delta\Lambda_i + \Lambda_i^0 - \Lambda_i \tag{9.73}$$

If height difference is observed, the parametric equation for the height difference ($\mathrm{d}h_{ij}$) can be added to the linearized model as

$$r_{\mathrm{d}h} = -\delta h_i + \delta h_j + \mathrm{d}h_{ij}^0 - \mathrm{d}h_{ij} \tag{9.74}$$

In adjusting horizontal networks in three dimensions, only approximate geodetic heights of the network points are needed in the model; note also that accurate vertical angles (or zenith angles) and astronomic latitudes and longitudes are not usually measured in horizontal networks.

9.5 Parametric Model Equations in Local Astronomic System

In order to allow easier interpretation of parameters, the geodetic coordinate differences ($\mathrm{d}\phi$, $\mathrm{d}\lambda$, $\mathrm{d}h$) in Equation (9.6) can be transformed into local Cartesian coordinate differences ($\mathrm{d}n$, $\mathrm{d}e$, $\mathrm{d}u$) (in local geodetic coordinate system) by combining Equations (9.6) and (9.13) as follows. In this case, the coordinate differences in Equation (9.13) relate to a particular point and not to two points. Note that Φ_i and Λ_i may be replaced by the corresponding geodetic latitude (ϕ_i) and geodetic longitude (λ_i) in Equation (9.13) without losing accuracy. For a particular point i, the following is obtained:

$$\begin{bmatrix} \delta n_i \\ \delta e_i \\ \delta u_i \end{bmatrix} = \begin{bmatrix} M_i + h_i & 0 & 0 \\ 0 & (N_i + h_i)\cos\phi_i & 0 \\ 0 & 0 & 1 \end{bmatrix} \begin{bmatrix} \delta\phi_i \\ \delta\lambda_i \\ \delta h_i \end{bmatrix} \tag{9.75}$$

or

$$\begin{bmatrix} \delta n_i \\ \delta e_i \\ \delta u_i \end{bmatrix} = H_i \begin{bmatrix} \delta\phi_i \\ \delta\lambda_i \\ \delta h_i \end{bmatrix} \tag{9.76}$$

where $(\delta n_i, \delta e_i, \delta u_i)$ are linear coordinate differences at point i in the directions of north, east, and plumb line, respectively, and the matrix H is evaluated for the station i with latitude ϕ_i. From Equation (9.76), the following is obtained:

$$
\begin{bmatrix} \delta\phi_i \\ \delta\lambda_i \\ \delta h_i \end{bmatrix} = H_i^{-1} \begin{bmatrix} \delta n_i \\ \delta e_i \\ \delta u_i \end{bmatrix}
\tag{9.77}
$$

Equation (9.77) converts local Cartesian coordinate differences to geodetic curvilinear coordinates at point i. Forming similar equation for point j, the following equation is obtained:

$$
\begin{bmatrix} \delta\phi_i \\ \delta\lambda_i \\ \delta h_i \\ \delta\phi_j \\ \delta\lambda_j \\ \delta h_j \end{bmatrix} = \begin{bmatrix} H_i^{-1} & 0 \\ 0 & H_j^{-1} \end{bmatrix} \begin{bmatrix} \delta n_i \\ \delta e_i \\ \delta u_i \\ \delta n_j \\ \delta e_j \\ \delta u_j \end{bmatrix}
\tag{9.78}
$$

By combining Equations (9.44) and (9.78) the following is obtained:

$$
r = \begin{bmatrix} A_i J_i H_i^{-1} & 0 \\ 0 & A_j J_j H_j^{-1} \end{bmatrix} \begin{bmatrix} \delta n_i \\ \delta e_i \\ \delta u_i \\ \delta n_j \\ \delta e_j \\ \delta u_j \end{bmatrix}
\tag{9.79}
$$

or

$$
\begin{bmatrix} r_s \\ r_A \\ r_v \end{bmatrix} = \begin{bmatrix} c_{11} & c_{12} & c_{13} & c_{14} & c_{15} & c_{16} \\ c_{21} & c_{22} & c_{23} & c_{24} & c_{25} & c_{26} \\ c_{31} & c_{32} & c_{33} & c_{34} & c_{35} & c_{36} \end{bmatrix} \begin{bmatrix} \delta n_i \\ \delta e_i \\ \delta u_i \\ \delta n_j \\ \delta e_j \\ \delta u_j \end{bmatrix} + \begin{bmatrix} s_{ij}^0 - s_{ij} \\ Az_{ij}^0 - Az_{ij} \\ v_{ij}^0 - v_{ij} \end{bmatrix}
\tag{9.80}
$$

Equation (9.80) can be expressed in matrix form in Equation (9.38) or in long form (with additional terms added as needed) as follows.

For *spatial distance measurement,* the parametric equation can be given as

$$
r_s = c_{11}\delta n_i + c_{12}\delta e_i + c_{13}\delta u_i + c_{14}\delta n_j + c_{15}\delta e_j + c_{16}\delta n_j + s_{ij}^0 - s_{ij}
\tag{9.81}
$$

For astronomical azimuth measurement, the parametric equation can be given as

$$r_A = c_{21}\delta n_i + c_{22}\delta e_i + c_{23}\delta u_i + c_{24}\delta n_j + c_{25}\delta e_j + c_{26}\delta u_j + c_{27}\delta\Phi_i + c_{28}\delta\Lambda_i + Az^0_{ij} - Az_{ij}$$

(9.82)

where the astronomical coordinates (Φ_i, Λ_i) of the setup station are treated as unknown with the corrections as $\delta\Phi_i$ and $\delta\Lambda_i$ to be determined. For the vertical angle measurement, the parametric equation can be given as

$$r_v = c_{31}\delta n_i + c_{32}\delta e_i + c_{33}\delta u_i + c_{34}\delta n_j + c_{35}\delta e_j + c_{36}\delta u_j + c_{37}\delta\Phi_i + c_{38}\delta\Lambda_i - \delta v + v^0_{ij} - v_{ij}$$

(9.83)

where δv is the unknown residual vertical angle refraction correction and the astronomical coordinates are considered unknown. The coefficients of Equations (9.81)–(9.83) are given in Vincenty and Bowring (1978), Vincenty (1979), and Vanicek and Krakiwsky (1986) as follows:

For *spatial distance measurement*, for example, the coefficients in Equation (9.81) can be determined by taking the partial derivatives of Equation (9.10) with respect to the unknowns, for example, with respect to n_i:

$$c_{11} = \frac{\partial s_{ij}}{\partial n_i} = \frac{-dn_{ij}}{s_{ij}}$$

(9.84)

Substituting Equation (9.7) into Equation (9.84) gives

$$c_{11} = -\cos v_{ij}\cos Az_{ij}$$

(9.85)

Similarly,

$$c_{12} = \frac{\partial s_{ij}}{\partial e_i} = -\cos v_{ij}\sin Az_{ij}$$

(9.86)

$$c_{13} = \frac{\partial s_{ij}}{\partial u_i} = -\sin v_{ij}$$

(9.87)

$$c_{14} = \frac{\partial s_{ij}}{\partial n_j} = -\cos v_{ji}\cos Az_{ji}$$

(9.88)

$$c_{15} = \frac{\partial s_{ij}}{\partial e_j} = -\cos v_{ji}\sin Az_{ji}$$

(9.89)

$$c_{16} = \frac{\partial s_{ij}}{\partial u_j} = -\sin v_{ji}$$

(9.90)

For *astronomic azimuth measurement*, the coefficients in Equation (9.82) can be determined by taking the partial derivatives of Equation (9.11) with respect to the unknowns, for example, with respect to n_i:

$$c_{21} = \frac{\partial Az_{ij}}{\partial n_i} = \frac{de_{ij}}{d\,e^2_{ij} + d\,n^2_{ij}}$$

(9.91)

Substituting Equations (9.7) and (9.8) into Equation (9.91) gives

$$c_{21} = \frac{s_{ij} \cos v_{ij} \sin Az_{ij}}{s_{ij}^2 \cos^2 v_{ij} \left(\sin^2 Az_{ij} + \cos^2 Az_{ij} \right)} \tag{9.92}$$

which can be simplified to

$$c_{21} = \frac{\sin Az_{ij}}{s_{ij} \cos v_{ij}} \tag{9.93}$$

Similarly, the following are obtained:

$$c_{22} = \frac{-\cos Az_{ij}}{s_{ij} \cos v_{ij}} \tag{9.94}$$

$$c_{23} = 0 \tag{9.95}$$

$$c_{24} = \frac{-\sin Az_{ij}}{s_{ij} \cos v_{ij}} \left[\cos \phi_i \cos \phi_j + \sin \phi_i \sin \phi_j \cos \left(\lambda_j - \lambda_i \right) + \right.$$
$$\left. \sin \phi_j \sin \left(\lambda_j - \lambda_i \right) \cot Az_{ij} \right] \tag{9.96}$$

$$c_{25} = \frac{\cos Az_{ij}}{s_{ij} \cos v_{ij}} \left[\cos \left(\lambda_j - \lambda_i \right) - \sin \phi_i \sin \left(\lambda_j - \lambda_i \right) \tan Az_{ij} \right] \tag{9.97}$$

$$c_{26} = \frac{-\sin Az_{ij}}{s_{ij} \cos v_{ij}} \left[\sin \left(\phi_j - \phi_i \right) + \cos \phi_j \sin \left(\lambda_j - \lambda_i \right) \cot Az_{ij} \right] \tag{9.98}$$

$$c_{27} = \sin Az_{ij} \tan v_{ij} \tag{9.99}$$

$$c_{28} = \sin \phi_i - \cos \phi_i \cos Az_{ij} \tan v_{ij} \tag{9.100}$$

For *vertical angle measurement*, the coefficients in Equation (9.83) can be determined by taking the partial derivatives of Equation (9.12) with respect to the unknowns, for example, with respect to n_i:

$$c_{31} = \frac{\partial v_{ij}}{\partial n_i} = \frac{du_{ij} \times dn_{ij}}{\left(s_{ij}^2 - d\, u_{ij}^2 \right)^{1/2} s_{ij}^2} \tag{9.101}$$

Substituting Equations (9.7) and (9.9) into Equation (9.101) (taking note that $\left(s_{ij}^2 - du_{ij}^2 \right)^{1/2}$ is the same as $s_{ij} \cos v_{ij}$) gives

$$c_{31} = \frac{s_{ij}^2 \cos v_{ij} \cos Az_{ij} \sin v_{ij}}{s_{ij} \cos v_{ij} \times s_{ij}^2} \tag{9.102}$$

which can be simplified to

$$c_{31} = \frac{\cos Az_{ij} \sin v_{ij}}{s_{ij}} \tag{9.103}$$

Similarly, the following are obtained:

$$c_{32} = \frac{\sin Az_{ij} \sin v_{ij}}{s_{ij}} \qquad (9.104)$$

$$c_{33} = \frac{-\cos v_{ij}}{s_{ij}} \qquad (9.105)$$

$$c_{34} = \frac{-\cos\phi_i \sin\phi_j \cos\left(\lambda_j - \lambda_i\right) + \sin\phi_i \cos\phi_j + \sin v_{ij} \cos v_{ji} \cos Az_{ji}}{s_{ij} \cos v_{ij}}$$

$$(9.106)$$

$$c_{35} = \frac{-\cos\phi_i \sin\left(\lambda_j - \lambda_i\right) + \sin v_{ij} \cos v_{ji} \cos Az_{ji}}{s_{ij} \cos v_{ij}} \qquad (9.107)$$

$$c_{36} = \frac{\cos\phi_i \cos\phi_j \cos\left(\lambda_j - \lambda_i\right) + \sin\phi_i \sin\phi_j + \sin v_{ij} \sin v_{ji}}{s_{ij} \cos v_{ij}} \qquad (9.108)$$

$$c_{37} = \cos Az_{ij} \qquad (9.109)$$

$$c_{38} = \cos\phi_i \sin Az_{ij} \qquad (9.110)$$

9.6 General Comments on Three-dimensional Adjustment

If the weight matrix (P) of measurements is expressed as usual, the parametric least squares adjustment solution for the corrections to the unknown parameters can be given as

$$\delta = -\left(A^T P A\right)^{-1} A^T P w \qquad (9.111)$$

where the first design matrix A and the vector of misclosures w can be formulated from Equations (9.37), (9.45), or (9.80) with the associated parameters. The adjusted parameters ($\hat{x}$) and the adjusted observations ($\hat{\ell}$) will be given as follows:

$$\hat{x} = x^0 + \delta \qquad (9.112)$$

$$\hat{\ell} = \ell + r \qquad (9.113)$$

where

$$r = A\delta + w \qquad (9.114)$$

x^0 is a vector of the approximate values of the parameters, ℓ is a vector of measurements, and r is the vector of residuals expressed by Equation (9.38).

If the azimuth (Equation (9.82)) is used for the total station direction measurement, an orientation parameter with coefficient of -1 can be added to the equation. In the least squares adjustment process, the adjusted positions of the previous iteration must be used at the current point of expansion. This is required irrespective of whether the partial derivatives are expressed in terms of Cartesian coordinates, geodetic latitudes, longitudes, and ellipsoidal heights or using azimuths and vertical angle measurements.

Note also that observations are not reduced to the marks on the ground, but to the line in space between the instrument and the target at the time of measurement. After the adjustment the reduction to the marks is determined indirectly "by applying the residuals, refraction corrections, and scale corrections (with the signs reversed) to the values computed by the inverse formula from adjusted coordinates of the marked point" (Vincenty 1979). The final values obtained will be identical to what would have been observed if the heights of the instruments and the target were zero.

In three-dimensional parametric adjustment, the provisional positions (usually, the geodetic coordinates) are required as input. Typically, the astronomic coordinates Φ_i and Λ_i are used to define the direction of plumb line, and their geodetic counterparts ϕ_i and λ_i with ellipsoidal height (h_i) are the true point coordinates that are essentially equivalent to the Cartesian X, Y, Z coordinates. If the astronomic coordinates are unknown, the geodetic values may be used instead (Vincenty 1979). It is also to be known that the same results will be obtained without using the geodetic latitudes, longitudes, and ellipsoidal heights anywhere in the computations since the ellipsoid is not considered at all in three-dimensional computations.

Note that since traditional observations are taken to some elevated target by an instrument at a height (HI) above the setup station, the geodetic heights of each station must be increased by the instrument heights (HIs) when computing geodetic coordinates. The approximate geodetic coordinates may be determined by first adjusting measurements in the map projection plane and converting the coordinates to geodetic equivalents later. The average geoidal heights for the region may also be used in order to determine the approximate ellipsoidal heights of points.

It should also be mentioned that vertical angles are subject to large systematic errors due to deflection of the vertical refraction and should not be used in an adjustment on a regular basis. If they must be used, the systematic errors must be corrected for or another terms to take care of the errors introduced in the model as unknown. This must be done with an understanding of the risk of over-parameterization by not adding too many unknowns to the model. Reciprocal distance measurements may be difficult to make because of heights of instrument and target changes at both ends; in this case, the forward and backward distance measurements should be treated as different. One choice for minimal constraint adjustment is to fix the coordinates (ϕ, λ, h) or (X, Y, Z) of one station, the azimuth or the longitude of another station, and the heights of two additional stations.

Differential Leveling Observations: Orthometric height differences obtained from differential leveling procedure can be included in three-dimensional geodetic network model. This will require, however, that the height differences be corrected for geoid undulation differences (dN_{ij}) between points i and j that are being considered. The adjustment parametric model can be given as

$$r_{dH} = \delta u_j - \delta u_i + dH_{ij} + dN_{ij} - dh_{ij} \tag{9.115}$$

where dH_{ij} is the elevation difference between the stations and dh_{ij} is the change in the ellipsoidal heights between the stations. It can be seen from Equation (9.115) that orthometric height difference equation cannot be formulated without a reference ellipsoid.

9.7 Adjustment Examples

The simple examples given in this section are mainly for the purpose of illustrating how to implement the equations discussed in this chapter. The examples can be solved using Microsoft Excel and MATLAB software applications or other programming environments. The author, however, used the MATLAB software application in processing the given measurements.

9.7.1 Adjustment in Cartesian Geodetic System

Three-dimensional data with regard to the network of points P1, P2, and P3 in Figure 9.4 were collected as shown in Table 9.1. Point P1 is a control point that is to be kept fixed during adjustment; the geodetic coordinates of the point and the approximate geodetic coordinates of P2 and P3 are provided in Table 9.2.

The adjustment in Cartesian system process starts with the use of Equations (9.1)–(9.5) with the results summarized in Tables 9.3 and 9.4 (note that the negative values of longitudes are used since the points are in the "West" region).

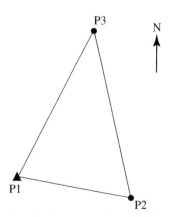

Figure 9.4 Sample 3D geodetic network.

Table 9.1 Field measurements.

Leg	Distance (m)	Zenith angle	Bearing
P1–P2	330.305 ± 0.003	89°53′10″ ± 5″	103°30′10″ ± 5″
P1–P3	584.140 ± 0.004	90°18′15″ ± 5″	21°14′30″ ± 5″
P2–P3	631.160 ± 0.003	90°20′40″ ± 5″	350°00′20″ ± 5″

Table 9.2 Initial geodetic coordinates of network points.

Point	Latitude (ϕ)	Longitude (λ)	Ellipsoidal height (m)
P1	49°05′24.73726″N	127°23′56.95384″W	291.895
P2	49°05′22.24485″N	127°23′41.12441″W	292.448
P3	49°05′42.33153″N	127°23′46.52178″W	289.560

Table 9.3 Radii of curvature calculations based on GRS80 ellipsoid.

Point	N (m)	M (m)
P1	6 390 365.3638	6 371 948.7421
P2	6 390 365.1070	6 371 947.9739
P3	6 390 367.1767	6 371 954.1651

Table 9.4 Initial Cartesian geodetic coordinates.

Point	X (m)	Y (m)	Z (m)
P1	−2 541 849.080 2	−3 324 702.344 2	4 797 354.789 3
P2	−2 541 629.586 0	−3 324 944.056 7	4 797 304.782 3
P3	−2 541 429.789 9	−3 324 502.379 0	4 797 708.960 7

The network in Figure 9.4 with the measurements in Table 9.1 is to be adjusted three-dimensionally using the method of least squares by fixing the three-dimensional Cartesian coordinates (X, Y, Z) of point P1 (by assigning standard deviation of 0.000 01 m to each coordinate) and the

Z coordinate of point P2 (by assigning standard deviation of 0.05 m to the coordinate).

9.7.1.1 Solution Approach

In adjusting the network, Equation (9.37) or Equation (9.38) will be used in order to be able to constrain the X, Y, Z of point P1 and Z of point P3. The first design matrix (A) is derived from Equations (9.37) and (9.38) with additional constraint equations due to the fixing of the X_1, Y_1, Z_1 coordinates of point P1 and of the Z_2 coordinate of point P2. These additional constraint equations, which constitute the 10th to 13th observations, can be expressed as follows:

$$r_{10} = \delta X_1 \tag{9.116}$$

$$r_{11} = \delta Y_1 \tag{9.117}$$

$$r_{12} = \delta Z_1 \tag{9.118}$$

$$r_{13} = \delta Z_2 \tag{9.119}$$

where the misclosures are zero (since the corresponding coordinates are the measurements); r_{10}, r_{11}, r_{12}, and r_{13} are the residuals of the observations; and δX_1, δY_1, δZ_1, and δZ_2 are the coordinate changes. The observations in Equations (9.116)–(9.118) are given very small standard deviations (0.001 mm) to ensure they are fixed after adjustment, and the observation in Equation (9.119) is given a standard deviation of 0.05 m since it is not well known. The first nine equations formulated from Equations (9.25)–(9.27) are based on three distances, three bearings, and three zenith angles. In formulating the A-matrix, the elements of the matrix due to the three distance measurements are derived from Equations (9.28)–(9.30), those due to the bearing measurements are derived from Equations (9.31)–(9.33), and those due to the zenith angles are derived from Equations (9.34)–(9.36), remembering that 90° minus zenith angle will give the vertical angle. The size of A-matrix is 13 observations by 9 unknown parameters (with the fixed parameters constrained by Equations (9.116)–(9.119) and to be highly weighted using some specified variances). The elements of A-matrix must also correspond with the appropriate parameters. For example, distance P1–P2 will have a_{11}, a_{12}, a_{13} values in the columns corresponding to δX_1, δY_1, δZ_1 and a_{14}, a_{15}, a_{16} in columns corresponding to δX_2, δY_2, and δZ_2, respectively; distance P1–P3 will have a_{11}, a_{12}, a_{13} values in the columns corresponding to X_1, Y_1, Z_1 and a_{14}, a_{15}, a_{16} in columns corresponding to δX_3, δY_3, and δZ_3, respectively; constraint Equation (9.116) will have 1.0 in column corresponding to δX_1 in row

10 corresponding to the order of the equation; etc. The computed A-matrix for this problem is given as follows:

$$A = \begin{bmatrix}
-0.665 & 0.732 & 0.151 & 0.665 & -0.732 & -0.151 & 0.0 & 0.0 & 0.0 \\
-0.718 & -0.342 & -0.606 & 0.0 & 0.0 & 0.0 & 0.718 & 0.342 & 0.606 \\
0.0 & 0.0 & 0.0 & -0.317 & -0.699\ 8 & -0.640\ 4 & 0.316\ 6 & 0.699\ 8 & 0.640\ 4 \\
-0.002 & -0.001 & -0.002 & 0.002 & 0.001\ 3 & 0.001\ 9 & 0.0 & 0.0 & 0.0 \\
-0.001 & 0.001 & 0.000\ 4 & 0.0 & 0.0 & 0.0 & 0.001\ 0 & -0.001\ 3 & -0.000\ 4 \\
0.0 & 0.0 & 0.00 & -0.001\ 4 & 0.000\ 8 & -0.000\ 18 & 0.001\ 4 & -0.000\ 8 & 0.000\ 2 \\
0.001 & 0.002 & -0.002\ 2 & -0.001\ 2 & -0.001\ 6 & 0.002\ 3 & 0.0 & 0.0 & 0.0 \\
0.001 & 0.001 & -0.001\ 3 & 0.0 & 0.0 & 0.0 & -0.000\ 7 & -0.000\ 9 & 0.001\ 3 \\
0.0 & 0.0 & 0.0 & 0.000\ 6 & 0.000\ 8 & -0.001\ 2 & -0.000\ 6 & -0.000\ 8 & 0.001\ 2 \\
1.000 & 0.0 & 0.0 & 0.0 & 0.0 & 0.0 & 0.0 & 0.0 & 0.0 \\
0.0 & 1.000 & 0.0 & 0.0 & 0.0 & 0.0 & 0.0 & 0.0 & 0.0 \\
0.0 & 0.0 & 1.000 & 0.0 & 0.0 & 0.0 & 0.0 & 0.0 & 0.0 \\
0.0 & 0.0 & 0.0 & 0.0 & 0.0 & 1.00 & 0.0 & 0.0 & 0.0
\end{bmatrix}$$

The least squares corrections to the approximate coordinates determined using the given geodetic coordinates are as follows:

$$\delta = -\left(A^T P A\right)^{-1} A^T P w$$

$$= \begin{bmatrix} 0.0 & 0.0 & 0.0 & -0.0007 & 0.0009 & 0.0067 & -0.0303 & 0.0028 & 0.0236 \end{bmatrix}^T$$

The adjusted geodetic Cartesian coordinates are

$$x^0 + \delta = \begin{bmatrix} \hat{X}_1 \\ \hat{Y}_1 \\ \hat{Z}_1 \\ \hat{X}_2 \\ \hat{Y}_2 \\ \hat{Z}_2 \\ \hat{X}_3 \\ \hat{Y}_3 \\ \hat{Z}_3 \end{bmatrix} = \begin{bmatrix} -2\ 541\ 849.080\ 2 \\ -3\ 324\ 702.344\ 2 \\ 4\ 797\ 354.789\ 3 \\ -2\ 541\ 629.586\ 8 \\ -3\ 324\ 944.055\ 8 \\ 4\ 797\ 304.789\ 0 \\ -2\ 541\ 429.820\ 2 \\ -3\ 324\ 502.376\ 1 \\ 4\ 797\ 708.984\ 3 \end{bmatrix}$$

where the approximate coordinates (x^0) were determined from the given geodetic coordinates and given in Table 9.4. The adjusted observations are given in Table 9.5 (note that the parameters adjusted as observations gave the same results as the adjusted parameters).

Table 9.5 Adjusted observations.

Leg	Distance (m)	Zenith angle	Bearing
P1–P2	330.3057 ± 0.0028	89°53′08.2″ ± 4.4″	103°30′08.0″ ± 3.3″
P1–P3	584.1406 ± 0.0036	90°18′18.4″ ± 3.7″	21°14′23.9″ ± 3.0″
P2–P3	631.1588 ± 0.0036	90°20′36.4″ ± 3.5″	350°00′24.1″ ± 2.9″

Table 9.6 Computed corrections to the geodetic coordinates of network points.

Point	Latitude ($\delta\phi$)	Longitude ($\delta\lambda$)	Ellipsoidal height (δh) (m)
P1	0.000 00″	0.000 00″	0.000 0
P2	+0.000 15″	−0.000 06″	+0.004 9
P3	+0.000 11″	−0.001 27″	+0.028 4

The covariance matrix ($C_{\hat{\ell}}$) of the adjusted observations is obtained using the following formula:

$$C_{\hat{\ell}} = s_0^2 \left[A \left(A^T P A \right)^{-1} A \right] \tag{9.120}$$

where the a posteriori variance factor of unit weight (s_0^2) was determined after the adjustment as 0.9101 with the number of degrees of freedom as 4. By using the transformation formula in Equation (9.42), three block diagonal submatrices of the coefficient matrix J can be formed for this problem; each block is a 3×3 submatrix formed using Equation (9.39), giving the overall J matrix of size 9×9. By inversion of the formula in Equation (9.42), the corrections to the given geodetic coordinates (ϕ, λ, h) in Table 9.2 are obtained, and the results are given in Table 9.6.

9.7.2 Adjustment in Curvilinear Geodetic System

Continuing from the example given in Section 9.7.1, assume that the geodetic network is to be adjusted in curvilinear (ϕ, λ, h) coordinate system (i.e. adjustment in curvilinear system) as discussed in Section 9.4. In this case, Equations (9.45)–(9.70) will be used. Assume the latitude, longitude, and the ellipsoidal height of point P1 and the ellipsoidal height of point P2 will be fixed to constraint the three-dimensional least squares adjustment of the measurements in Table 9.1. The constraint equations to be used with the measurement equations in Equation (9.45) can be expressed as follows:

$$r_{10} = \delta\phi_1 \tag{9.121}$$

$$r_{11} = \delta\lambda_1 \tag{9.122}$$

$$r_{12} = \delta h_1 \tag{9.123}$$

$$r_{13} = \delta h_2 \tag{9.124}$$

where the misclosures are zero (since the coordinates are directly measured) and $\delta\phi_1$, $\delta\lambda_1$, δh_1, and δh_2 are the geodetic coordinate changes. The observations in Equations (9.121)–(9.122) are given very small standard deviations (0.001″), the observation in Equation (9.123) is given 0.01 mm standard deviation, and the observation in Equation (9.124) is given a standard deviation of 0.05 m since it is not well known. The first design matrix A has a size of 13 equations by 9 unknown parameters; the unknown parameters include the fixed coordinates of point P1. The resulting A-matrix is given as follows; the adjustment results are given in Tables 9.7–9.9.

$$
A = \begin{bmatrix}
-1.49\text{E}6 & 4.07\text{E}6 & -1.98\text{E}-3 & 1.49\text{E}6 & -4.07\text{E}6 & -1.98\text{E}-3 & 0.0 & 0.0 & 0.0 \\
-5.94\text{E}6 & -1.52\text{E}6 & 5.37\text{E}-3 & 0.0 & 0.0 & 0.0 & 5.94\text{E}6 & 1.52\text{E}6 & 5.37\text{E}-3 \\
0.0 & 0.0 & 0.0 & -6.28\text{E}6 & 7.26\text{E}5 & 6.03\text{E}-3 & 6.28\text{E}6 & -7.26\text{E}5 & 6.03\text{E}-3 \\
-1.88\text{E}4 & -2.96\text{E}3 & 0.0 & 1.88\text{E}4 & 2.96\text{E}3 & 0.0 & 0.0 & 0.0 & 0.0 \\
3.95\text{E}3 & -6.68\text{E}3 & 0.0 & 0.0 & 0.0 & 0.0 & -3.95\text{E}3 & 6.68\text{E}3 & 0.0 \\
0.0 & 0.0 & 0.0 & -1.75\text{E}3 & -6.53\text{E}3 & 0.0 & 1.75\text{E}3 & 6.53\text{E}3 & 0.0 \\
8.9 & -24.4 & -3.03\text{E}-3 & -8.69\text{E}3 & 23.8 & 3.03\text{E}-3 & 0.0 & 0.0 & 0.0 \\
-55.0 & -13.9 & -1.71\text{E}-3 & 0.0 & 0.0 & 0.0 & 53.71 & 13.71 & 1.71\text{E}-3 \\
0.0 & 0.0 & 0.0 & -59.97 & 6.94 & -1.58\text{E}-3 & 58.99 & -6.82 & 1.58\text{E}-3 \\
1.0 & 0.0 & 0.0 & 0.0 & 0.0 & 0.0 & 0.0 & 0.0 & 0.0 \\
0.0 & 1.0 & 0.0 & 0.0 & 0.0 & 0.0 & 0.0 & 0.0 & 0.0 \\
0.0 & 0.0 & 1.0 & 0.0 & 0.0 & 0.0 & 0.0 & 0.0 & 0.0 \\
0.0 & 0.0 & 0.0 & 0.0 & 0.0 & 1.0 & 0.0 & 0.0 & 0.0
\end{bmatrix}
$$

The a posteriori variance factor of unit weight for the adjustment is 0.6936 with the number of degrees of freedom as 4.

Table 9.7 Computed corrections to the geodetic coordinates of network points.

Point	Latitude ($\delta\phi$)	Longitude ($\delta\lambda$)	Ellipsoidal height (δh) (m)
P1	0.000 00″	0.000 00″	0.000 0
P2	+0.000 12″	+0.000 22″	+0.004 9
P3	+0.000 06″	−0.001 13″	+0.028 5

Table 9.8 Adjusted geodetic coordinates of network points.

Point	Latitude (ϕ)	Longitude (λ)	Ellipsoidal height (m)
P1	49°05′24.73726″N	127°23′56.95384″W	291.8950
P2	49°05′22.24497″N	127°23′41.12463″W	292.4529
P3	49°05′42.33159″N	127°23′46.52065″W	289.5885

Table 9.9 Adjusted observations.

Leg	Distance (m)	Zenith angle	Bearing
P1–P2	330.3044 ± 0.0025	89°53′08.2″ ± 3.8″	103°30′08.3″ ± 2.9″
P1–P3	584.1405 ± 0.0031	90°18′18.4″ ± 3.2″	21°14′25.1″ ± 2.6″
P2–P3	631.1590 ± 0.0031	90°20′36.3″ ± 3.0″	350°00′23.2″ ± 2.6″

9.7.3 Adjustment in Local System

Continuing from the example given in Section 9.7.1, assume that the geodetic network is to be adjusted in the LA (n, e, u) coordinate system (i.e. adjustment in local system) as discussed in Section 9.5. In this case, Equations (9.80)–(9.110) will be used. Assume the local coordinates (n, e, u) of point P1 and the u-coordinate of points P2 will be fixed to constraint the three-dimensional least squares adjustment of the measurements in Table 9.1. The constraint equations to be used with the measurement equations in Equation (9.80) can be expressed as follows:

$$r_{10} = \delta n_1 \tag{9.125}$$
$$r_{11} = \delta e_1 \tag{9.126}$$
$$r_{12} = \delta u_1 \tag{9.127}$$
$$r_{13} = \delta u_2 \tag{9.128}$$

where the misclosures are zero (since the coordinates are measured directly) and δn_1, δe_1, δu_1, and δu_2 are the coordinate changes. The constraint measurements in Equations (9.125)–(9.127) are assigned very small standard deviations of 0.01 mm, and the u-coordinate value for point P2 in Equation (9.128) is assigned a standard deviation of 0.05 m since it is not well known. The first design matrix A formed from Equation (9.80) with the constraint Equations (9.125)–(9.128) is given as a matrix of 13 rows (for the number of equations) by 9 columns (for the number of unknown coordinate corrections) as follows:

$$A = \begin{bmatrix}
-0.233 & 0.972 & 1.98E-3 & 0.233 & -0.972 & -2.03E-3 & 0.0 & 0.0 & 0.0 \\
-0.932 & -0.362 & 5.37E-3 & 0.0 & 0.0 & 0.0 & 0.932 & 0.362 & -5.31E-3 \\
0.0 & 0.0 & 0.0 & -0.985 & 0.173 & 6.0E-3 & 0.985 & -0.173 & -6.0E-3 \\
-2.9E-3 & -7.1E-4 & 0.0 & 2.9E-3 & 7.1E-4 & -7.1E-8 & 0.0 & 0.0 & 0.0 \\
6.2E-4 & -1.6E-3 & 0.0 & 0.0 & 0.0 & 0.0 & -6.2E-4 & 1.6E-3 & -1.1E-7 \\
0.0 & 0.0 & 0.0 & -2.8E-4 & -1.6E-3 & 0.0 & 2.75E-4 & 1.6E-3 & 5.4E-8 \\
1.4E-6 & -5.8E-6 & -3.0E-3 & 1.4E-6 & -1.6E-6 & 1.0 & 0.0 & 0.0 & 0.0 \\
-8.6E-6 & -3.3E-6 & -1.7E-3 & 0.0 & 0.0 & 0.0 & -8.7E-6 & 8.5E-6 & 1.0 \\
0.0 & 0.0 & 0.0 & -9.4E-6 & 1.66E-6 & -1.6E-3 & -9.6E-6 & 9.4E-6 & 1.0 \\
1.0 & 0.0 & 0.0 & 0.0 & 0.0 & 0.0 & 0.0 & 0.0 & 0.0 \\
0.0 & 1.0 & 0.0 & 0.0 & 0.0 & 0.0 & 0.0 & 0.0 & 0.0 \\
0.0 & 0.0 & 1.0 & 0.0 & 0.0 & 0.0 & 0.0 & 0.0 & 0.0 \\
0.0 & 0.0 & 0.0 & 0.0 & 0.0 & 1.0 & 0.0 & 0.0 & 0.0
\end{bmatrix}$$

The least squares computed corrections of δn, δe, and δu are given as

$$\delta = -\left(A^T P A\right)^{-1} A^T P w$$
$$= \begin{bmatrix} 0.0 & 0.0 & 0.0 & 0.0037 & 0.0044 & 0.0 & 0.0019 & -0.023 & 0.0 \end{bmatrix}^T$$

The computed corrections (δn, δe, δu) are for each instrument setup points. In order to interpret the vector of corrections determined in this section, one should take note that LA system is not a unified coordinate system since each instrument setup point is a separate coordinate system with its separate origin. This is the reason why the initial coordinates of the instrument points must be given in unified coordinate system such as the Cartesian geodetic (X, Y, Z) coordinate system or the curvilinear geodetic (ϕ, λ, h) coordinate system. Since the computed corrections actually do not relate to the same coordinate system, the geodetic coordinates must be computed to represent the coordinates of the instrument points (in a unified coordinate system). In this case, the corrections must be transformed into changes in the geodetic coordinates using the inverse formula of Equation (9.78).

The H-matrix based on Equations (9.75)–(9.78) is based on three diagonal submatrix blocks formed for each geodetic point using Equation (9.75); this results in a matrix of size 9×9. The determined H-matrix is given as follows:

$$H = \text{diag}([6.372E6 \quad 4.185E6 \quad 1.0 \quad 6.372E6 \quad 4.185E6 \quad 1.0 \quad 6.372E6 \quad 4.185E6 \quad 1.0])$$

Table 9.10 Adjusted observations.

Leg	Distance (m)	Zenith angle	Bearing
P1–P2	330.3044 ± 0.0027	89°53′10.0″ ± 4.6″	103°30′08.3″ ± 3.2″
P1–P3	584.1406 ± 0.0035	90°18′19.7″ ± 3.3″	21°14′25.0″ ± 2.9″
P2–P3	631.1589 ± 0.0035	90°20′35.3″ ± 3.3″	350°00′23.2″ ± 2.8″

The computed changes in the curvilinear geodetic coordinates ($\delta\phi$, $\delta\lambda$, δh) are given as follows:

$$
\begin{bmatrix} \delta\phi_1 \\ \delta\lambda_1 \\ \delta h_1 \\ \delta\phi_2 \\ \delta\lambda_2 \\ \delta h_2 \\ \delta\phi_3 \\ \delta\lambda_3 \\ \delta h_3 \end{bmatrix} = H^{-1} \begin{bmatrix} \delta n_1 \\ \delta e_1 \\ \delta u_1 \\ \delta n_2 \\ \delta e_2 \\ \delta u_2 \\ \delta n_3 \\ \delta e_3 \\ \delta u_2 \end{bmatrix} \rightarrow \begin{bmatrix} \delta\phi_1 \\ \delta\lambda_1 \\ \delta h_1 \\ \delta\phi_2 \\ \delta\lambda_2 \\ \delta h_2 \\ \delta\phi_3 \\ \delta\lambda_3 \\ \delta h_3 \end{bmatrix} = \begin{bmatrix} 0.000\ 00″ \\ 0.000\ 00″ \\ 0.000\ 0\ \text{m} \\ +0.000\ 12″ \\ +0.000\ 22″ \\ +0.000\ 1\ \text{m} \\ +0.000\ 06′ \\ -0.001\ 13″ \\ 0.000\ 0\ \text{m} \end{bmatrix}
$$

The adjusted observations and their corresponding standard deviations are given in Table 9.10.

The computed a posteriori variance factor of unit weight is 0.8509 with the number of degrees of freedom as 4.

10

Nuisance Parameter Elimination and Sequential Adjustment

CHAPTER MENU

OBJECTIVES

After studying this chapter, you should be able to:

1) Explain the concept of nuisance parameter elimination, including the applications.
2) Formulate nuisance parameter elimination models.
3) Perform least squares adjustment with nuisance parameters eliminated.
4) Explain the concept of sequential least squares adjustment, including the applications.
5) Formulate sequential least squares adjustment models.
6) Perform sequential least squares adjustment to solve some geomatics problems.

10.1 Nuisance Parameters

Nuisance parameters are quantities in a mathematical model that are required in completing the formulation of the model but do not need to be solved for in an adjustment since the knowledge of their values is not actually of any importance.

Understanding Least Squares Estimation and Geomatics Data Analysis, First Edition.
John Olusegun Ogundare.
© 2019 John Wiley & Sons, Inc. Published 2019 by John Wiley & Sons, Inc.
Companion website: www.wiley.com/go/ogundare/Understanding-lse-and-gda

In this case, if the knowledge of some parameters is not important, then only those parameters that are important in the formulated model will need to be solved for in a least squares adjustment.

In some practical method of positioning, parametric equations may be set up to include two types of parameters: those that are required to be estimated and those that need not necessarily be estimated in the current problem but have to be included in the parametric equations for the sake of completing the equations (nuisance parameters). Typical nuisance parameters in *photogrammetric aerial triangulation*, for example, will be camera station coordinates (X_0, X_0, Z_0) and the rotation elements (Ω, Φ, K) of the camera axes; in *satellite positioning*, the nuisance parameters could be parameters for modeling the effects of the ionosphere and troposphere and also the orbital errors; in traditional surveying, the nuisance parameters could be convergence of meridians, convergence of meridians when transforming geodetic azimuths to map projection grid bearings, scale change when using different EDM equipment in a network distance measurements, scale factors and system constants of EDM, and orientation parameters in total station direction measurements. For example, when different EDMs are used in measuring distances in a survey network, the distances are used in the adjustment for the unknown network coordinates, while the scale factors and system constants of the EDMs are not directly solved for even though they are part of the model formulation. With regard to horizontal control survey adjustment, if total station directions are measured to some targets in a setup, the azimuth of the direction of zero graduation of the total station at that setup is considered as nuisance (or orientation) parameter when relating all the measured directions to their corresponding azimuths (or bearings). The number of this type of orientation parameters depends on the number of times the total station is set up (recentered and releveled) over a survey point (or over different survey points) from which a set of direction measurements are made. Each setup of a total station will contribute one unknown orientation parameter; for example, for a total of four setups (where directions are measured to targets), there will be a total of four orientation unknowns in addition to the unknown coordinates of the network points to be determined. The formulation of parametric models with nuisance (orientation) parameters is discussed in Examples 5.5 and 5.6.

10.2 Needs to Eliminate Nuisance Parameters

Some of the needs for nuisance parameter elimination include:

1) Reducing size of the matrix of the coefficients of normal equations (N) in least squares adjustment so that the inversion of the matrix is manageable. In this case, measurements are divided into groups, and the parameters can

be estimated in steps known as *step-by-step* or *phased estimation* method. This method is useful in computing large continental control networks where thousands of parameters are involved; simultaneous solution for all the parameters may be inconvenient for political reasons or from the data processing point of view. In this case, the network is broken into blocks, and the parameters are estimated in steps; those parameters that do not need to be estimated are considered as nuisance parameters and eliminated. For example, if blocks I and II are created and assuming measurements in one block are independent of those in the other block, only the parameters common to the two blocks will have to be solved for. In this case, the reduced normal equations for the common parameters will be formed and others treated as nuisance. The method is known as *Helmert–Wolf* method. The step-by-step method, however, requires that the approximate parameters be sufficiently accurate so as to avoid iterative solution. The solution obtained using the method, however, will be identical to what would have been obtained if all the data had been processed simultaneously. This is no longer an important issue due to the power of computers today.

2) Eliminating effects of orientation parameters due to direction measurements from an adjustment done for the purpose of deformation analysis. If total station direction measurements are made in deformation survey, the orientation (nuisance) parameters as well as the desired coordinates of survey points will constitute the overall parameters to be solved for in an adjustment. Most importantly, the variance–covariance matrix of the combined parameters will be constructed after the adjustment. Since the adjusted coordinates of survey points and their variance–covariance matrix are only needed in the subsequent deformation analysis, effort must be made to ensure that the effects of nuisance parameters and their covariance matrix are eliminated from the adjustment results before the deformation analysis stage. This constitutes an important advantage of nuisance parameter elimination.

10.3   Nuisance Parameter Elimination Model

In least squares adjustment, the orientation unknowns for sets of observed directions (from the use of total station equipment), for example, can be eliminated, thereby reducing sizes of matrices involved in the adjustment. One way of eliminating such nuisance parameters is to transform direction measurements into angle equivalents by subtracting one direction from the other. The other way will be to formulate each direction in the form of azimuth model and apply appropriate nuisance parameter as discussed in Section 5.1.2 on parametric model equation formulation. The nuisance parameters are then eliminated mathematically from the final adjusted solution model. In this section, matrix algebra expressions that will allow the mathematical elimination of nuisance parameters will be derived.

One of the methods of eliminating orientation parameters (nuisance parameters) is by partitioning the first design matrix (A) for least squares adjustment. The A-matrix may be partitioned so that one partition (A_1) consists of coefficients pertaining to the desired portion of the solution vector and the second partition (A_2) consists of the remaining coefficients pertaining to the nuisance parameters. It is necessary that the vector of unknowns be partitioned to be conformable for matrix manipulations with the partitioned A-matrix. To illustrate this, consider a case where an original parametric model is given as $\hat{\ell} = f(\hat{x})$, where $\hat{\ell}$ is a vector of adjusted observations and $\hat{x}$ is a vector of adjusted parameters consisting of both the desired parameters ($\hat{x}_1$) and the nuisance parameters ($\hat{x}_2$). This model can be partitioned into two according to the parameters $\hat{x}_1$ and $\hat{x}_2$, assuming a weight matrix (P) of observations is associated with the model. This is represented mathematically as follows. The general parametric least squares equations can be given as

$$\hat{\ell} = f(\hat{x}) \quad P \tag{10.1}$$

and the partitioned parametric model can be given as

$$\hat{\ell} = f(\hat{x}_1, \hat{x}_2) \quad P \tag{10.2}$$

where $\hat{\ell} = \ell + v$ is the vector of adjusted observations with v as residual corrections to be applied to the original observations (ℓ), $\hat{x}_1 = x_1^0 + \delta_1$ is the vector of adjusted main (desired) parameters with δ_1 as the corrections to be applied to the approximate values of parameters (x_1^0), and $\hat{x}_2 = x_2^0 + \delta_2$ is the vector of the adjusted nuisance (not desired) parameters with δ_2 as the corrections to be applied to the approximate values of parameters (x_2^0).

Applying Taylor series expansion to Equation (10.2) gives the following linearized form:

$$\ell + v = f\left(x_1^0, x_2^0\right) + \frac{\partial f}{\partial x_1}\delta_1 + \frac{\partial f}{\partial x_2}\delta_2 \tag{10.3}$$

If the misclosure vector $w = f\left(x_1^0, x_2^0\right) - \ell$, $A_1 = \dfrac{\partial f}{\partial x_1}$, and $A_2 = \dfrac{\partial f}{\partial x_2}$, then Equation (10.3) can be rewritten as

$$v = A_1\delta_1 + A_2\delta_2 + w \tag{10.4}$$

where A_1 and A_2 are the partitioned first design matrices corresponding to the desired parameters and the nuisance parameters, respectively. Equation (10.3) or Equation (10.4) is the linearized form of the model in Equation (10.2).

By imposing the least squares criterion ($v^T Pv$ = minimum) on Equation (10.4), the following variation function (otherwise known as Lagrange function) is obtained:

$$\varphi = v^T Pv - 2k^T (A_1 \delta_1 + A_2 \delta_2 + w - v) = \text{minimum} \tag{10.5}$$

where k is the vector of Lagrangian multipliers (or vector of correlates) and P is the weight matrix of observations (derived from the variances of the observations). The partitioned normal equations can be derived from Equation (10.5) by taking its partial derivatives with respect to the unknown quantities in the equation (i.e. v, δ_1, δ_2, k^T):

$$\frac{\partial \varphi}{\partial v} = 2v^T P + 2k^T = 0 \quad \text{or} \quad v = -P^{-1}k \tag{10.6}$$

$$\frac{\partial \varphi}{\partial \delta_1} = -2k^T A_1 = 0 \quad \text{or} \quad A_1^T k = 0 \tag{10.7}$$

$$\frac{\partial \varphi}{\partial \delta_2} = -2k^T A_2 = 0 \quad \text{or} \quad A_2^T k = 0 \tag{10.8}$$

$$\frac{\partial \varphi}{\partial k^T} = -2(A_1 \delta_1 + A_2 \delta_2 + w - v) = 0 \tag{10.9}$$

Equations (10.6)–(10.9) are the normal equation system in the most expanded form. These equations can be reduced by putting Equation (10.6) into Equation (10.9), giving

$$A_1 \delta_1 + A_2 \delta_2 + w + P^{-1} k = 0 \tag{10.10}$$

or

$$PA_1 \delta_1 + PA_2 \delta_2 + Pw + k = 0 \tag{10.11}$$

Pre-multiplying Equation (10.11) by A_1^T gives

$$A_1^T PA_1 \delta_1 + A_1^T PA_2 \delta_2 + A_1^T Pw + A_1^T k = 0 \tag{10.12}$$

Pre-multiplying Equation (10.11) again by A_2^T gives

$$A_2^T PA_1 \delta_1 + A_2^T PA_2 \delta_2 + A_2^T Pw + A_2^T k = 0 \tag{10.13}$$

Referring to Equations (10.7) and (10.8), we can set $A_1^T k = 0$ and $A_2^T k = 0$ in Equations (10.12) and (10.13) to obtain the following:

$$A_1^T PA_1 \delta_1 + A_1^T PA_2 \delta_2 + A_1^T Pw = 0 \tag{10.14}$$

$$A_2^T PA_1 \delta_1 + A_2^T PA_2 \delta_2 + A_2^T Pw = 0 \tag{10.15}$$

Solve for the correction vector to nuisance parameters from Equation (10.15):

$$\delta_2 = -\left(A_2^T P A_2\right)^{-1}\left(A_2^T P A_1 \delta_1 + A_2^T P w\right) \tag{10.16}$$

Put Equation (10.16) into Equation (10.14) and solve for the corrections to the required parameters:

$$\delta_1 = -\left(A_1^T P^* A_1\right)^{-1} A_1^T P^* w \tag{10.17}$$

where $P^* = P\left(I - A_2\left(A_2^T P A_2\right)^{-1} A_2^T P\right)$ and Equation (10.17) is the solution vector for problem. It can be seen from Equation (10.17) that the matrix size inverted is the size of the design matrix A_1 (with the effects of nuisance parameters eliminated). The adjusted parameters (without the nuisance parameters) will be given as

$$\hat{x}_1 = x_1^0 + \delta_1 \tag{10.18}$$

The residual vector can be given in partitioned mode from Equation (10.4) as

$$v = \begin{bmatrix} A_1 & A_2 \end{bmatrix}\begin{bmatrix} \delta_1 \\ \delta_2 \end{bmatrix} + w \tag{10.19}$$

and the a posteriori variance factor of unit weight can be given as usual as

$$s_0^2 = \frac{v^T P v}{n - u} \tag{10.20}$$

where n is the number of observations and u is the number of unknown parameters (the sum of the number of required parameters and the number of nuisance parameters). The cofactor of the adjusted parameters with no nuisance parameters will be given from Equation (10.17) as

$$Q_{\hat{x}_1} = \left(A_1^T P^* A_1\right)^{-1} \tag{10.21}$$

The variance–covariance matrix of the adjusted parameters (without the nuisance parameters) will be given as

$$C_{\hat{x}_1} = s_0^2\left(A_1^T P^* A_1\right)^{-1} \tag{10.22}$$

10.3.1 Nuisance Parameter Elimination Summary

The summary of the models for computing the adjusted parameters and the associated variance–covariance matrix with the effects of nuisance parameters eliminated is given as follows:

1) The adjusted parameters with nuisance eliminated:

$$\hat{x}_1 = x_1^0 + \delta_1 \tag{10.23}$$

where

$$\delta_1 = -\left(A_1^T P^* A_1\right)^{-1} A_1^T P^* w \tag{10.24}$$

and

$$P^* = P\left(I - A_2\left(A_2^T P A_2\right)^{-1} A_2^T P\right)$$

2) Variance–covariance matrix of the adjusted parameters (with nuisance parameters eliminated):

$$C_{\hat{x}_1} = s_0^2 \left(A_1^T P^* A_1\right)^{-1} \tag{10.25}$$

where

$$s_0^2 = \frac{v^T P v}{n - u} \tag{10.26}$$

$$v = \begin{bmatrix} A_1 & A_2 \end{bmatrix} \begin{bmatrix} \delta_1 \\ \delta_2 \end{bmatrix} + w \tag{10.27}$$

$$\delta_2 = -\left(A_2^T P A_2\right)^{-1} \left(A_2^T P A_1 \delta_1 + A_2^T P w\right) \tag{10.28}$$

where n is the number of observations and u is the number of unknown parameters (the sum of the number of required parameters and the number of nuisance parameters). The following example illustrates the steps involved in nuisance parameter elimination.

10.3.2 Nuisance Parameter Elimination Example

Consider a simple resection problem in Figure 10.1, where point O (x, y) is to be resected from the given control points A $(x_A = 1000.000 \text{ m},$

Figure 10.1 Simple resection problem to illustrate orientation parameter.

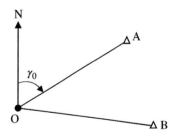

y_A = 1000.000 m) and B (x_B = 1500.000 m, y_B = 500.000 m). The measurements are azimuth O–A as ℓ_1 = 30°00′00″ ± 30″; the total station directions O–A and O–B as ℓ_2 = 0°00′00″ ± 30″ and ℓ_3 = 65°00′00″ ± 30″, respectively; and the distances O–B and O–A as ℓ_4 = 753.670 ± 0.010 m and ℓ_5 = 501.390 ± 0.010 m, respectively. Assume the approximate coordinates of point O are x = 749 m, y = 566 m and perform the following tasks.

a) Determine the adjusted coordinates of point O and the adjusted value of the orientation parameter and their covariance matrix.

Solution:

Formulate the parametric model $\hat{\ell} = f(\hat{x})$.

The unknown parameters are $x = \begin{bmatrix} x \\ y \\ \gamma_0 \end{bmatrix}$

Let the approximate values of parameters be $x^0 =$
$\begin{bmatrix} x^0 = 749.0 \\ y^0 = 566.0 \\ \gamma_0^0 = 30° \text{ or } 0.523\ 598\ 8 \text{ rad} \end{bmatrix}$

$$\hat{\ell}_1 = a\tan\left(\frac{x_A - \hat{x}}{y_A - \hat{y}}\right) \tag{10.29}$$

$$\hat{\ell}_2 = a\tan\left(\frac{x_A - \hat{x}}{y_A - \hat{y}}\right) - \hat{\gamma}_0 \tag{10.30}$$

$$\hat{\ell}_3 = a\tan\left(\frac{x_B - \hat{x}}{y_B - \hat{y}}\right) - \hat{\gamma}_0 + 180° \tag{10.31}$$

$$\hat{\ell}_4 = \left[(x_B - \hat{x})^2 + (y_B - \hat{y})^2\right]^{1/2} \tag{10.32}$$

$$\hat{\ell}_5 = \left[(x_A - \hat{x})^2 + (y_A - \hat{y})^2\right]^{1/2} \tag{10.33}$$

Form the A-matrix as follows (with the partials evaluated at the approximate values of the parameters):

$$\frac{\partial \hat{\ell}_1}{\partial \hat{x}} = \frac{1}{1 + \left(\frac{x_A - \hat{x}}{y_A - \hat{y}}\right)^2} \frac{-1}{(y_A - \hat{y})} \rightarrow \frac{-(y_A - \hat{y})}{(x_A - \hat{x})^2 + (y_A - \hat{y})^2} = -1.726\ 63\text{E} - 3$$

$$\frac{\partial \hat{\ell}_1}{\partial \hat{\gamma}} = \frac{1}{1 + \left(\dfrac{x_A - \hat{x}}{y_A - \hat{y}}\right)^2} \frac{(x_A - \hat{x})}{(y_A - \hat{y})^2} \rightarrow \frac{(x_A - \hat{x})}{(x_A - \hat{x})^2 + (y_A - \hat{y})^2} = 9.985\ 797\text{E}-4$$

$$\frac{\partial \hat{\ell}_1}{\partial \hat{\gamma}_0} = 0.0$$

$$\frac{\partial \hat{\ell}_2}{\partial \hat{x}} = \frac{1}{1 + \left(\dfrac{x_A - \hat{x}}{y_A - \hat{y}}\right)^2} \frac{-1}{(y_A - \hat{y})} \rightarrow \frac{-(y_A - \hat{y})}{(x_A - \hat{x})^2 + (y_A - \hat{y})^2} = -1.726\ 63\text{E}-3$$

$$\frac{\partial \hat{\ell}_2}{\partial \hat{y}} = \frac{1}{1 + \left(\dfrac{x_A - \hat{x}}{y_A - \hat{y}}\right)^2} \frac{(x_A - \hat{x})}{(y_A - \hat{y})^2} \rightarrow \frac{(x_A - \hat{x})}{(x_A - \hat{x})^2 + (y_A - \hat{y})^2} = 9.985\ 797\text{E}-4$$

$$\frac{\partial \hat{\ell}_2}{\partial \hat{\gamma}_0} = -1.0$$

$$\frac{\partial \hat{\ell}_3}{\partial \hat{x}} = \frac{1}{1 + \left(\dfrac{x_B - \hat{x}}{y_B - \hat{y}}\right)^2} \frac{-1}{(y_B - \hat{y})} \rightarrow \frac{-(y_B - \hat{y})}{(x_B - \hat{x})^2 + (y_B - \hat{y})^2} = 1.161\ 242\text{E}-4$$

$$\frac{\partial \hat{\ell}_3}{\partial \hat{y}} = \frac{1}{1 + \left(\dfrac{x_B - \hat{x}}{y_B - \hat{y}}\right)^2} \frac{(x_B - \hat{x})}{(y_B - \hat{y})^2} \rightarrow \frac{(x_B - \hat{x})}{(x_B - \hat{x})^2 + (y_B - \hat{y})^2} = 1.321\ 353\text{E}-3$$

$$\frac{\partial \hat{\ell}_3}{\partial \hat{\gamma}_0} = -1.0 \qquad \frac{\partial \hat{\ell}_4}{\partial \hat{x}} = \frac{-(x_B - \hat{x})}{\sqrt{(x_B - \hat{x})^2 + (y_B - \hat{y})^2}} = -0.996\ 161$$

$$\frac{\partial \hat{\ell}_4}{\partial \hat{x}} = \frac{-(x_B - \hat{x})}{\sqrt{(x_B - \hat{x})^2 + (y_B - \hat{y})^2}} = 0.996\ 161$$

$$\frac{\partial \hat{\ell}_4}{\partial \hat{y}} = \frac{-(y_B - \hat{y})}{\sqrt{(x_B - \hat{x})^2 + (y_B - \hat{y})^2}} = 0.087\ 545\ 4 \qquad \frac{\partial \hat{\ell}_4}{\partial \hat{\gamma}_0} = 0.0$$

$$\frac{\partial \hat{\ell}_5}{\partial \hat{x}} = \frac{-(x_A - \hat{x})}{\sqrt{(x_A - \hat{x})^2 + (y_A - \hat{y})^2}} = -0.500\ 643\ 1$$

$$\frac{\partial \hat{\ell}_5}{\partial \hat{y}} = \frac{-(y_A - \hat{y})}{\sqrt{(x_A - \hat{x})^2 + (y_A - \hat{y})^2}} = -0.865\ 653\ 8 \qquad \frac{\partial \hat{\ell}_5}{\partial \hat{\gamma}_0} = 0.0$$

The first design matrix (A) can be given from the above partial derivatives as

$$A = \begin{bmatrix} -1.726\ 63\text{E}-3 & 9.985\ 797\text{E}-4 & 0.0 \\ -1.726\ 63\text{E}-3 & 9.985\ 797\text{E}-4 & -1.0 \\ 1.161\ 242\text{E}-4 & 1.321\ 352\text{E}-3 & -1.0 \\ -9.961\ 61\text{E}-1 & 8.754\ 54\text{E}-2 & 0.0 \\ -5.006\ 431\text{E}-1 & -8.656\ 538\text{E}-1 & 0.0 \end{bmatrix} \tag{10.34}$$

The weight matrix (P) formed from the variances of the observations:

$$P = \begin{bmatrix} 4.731\ 835\text{E}7 & & & & \\ & 4.731\ 835\text{E}7 & & & \\ & & 4.731\ 835\text{E}7 & & \\ & & & 100\ 00 & \\ & & & & 100\ 00 \end{bmatrix}$$

The misclosure vector (w) and the inverse of the matrix of the coefficients of the normal equations (N^{-1}):

$$w = f(x^0) - \ell = \begin{bmatrix} 7.427\ 40\text{E}-4 \\ 7.427\ 40\text{E}-4 \\ 3.911\ 52\text{E}-4 \\ 0.224\ 555\ 0 \\ -0.034\ 836\ 5 \end{bmatrix} \quad A^T P w = \begin{bmatrix} -2\ 181.737\ 84 \\ 592.798\ 120 \\ -53\ 653.877\ 4 \end{bmatrix} \tag{10.35}$$

$$N^{-1} = (A^T P A)^{-1} = \begin{bmatrix} 8.977\ 26\text{E}-5 & -3.998\ 88\text{E}-5 & -1.186\ 752\text{E}-7 \\ -3.998\ 88\text{E}-5 & 1.490\ 489\ 1\text{E}-4 & 2.050\ 927\ 4\text{E}-7 \\ -1.186\ 752\text{E}-7 & 2.050\ 927\ 4\text{E}-7 & 1.090\ 018\ 97\text{E}-8 \end{bmatrix}$$

The corrections to be applied to the approximate parameters x^0 are

$$\delta = -(A^T P A)^{-1} A^T P w = \begin{bmatrix} 0.213\ 2 \\ -0.164\ 6 \\ 0.000\ 204\ 3 \end{bmatrix}$$

The adjusted parameters (including the nuisance parameter):

$$\hat{x} = x^0 + \delta = \begin{bmatrix} 749.21 \\ 565.84 \\ 30°01' \text{ or } 0.523\ 803\ 1 \text{ rad} \end{bmatrix}$$

Residual vector:

$$v = A\delta + w = \begin{bmatrix} 2.102\ 6E-4 \\ 5.921\ 83E-6 \\ -5.921\ 83E-6 \\ -2.234\ 3E-3 \\ 9.112\ 95E-5 \end{bmatrix}$$

The a posteriori variance factor of unit weight:

$$s_0^2 = \frac{v^T P v}{n-u} \rightarrow s_0^2 = \frac{2.151\ 48}{5-3} = 1.075\ 7$$

The cofactor matrix of the adjusted parameters (including the nuisance parameter):

$$Q_{\hat{x}} = \left(A^T P A\right)^{-1} = \begin{bmatrix} 8.977\ 26E-5 & -3.998\ 88E-5 & -1.186\ 752E-7 \\ -3.998\ 88E-5 & 1.490\ 489\ 1E-4 & 2.050\ 927\ 4E-7 \\ -1.186\ 752E-7 & 2.050\ 927\ 4E-7 & 1.090\ 018\ 97E-8 \end{bmatrix}$$

The covariance matrix of the adjusted parameters (including the nuisance parameter):

$$C_{\hat{x}} = s_0^2 Q_{\hat{x}} = \begin{bmatrix} 9.666\ 3E-5 & -4.305\ 8E-5 & -1.277\ 84E-7 \\ -4.305\ 8E-5 & 1.604\ 89E-4 & 2.208\ 34E-7 \\ -1.277\ 84E-7 & 2.208\ 34E-7 & 1.173\ 68E-8 \end{bmatrix}$$

The MATLAB code for performing the adjustment in this question (without nuisance elimination) is given in Table 10.1.

Table 10.1 MATLAB code for adjustment without nuisance parameter elimination.

```
>> % Use syms to define the variables x, y gamma (gam)
>> syms x y gam
>> % form the parametric Equations (10.29) - (10.33)
>> L1=atan((1000-x)/(1000-y));
>> L2=atan((1000-x)/(1000-y))-gam;
>> L3=atan((1500-x)/(500-y))-gam +pi;
>> L4=sqrt((1500-x)^2+(500-y)^2);
>> L5=sqrt((1000-x)^2+(1000-y)^2);
>>% Form Jacobian wrt x, y gam and evaluate as A matrix
>> J=jacobian([L1;L2;L3;L4;L5],[x y gam]);
```

(Continued)

Table 10.1 (Continued)

```
>> x=749.0; y=566.0;gam=0.5235988;
>> format long
>> A=eval(J);
>>% form weight matrix P, computed observations L0, given
observations L and misclosure w
>> P=inv(diag([(30/206265)^2 (30/206265)^2 (30/206265)^2
0.010^2 0.010^2]));
>> L0=eval([L1;L2;L3;L4;L5]);
>> L=[30*pi/180; 0.0; 65*pi/180; 753.670; 501.390];
>> w=L0-L;
>>% determine solution d and adjusted parameters xa
>> u=A'*P*w;
>> N=A'*P*A;
>> Ninv=inv(N);
>> d=-Ninv*u;
>> x0=[x;y;gam];
>> xa=x0+d
xa =
    1.0e+02 *
    7.492131962259725
    5.658354051936339
    0.005238031203358
>>% determine residuals v, aposteriori variance s02,
covariance matrix Cx
>> v=A*d+w;
>> s02=v'*P*v/2
s02 =
    1.075739194472152
>> Qx=Ninv;
>> Cx=s02*Qx;
>>
```

b) Determine the adjusted coordinates of point O with the orientation parameter eliminated (using the Equation (10.24)) and the covariance matrix of the
   adjusted coordinates without the orientation parameter (Equation (10.25)).

   Solution:

   Using the parametric equations (10.29)–(10.33) above,

$$\text{Required parameters } x_1 = \begin{bmatrix} x \\ y \end{bmatrix} \quad \text{Nuisance parameter} : x_2 = [\gamma_0]$$

Partial derivatives of the parametric equations (10.29)–(10.33) with respect to parameters x_1, which are the first two columns in Equation (10.34):

$$A_1 = \begin{bmatrix} -1.726\ 63\text{E}-3 & 9.985\ 797\text{E}-4 \\ -1.726\ 63\text{E}-3 & 9.985\ 797\text{E}-4 \\ 1.1612\ 42\text{E}-4 & 1.321\ 352\text{E}-3 \\ -9.961\ 61\text{E}-1 & 8.754\ 54\text{E}-2 \\ -5.006\ 431\text{E}-1 & -8.656\ 538\text{E}-1 \end{bmatrix}$$

Partial derivatives of the parametric equations (10.29)–(10.33) with respect to x_2:

$$A_2 = \begin{bmatrix} 0.0 \\ -1.0 \\ -1.0 \\ 0.0 \\ 0.0 \end{bmatrix} \quad P = \begin{bmatrix} 4.731\ 835\text{E}7 & & & & \\ & 4.731\ 835\text{E}7 & & & \\ & & 4.731\ 835\text{E}7 & & \\ & & & 10\ 000 & \\ & & & & 10\ 000 \end{bmatrix}$$

The misclosure vector is the same as that in Equation (10.35):

$$w = f\left(x^0\right) - \ell = \begin{bmatrix} 7.427\ 40\text{E}-4 \\ 7.427\ 40\text{E}-4 \\ 3.911\ 52\text{E}-4 \\ 0.224\ 555\ 0 \\ -0.034\ 836\ 5 \end{bmatrix}$$

$$\left(A_2^T P A_2\right)^{-1} = [1.056\ 672\ 6\text{E}-8]$$

$$P^* = P\left(I - A_2\left(A_2^T P A_2\right)^{-1} A_2^T P\right)$$

$$= \begin{bmatrix} 4.731\ 83\text{E}7 & & & & \\ & 2.365\ 92\text{E}7 & -2.365\ 92\text{E}7 & & \\ & -2.365\ 92\text{E}7 & 2.365\ 92\text{E}7 & & \\ & & & 10\ 000 & \\ & & & & 10\ 000 \end{bmatrix}$$

Table 10.2 MATLAB code for nuisance parameter elimination (continuation of code in (a)).

```
>>% continue from Table 10.2, form A1 matrix for x, y
parameters
>> A1=[A(1,1) A(1,2);A(2,1) A(2,2);A(3,1) A(3,2);A(4,1) A
(4,2);A(5,1) A(5,2)]
A1 =
   -0.001726627863954    0.000998579709338
   -0.001726627863954    0.000998579709338
    0.000116124196588    0.001321352600566
   -0.996160530750547    0.087545399506706
   -0.500643093474549   -0.865653795091452
>>% Form A1 matrix for the nuisance parameter gam
>> A2=[A(1,3);A(2,3);A(3,3);A(4,3);A(5,3)]
A2 =
    0
   -1
   -1
    0
    0
>> % solve for solution d1 and adjusted parameters x1, y1
>> N2=A2'*P*A2;
>> I=eye(5);
>> Ps=P*(I-A2*inv(N2)*A2'*P);
>> N1=A1'*Ps*A1;
>> u1=A1'*Ps*w;
>> d1=-inv(N1)*u1
d1 =
    0.213196225972572
   -0.164594806366143
>> x1=x+d1(1)
x1 =
      7.492131962259725e+02
>> y1=y+d1(2)
y1 =
      5.658354051936339e+02
>>% determine covariance matrix Cx1 for adjusted parameters
>> Qx1=inv(N1);
>> Cx1=s02*Qx1
Cx1 =
    1.0e-03 *
    0.096574351812482 -0.043019709276996
   -0.043019709276996 0.160340110920531
>>
```

$$\left(A_1^T P^* A_1\right)^{-1} = \begin{bmatrix} 8.97726E-5 & -3.99888E-5 \\ -3.99888E-5 & 1.4904891E-4 \end{bmatrix}$$

$$A_1^T P^* w = \begin{bmatrix} -2138.5330 \\ 530.5614 \end{bmatrix}$$

The corrections to be applied to the approximate parameters x^0 are

$$\delta_1 = -\left(A_1^T P^* A_1\right)^{-1} A_1^T P^* w = \begin{bmatrix} 0.2132 \\ -0.1646 \end{bmatrix}$$

The adjusted parameters (without the nuisance parameter):

$$\hat{x}_1 = x_1^0 + \delta_1 = \begin{bmatrix} 749.21 \\ 565.84 \end{bmatrix}$$

The cofactor matrix of the adjusted parameters:

$$Q_{\hat{x}_1} = \left(A_1^T P^* A_1\right)^{-1} = \begin{bmatrix} 8.977\ 26E-5 & -3.998\ 88E-5 \\ -3.998\ 88E-5 & 1.490\ 489\ 1E-4 \end{bmatrix}$$

The covariance matrix will be

$$C_{\hat{x}_1} = s_0^2 \left(A_1^T P^* A_1\right)^{-1} = \begin{bmatrix} 9.6663E-5 & -4.3058E-5 \\ -4.3058E-5 & 1.60489E-4 \end{bmatrix}$$

The a posteriori variance factor of unit weight is the same as in part (a): $s_0^2 = 1.0767$. The MATLAB code for completing the task in this question is given in Table 10.2 as a continuation of the code given in Table 10.1.

10.4 Sequential Least Squares Adjustment

There are certain times when all the measurements for a given geodetic network to be adjusted are not all available at the same time as a single batch; for example, a large portion of the measurements (ℓ_1) may be available as *batch I* and the remaining measurements (ℓ_2) to come at a much later date as *batch II*. Since the adjusted coordinates ($\hat{x}$) of the network are required based on batch I, these available measurements (ℓ_1) must be used to obtain initial coordinates of the network as $\hat{x}_1$; when batch II measurements (ℓ_2) are available, instead of readjusting the combined observations ($\ell_1 + \ell_2$) in order to determine the final adjusted coordinates ($\hat{x}$) of the network, the batch II observations (ℓ_2) alone are used to provide an update (Δx) to be applied to the adjusted coordinates ($\hat{x}_1$) based on batch I measurements. The final adjusted coordinates are then obtained as

$$\hat{x} = \hat{x}_1 + \Delta x \qquad (10.36)$$

Similarly, the covariance matrix ($C_{\hat{x}}$) of the final adjusted coordinates is obtained as

$$C_{\hat{x}} = C_{\hat{x}_1} + \Delta C_{x_1} \qquad (10.37)$$

where $C_{\hat{x}_1}$ is the covariance matrix of the adjusted coordinates from batch I and ΔC_{x_1} is the update due to batch II measurements (ℓ_2). The above procedure is considered as sequential adjustment, in which initial coordinates are updated sequentially as new observations are available, without having to readjust all of the observations as a single batch with no need to store all the observations. This sequential adjustment procedure will provide results that are identical to those expected when combined measurements ($\ell_1 + \ell_2$) are adjusted together as a single batch. The adjustment of combined measurements, however, may be more expensive compared with sequential adjustment. The simplest form of sequential adjustment is the computation of moving averages, where a new improved mean is calculated sequentially as new values are available. Two possible cases of sequential adjustment can be given as follows:

a) When the number of parameters is fixed, e.g. no stations are added or deleted from one batch to another. This case is only considered in this book and derived in Section 10.5.
b) When the number of parameters varies from one batch to another.

Some of the applications of sequential least squares adjustment are as follows:

1) Real-time data acquisition where estimates of a number of stationary parameters are required. If new parameters are being "generated" by the new data, one must use step-by-step method, and if the parameters are changing with time, a filtering technique must be applied. Specific sequential adjustment problems include fixing stationary oil rigs by continuous interrogation of a number of different navigation systems.
2) Coordinates update of datum points. National and international mapping organizations are continuously undertaking new measurements, requiring that they continuously update the previously adjusted coordinates of their survey points. This is particularly relevant in updating the International Terrestrial Reference Frame (ITRF) and the Active Control Systems (ACS).
3) Design of position fixes. Computer simulation method of designing position fixes involves continuously altering a proposed position fixing configuration to search for the one that fulfills the design criteria most economically. In this case, the computed corrections to the precisions of the previous adjustment can be separately analyzed.

10.4.1 Sequential Adjustment in Simple Form

As a simple illustration of sequential adjustment, consider designing an algorithm for real-time calculation of running averages in which previous measurements will not have to be used when new measurement (having equal weight as the previous measurements) becomes available later. Let the old measurements up to n times be given as $\ell_1, \ell_2, \ell_3, ..., \ell_n$ with its average as $\hat{x}_n$; the new measurement to be added later is ℓ_{n+1}; and the average of $\ell_1, \ell_2, \ell_3, ..., \ell_n, \ell_{n+1}$ is to be $\hat{x}_{n+1}$. The average of the old set of measurements can be given as

$$\hat{x}_n = \frac{1}{n}\sum_{i=1}^{n}\ell_i \tag{10.38}$$

and the average of all the old and the new measurements combined can be given as

$$\hat{x}_{n+1} = \frac{1}{n+1}\left[\left(\sum_{i=1}^{n}\ell_i\right) + \ell_{n+1}\right] \tag{10.39}$$

From Equation (10.38), Equation (10.39) can be rewritten as

$$\hat{x}_{n+1} = \frac{1}{n+1}\left[n\hat{x}_n + \ell_{n+1}\right] \tag{10.40}$$

or

$$\hat{x}_{n+1} = \left(\frac{n}{n+1}\right)\hat{x}_n + \frac{\ell_{n+1}}{n+1} + \frac{\hat{x}_n}{n+1} - \frac{\hat{x}_n}{n+1} \tag{10.41}$$

where the last two terms will cancel out; they are added for the sake of manipulating the equation further. Equation (10.41) can be rearranged and simplified to give

$$\hat{x}_{n+1} = \hat{x}_n + \frac{1}{n+1}(\ell_{n+1} - \hat{x}_n) \tag{10.42}$$

Equation (10.42) can be simplified as

$$\hat{x}_{n+1} = \hat{x}_n + \Delta x \tag{10.43}$$

where Δx is an update to be sequentially applied to the average value ($\hat{x}_n$) determined from the previous batch of observations; this update can be given from Equation (10.42) as

$$\Delta x = \frac{1}{n+1}(\ell_{n+1} - \hat{x}_n) \tag{10.44}$$

It can be seen from Equation (10.44) that the update (Δx) is based only on the previous adjusted parameter ($\hat{x}_n$) and the new observation (ℓ_{n+1}). By applying the laws of variance–covariance propagation on Equation (10.38), the following is obtained:

$$\sigma_{\hat{x}_n}^2 = \left(\frac{1}{n}\right)^2 \left[n\sigma_\ell^2\right] \tag{10.45}$$

Rearranging Equation (10.42) gives the following:

$$\hat{x}_{n+1} = k\ell_{n+1} + (1-k)\hat{x}_n \tag{10.46}$$

where

$$k = \frac{1}{n+1} \tag{10.47}$$

By applying the laws of variance–covariance propagation on Equation (10.46), the following is obtained:

$$\sigma_{\hat{x}_{n+1}}^2 = k^2\sigma_{\ell_{n+1}}^2 + \sigma_{\hat{x}_n}^2 - 2k\sigma_{\hat{x}_n}^2 + k^2\sigma_{\hat{x}_n}^2 \tag{10.48}$$

Equation (10.48) can be rearranged to give

$$\sigma_{\hat{x}_{n+1}}^2 = k^2\left(\sigma_{\ell_{n+1}}^2 + \sigma_{\hat{x}_n}^2\right) + \sigma_{\hat{x}_n}^2 - 2k\sigma_{\hat{x}_n}^2 \tag{10.49}$$

Noting that all the measurements have equal weights (or equal standard deviations), then $\sigma_{\ell_1}^2 = \sigma_{\ell_2}^2 \cdots \sigma_{\ell_n}^2 = \sigma_{\ell_{n+1}}^2 = \sigma_\ell^2$; and from Equation (10.45), $\sigma_{\hat{x}_n}^2 = \frac{\sigma_\ell^2}{n}$; substituting these values and the value of k in Equation (10.47) into Equation (10.49) gives the following:

$$\sigma_{\hat{x}_{n+1}}^2 = \left(\frac{1}{n+1}\right)^2\left(\sigma_\ell^2 + \frac{\sigma_\ell^2}{n}\right) + \sigma_{\hat{x}_n}^2 - 2\left(\frac{1}{n+1}\right)\sigma_{\hat{x}_n}^2 \tag{10.50}$$

Equation (10.50) can be simplified to give the following:

$$\sigma_{\hat{x}_{n+1}}^2 = \left(\frac{1}{n+1}\right)\frac{\sigma_\ell^2}{n} + \sigma_{\hat{x}_n}^2 - 2\left(\frac{1}{n+1}\right)\sigma_{\hat{x}_n}^2 \tag{10.51}$$

Substituting $\sigma_{\hat{x}_n}^2 = \frac{\sigma_\ell^2}{n}$ back into Equation (10.51) gives the following simplification:

$$\sigma_{\hat{x}_{n+1}}^2 = \left(\frac{1}{n+1}\right)\sigma_{\hat{x}_n}^2 + \sigma_{\hat{x}_n}^2 - 2\left(\frac{1}{n+1}\right)\sigma_{\hat{x}_n}^2 \tag{10.52}$$

or

$$\sigma_{\hat{x}_{n+1}}^2 = \sigma_{\hat{x}_n}^2 - \left(\frac{1}{n+1}\right)\sigma_{\hat{x}_n}^2 \tag{10.53}$$

which can be rewritten in sequential form as

$$\sigma_{\hat{x}_{n+1}}^2 = \sigma_{\hat{x}_n}^2 + \Delta C \tag{10.54}$$

where

$$\Delta C = -\left(\frac{1}{n+1}\right)\sigma_{\hat{x}_n}^2 \tag{10.55}$$

Equations (10.43) and (10.54) are referred to as sequential adjustment solutions.

10.5 Sequential Least Squares Adjustment Model

Consider a case where a batch of measurements (ℓ_1) constituting a batch I measurements are taken for a given network (for the purpose of determining the coordinates of the network points). Let the batch of measurements be represented in a parametric form as follows:

$$\hat{\ell}_1 = f_1(\hat{x}) \quad P_1, \quad n_1 \tag{10.56}$$

where P_1 is the weight matrix of n_1 observations in the given batch I, $\hat{\ell}_1$ is a vector of the adjusted observations for the batch, and $\hat{x}$ is a vector of the adjusted parameters. At a later date, another batch (smaller than the previous batch) of measurements (ℓ_2) constituting batch II measurements were made to strengthen the previous network. Let the parametric model for the batch II measurements be given as

$$\hat{\ell}_2 = f_2(\hat{x}) \quad P_2, \quad n_2 \tag{10.57}$$

where P_2 is the weight matrix of n_2 observations in batch II and $\hat{\ell}_2$ is a vector of the adjusted observations in the batch. Formulate the sequential models for adding batch II measurements (ℓ_2) to batch I measurements (ℓ_1), without having to readjust the two batches of measurements together as a single batch. The outcome of the adjustment must give the same result as if the two batches of measurements are adjusted simultaneously. Sequential adjustment is about applying effects of new measurements to previous ones without directly readjusting the previous and the current measurements together. It should be noted that in the steps developed below, the same approximate values of parameters (x^0) used in the adjustment of batch I observations will be used again in the sequential adjustment of subsequent batch II observations. The sequential steps are formulated as follows, with regard to Equations (10.56) and (10.57). Assume that Equation (10.56) is the parametric model for batch I observations and Equation (10.57) is the model for the batch II observations. Equations (10.56) and (10.57) can be rewritten as follows:

$$\ell_1 + v_1 = f_1\left(x^0 + \delta\right) \tag{10.58}$$

$$\ell_2 + v_2 = f_2\left(x^0 + \delta\right) \tag{10.59}$$

where v_1 and v_2 are vectors of residuals of the first and second batches of observations, respectively, and δ is a vector of corrections to be applied to the approximate values (x^0) of the parameters. Linearize Equations (10.58) and (10.59) by Taylor series expansion as follows:

$$\ell_1 + v_1 = f_1\left(x^0\right) + \frac{\partial f_1}{\partial x}\delta \tag{10.60}$$

$$\ell_2 + v_2 = f_2\left(x^0\right) + \frac{\partial f_2}{\partial x}\delta \tag{10.61}$$

Equations (10.60) and (10.61) can be simplified as

$$A_1\delta + w_1 - v_1 = 0 \tag{10.62}$$

$$A_2\delta + w_2 - v_2 = 0 \tag{10.63}$$

where

$$A_1 = \frac{\partial f_1}{\partial x}; w_1 = f_1\left(x^0\right) - \ell_1; \tag{10.64}$$

$$A_2 = \frac{\partial f_2}{\partial x}; w_2 = f_2\left(x^0\right) - \ell_2; \tag{10.65}$$

The variation function (φ) is formulated from Equations (10.62) and (10.63) by imposing the least squares criterion on them as follows:

$$\varphi = v_1^T P_1 v_1 + v_2^T P_2 v_2 - 2k_1^T\left(A_1\delta + w_1 - v_1\right) - 2k_2^T\left(A_2\delta + w_2 - v_2\right) = \text{minimum} \tag{10.66}$$

where k_1 and k_2 are the Lagrangian multipliers (or vectors of correlates) and P_1 and P_2 are the weight matrices of the first and second batches of observations, respectively. The normal equations can be derived by finding the partial derivatives of the variation function φ with respect to the unknown quantities v_1, v_2, δ, k_1, k_2:

$$\frac{\partial \varphi}{\partial v_1} = 2v_1^T P_1 + 2k_1^T = 0 \tag{10.67}$$

$$\frac{\partial \varphi}{\partial v_2} = 2v_2^T P_2 + 2k_2^T = 0 \tag{10.68}$$

$$\frac{\partial \varphi}{\partial \delta} = -2k_1^T A_1 - 2k_2^T A_2 = 0 \tag{10.69}$$

$$\frac{\partial \varphi}{\partial k_1^T} = -2(A_1\delta + w_1 - v_1) = 0 \tag{10.70}$$

$$\frac{\partial \varphi}{\partial k_2^T} = -2(A_2\delta + w_2 - v_2) = 0 \tag{10.71}$$

Simplify Equations (10.67)–(10.71) to give the following normal equations in the most expanded forms:

$$v_1^T P_1 + k_1^T = 0 \tag{10.72}$$

$$v_2^T P_2 + k_2^T = 0 \tag{10.73}$$

$$k_1^T A_1 + k_2^T A_2 = 0 \tag{10.74}$$

$$A_1\delta + w_1 - v_1 = 0 \tag{10.75}$$

$$A_2\delta + w_2 - v_2 = 0 \tag{10.76}$$

Transpose Equations (10.72) and (10.73) and make v_1 and v_2 as subjects as follows:

$$v_1 = -P_1^{-1}k_1 \tag{10.77}$$

$$v_2 = -P_2^{-1}k_2 \tag{10.78}$$

Substitute v_1 and v_2 into Equations (10.75) and (10.76):

$$A_1\delta + w_1 + P_1^{-1}k_1 = 0 \tag{10.79}$$

$$A_2\delta + w_2 + P_2^{-1}k_2 = 0 \tag{10.80}$$

Make k_1 the subject from Equation (10.79):

$$k_1 = -P_1(A_1\delta + w_1) \tag{10.81}$$

Transpose Equation (10.74), substitute Equation (10.81) into it, and simplify to give the following:

$$-A_1^T P_1(A_1\delta + w_1) + A_2^T k_2 = 0 \tag{10.82}$$

or

$$A_1^T P_1 A_1\delta + A_1^T P_1 w_1 - A_2^T k_2 = 0 \tag{10.83}$$

The remaining un-manipulated equation is Equation (10.80). Putting Equations (10.83) and (10.80) in matrix form gives the following:

$$\begin{bmatrix} A_1^T P_1 A_1 & A_2^T \\ A_2 & -P_2^{-1} \end{bmatrix} \begin{bmatrix} \delta \\ -k_2 \end{bmatrix} = \begin{bmatrix} -A_1^T P_1 w_1 \\ -w_2 \end{bmatrix} \tag{10.84}$$

The solution of the matrix (10.84) can be given as follows:

$$
\begin{bmatrix} \delta \\ -k_2 \end{bmatrix} = \begin{bmatrix} N_1 & A_2^T \\ A_2 & -P_2^{-1} \end{bmatrix}^{-1} \begin{bmatrix} -u_1 \\ -w_2 \end{bmatrix}
$$

(10.85)

where $N_1 = A_1^T P_1 A_1$ and $u_1 = A_1^T P_1 w_1$. Matrix product in Equation (10.85) can be simplified further as follows:

$$
\begin{bmatrix} \delta \\ -k_2 \end{bmatrix} = \begin{bmatrix} Q_{11} & Q_{12} \\ Q_{21} & Q_{22} \end{bmatrix} \begin{bmatrix} -u_1 \\ -w_2 \end{bmatrix}
$$

(10.86)

Use the matrix inversion identity (E.2) on the matrix inverse of Equation (10.85) above, assuming $T = N_1$, $U = A_2^T$, $V = A_2$, $W = -P_2^{-1}$. The following results are obtained:

$$
Q_{11} = N_1^{-1} + N_1^{-1} A_2^T Q^{-1} A_2 N_1^{-1}
$$

(10.87)

$$
Q_{12} = -N_1^{-1} A_2^T Q^{-1}
$$

(10.88)

$$
Q_{21} = -Q^{-1} A_2 N_1^{-1}
$$

(10.89)

$$
Q_{22} = Q^{-1}
$$

(10.90)

where

$$
Q = -\left[P_2^{-1} + A_2 N_1^{-1} A_2^T \right]
$$

(10.91)

Let the following representation be made:

$$
M = -Q^{-1}
$$

(10.92)

or

$$
M = \left[P_2^{-1} + A_2 N_1^{-1} A_2^T \right]^{-1}
$$

(10.93)

Substituting M into Equations (10.87)–(10.90) gives the following elements of the inverse matrix in Equation (10.86):

$$
Q_{11} = N_1^{-1} - N_1^{-1} A_2^T M A_2 N_1^{-1}
$$

(10.94)

$$
Q_{12} = N_1^{-1} A_2^T M
$$

(10.95)

$$
Q_{21} = M A_2 N_1^{-1}
$$

(10.96)

$$
Q_{22} = -M
$$

(10.97)

From Equation (10.86), the solution vectors can be given as

$$
\delta = -(Q_{11} u_1 + Q_{12} w_2)
$$

(10.98)

Substituting Equations (10.94) and (10.95) into Equation (10.98) gives

$$
\delta = -N_1^{-1} u_1 + N_1^{-1} A_2^T M A_2 N_1^{-1} u_1 - N_1^{-1} A_2^T M w_2
$$

(10.99)

The standard least squares correction vector to the approximate vector of parameters (x^0) from the first batch of observations (ℓ_1) can be taken as the first term in Equation (10.99), which can be expressed as

$$\delta' = -N_1^{-1} u_1 \tag{10.100}$$

By considering the final correction vector (δ) to the same approximate vector of parameters (x^0) in Equation (10.99), the following can be deduced:

$$\delta = \delta' + \Delta\delta \tag{10.101}$$

where

$$\Delta\delta = N_1^{-1} A_2^T M A_2 N_1^{-1} u_1 - N_1^{-1} A_2^T M w_2 \tag{10.102}$$

or

$$\Delta\delta = N_1^{-1} A_2^T M \left(A_2 N_1^{-1} u_1 - w_2 \right) \tag{10.103}$$

Substitute Equation (10.100) into Equation (10.103):

$$\Delta\delta = -N_1^{-1} A_2^T M (A_2 \delta' + w_2) \tag{10.104}$$

By variance–covariance propagation on Equation (10.99), the following cofactor matrix of the vector of corrections to the approximate parameters can be given as

$$Q_\delta = N_1^{-1} - N_1^{-1} A_2^T M A_2 N_1^{-1} \tag{10.105}$$

Equation (10.105) can be rewritten as

$$Q_\delta = Q_{\delta'} + \Delta Q_\delta \tag{10.106}$$

where $Q_{\delta'} = N_1^{-1}$ is the cofactor matrix of the vector of corrections to the approximate parameters based on the first batch of observations (ℓ_1), which is the same as the cofactor of the adjusted parameters $(Q_{\hat{x}_1})$; $\Delta Q_\delta = -N_1^{-1} A_2^T M A_2 N_1^{-1}$ is the correction matrix to be applied to the previous cofactor, due to batch II observations vector (ℓ_2); and $Q_\delta = Q_{\hat{x}}$ is the cofactor matrix of the overall adjusted parameters.

Substitute Equation (10.101) into Equation (10.75) and rearrange to obtain

$$v_1 = A_1(\delta' + \Delta\delta) + w_1 \tag{10.107}$$

or

$$v_1 = v_1' + A_1 \Delta\delta \tag{10.108}$$

where $v_1' = A_1 \delta' + w_1$ is the vector of residuals computed based on the first batch of observations (ℓ_1). From the solution vector in Equation (10.86), the vector of correlates (k_2) can be given as

$$k_2 = Q_{21} u_1 + Q_{22} w_2 \tag{10.109}$$

Substituting Equations (10.96) and (10.97) into Equation (10.109) gives

$$k_2 = MA_2N_1^{-1}u_1 - Mw_2 \qquad (10.110)$$

Substitute Equation (10.100) into Equation (10.110), giving

$$k_2 = -M(A_2\delta' + w_2) \qquad (10.111)$$

Substitute Equation (10.111) into Equation (10.78), giving

$$v_2 = P_2^{-1}M(A_2\delta' + w_2) \qquad (10.112)$$

or from Equation (10.76)

$$v_2 = A_2\delta + w_2 \qquad (10.113)$$

Remember that the sequential approach given above assumes that one uses the same approximate parameters (x^0) in the adjustment of the first and the second batches of observations.

10.5.1 Summary of Sequential Least Squares Adjustment Steps

The steps for the sequential least squares adjustment can be summarized as follows.

1) *Batch I adjustment*: Perform the initial adjustment with n_1 number of batch I observations as follows.
   a) Formulate $\hat{\ell}_1 = f_1(\hat{x})$.
   b) Find $A_1 = \dfrac{\partial f_1}{\partial x}$, P_1, $w_1 = f_1(x^0) - \ell_1$.
   c) Calculate the corrections to approximate parameters (x^0):

   $$\delta' = -N_1^{-1}u_1 \qquad (10.114)$$

   where $N_1 = A_1^T P_1 A_1$ and $u_1 = A_1^T P_1 w_1$.
   d) Determine the cofactor matrix of the adjusted parameters for batch I observations:

   $$Q_{\hat{x}_1} = N_1^{-1} \qquad (10.115)$$

   e) Calculate the residual vector to be applied to batch I observations:

   $$v_1' = A_1\delta' + w_1 \qquad (10.116)$$

   f) Calculate the a posteriori variance factor of unit weight for the batch I observations:

   $$s_0'^2 = \frac{v_1'^T P_1 v_1'}{n_1 - u} \qquad (10.117)$$

g) Determine the covariance matrix of the adjusted parameters from the batch I observations:

$$C_{\hat{x}_1} = s_0'^2 Q_{\hat{x}_1} \tag{10.118}$$

h) Calculate the cofactor of residual vector determined from the batch I observations:

$$Q_{V_1} = Q_{\ell_1} - A_1 N_1^{-1} A_1^T \tag{10.119}$$

2) *Batch II adjustment*: When new observations are available, perform the following initial adjustment steps with n_2 number of observations. The batch II observations are sequentially used to correct batch I observations in order to determine the updated values of the adjusted parameters and the updated values of the precisions of the adjusted parameters. The batch II adjustment steps are as follows:

a) Formulate $\hat{\ell}_2 = f_2(\hat{x})$.

It should be mentioned here that the mathematical model equations formed here will depend on the number of observations involved, e.g. for one observation, there will be only one equation relating only to the relevant parameters; however, the partial derivatives will still be done with respect to all the parameters identified in phase I above.

b) Find $A_2 = \frac{\partial f_2}{\partial x}$, P_2,

$$w_2 = f_2(x^0) - \ell_2, \tag{10.120}$$
$$N_2 = A_2^T P_2 A_2,$$
$$u_2 = A_2^T P_2 w_2$$

The misclosure vector (w_2) is determined by substituting into the equations formed in (a) above the approximate values of the parameters and the values of the observations.

c) Determine the changes in corrections ($\Delta\delta$) due to adjustment of batch II observations:

$$\Delta\delta = -N_1^{-1} A_2^T M (A_2 \delta' + w_2), \tag{10.121}$$
$$M = \left[P_2^{-1} + A_2 N_1^{-1} A_2^T \right]^{-1} \tag{10.122}$$

d) Determine the final correction vector:

$$\delta = \delta' + \Delta\delta \tag{10.123}$$

e) Determine the final cofactor matrix:

$$Q_{\hat{x}} = Q_{\hat{x}_1} + \Delta Q_{\hat{x}_1} \tag{10.124}$$

where

$$\Delta Q_{\hat{x}_1} = -N_1^{-1}A_2^T MA_2 N_1^{-1} \tag{10.125}$$

f) Determine the residual vectors:

$$v_1 = A_1\delta + w_1 \tag{10.126}$$

$$v_2 = A_2\delta + w_2 \tag{10.127}$$

or in sequential form

$$v_1 = v_1' + \Delta v_1 \tag{10.128}$$

where $\Delta v_1 = A_1\Delta\delta$ is the vector of corrections to be made to the previously computed residuals (v') and the residuals for the batch II observations can be given as

$$v_2 = P_2^{-1}M(A_2\delta' + w_2) \tag{10.129}$$

If $k_2 = -M(A_2\delta' + w_2)$, then Equation (10.129) will become

$$v_2 = -P_2^{-1}k_2 \tag{10.130}$$

g) Determine the final a posteriori variance factor of unit weight:

$$s_0^2 = \frac{v_1^T P_1 v_1 + v_2^T P_2 v_2}{n_1 + n_2 - u} \tag{10.131}$$

where n_1 and n_2 are the numbers of observations in batches I and II, respectively, and u is the number of unknown parameters (which is the same in both batches).

h) Determine the final covariance matrix of the adjusted parameters:

$$C_{\hat{x}} = s_0^2 Q_{\hat{x}} \tag{10.132}$$

i) Determine the cofactor matrix of the corrected residual vector (v_1) for batch *I* observations:

$$Q_{v_1} = Q_{v_1'} + \Delta Q_{v_1} \tag{10.133}$$

where

$$\Delta Q_{v_1} = A_1 N_1^{-1}A_2^T MA_2 N_1^{-1}A_1^T \tag{10.134}$$

j) Determine the cofactor matrix of the residual vector (v_2) for batch II observations:

$$Q_{v_2} = P_2^{-1}MP_2^{-1} \tag{10.135}$$

Examples of sequential least squares adjustment are given as follows.

Example 10.1   Given the following mathematical models

$$\lambda_1 = f_1(x_1, x_2) \quad C_{\lambda_1} \tag{10.136}$$

$$\lambda_2 = f_2(x_1, x_2) \quad C_{\lambda_2} \tag{10.137}$$

where f_1 and f_2 are vectors of mathematical models, x_1 and x_2 are vectors of unknown parameters, λ_1 and λ_2 are vectors of observations, and C_{λ_1} and C_{λ_2} are covariance matrices.

a) Formulate the variation function.

Solution:

Rewrite Equations (10.136) and (10.137) in terms of their respective residual vectors v_1 and v_2, vectors of adjusted observations $\hat{\lambda}_1$ and $\hat{\lambda}_2$, vectors of approximate values of parameters x_1^0 and x_2^0, and vectors of corrections to approximate parameters δ_1 and δ_2:

$$\lambda_1 + v_1 = f_1\left(x_1^0 + \delta_1, x_2^0 + \delta_2\right) \tag{10.138}$$

$$\lambda_2 + v_2 = f_2\left(x_1^0 + \delta_1, x_2^0 + \delta_2\right) \tag{10.139}$$

Linearize Equations (10.138) and (10.139) by Taylor series method:

$$v_1 = f_1\left(x_1^0, x_2^0\right) + \frac{\partial f_1}{\partial x_1}\delta_1 + \frac{\partial f_1}{\partial x_2}\delta_2 - \lambda_1 \tag{10.140}$$

$$v_2 = f_2\left(x_1^0, x_2^0\right) + \frac{\partial f_2}{\partial x_1}\delta_1 + \frac{\partial f_2}{\partial x_2}\delta_2 - \lambda_2 \tag{10.141}$$

or

$$v_1 = A_{11}\delta_1 + A_{12}\delta_2 + w_1 \tag{10.142}$$

$$v_2 = A_{21}\delta_1 + A_{22}\delta_2 + w_2 \tag{10.143}$$

where $w_1 = f_1\left(x_1^0, x_2^0\right) - \lambda_1$ and $w_2 = f_2\left(x_1^0, x_2^0\right) - \lambda_2$ are the corresponding misclosure vectors and $A_{11} = \dfrac{\partial f_1}{\partial x_1}$, $A_{12} = \dfrac{\partial f_1}{\partial x_2}$, and $A_{21} = \dfrac{\partial f_2}{\partial x_1}$, $A_{22} = \dfrac{\partial f_2}{\partial x_2}$ are the first design matrices.

Impose the least squares criterion $v_1^T P_1 v_1 + v_2^T P_2 v_2 = \text{minimum}$ and create the variation function as follows:

$$\varphi = v_1^T P_1 v_1 + v_2^T P_2 v_2 - 2k_1^T\left(A_{11}\delta_1 + A_{12}\delta_2 + w_1 - v_1\right)$$
$$- 2k_2^T\left(A_{21}\delta_1 + A_{22}\delta_2 + w_2 - v_2\right) = \text{minimum} \tag{10.144}$$

where k_1 and k_2 are the vectors of Lagrangian multipliers (or vectors of correlates). Since the covariance matrices C_{λ_1} and C_{λ_2} are given, the equations must be expressed with respect to them; hence $P_1 = C_{\lambda_1}^{-1}$ and $P_2 = C_{\lambda_2}^{-1}$ (for $\sigma_0^2 = 1$). The variation function becomes

$$
\begin{aligned}
\varphi &= v_1^T C_{\lambda_1}^{-1} v_1 + v_2^T C_{\lambda_2}^{-1} v_2 - 2k_1^T (A_{11}\delta_1 + A_{12}\delta_2 + w_1 - v_1) \\
&\quad - 2k_2^T (A_{21}\delta_1 + A_{22}\delta_2 + w_2 - v_2) = \text{minimum}
\end{aligned}
\tag{10.145}
$$

b) Derive the most expanded form of the least squares normal equation system.

Solution:

Find the partials of φ with respect to the unknown quantities in the function, i.e. v_1, v_2, δ_1, δ_2, k_1^T, k_2^T:

$$
\frac{\partial \varphi}{\partial v_1} = 2v_1^T C_{\lambda_1}^{-1} + 2k_1^T = 0
\tag{10.146}
$$

$$
\frac{\partial \varphi}{\partial v_2} = 2v_2^T C_{\lambda_2}^{-1} + 2k_2^T = 0
\tag{10.147}
$$

$$
\frac{\partial \varphi}{\partial \delta_1} = -2k_1^T A_{11} - 2k_2^T A_{21} = 0
\tag{10.148}
$$

$$
\frac{\partial \varphi}{\partial \delta_2} = -2k_1^T A_{12} - 2k_2^T A_{22} = 0
\tag{10.149}
$$

$$
\frac{\partial \varphi}{\partial k_1^T} = -2(A_{11}\delta_1 + A_{12}\delta_2 + w_1 - v_1) = 0
\tag{10.150}
$$

$$
\frac{\partial \varphi}{\partial k_2^T} = -2(A_{21}\delta_1 + A_{22}\delta_2 + w_2 - v_2) = 0
\tag{10.151}
$$

In summary, the normal equation system (Equations (10.146)–(10.151)) in the most expanded form can be given as

$$
v_1^T C_{\lambda_1}^{-1} + k_1^T = 0
\tag{10.152}
$$

$$
v_2^T C_{\lambda_2}^{-1} + k_2^T = 0
\tag{10.153}
$$

$$
k_1^T A_{11} + k_2^T A_{21} = 0
\tag{10.154}
$$

$$
k_1^T A_{12} + k_2^T A_{22} = 0
\tag{10.155}
$$

$$
A_{11}\delta_1 + A_{12}\delta_2 + w_1 - v_1 = 0
\tag{10.156}
$$

$$
A_{21}\delta_1 + A_{22}\delta_2 + w_2 - v_2 = 0
\tag{10.157}
$$

10.5.2 Sequential Least Squares Adjustment Example

To illustrate the sequential least squares adjustment steps given in Section 10.5.1, consider five height differences observed between four survey stations as shown in Table 10.3 and Figure 10.2. The height of station A is

Table 10.3 Field measurements.

Stations		Observed height differences (m)	Approx. distance (km)
From	**To**		
A	1	61.478	10
1	2	16.994	15
2	3	−25.051	9
3	A	−53.437	18
A	2	78.465	20

Figure 10.2 Leveling network.

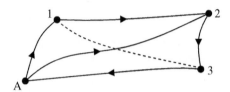

214.880 m above the datum, and the standard deviation of an observed height difference can be determined using 0.005 m $\sqrt{k}$ (where k is the length of the level route in km).

Perform the following tasks:

a) Compute the least squares adjusted estimates of the heights of points 1, 2, and 3 and their standard errors.

b) Now if, at later date, the height difference from 1 to 3 is observed to be −8.070 m (length 22 km), determine new sets of least squares estimates and standard deviations of points 1, 2, and 3 using sequential parametric least squares method.

Solution:

a) Least squares adjusted estimates of the heights of points 1, 2, and 3 and their standard errors are determined as follows. Form the parametric equations: $\hat{\ell}_1 = f_1(\hat{x})$

$$\hat{\ell}_1 = \hat{x}_1 - 214.880 \tag{10.158}$$

$$\hat{\ell}_2 = \hat{x}_2 - \hat{x}_1 \tag{10.159}$$

$$\hat{\ell}_3 = \hat{x}_3 - \hat{x}_2 \tag{10.160}$$

$$\hat{\ell}_4 = 214.880 - \hat{x}_3 \tag{10.161}$$

$$\hat{\ell}_5 = \hat{x}_2 - 214.880 \tag{10.162}$$

where $\hat{x}_1$, $\hat{x}_2$, and $\hat{x}_3$ represent the adjusted estimates of the heights of points 1, 2, and 3, respectively, and $\hat{\ell}_1$, $\hat{\ell}_2$, $\hat{\ell}_3$, $\hat{\ell}_4$, and $\hat{\ell}_5$ represent the adjusted elevation differences as given in Table 10.3, respectively. Let the approximate heights of points 1, 2, and 3 be given as follows:

$$\begin{bmatrix} x_1^0 \\ x_2^0 \\ x_3^0 \end{bmatrix} = \begin{bmatrix} 276.358 \\ 293.352 \\ 268.301 \end{bmatrix} \quad \ell_1 = \begin{bmatrix} 61.478 \\ 16.994 \\ -25.051 \\ -53.437 \\ 78.465 \end{bmatrix}$$

Form the A-matrix from Equations (10.158)–(10.162) and the misclosure vector (w_1):

$$A - \text{matrix}: A_1 = \begin{bmatrix} 1 & 0 & 0 \\ -1 & 1 & 0 \\ 0 & -1 & 1 \\ 0 & 0 & -1 \\ 0 & 1 & 0 \end{bmatrix} \quad \text{Misclosure vector}: w_1 = f_1(x^0) - \ell_1 = \begin{bmatrix} 0.000 \\ 0.000 \\ 0.000 \\ 0.016 \\ 0.007 \end{bmatrix}$$

Determine the standard deviation of each elevation difference from 0.005 m $\sqrt{k}$ (where k is the length of the level route in km); then form the covariance matrix and the weight matrix of observations P:

$$\text{Weigh matrix}: P_1 = \begin{bmatrix} 4000 & & & & \\ & 2667.67 & & & \\ & & 4444.44 & & \\ & & & 2222.22 & \\ & & & & 2000 \end{bmatrix}$$

Form the matrix of coefficients of normal equations (N) and the u vector:

$$N_1 = A_1^T P_1 A_1 = \begin{bmatrix} 6666.7 & -2666.7 & 0.0 \\ -2666.7 & 9111.1 & -4444.4 \\ 0.0 & -4444.4 & 6666.7 \end{bmatrix}$$

$$u_1 = A_1^T P_1 w_1 = \begin{bmatrix} 0.000 \\ 14.000 \\ -35.556 \end{bmatrix}$$

$$N_1^{-1} = \begin{bmatrix} 1.8149E-4 & 7.872E-2 & 5.248E-5 \\ 7.872E-2 & 1.968E-4 & 1.312E-4 \\ 5.248E-5 & 1.312E-4 & 2.375E-4 \end{bmatrix}$$

Compute the corrections to the approximate heights from Equation (10.114):

$$\delta' = -N_1^{-1} u_1$$

$$\begin{bmatrix} \delta_1 \\ \delta_2 \\ \delta_3 \end{bmatrix} = \begin{bmatrix} 0.0008 \\ 0.0019 \\ 0.0066 \end{bmatrix}$$

Least squares adjusted estimates $\hat{x} = x^0 + \delta'$ are given in Table 10.4.

Table 10.4 Least squares adjusted heights based on batch I observations.

Station	Adjusted heights (m)
1	276.359
2	293.354
3	268.308

Compute the observation residual vector from Equation (10.116):

$$v' = A_1 \delta' + w_1$$

$$v' = \begin{bmatrix} 7.64E-4 \\ 1.145E-3 \\ 4.700E-3 \\ 9.394E-3 \\ 8.910E-3 \end{bmatrix}$$

Compute the a posteriori variance factor of unit weight from Equation (10.117):

$$s'^2_0 = \frac{v'^T P_1 v'}{df} \rightarrow \frac{0.458\ 73}{2} = 0.229\ 4$$

Cofactor matrix of the adjusted heights is given by Equation (10.115):

$$Q_{\hat{x}_1} = \begin{bmatrix} 1.8149\text{E}-4 & 7.872\text{E}-5 & 5.248\text{E}-5 \\ 7.872\text{E}-5 & 1.968\text{E}-4 & 1.312\text{E}-4 \\ 5.248\text{E}-5 & 1.312\text{E}-4 & 2.375\text{E}-4 \end{bmatrix}$$

Compute covariance of the adjusted heights from Equations (10.115) and (10.118):

$$C_{\hat{x}_1} = s'^2_0 Q_{\hat{x}_1}$$

$$C_{\hat{x}_1} = \begin{bmatrix} 4.163\text{E}-5 & 1.805\text{E}-5 & 1.204\text{E}-5 \\ 1.805\text{E}-5 & 4.514\text{E}-5 & 3.009\text{E}-5 \\ 1.204\text{E}-5 & 3.009\text{E}-5 & 5.447\text{E}-5 \end{bmatrix}$$

Least squares adjusted heights and their standard errors (square root of the diagonal elements of the covariance matrix of the adjusted heights) are given in Table 10.5.

Table 10.5 Adjusted heights and their standard deviations.

Station	Adjusted heights (m)	Standard error (m)
1	276.359	0.0064
2	293.354	0.0067
3	268.308	0.0074

The MATLAB code for the least squares adjustment of batch I observations is given in Table 10.6.

Table 10.6 MATLAB code for the adjustment of batch I observations.

```
>> % define the variables x1, x2, x3 using syms
>> syms x1 x2 x3
>> % form parametric Equations (10.158)-(10.162)
>> L1=x1-214.880;
>> L2=x2-x1;
>> L3=x3-x2;
>> L4=214.880-x3;
>> L5=x2-214.880;
>>% determine Jacobian of the equations wrt to x1,x2, x3 and
evaluate as A1 matrix
>> J=jacobian([L1;L2;L3;L4;L5],[x1 x2 x3]);
>> x1=276.358;x2=293.352;x3=268.301;
>> A1=eval(J);
>>% Determine computed observations L0, given observations
L, misclosure w1
>> L0=eval([L1;L2;L3;L4;L5]);
>> L=[61.478;16.994;-25.051;-53.437;78.465];
>> w1=L0-L;
>> % Form weight matrix of observations P1, solution d1 and
adjusted parameters xa1
>> P1=inv(diag(0.005^2*[10 15 9 18 20]));
>> format long
>> N1=A1'*P1*A1;
>> u1=A1'*P1*w1;
>> d1=-inv(N1)*u1;
>> x0=[x1;x2;x3];
>> xa1=x0+d1;
>>% determine residuals vp, aposteriori variance s012
>> vp=A1*d1+w1;
>> s012=vp'*P1*vp/2
s012 =
   0.229364431486732
>>% determine covariance matrix of parameters Cx1
>> Qx1=inv(N1);
>> Cx1=s012*inv(N1)
Cx1 =
   1.0e-04 *
   0.416266351604929   0.180549260937078
0.120366173958052
   0.180549260937078   0.451373152342694
0.300915434895130
   0.120366173958052   0.300915434895130
0.544656937160184
>>
```

b) Implement the sequential least squares procedure. The vector of initial values of the parameters (x^0) used here should be the same as that used in question (a). Add the sixth observation equation as follows (with $\hat{\ell}_6$ as the adjusted elevation difference newly added). Form the parametric equation for the new observation: $\hat{\ell}_2 = f_2(\hat{x})$

$$\hat{\ell}_6 = \hat{x}_3 - \hat{x}_1 \tag{10.163}$$

Form the A_2-matrix from Equation (10.163) with respect to all the parameters (x_1, x_2, x_3) as in batch I, and form the misclosure vector w_2:

$$f_2(x^0) = 268.301 - 276.358 = -8.057 \text{ m} \quad \text{and} \quad \ell_2 = -8.070 \text{ m}$$

$$A_2 = [-1 \ 0 \ 1] \quad w_2 = f_2(x^0) - \ell_2 = [0.013]$$

Use the distance to construct the weight matrix P_2:

$$Q_2 = P_2^{-1} = \left[\left(0.005\sqrt{22}\right)^2\right] \quad Q_2 = P_2^{-1} = [0.000 \ 55]$$

$$A_2 N_1^{-1} A_2^T = 0.000 \ 314$$

Calculate M from Equation (10.122):

$$P_2^{-1} + A_2 N_1^{-1} A_2^T = 0.000 \ 864$$

$$M = \left[P_2^{-1} + A_2 N_1^{-1} A_2^T\right]^{-1} = 1 \ 157.415 \ 22$$

$$v_2 = P_2^{-1} M (A_2 \delta' + w_2)$$

$$A_2 \delta' = 0.005 \ 843$$

$$A_2 \delta' + w_2 = 0.018 \ 843$$

$$N_1^{-1} A_2^T = \begin{bmatrix} -0.000 \ 129 \\ 0.000 \ 052 \\ 0.000 \ 185 \end{bmatrix}$$

Calculate the corrections to initial correction vector from Equation (10.121):

$$\Delta\delta = -N_1^{-1} A_2^T M (A_2 \delta' + w_2) = \begin{bmatrix} 0.002 \ 81 \\ -0.001 \ 14 \\ -0.004 \ 03 \end{bmatrix}$$

Compute the new values of δ obtained from Equation (10.123): $\delta = \delta' + \Delta\delta$.

$$\delta = \begin{bmatrix} 0.000 \ 76 \\ 0.001 \ 91 \\ 0.006 \ 61 \end{bmatrix} + \begin{bmatrix} 0.002 \ 81 \\ -0.001 \ 14 \\ -0.004 \ 03 \end{bmatrix} = \begin{bmatrix} 0.003 \ 577 \\ 0.000 \ 765 \\ 0.002 \ 572 \end{bmatrix}$$

Compute the correction to the cofactor matrix from Equation (10.125):

$$\Delta Q_{\hat{x}_1} = -N_1^{-1} A_2^T M A_2 N_1^{-1}$$

$$= - \begin{bmatrix} -0.000\ 129 \\ 0.000\ 052 \\ 0.000\ 185 \end{bmatrix} [1\ \ 157][-0.000\ 129\ \ 0.000\ 052\ \ 0.000\ 185]$$

$$= \begin{bmatrix} -0.000\ 019 & 0.000\ 008 & 0.000\ 028 \\ 0.000\ 008 & -0.000\ 003 & -0.000\ 011 \\ 0.000\ 028 & -0.000\ 011 & -0.000\ 040 \end{bmatrix}$$

Compute the new cofactor matrix of adjusted parameters from Equation (10.124):

$$Q_{\hat{x}} = Q_{\hat{x}_1} + \Delta Q_{\hat{x}}$$

$$Q_{\hat{x}} = \begin{bmatrix} 0.000\ 162 & 0.000\ 087 & 0.000\ 080 \\ 0.000\ 087 & 0.000\ 194 & 0.000\ 120 \\ 0.000\ 080 & 0.000\ 120 & 0.000\ 198 \end{bmatrix}$$

Compute the corrections to the phase I observation residuals from Equation (10.128):

$$\Delta v_1 = A_1 \Delta \delta = \begin{bmatrix} 0.0028 \\ -0.0040 \\ -0.0029 \\ 0.0040 \\ -0.0011 \end{bmatrix}$$

Compute the adjusted phase I observation residuals from Equation (10.128):

$$v_1 = v_1' + \Delta v_1 = \begin{bmatrix} 0.0036 \\ -0.0028 \\ 0.0018 \\ 0.0134 \\ 0.0078 \end{bmatrix}$$

Compute the batch II observation residuals from Equation (10.129):

$$v_2 = Q_2 M (A_2 \delta + w_2) = [0.011\ 995]$$

Compute the overall a posteriori variance factor of unit weight from Equation (10.131):

$$s_0^2 = \frac{v_1^T P_1 v_1 + v_2^T P_2 v_2}{n_1 + n_2 - u} \rightarrow s_0^2 = \frac{0.608\ 07 + 0.261\ 590}{5 + 1 - 3}$$

$$s_0^2 = 0.289\ 887$$

Compute the covariance matrix of the adjusted parameters from Equation (10.132):

$$C_{\hat{x}} = s_0^2 Q_{\hat{x}} = \begin{bmatrix} 0.000\ 047 & 0.000\ 025 & 0.000\ 023 \\ 0.000\ 025 & 0.000\ 056 & 0.000\ 035 \\ 0.000\ 023 & 0.000\ 035 & 0.000\ 057\ 4 \end{bmatrix}$$

From the combination of all of the six observations, the least squares adjusted parameters and their standard errors are given in Table 10.7.

Table 10.7 Results of sequential adjustment.

Station	Adjusted heights (m)	Standard error (m)
1	276.361	0.0069
2	293.353	0.0075
3	268.303	0.0076

The above solution should be identical to one derived from simultaneous computation with all six observations. The MATLAB code for the sequential adjustment of batch II observations to the batch I observations (as a continuation of the code in Table 10.6) is given in Table 10.8.

Table 10.8 MATLAB code for the sequential adjustment (continued from code in Table 10.6).

```
>>% Continuing from Table 10.7 form Equation (10.163) as L6
and Jacobian wrt x1, x2, x3 and evaluate as A2
>> syms x1 x2 x3
>> L6=x3-x1;
>> J2=jacobian([L6],[x1 x2 x3]);
>> x1=276.358;x2=293.352;x3=268.301;
>> A2=eval(J2)
A2 =
    -1    0    1
>>% form misclosure w2 cofactor Q2 of L6
>> w2=eval(L6)-(-8.070)
w2 =
   0.012999999999984
>> Q2=0.005^2*22
Q2 =
    5.500000000000000e-04
>>% Solve corrections (Dd) to parameters d1
>> M=inv(Q2+A2*inv(N1)*A2')
M =
    1.157415218491648e+03
>> Dd=-inv(N1)*A2'*M*(A2*d1+w2);
>>% adjust parameters d1 with Dd
>> d=d1+Dd;
>>% find corrections to cofactor of parameters DQx and
correct cofactor Qx1 to give adjusted cofactor Qxa
>> DQx=-inv(N1)*A2'*M*A2*inv(N1);
>> Qxa=Qx1+DQx;
>>% find corrections (Dv1)to residuals and corrected
residuals v1
>> Dv1=A1*Dd;
>> v1=vp+Dv1;
>>% find residuals (v2)of second batch of observations
>> v2=Q2*M*(A2*d1+w2)
v2 =
   0.011994769697992
>>% find aposteriori variance s02 and covariance
matrix Cx of adjusted parameters
>> s02=(v1'*P1*v1+v2'*inv(Q2)*v2)/3
s02 =
   0.289886733029535
>> Cx=s02*Qxa;
>>
```

Example 10.2 Two offshore rigs B (x_B, y_B) and D (x_D, y_D) were previously coordinated using classical survey method with total station equipment used. In order to improve on the coordinates of the rigs by sequential least squares approach, the horizontal coordinates (with their standard deviations) of points B and D were determined again using GNSS, and the distance (S_{BD}) and bearing (α_{BD}) from B to D were measured with a total station instrument that is more precise than the one used in the previous survey. Points A and C are fixed control points with known (map projection) coordinates as shown in Figure 10.3 (taking the direction AN as the north direction).

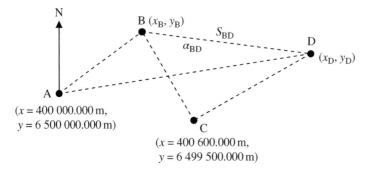

Figure 10.3 GPS and distance measurements (not to scale).

Formulate the parametric model relating the new measurements with the unknown coordinates.

Solution:

Parametric model equations are as follows:

$$\hat{\ell}_{x_B} = \hat{x}_B \tag{10.164}$$

$$\hat{\ell}_{y_B} = \hat{y}_B \tag{10.165}$$

$$\hat{\ell}_{x_D} = \hat{x}_D \tag{10.166}$$

$$\hat{\ell}_{y_D} = \hat{y}_D \tag{10.167}$$

$$\hat{S}_{BD} = \sqrt{(\hat{x}_D - \hat{x}_B)^2 + (\hat{y}_D - \hat{y}_B)^2} \tag{10.168}$$

$$\hat{\alpha}_{BD} = a\tan\left(\frac{(\hat{x}_D - \hat{x}_B)}{(\hat{y}_D - \hat{y}_B)}\right) + 180° \tag{10.169}$$

Problems

10.1 Given the following mathematical models

$$\ell_1 = f_1(x_1, x_2) \quad C_{\ell_1}$$
$$\ell_2 = f_2(x_1) \qquad C_{\ell_2}, C_{x_1}$$

where f_1 and f_2 are vectors of mathematical models, x_1 is a vector of unknown useful parameters, x_2 is a vector of unknown nuisance parameters, ℓ_1 and ℓ_2 are vectors of observations with their corresponding covariance matrices C_{ℓ_1} and C_{ℓ_2}, and C_{x_1} is the covariance matrix of the parameters, formulate the variation function.

10.2 Five distances and their corresponding standard errors were observed among 4 stations (A, B, R1, R2) in order to fix two nearby offshore rigs (R1 and R2) as shown in Figure P10.2. The field measurements, the known coordinates of points A and B, and the approximate coordinates of points R1 and R2 are given in the following tables.

Figure P10.2

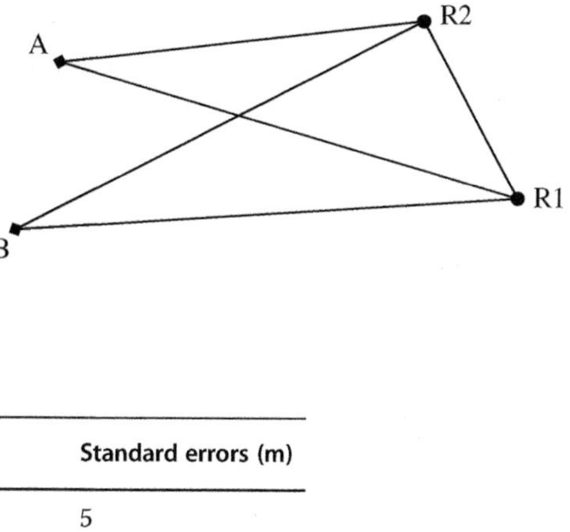

Field measurements

	Distances (m)	Standard errors (m)
A-R2	114 737.8	5
A-R1	114 948.5	5
B-R2	146 667.7	5
B-R1	147 802.4	5
R1-R2	1 981.81	0.05

Known and approximate coordinates of stations.

Point		
A	163 304.56	894 962.77
B	108 791.23	943 117.05
R1	255 086.5	964 173.1
R2	253 717.3	965 605.9

Perform the following tasks:
a) Compute the least squares adjusted estimates of the coordinates of points R1 and R2 and their standard errors.
b) Now if, at a later date, the coordinates of point R1 were measured using GPS as easting = 255 086.5 m and northing = 964 173.1 m with corresponding standard errors of ±3 m, determine new sets of least squares estimates and standard deviations of points R1 and R2 using the sequential parametric least squares method.

10.3 In order to determine the coordinates of point B (x, y), two total station directions (d_{BC}, d_{BA}) and two distances (S_{BC}, S_{BA}) were measured as shown in Figure P10.3. The field measurements, the coordinates of the fixed points A and C, and the approximate coordinates of the unknown point B are given in the following tables.

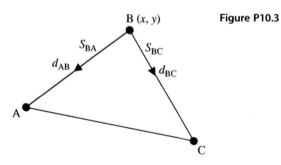

Figure P10.3

Field measurements

Line	Measurement	Standard deviation
d_{BC}	0°00′00″	±5″
d_{BA}	112°58′00″	±5″
S_{BC}	600.045 m	±0.010 m
S_{BA}	89.420 m	± 0.005 m

Coordinates

Station	X coordinate (m)	Y coordinate (m)	Comments
A	1000.000	900.000	Fixed point
B	1083.5	932.000	Approximate
C	1500.020	500.105	Fixed point

In order to perform the least squares adjustment of this problem, the following were decided:

Coordinate parameters, $x_1 = \begin{bmatrix} x_B \\ y_B \end{bmatrix}$; orientation parameters, $x_2 = [z]$;

observations, $\ell = [d_{BC} \quad d_{BA} \quad S_{BC} \quad S_{BA}]^T$

Parametric equations $[\hat{\ell} = f(\hat{x})]$:

$$\hat{d}_{BC} = a\tan\left(\frac{1500.020 - \hat{x}_B}{500.105 - \hat{y}_B}\right) + 180° - \hat{z}$$

$$\hat{d}_{BA} = a\tan\left(\frac{1000.000 - \hat{x}_B}{900.000 - \hat{y}_B}\right) + 180° - \hat{z}$$

$$\hat{S}_{BC} = \sqrt{(1500.020 - \hat{x}_B)^2 + (500.105 - \hat{y}_B)^2}$$

$$\hat{S}_{BA} = \sqrt{(1000 - \hat{x}_B)^2 + (900 - \hat{y}_B)^2}$$

Assumed approximate coordinates: $x^0 = \begin{bmatrix} x_B^0 = 1083.5 \\ y_B^0 = 932.0 \end{bmatrix}$ and

Assumed approximate orientation parameters: $z^0 = 136°02'18''$

A-matrices (evaluated at the approximate parameters):

$$A_1 = \frac{\partial f}{\partial x} = \begin{bmatrix} 0.001\ 199\ 634\ 351 & 0.001\ 156\ 928\ 652 \\ 0.004\ 001\ 875\ 879 & -0.010\ 442\ 394\ 87 \\ -0.694\ 178\ 595\ 5 & 0.719\ 802\ 805\ 5 \\ 0.933\ 777\ 260\ 0 & 0.357\ 854\ 758\ 4 \end{bmatrix} \qquad A_2 = \frac{\partial f}{\partial z} = \begin{bmatrix} -1 \\ -1 \\ 0 \\ 0 \end{bmatrix}$$

Using all of the above information as needed, compute the adjusted coordinates of point B with eliminated orientation parameters using the parametric method of least squares. Your work must clearly and explicitly show the elimination of the orientation parameter as discussed in the lecture notes.

11

Post-adjustment Data Analysis and Reliability Concepts

CHAPTER MENU

OBJECTIVES

After studying this chapter, you should be able to:

1) Explain the steps involved in post-adjustment data analysis.
2) Perform global tests for outlier detection using one-tailed and two-tailed approaches.
3) Perform local tests for outlier identification and elimination using Pope's and Baarda's approaches.
4) Discuss the concept of redundancy numbers and the applications to geomatics.
5) Discuss the concept of reliability, including internal and external reliability measures and sensitivity measures and the applications to geomatics.

Understanding Least Squares Estimation and Geomatics Data Analysis, First Edition.
John Olusegun Ogundare.
© 2019 John Wiley & Sons, Inc. Published 2019 by John Wiley & Sons, Inc.
Companion website: www.wiley.com/go/ogundare/Understanding-lse-and-gda

11.1 Introduction

Statistical testing of agreement of measurements and their residuals with some specified tolerance or standard is referred to as post-adjustment analysis in this chapter. Conclusions about measurements and the quality of their agreement are based on least squares adjustment of the measurements and the statistical evidence. In this chapter, the concepts of data analysis by Baarda (1968) and Pope (1976) will be considered. It should be remembered that errors are the differences between true values of quantities and their measured values, knowing that true values cannot be determined. In the process of collecting measured values, different error sources are possible. The following definitions are to help understand the important error sources that are relevant to data analysis discussed in this chapter:

- *Random errors* are accidental errors, which follow statistical and probability rules. In other words, the errors are considered as stochastic, and cannot be predicted precisely, requiring probability distributions, as discussed in Chapters 2 and 3.
- *Systematic errors* are non-stochastic errors, which can be explained with regard to least squares adjustment, as differences between predicted values based on functional models describing observations and the true values of the observables. The differences (known as systematic errors) can be due to environmental conditions, such as the effects of refraction, temperature changes, etc.; *instrumental errors*, such as faulty construction and poor calibration of instrument, resulting in instrument constant, collimation errors, tilting axis errors, etc.; and *geometric reductions*, which may be due to slope reduction, map projection, nuisance parameters, etc. These types of errors can be determined mathematically, based on some physical laws.
- *Gross errors* are errors due to the malfunctioning of instrument or oversight of the surveyor while making measurements. Gross errors are usually considered as mistakes in measurements. In least squares adjustment, some of the gross errors are likely to go to the residuals as outliers, and the rest spread to other measurements, sometimes, making some of these measurements appear as outliers. Gross errors and systematic errors can only be detected when there are redundant measurements to provide the checks.
- *Residuals* are errors, which can be explained with regard to least squares adjustment, as differences between predicted values based on functional models describing observations and the actual observations (not the true values of the observables). In other words, a residual is the difference between the simulated measurement (based on some mathematical models) and the actual measurement. This type of error will most likely have a mixture of all possible types of errors, such as random, systematic, and gross errors.

- *Misclosure* can be considered as the discrepancy between the actual measurement (or a given value) and the derived measurement (or derived value). The derived value in terms of the misclosure of a measured forward bearing in a traverse will be the calculated bearing based on the initial bearing and the measured included traverse angles; the derived measurement in the case of parametric least squares adjustment will be the one calculated from the mathematical model using approximate coordinates. The least squares adjustment is to remove misclosures so that at the end of the adjustment (with the residuals added to the observations to make them fit the mathematical model), the misclosures are zero.
- *Outlier* is a data value that does not match the trend of the other data values in a given data set. It is an observation that seems to be inconsistent with the rest. It has a residual that, according to some test rule, is in disagreement with the assumption made (detected outliers are likely caused by gross errors). If outlying residual is detected, it is usual to try to find evidence of gross errors. Generally, an observation will be considered an outlier if its residual is more than about 2.5–3 times its standard deviation.
- *Quality* is about efficiently producing a product or a service that is of high standard (free of errors) within an acceptable economy without wasting much time.
- *Quality assurance* is an approach (action plan) aimed at preventing problems from occurring during a survey work. It considers the work processes or methods and the people doing the work as well as the end product to be delivered to the client. An example of quality assurance is an idea of testing different computer software for accuracy before it is used for data processing.
- *Quality control* is about checking a survey work (or the product) after its completion to detect errors and correct them. This process, however, is not as efficient as the quality assurance measure since it is to correct the error that had already been made in producing a product.

11.2 Post-adjustment Detection and Elimination of Non-stochastic Errors

One important disadvantage of least squares method of adjustment is that when a non-stochastic error is present in one measurement and the measurement is adjusted together with others, the non-stochastic error is usually distributed among the other adjusted measurements so that many residuals are somehow contaminated with a portion of the error. The least squares method usually spread the non-stochastic error in order to satisfy the least squares criterion of making the weighted sum of corrections to all observations as small as possible. Because of this, many residuals detected as outliers in an adjustment may not necessary be outliers, so that some cautions are required in rejecting the associated measurements. The presence and localization of non-stochastic errors are determined after least squares adjustment by using statistical tests. Examples of non-stochastic errors are model errors, systematic errors, large errors, or blunders. Other type of error detected is precision error. Precision errors are caused by under- or overestimating precisions of measurements, e.g. assigning precision of $1''$ to vertical angle observation in poor atmospheric condition.

Probability distribution function does not exist for non-stochastic errors, and their presence in measurements prevents the measurements from being normally distributed. Statistical tests to detect and localize non-stochastic and precision errors are based on confidence limits and critical values so that if test value falls within the range described by the critical values, then the test is passed, and the measurements are assumed to be free of errors. The usual null hypothesis (H_0) to be tested is that the model is correct and complete and the distributional assumptions meet the reality or are normal. Two possible outcomes for the testing of the H_0 hypothesis are to accept or to reject it. There is usually no specific alternative hypothesis apart from rejecting or accepting the null hypothesis with the critical values to establish a window of acceptance or cutoff points for rejecting the hypothesis. These critical values are determined by the choice of a level of significance level α (the probability of an incorrect rejection). The statistical testing procedure is given in Figure 11.1 and summarized as follows:

1) Perform *global test* of the variance factor of unit weight (a good indicator for the presence of any type of error). If this test fails, it may be interpreted that there are gross, systematic, or precision errors or a combination of the errors.
2) Perform *local test* on individual measurements to identify possible outliers. This requires testing standardized residuals of individual measurements against critical factors derived from an appropriate statistical distribution.
3) Measurements that have residuals that fail local test are likely to be outliers. To identify the actual outlier, the following steps can be taken:
   - If many measurements are flagged as outliers, remove the measurement with the largest standardized residual as the only outlier for now. It is advisable to give the outlying measurement a large input standard deviation so that it has little weight effect on the solution rather than just deleting it outrightly from the data set. With this, the new residuals may provide a better picture of the true size of the errors in the measurements.
   - Readjust the measurements without the outlying measurement or with it, but with larger input standard deviation.
   - After the adjustment, return to step 1 and start the testing procedure all over.

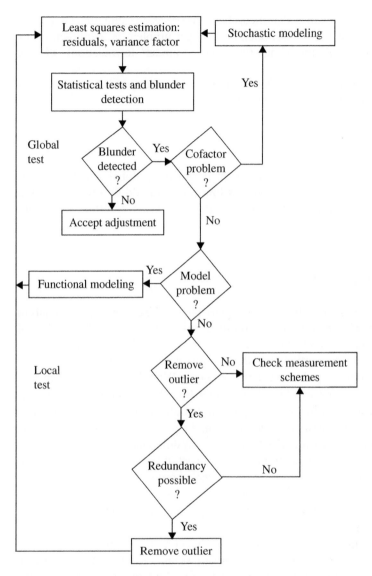

Figure 11.1 Schematic diagram of statistical tests and blunder detection procedure in least squares estimation.

4) If no measurements fail the local test but the global test fails, do the following:
   • Review the measurements for precision (cofactor problem) or systematic (model problem) errors.

- If cofactor problem is identified, check the precision estimation (stochastic modeling) of the measurements; if model problem is identified, check the functional modeling (parametric model formulation) of the measurements.
- After correcting for the effects of the systematic or precision errors, readjust the corrected measurements.
- Return to step 1 and start the testing procedure all over
5) If no measurements fail the local test, and the global test passes, the network adjustment is considered satisfactory.

Different approaches are possible in performing global and local tests. Some of the approaches are given as follows. It should be mentioned that global and local tests should only be performed for minimal constraint adjustments.

11.3   Global Tests

Global or variance factor test is a chi-square test on the observation residuals to infer whether all the standardized residuals involved are normally distributed with a mean of zero and a standard deviation of one. The concept of the test is based on the assumption that the a priori variance factor of unit weight $(\sigma_0^2 = 1)$ is from the population and is *well known*. With this idea, the a posteriori variance factor of unit weight (s_0^2) computed after the adjustment procedure is statistically checked whether it is significantly different from the a priori value $(\sigma_0^2 = 1)$ at a given confidence level $(1 - \alpha)$ and a power of the test $(1 - \beta)$. If the two variance factors are found to be significantly different by this test, then there is an indication that at least an outlier may exist in the measured quantities. It should be mentioned that the purpose of a global test will be defeated if σ_0^2 is not well known. When properly used, the global test usually reflects the true quality of survey measurements. In some cases, if s_0^2 is too small compared with σ_0^2, the input standard deviations of the observations are likely too large; if s_0^2 is too large, then the input standard deviations of the observations are too small. Some computer software packages allow automatic scaling of covariance matrices of parameters and observations to account for the effect of the a posteriori variance factor of unit weight, while some do not. If the input variances of the observations are known to be well estimated, the global test should be done to check if the a posteriori variance factor of unit weight is close to unity; if the test fails, the cause of the failure must be investigated; if the test passes, the estimated covariance matrices of the parameters are not to be scaled with s_0^2. If, however, the input variances of the observations are not well estimated, different approaches exist, such as scaling the covariance matrices of the parameters by s_0^2, or changing the input variances of the observations until the test passes if large redundancies are involved; and for small redundant observations, it is usually not obvious what should be done. Effort, however, should be made not to manipulate measurement weights (P) in order to satisfy global tests.

11.3.1 Standard Global Test

It has become a standard procedure in outlier detection for least squares adjustment packages to first perform global test to indicate whether outlier exists and then local test to identify the outlier. In commercial software packages, the hypotheses for global test (which is considered as standard approach in this book) can be formulated as

$$H_0 : s_0^2 = \sigma_0^2 \text{ against } H_A : s_0^2 \neq \sigma_0^2$$

which is two-tailed test. The commonly used standard test statistic (with the usual notations for the lower-tail areas of the chi-square used) is given as follows (refer to Chapter 3 and Table B.3):

$$\frac{(\text{df}) \, s_0^2}{\chi_{1-\alpha/2,\text{df}}^2} < \sigma_0^2 < \frac{(\text{df}) \, s_0^2}{\chi_{\alpha/2,\text{df}}^2} \tag{11.1}$$

or

$$\chi_{\alpha/2,\text{df}}^2 < \chi^2 < \chi_{1-\alpha/2,\text{df}}^2 \tag{11.2}$$

where n is the number of observations, α is the significance level (e.g. 0.05), df is the number of degrees of freedom (e.g. $n - 1$ for single observable), $\sigma_0^2 = 1.0$ is the a priori variance factor of unit weight, $\chi_{\alpha/2,\text{df}}^2$ and $\chi_{1-\alpha/2,\text{df}}^2$ are the chi-square values based on the lower area type of chi-square distribution (refer to Table B.3), s_0^2 is the a posteriori variance factor of unit weight (computed after least squares adjustment), and

$$\chi^2 = \frac{(\text{df}) \, s_0^2}{\sigma_0^2} \tag{11.3}$$

Equation (11.1) or Equation (11.2) will be referred to as the *standard global test*. The failure of the above test on the *low end* of confidence region (i.e. $\chi^2 < \chi_{\alpha/2,\text{df}}^2$) indicates an incorrect scaling of covariance matrix of observation, while the failure on the *high end* (i.e. $\chi^2 > \chi_{1-\alpha/2,\text{df}}^2$) means a possible presence of blunders or model errors in the data, either through an incorrect scaling of observation variances or by the choice of mathematical model. The following example illustrates how standard global test is done.

Example 11.1 In order to determine the coordinates of a new control point, five measurements were made from two control points whose positions are known and fixed. After the parametric least squares adjustment of the two-dimensional survey network, the computed a posteriori variance factor of unit weight is $s_0^2 = 0.213$. Perform two-tailed (standard) global test at 95% confidence level on this adjustment to find out if outliers are suspected in the measurements.

Solution:

Number of observations, $n = 5$; number of unknown (x, y) coordinates, $u = 2$; degrees of freedom, df = 5 − 2 (or 3); and $\alpha = 0.05$. Equation (11.1) will be used as follows:

$$\frac{(df) s_0^2}{\chi^2_{1-\alpha/2, df}} < \sigma_0^2 < \frac{(df) s_0^2}{\chi^2_{\alpha/2, df}}$$

Based on the standard chi-square table, $\chi^2_{0.025, df=3} = 0.216$, $\chi^2_{0.975, df=3} = 9.348$, $s_0^2 = 0.213$, and $\sigma_0^2 = 1$ (usually adopted value),

$$\frac{(3)0.213}{9.348} < 1.0 < \frac{(3)0.213}{0.216}$$

$$0.0684 < 1.0 < 2.958$$

Since the above condition is satisfied at 95% confidence level, the hypothesis that $s_0^2 = 0.213$ is statistically equal to $\sigma_0^2 = 1.0$ is accepted; the global test passes and there is no indication that the measurements contain outliers.

11.3.2 Global Test by Baarda

The concept of global tests flows directly from Baarda (1968); it is as discussed in Section 11.3.1 except that Baarda considers the failure in the high end to be more critical. The hypotheses for Baarda's global test can be formulated as

$$H_0 : s_0^2 = \sigma_0^2 \text{ against } H_A : s_0^2 > \sigma_0^2$$

which is one-tailed test. The test statistics used by Baarda to check the failure in the high end is as follows:

$$\frac{(df) s_0^2}{\sigma_0^2} < \chi^2_{1-\alpha, df} \tag{11.4}$$

The null hypothesis for the above test is that there are no blunders or model errors in the observations (presence of errors in the data would make the global test fail on the upper side of the confidence interval as presented in Equation (11.2)). Some literatures called the above procedure *B-method*. The following example illustrates how the B-method or Baarda's global (one-tailed) test is applied.

Example 11.2 In order to determine the coordinates of a new control point, five measurements were made from two control points whose positions are known and fixed. After the parametric least squares adjustment of the two-dimensional survey network, the computed a posteriori variance factor of unit weight is $s_0^2 = 0.213$. Perform the one-tailed (Baarda's) global test at 95% confidence level on this adjustment to find out if outliers are suspected in the measurements.

Solution:

Number of observations, $n = 5$; number of unknown (x, y) coordinates, $u = 2$; degrees of freedom, df = 5 – 2 (or 3); and $\alpha = 0.05$. Equation (11.4) will be used as follows:

$$\frac{(df)\, s_0^2}{\sigma_0^2} < \chi^2_{1-\alpha,\, df}$$

Based on the standard chi-square table, $\chi^2_{0.95,\, df = 3} = 7.815$ and $\sigma_0^2 = 1$ (usually adopted value),

$$\frac{(3)0.213}{1.0} < 7.815$$
$$0.639 < 7.815$$

Since the above condition is satisfied at 95% confidence level, the hypothesis that $s_0^2 = 0.213$ is statistically equal to $\sigma_0^2 = 1.0$ is accepted; the global test passes and there is no indication that the measurements contain outliers.

11.4 Local Tests

Local tests are performed on standardized residuals to flag for rejection of observations containing blunders. The test checks if the standardized residuals have a mean of zero and standard deviation of one; otherwise, the observation is rejected as containing outlier (or blunder). Two types of tests can be conducted on the standardized residuals: *in-context* and *out-of-context* tests.

- *In-context test* relates to many points simultaneously, rather than the conventional notion of only one point without regard to any others. It is about testing individual data value while regarding the possibility of influence from all of the other data values associated with the one that is being tested (this requires using smaller significance level than in the case of out-of-context test). The significance level (α_0) for local test based on in-context testing of each of the n number of observations can be derived from the significance level (α) for the global test as follows:

$$\alpha_0 = 1 - (1-\alpha)^{1/n} \tag{11.5}$$

or using approximate value given as

$$\alpha_0 = \frac{\alpha}{n} \tag{11.6}$$

Although Baarda's approach to local test is essentially in context, it should be mentioned here that Baarda's approach does not use Equations (11.5) and (11.6); instead, the values for α_0 are determined in a different way as discussed in Section 11.4. The in-context test assumes that all quantities can be tested simultaneously with a joint probability of $1 - \alpha$. This means that if each quantity is tested with α_0 in Equation (11.6) and all the quantities being tested are uncorrelated, the significance level for the whole quantities will be α. If the quantities are correlated, however, their combined significance level will be less than α.

- *Out-of-context test* is conducted on a piece of data without any regard for the remaining data in the set (uses larger significance level than in-context test). The significance level used in global test is usually the same as the one used in the local out-of-context test, such that $\alpha_0 = \alpha$. In performing out-of-context test, one is deliberately disregarding the existence of other members of the set of quantities being tested.

Generally, *in-context confidence regions* are not commonly computed as the out-of-context regions.

11.5 Pope's Approach to Local Test

Pope's approach to local test (1976) is known as Tau test and is a method for *in-context local testing*. It also assumes (like Baarda's approach) that only one outlier occurs at a time and that all the observations are uncorrelated, but it does not expect the global test to be performed at all. The steps of Pope's Tau test can be given as follows:

1) The variance factor $\sigma_0^2 = 1$ is assumed unknown (meaning that the covariance matrix of observations C_ℓ is assumed unreliably known), and the estimated factor s_0^2 is used to compute the covariance matrix of residuals as follows:

$$\hat{C}_v = s_0^2 \left(Q_\ell - A Q_{\hat{x}} A^T \right) \tag{11.7}$$

where A is the first design matrix, Q_ℓ is the cofactor matrix of the observations, and $Q_{\hat{x}}$ is the cofactor matrix of the adjusted parameters.

2) Compute the standardized residuals of all the observations as follows:

$$\tau_i = \left| \frac{v_i}{\sigma_{v_i}} \right| \tag{11.8}$$

where v_i is the residual of a given observation i and σ_{v_i} is the standard deviation of the residual of the observation number i, determined from the covariance matrix given in Equation (11.7). The standardized residual in Equation (11.8) is distributed as Pope's Tau distribution. In some cases, standardized residuals are computed as

$$\tau_i = \left| \frac{v_i}{\sigma_{\ell_i}} \right| \tag{11.9}$$

where σ_{ℓ_i} is the standard deviation of the given measurement. It should be noted that Equations (11.8) and (11.9) are not the same since $\sigma_{v_i} < \sigma_{\ell_i}$ so that the values determined from Equation (11.9) are usually smaller than those from Equation (11.8) and there will be less failures of the test using Equation (11.9). In this case, Equation (11.9) may not detect all the outliers.

3) Perform the local test as follows: Observation ℓ_i is a potential outlier if $\tau_i > \tau_{\alpha_0, n, n-u}$, where α_0 is the significance level for in-context testing given in Equation (11.5) or Equation (11.6) and $\tau_{\alpha_0, n, n-u}$ is Tau distribution critical value with n number of observations and $n - u$ degrees of freedom. According to Bonferroni's inequality (Vanicek and Krakiwsky 1986), if the observations (and the corresponding residuals) are uncorrelated, the significance level for the whole vector of observations (or residuals) is α, but if they are correlated, the significance level will be less than α. By considering Bonferroni's inequality, it can be stated that by performing all the individual tests of the residuals at α_0 significance level, the confidence level for the whole vector of residuals (or the whole vector of corresponding observations) will be at least $1 - \alpha$. Pope (1976) provided a subroutine or a table that provides the critical values for the desired probability of type I error (α) with Equation (11.5) or Equation (11.6) considered. In this case, when the value of α is provided with the number of observations (n) and the degrees of freedom $n - u$, the critical value $(\tau_{\alpha, n, n-u})$ obtained is the same as $\tau_{\alpha_0, n, n-u}$ since α_0 is already internally computed before the critical value is determined (Pope 1976). The Tau critical values provided in the table in Appendix C are based on Pope's subroutine; the table should be used with α instead of α_0, where from Equation (11.6), $\alpha = n \times \alpha_0$.

In summary, the following statements are consistent in both Pope's and Baarda's approaches:

- Significance level (α_0) of the local test is not the same as that of the global test, which is α.
- Global test requires a larger significance level than the local test.
- If global test is performed with α, local test must be performed with $\alpha_0 = \alpha/n$ according to Pope for in-context testing, and if $\alpha_0 = \alpha$ is used, the testing is considered out of context.

- Baarda's approach for determining α_0 is different; the approach is discussed in Section 11.7.1.
- It is assumed in both methods that only one observation is in error, so the largest outlier test value should be eliminated and new adjustment performed on the remaining observations; the process is done iteratively until no outlier is detected any more. This iterative testing procedure is considered (Stefanovic 1980) as providing the best results.
- Since Tau test does not consider the type II error, the reliability statements on the sensitivity of the detection cannot be made as in Baarda's approach. Pope's approach may be less sensitive since the a posteriori variance factor used in scaling the cofactor of the residuals is also affected by the possible presence of outliers.

Usually, the commercial software may use $\alpha = 0.05$ for global test and $\alpha_0 = 0.01$ (for smaller redundancy) and $\alpha_0 = 0.001$ (for larger redundancy) in local test. The following example illustrates Pope's approach to local test.

Example 11.3   Perform Pope's Tau test at $\alpha_0 = 0.7\%$ on the following residual vector computed in a least squares adjustment using seven independent measurements (with the degrees of freedom as 4): $v^T = [0.50, -0.15, 0.00, 0.05, -0.20, 0.10, -0.40]$ m. The standard deviation of each measurement is 0.25 m, and the standard deviation of each residual is $\sigma_{v_i} = 0.15$ m.

Solution:

The standardized residual according to Pope is given in Equation (11.8) as

$$\tau_i = \left| \frac{v_i}{\sigma_{v_i}} \right|$$

For the whole measurements, $\tau_i = [\ 3.33, 1.00, 0.00, 0.33, 1.33, 0.67, 2.67]$.

$\alpha_0 = 0.7\%$ for $n = 7$ observations; $\alpha = n \times \alpha_0$ or $\alpha = 7 \times 0.007$ or 0.05.

The desired probability of type I error is $\alpha = 0.05$ for this problem; from the Tau table (in Appendix C), $\tau_{0.05,\ 7,\ 4} = 1.933$; by checking if $\tau_i > \tau_{0.05,\ 7,\ 4}$ is satisfied for each of the standardized residuals, it can be seen that the first and the last observations $(3.33 > 1.933)$ and $(2.67 > 1.933)$ qualify as outliers, respectively.

11.6   Concepts of Redundancy Numbers

Baarda's approach to data analysis involves understanding the concepts of gross errors and how they distort mathematical models in adjustment. Gross errors in observations will result in distorted computed corrections to parameters (δ^*)

and distorted residuals ($v*$), which can be expressed as follows. From Equation (5.114), the cofactor matrix for distorted residuals can be given as

$$Q_{v*} = P^{-1} - AN^{-1}A^T \tag{11.10}$$

or

$$Q_{v*}P = I - AN^{-1}A^T P \tag{11.11}$$

where I is the identity matrix with all the diagonal elements being equal to one and the off-diagonal elements equal to zero, A is the first design matrix, $N = A^TPA$ is the matrix of the coefficients of the normal equations (in the parametric least squares adjustment), Q_{v*} is the cofactor matrix for the distorted residuals, and P is the weight matrix of measurements. If Equation (11.11) is post-multiplied by the distorted misclosure vector $w*$, the following equation is obtained:

$$Q_{v*}Pw^* = w^* - AN^{-1}A^T Pw^* \tag{11.12}$$

Remember that the parametric least squares adjustment solution of the distorted corrections (δ^*) to the approximate parameters is usually given as

$$\delta^* = -N^{-1}A^T Pw^* \tag{11.13}$$

If Equation (11.13) is substituted into Equation (11.12), the following is obtained:

$$Q_{v*}Pw^* = w^* + A\delta^* \tag{11.14}$$

Remember also that the usual model for residual (from parametric adjustment) is given as

$$v^* = w^* + A\delta^* \tag{11.15}$$

It can then be deduced from Equations (11.14) and (11.15) that

$$v^* = Q_{v*}Pw^* \tag{11.16}$$

The diagonal elements of the product ($Q_{v*}P$) in Equation (11.16) are known as the redundancy numbers (r_i) of the corresponding observations (Förstner 1979). The redundancy number expresses the contribution of each single observation to the total redundancy $r = n - u$ with n as the number of observations and u as the number of unknown parameters; r can also be given as the trace ($Q_{v*}P$) = $n - u$, where trace() simply means the sum of the diagonal elements of the matrix in the bracket. The redundancy number r_i for the ith observation can be given as $r_i = (Q_{v*}P)_{ii}$.

Each observation in a network has a redundancy number. Practically, the redundancy number of an observation should fall within the range $0 \le r_i \le 1$; the redundancy number that is closer to one is better, meaning that the observation is strongly checked by the other observations and any error in that observation will probably be detected. For example, it will be very difficult or impossible to detect gross errors in observations with redundancy numbers that are close to zero, while it will be easy in the case of observations with redundancy numbers that are close to one. The average redundancy number expected for geometrically strong and well-measured control networks (which should be equal to or >0.5) can be expressed as

$$\bar{r} = \frac{1}{n}\sum_{i=1}^{n} r_i \approx 0.5 \tag{11.17}$$

The following example illustrates how redundancy numbers are determined.

Example 11.4   Given the following cofactor matrix of residuals Q_v and the weight matrix of observations P, compute the redundancy numbers of each measurement in the survey network and the average redundancy number for the network:

$$Q_v = \begin{bmatrix} 17.6 & -6.6 & 9.9 & 3.3 & 3.3 \\ -6.6 & 13.2 & 6.6 & -6.6 & -6.6 \\ 9.9 & 6.6 & 17.6 & -3.3 & -3.3 \\ 3.3 & -6.6 & -3.3 & 17.6 & -9.9 \\ 3.3 & -6.6 & -3.3 & -9.9 & 17.6 \end{bmatrix} \times \text{E}-4 \quad P = \begin{bmatrix} 400 & & & & \\ & 400 & & & \\ & & 400 & & \\ & & & 400 & \\ & & & & 400 \end{bmatrix}$$

Solution:

The redundancy numbers of the observations can be determined from

$$R = Q_v P = \begin{bmatrix} 17.6 & -6.6 & 9.9 & 3.3 & 3.3 \\ -6.6 & 13.2 & 6.6 & -6.6 & -6.6 \\ 9.9 & 6.6 & 17.6 & -3.3 & -3.3 \\ 3.3 & -6.6 & -3.3 & 17.6 & -9.9 \\ 3.3 & -6.6 & -3.3 & -9.9 & 17.6 \end{bmatrix} \times \text{E}-4 \times \begin{bmatrix} 400 & & & & \\ & 400 & & & \\ & & 400 & & \\ & & & 400 & \\ & & & & 400 \end{bmatrix}$$

$$R = \begin{bmatrix} 0.704 & -0.264 & 0.396 & 0.132 & 0.132 \\ -0.264 & 0.528 & 0.264 & -0.264 & -0.264 \\ 0.396 & 0.264 & 0.704 & -0.132 & -0.132 \\ 0.132 & -0.264 & -0.132 & 0.704 & -0.396 \\ 0.132 & -0.264 & -0.132 & -0.396 & 0.704 \end{bmatrix}$$

The redundancy numbers of the corresponding observations are the diagonal elements $r_i = 0.704, 0.528, 0.704, 0.704, 0.704$. The average redundancy number for the network can be given as

$$\bar{r} = \frac{1}{n}\sum_{i=1}^{n} r_i \quad \text{(where the diagonal elements are the } r_i)$$

$$\bar{r} = \frac{1}{5}(3.344) = 0.67$$

The weakest measurement has the redundancy number of 0.528 with the remaining measurements having higher values of 0.704. The higher the value of redundancy number, the more reliable the measurement becomes. Usually, the observation between two fixed points will have redundancy number of 1, meaning that all the errors can be recovered.

11.7 Baarda's Data Analysis Approach

According to the least squares procedure, it is possible to predict how a specific systematic error will propagate into adjusted measurements. If the systematic error $\nabla \ell_i$ is present in a particular measurement ℓ_i with the other measurements in the vector of measurements ℓ free of any systematic error, a systematic error vector $\nabla \ell$ can be formed containing zero elements for other measurements with systematic error $\nabla \ell_i$ for measurement ℓ_i. The systematic error in all the residuals (∇v) of all the measurements due to the single systematic error in measurement ℓ_i can be predicted from Equation (5.109) as

$$\nabla v = \left(A^T\left(A^T PA\right)^{-1}A^T P - I\right)\nabla \ell \tag{11.18}$$

where A is the first design matrix, P is the weight matrix of measurements, and I is the identity matrix. According to Baarda (1968), a gross error or a systematic error of an observation must reach a certain value (a multiple of the standard deviation of the corresponding observation) in order to be detected with a probability of $1 - \beta_0$, where β_0 is a specified value of type II error. Baarda's noncentrality parameter (λ) is related to the systematic errors in all the residuals as follows (Baarda 1968; Cooper 1987):

$$\lambda = \frac{\nabla v^T P \nabla v}{\sigma_0^2} \tag{11.19}$$

where σ_0^2 is the a priori variance factor of unit weight. The noncentrality parameter (λ) is shift of the probability distribution of null hypothesis (H_0) from that of the alternative (H_A) hypothesis due to a gross error in observations. This can be considered as the offset of the expected value of the test statistic that the sample value should attain in order that it may exceed the critical value at the given α_0 type I error and β_0 type II error. The boundary value of noncentrality parameter (λ_0) is provided in Baarda's nomogram so that the hypothesis that there is no non-stochastic error in the measurement is rejected if the following condition is satisfied:

$$\frac{\nabla v^T P \nabla v}{\sigma_0^2} > \lambda_0 \tag{11.20}$$

Equation (11.20) can be inverted to determine the smallest systematic error in any measurement ℓ_i that can just be detected (or the maximum error that will remain undetected) at a given α and β probability levels, assuming there is only one measurement that is contaminated with systematic error. Note that the noncentrality parameters in Equations (11.19) and (11.20) relate to multidimensional cases, in which Fisher's F-statistic and chi-square statistic are involved. In one-dimensional case, involving normal distribution function, the upper limit of the noncentrality parameter (λ_0) is the value that will lead to the acceptance of the null hypothesis at the α and β probability levels. This can be expressed as

$$\sqrt{\lambda_0} = z_{1-\alpha_0/2} + z_{1-\beta_0} \tag{11.21}$$

where z is the standard score from the standard normal distribution curve. For example, at $\alpha_0 = 5\%$ and $\beta_0 = 20\%$, using the Microsoft Excel *NORM.S.INV* function will give $\sqrt{\lambda_0} = 2.802$ based on Equation (11.21). This is considered as the number of standard deviations by which the error is off from the expected true value, which is zero. The noncentrality parameter is the degree to which a sampling distribution (Student's t, F, chi-square, etc.) is stretched to the right; the larger it is, the higher the departure from the null hypothesis.

The idea of global test is from Baarda, who believes that the a priori variance factor of unit weight ($\sigma_0^2 = 1$) is well known and should be tested against the computed a posteriori variance factor of unit weight (s_0^2) using Equation (11.4) given above. According to Baarda (1968), the significance level (α) used in global test must be consistent with the one (α_0) to be used in local test. In this case, global test will indicate correctly at probability α and β_0 if blunders actually exist provided the following equation is satisfied:

$$w_i = \left| \frac{v_i}{\sigma_{\ell_i} \sqrt{r_i}} \right| \geq \sqrt{\lambda_0} \tag{11.22}$$

where σ_{ℓ_i} is the standard deviation of the observation ℓ_i, λ_0 is the noncentrality parameter, v_i is the residual of the observation, r_i is the redundancy number of the observation, $z_{1-\alpha_0/2}$ and $z_{1-\beta_0}$ are the standard normal distribution values at $1 - \alpha_0/2$ and $1 - \beta_0$ probability levels, respectively, $\alpha_0/2$ is the two-sided probability of rejecting good data, and β_0 is the one-sided probability of accepting bad data. Usually, Baarda provided a formula for solving for λ_0 directly (refer to Table A.1 for some of the extracted values) as

$$\lambda_0 = \lambda(\alpha, \beta_0, \mathrm{df} = n - u) = \lambda(\alpha_0, \beta_0, \mathrm{df} = 1) \tag{11.23}$$

where df = $n - u$ is the number of degrees of freedom in the adjustment, n is the number of model equations, and u is the number of unknown parameters. The term on the right-hand side of Equation (11.23) requires choosing α_0 and β_0 quantities first, using them and the degrees of freedom of one to compute the noncentrality parameter (λ_0) for univariate case and then using the computed λ_0 to compute the significance level α for the multidimensional global test. This relationship can also be used to obtain α_0 to be used in the local test if α, β_0, and df are known. Since the explicit form of Equation (11.23) is complex, Baarda's nomogram is commonly used. The nomogram provides λ_0 and α_0 directly, given the β_0, α (used in the global test), and the degrees of freedom df. The sample values from Baarda's nomogram are given in Tables A.1–A.4.

11.7.1 Baarda's Approach to Local Test

Baarda's approach to local test (1968) is for *in-context local testing* in which local significance level (α_0) that corresponds to the one (α) used in global test is determined. The relationship between α_0 and α is determined in terms of the probability of both type I and II errors through the assumptions built into Baarda's reliability theory. According to Baarda (1968), the risk level α of the r-dimensional (with r representing the redundancy) global test must relate to one-dimensional test with type I error α_0 and type II error β_0 in such a way that a gross error Δ_{0i} causes the outlier test to fail with a probability of $1 - \beta_0$. Baarda's approach assumes that the covariance matrix of observations (C_ℓ) is well known or is properly scaled (i.e. variance factor $\sigma_0^2 = 1$ is well known), so that the covariance matrix of residuals (C_v) can be computed from the following:

$$C_v = C_\ell - A C_{\hat{x}} A^T \tag{11.24}$$

where A is the first design matrix and $C_{\hat{x}}$ is the covariance matrix of the adjusted parameters. Note that when σ_0^2 is assumed known, it means that the variances of measurements have been properly scaled so that there is no need of additional scaling using the a posteriori variance factor (s_0^2). The generalized test statistic utilized for testing residuals, or groups of residuals, is given by Baarda as w-statistic as follows:

$$w_i = \left| \frac{v_i}{\sigma_{\ell_i} \sqrt{r_i}} \right| < z_{1 - \alpha_0/2} \tag{11.25}$$

where σ_{ℓ_i} is the standard deviation of the observation ℓ_i, v_i is the residual of the observation, r_i is the redundancy number of the observation, and $z_{1-\alpha_0/2}$ is the standard normal distribution table value. Equation (11.25) is used when the covariance matrix of the observations is fully populated. For diagonal weight matrix (P) of observations (or diagonal covariance matrix of measurements C_ℓ with no correlation between the measurements), w_i-statistic in Equation (11.25) can be simplified to

$$w_i = \left| \frac{v_i}{\sigma_{v_i}} \right| < z_{1-\alpha_0/2} \tag{11.26}$$

where σ_{v_i} is the standard deviation of the residual for observation ℓ_i. Equation (11.26) is the same as that given in Pope's approach in Equation (11.8) except that the standard deviation of residual (σ_{v_i}) is not pre-scaled with the a posteriori variance factor of unit weight (s_0^2). Baarda's one-dimensional testing procedure expressed by Equations (11.25) and (11.26) is called (Baarda 1968) *data snooping* or *w-test*, which assumes that only one outlier occurs at a time; Equation (11.25) assumes all observations are correlated, while Equation (11.26) assumes they are uncorrelated. In general, Baarda (1968) suggested the use of global test (Equation (11.4)) for the detection of outliers and the data snooping test (Equations (11.25) and (11.26)) for the localization of outliers. The essential element of Baarda's approach is that the *B-method* of testing (or the χ^2-test or global test) and the *w-test* (local test) detect an outlier with the same probability with the decisions from the two tests being consistent. The following example illustrates Baarda's approach to local test.

Example 11.5 The residual vector computed in a least squares adjustment using seven independent measurements is given as $v^T = [0.50, -0.15, 0.00, 0.05, -0.20, 0.10, -0.40]$ m with the degrees of freedom as 4. If the standard deviation of each measurement is 0.25 m and the standard deviation of each residual is $\sigma_{v_i} = 0.15$ m, perform the B-method blunder detection test on the adjustment at $\alpha = 5\%$ and $\beta_0 = 20\%$.

Solution:

$v^T P v = 48.5$, df $= 4$, $s_0^2 = 12.125$, and $\sigma_0^2 = 1$ is assumed known.
   Global test:

$$\frac{(\text{df}) \, s_0^2}{\sigma_0^2} = \frac{v^T P v}{\sigma_0^2} = 48.5 \; \chi^2_{0.95, \text{df} = 4} = 9.49$$

Since $\dfrac{(\text{df}) \, s_0^2}{\sigma_0^2} < \chi^2_{0.95, \text{df} = 4}$ is not satisfied, the null hypothesis is not accepted, and the global test fails. The adjustment will be subjected to further data snooping as follows. Since the covariance matrix of the observations is diagonal (no correlation between measurements), the w-statistic can be given as

$$w_i = \left| \frac{v_i}{\sigma_{v_i}} \right| \tag{11.27}$$

For the whole measurements, $w^T = [3.33, 1.00, 0.00, 0.33, 1.33, 0.67, 2.67]$. Referring to Table A.3 for Baarda's nomogram value for α_0 (corresponding to $\alpha = 0.05$ used in the global test) to be used with type II error $\beta_0 = 0.20$, the extracted value at df $= 4$ is $100\alpha_0 = 0.9$ or $\alpha_0 = 0.009$. The standard normal distribution value is $z_{1 - 0.009/2} = 2.61$; by comparing w_i with 2.61 according to Equation (11.25), it can be seen that the first and the last observations failed the local tests (i.e. $3.33 < 2.61$ and $2.67 < 2.61$ are not satisfied). Usually the observation having the larger value (e.g. 3.33) is removed, and the adjustment is carried out again with the global and local test carried out again.

11.8 Concepts of Reliability Measures

Reliability is a measure of the closeness of some quantities to their true values or the ability a model or a network to facilitate gross error and systematic error detection and localization. It should be remembered that the concept of global tests (Section 11.3) is to detect outlier and that of local tests (Section 11.7.1) is to identify or locate the outlier. The concept of reliability is related to detection and rejection of outliers. Observations or surveying method with high reliability can easily detect gross and systematic errors. This is the hypothesis behind the concepts of B-method and data snooping for *blunder detection* in observations (Baarda 1968). Reliability of a survey network can be divided into *internal* and *external* reliability.

11.8.1 Internal Reliability Measures

Internal reliability of a network or a model is the ability of the network or model to allow the detection of systematic error or a blunder that is present in one of the measurements by tests of hypotheses made with specific confidence level ($1 - \alpha$) and power ($1 - \beta$). It is a means of analyzing the sensitivity of a network to the presence of an outlier with the chances of being able to detect and locate it; it is a measure of how self-checking the model or the network is. Internal reliability is usually quantified by what is usually referred to as *marginally detectable error* (MDE) or *marginally detectable bias* (MDB). The MDE is the magnitude of the error that can just be marginally detected as an outlier, and hence any error smaller will probably remain undetected. The concept of MDE gives an idea about the minimal magnitude of a gross error in a particular observation, but does not cover how to identify the error. MDE is then the smallest outlier that can be detected by the method of data snooping at a particular significance level α and at the power of test of $1 - \beta$. Remember that the significance level, α, is the probability of rejecting good data as an outlier or the probability of a good observation failing the outlier test, the power of test $(1 - \beta)$ is the probability of correctly detecting outlier if it occurs, and the probability of type II error, β, is the chance of incorrectly accepting bad data (i.e. the error in accepting an outlier as a good observation or the error of a true outlier passing the outlier test). For example, if the outlier test flags residual greater than 3σ as outlier, then this is the MDE and anything smaller than it will not be detected. If the power of the outlier detection test is $1 - \beta$, then the MDE is the smallest size of an outlier that is possible to detect at $1 - \beta$ probability level; this means there is a β chance that the outlier will be undetected. For uncorrelated observations, the MDE can be given (Baarda 1968) as

$$\nabla_0 \ell_i = \sigma_{\ell_i} \sqrt{\frac{\lambda_0}{r_i}} \tag{11.28}$$

where σ_{ℓ_i} is the standard deviation of the ith observation, r_i is the redundancy number, and λ_0 is the noncentrality parameter, which can be defined as a measure of the number of standard deviations between the mean (μ_1) of the population of good data and the mean (μ_2) of the nearest possible population of outlying data (i.e. the data containing the bad data). The values $\nabla_0 \ell_i$ for all observations may be calculated during the design stage of a network to give insight into the sensitivity of the network to the presence of non-stochastic errors. The residual bias value of the residual v_i or the amount of blunder that can be detected marginally $(\nabla_0 v_i)$ corresponding to the marginal error expressed in Equation (11.28) is given as follows based on the concept expressed by Equation (11.16):

$$\nabla_0 v_i = \nabla_0 \ell_i (r_i) \tag{11.29}$$

where r_i is the redundancy number for the observation. Note that MDE is often used to mean the same thing as *minimal detectable bias* (MDB).

Internal reliability measure can also be given as the ratio of the standard deviation of an adjusted measurement $(\sigma_{\hat{\ell}_i})$ to the standard deviation of the measurement (σ_{ℓ_i}) itself, expressed mathematically as follows (Ashkenazi 1980):

$$\rho_i = \frac{\sigma_{\hat{\ell}_i}}{\sigma_{\ell_i}} \tag{11.30}$$

The internal reliability measure expressed by Equation (11.30) represents the way in which networks increase precisions of measured quantities. For an example, a completely internally reliable measurement will have $\sigma_{\hat{\ell}_i} = 0$ so that the measure $\rho_i = 0$; for a completely internally unreliable measurement, the measure $\rho_i = 1$. This means that an internally reliable network results in high precision of the adjusted measurements compared with the precision of the original measurements themselves. If the standard deviations $(\sigma_{\hat{\ell}_i})$ of the adjusted measurements are the same as the standard deviations (σ_{ℓ_i}) of the corresponding measurements, it will mean that the smallest detectable error is infinitely large and the network is completely unreliable internally. This will be the case if there are no redundant measurements to check each other. A more internally reliable network can detect smaller error. In other words, a highly internally reliable network is one in which the MDEs are very small, and the overall measure of internal reliability may be taken as the largest of the MDEs.

Another form of internal reliability measure can also be expressed in terms of the local redundancy number (r_i), given as

$$r_i = (Q_v P)_{ii} \tag{11.31}$$

or

$$r_i = \frac{\sigma_{v_i}^2}{\sigma_{\ell_i}^2} \tag{11.32}$$

where Q_v is the covariance matrix of residuals, P is the weight matrix of observations, and $\sigma_{v_i}^2$ and $\sigma_{\ell_i}^2$ are the variances of the residuals and the corresponding measurements, respectively. The local redundancy number (r_i) represents the contribution of an observation to the total redundancy; the contribution, however, should be within zero and one $(0 \leq r_i \leq 1)$. Remember from Equation (11.16) that by multiplying a misclosure by a local redundancy number, the magnitude of the observation error is estimated; this means that the larger the r_i the better, since it will allow the largest residual to be considered as outlier.

Another form of internal reliability measure is the *global internal reliability measure*, which can be given as

$$\bar{r} = \frac{1}{n} \sum_{i=1}^{n} r_i \approx 0.5 \tag{11.33}$$

Equation (11.33) is a reasonable global measure of internal reliability if the r_i values are distributed evenly in the network. Global measure of internal reliability is about the probability of outlier detection, while the local measure of internal reliability is about the probability of locating the erroneous observations. For example, the greater the value of $\nabla_0 v_i$ in Equation (11.29) due to a single measurement error $\nabla_0 \ell_i$, the greater is the probability of localizing the erroneous observation ℓ_i. The internal reliability measures expressed by Equations (11.31) and (11.33) are means of checking how well observations check geometrically with each other. If the redundancy numbers of individual observations approach their maximum value of 1, the sizes of the undetected blunders in the network will be minimal. The measures also help to locate weak areas in a network, which can be identified as the regions where redundancy numbers are smaller compared with the global internal reliability measure.

11.8.2 External Reliability Measures

External reliability of a network is related to the maximum influence of undetected errors (outliers or MDE) on the adjusted parameters and also on the functions of such parameters; it measures the response of a survey network to undetected systematic and gross errors. In other words, external reliability is the effect of potential failure of blunder detection procedure (data snooping) on estimated parameters (such as coordinates). A measure of external reliability of an adjustment is obtained by calculating the effect of an undetected outlier on the estimated parameters, assuming that the magnitude of the outlier is equal to the MDE of the corresponding observation. For example, the effect of an error in the ith measurement on each of the parameters is called *local external reliability*. For each observation, there will be a vector of local external reliability, showing the impact of each marginally detectable blunder on the parameters. External reliability is of more practical value than internal reliability, but some indication of the largest undetected error for a given α and β probability levels is necessary in considering external reliability.

External reliability measure results through the propagation of the values for the individual minimally detectable errors $\nabla_0 \ell_i$ (or MDEs) through the least squares process. For each observation ℓ_i, an influence vector $\nabla_0 x_i$ is computed as follows:

$$\nabla_0 x_i = \left(A^T C_\ell^{-1} A \right)^{-1} A^T C_\ell^{-1} e \nabla_0 \ell_i \tag{11.34}$$

where $\nabla_0 x_i$ is a vector with each element corresponding to the shift in an adjusted coordinate value and e is a vector having its elements as zeroes for observations that are not containing any error and one for the only observation containing an error. From Equation (11.34), it can be seen that the value of the influence vector will be minimum if the sizes of redundancy numbers are increased. In practice, the parameters associated with the ℓ_i observation equation are expected to be more biased than others that are not directly connected with the observation equation. In this case, the effect of one error can be considered as local, and $\nabla_0 x_i$ can be considered as the effect of a gross error in one observation on the corresponding parameter. The often used *global measure of external reliability* can be given as

$$\delta_{0i}^2 = \left(\frac{1 - r_i}{r_i} \right) \lambda_0 \tag{11.35}$$

where r_i is the *redundancy number* and $1 - r_i$ is the *absorption number*. While the redundancy number r_i is an indication of how much the gross errors are likely to affect the estimated quantities, the absorption number indicates how much the solution will absorb the errors. With high redundancy numbers (values between 0.5 and 1.0), any undetected gross errors in the observations are most likely going to show up in the residuals, rather than being absorbed by the solution. Further discussions on the detection of outliers and the determination of reliability in geodetic networks can be found in Kavouras (1982).

11.9   Network Sensitivity

Network sensitivity is the ability of a network to allow specific small deformation of the network to be detected. According to Even-Tzur (2006), sensitivity of a geodetic monitoring network is the ability of the network to detect and measure movements and deformation in the area covered by the network. A sensitive network will not tolerate even a small observation error because its effect will be amplified on the parameter estimates. Large errors, however, may be tolerated if they have only a marginal effect on the network. Network sensitivity, however, is relevant only when the network is to be used to monitor deformation of a point or group of points. At the design stage of such a network, the point or the group of points may be assigned some specific predicted displacements based on some previous knowledge of the deformation activities in the region, which the network is expected to be able to detect if such amount of displacements occur. If such a network is able to detect very small displacements, the network is said to be very sensitive.

Example 11.6   A horizontal control network of five points has been measured by classical methods of triangulation and trilateration. The measured quantities consisting of 8 directions, 1 angle, and 10 distances are given in Table 11.1 and Figure 11.2. The fixed and approximate coordinates of the network points are given in Table 11.2. Assume the position of point A and the departure (X) of point B to be fixed (errorless) for minimal constraint least squares adjustment.

Perform the parametric least squares adjustment and complete the following:

a) Conduct a global test (at 95% confidence level) by standard method to decide if there exists outlier or not.
b) Conduct outlier detection tests by B-method to locate the outlier, and then compute the outlier-free variance factor of unit weight and the outlier-free coordinates of the unknown points (at $\alpha_0 = 0.1\%$, $\beta_0 = 0.20$).
c) Calculate the internal and external measures of reliability for the final adjustment of the network (at $\alpha_0 = 0.1\%$, $\beta_0 = 0.20$).

Table 11.1 Field measurements.

i	At	From	To	Observation (l_i)	Standard dev. (σ)
1		A	B	0°00′00″	3″
2		A	E	20°33′27.6″	3″
3		A	C	34°18′41.7″	3″
4		A	D	64°44′30.0″	3″
5		E	B	0°00′00″	3″
6		E	C	82°20′50.1″	3″
7		E	D	130°40′50.2″	3″
8		E	A	233°46′42.7″	3″
9	C	D	B	122°38′24.3″	5″
10		A	B	4509.711 m	0.009 m
11		A	E	3062.624 m	0.007 m
12		A	C	5728.087 m	0.011 m
13		A	D	5518.873 m	0.011 m
14		B	C	3236.519 m	0.007 m
15		B	D	5436.573 m	0.011 m
16		B	E	1962.939 m	0.004 m
17		C	D	2958.628 m	0.006 m
18		C	E	2847.985 m	0.006 m
19		D	E	3949.296 m	0.008 m

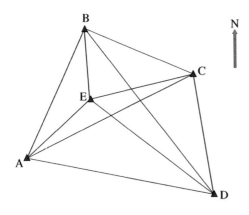

Figure 11.2 Control network (not to scale).

Table 11.2 Fixed and approximate coordinates of control stations.

Station	X (m)	Y (m)
A	**401 176.143**	**5 501 862.458**
B	**403 602.951**	5 505 663.515
C	406 443.716	5 504 112.681
D	406 650.278	5 501 161.279
E	403 625.708	5 503 700.754

Bolded coordinates are fixed.

Solution:

a) Global test (at 95% confidence level) by standard method (not by Baarda's approach):

$$s_0^2 = \frac{v^T P v}{10} = 2.6176 \text{ degrees of freedom}, \text{ df} = 10$$

The degrees of freedom (df = 10) is obtained as follows:

i) Total number of observations is 19 (8 directions plus 1 angle plus 10 distances).

ii) Out of the total 10 coordinates for the 2-D network (5 points), 3 coordinates (point A and X-departure of point B) are fixed, remaining 7 unknown coordinates to be determined.

iii) From Table 11.2, directions were measured from points A and E, creating additional two unknown orientation parameters.

iv) The total number of unknown parameters is the sum of seven unknown coordinates (from step (ii)) and two orientation unknowns (from step (iii)), giving nine unknown parameters.

v) The number of degrees of freedom is the number of observations (19) minus the number of unknown parameters (9), giving 10 degrees of freedom.

Using the significance level $\alpha = 0.05$ and the degrees of freedom (df = 10) in the Microsoft Excel 2013 program CHISQ.INV $(1 - \alpha, \text{df})$, the following are obtained:

$$\chi^2_{0.025, \text{df} = 10} = 3.25 \quad \chi^2_{0.975, \text{df} = 10} = 20.50$$

$$\chi^2 = \frac{(\text{df}) s_0^2}{\sigma_0^2} \rightarrow \sigma_0^2 = 1.0$$

$$\chi^2 = 26.176$$

$$\chi^2_{0.025, \text{df} = 10} < \chi^2 < \chi^2_{0.975, \text{df} = 10}? \quad 3.25 \le 26.18 \le 20.50$$

The H_0 is not accepted (global test fails) at 95% confidence level. Outlier(s) are suspected among the measurements.

b) Outlier detection by B-method:

i) Global test by B-method:

$$s_0^2 = \frac{v^T P v}{10} = 2.6176 \quad df = 10 \quad a_0 = 0.1\% \quad \beta_0 = 0.20$$

where $\alpha = 0.043$ or 4.3% (from nomogram extract in Table A.2) with $\lambda_0 = 17.0$. Extract the chi-square value from the Microsoft Excel 2013 program CHISQ.INV $(1 - \alpha, \ df = 10)$ to obtain the following: $\chi^2_{0.957, df = 10} = 18.791$

$$\chi^2 = \frac{(df) \, s_0^2}{\sigma_0^2} \quad (\text{for } \sigma_0^2 = 1.0) \quad \chi^2 = 26.176$$

$$\chi^2 < \chi^2_{0.957, df = 10}? \quad \text{or} \quad 26.176 < 18.791?$$

The H_0 is not accepted (global test fails) at 95.7% confidence level. Outlier(s) are suspected among the measurements.

ii) Local tests by data snooping:

For $\alpha_0 = 0.001$ and $\beta_0 = 0.20$, $\sqrt{\lambda_0} = 4.123$ for $\lambda_0 = 17.0$, $z_{1-\alpha_0/2} = 3.291$.

Table 11.3 gives the summary of the computations with the asterisked item corresponding with the measurement that failed the local test.

In Table 11.3, v_i is the residual, σ_i is the standard deviation of a measurement, r_i is the redundancy number of the measurement, and w_i is the test statistic for Baarda's approach; the measurement is accepted if $w_i \leq z_{0.9995} = 3.291$.

After removing the blunder and readjusting the network without the blunder, $\alpha_0 = 0.1\%$, $\beta_0 = 0.20$, $\alpha = 0.040$ or 4.0% (from Nomogram extract in Table A.2), and the global test is done again as follows:

$$s_0^2 = \frac{v^T P v}{9} = 1.135 \quad \chi^2_{0.96, df = 9} = 17.608 \, (\text{for } df = 9)$$

$$\chi^2 = \frac{(df) \, s_0^2}{\sigma_0^2} \quad (\text{for } \sigma_0^2 = 1.0) \quad \chi^2 = 10.215$$

$$\chi^2 < \chi^2_{0.96, df = 9}? \quad 10.215 < 17.608?$$

Table 11.3 Summary of computations.

i	v_i	σ_i (rad and m)	r_i	$w_i = \left\| \dfrac{v_i}{\sigma_i \sqrt{r_i}} \right\|$
1	−0.455 96E−5	1.454 44E−5	0.7446	0.363 49
2	0.112 03E−4	1.454 44E−5	0.7345	0.899 24
3	0.368 92E−5	1.454 44E−5	0.7460	0.293 81
4	0.120 74E−4	1.454 44E−5	0.7416	0.964 50
5	0.619 17E−5	1.454 44E−5	0.6991	0.509 41
6	−0.367 16E−5	1.454 44E−5	0.7275	0.296 11
7	0.575 32E−5	1.454 44E−5	0.7300	0.463 30
8	−0.827 32E−5	1.454 44E−5	0.6900	0.685 23
9	−0.954 50E−4	1.454 44E−5	0.9725	3.994 94*
10	0.110 04E−1	0.009 m	0.3312	2.124 49
11	−0.119 12E−1	0.007 m	0.3984	2.696 34
12	0.858 69E−2	0.011 m	0.6135	0.996 63
13	0.955 12E−2	0.011 m	0.2366	1.785 22
14	0.769 15E−3	0.007 m	0.2130	0.238 10
15	0.103 41E−1	0.011 m	0.6616	1.155 78
16	−0.242 34E−2	0.004 m	0.1050	1.870 12
17	−0.172 86E−2	0.006 m	0.0866	0.979 09
18	−0.587 94E−2	0.006 m	0.2268	2.057 41
19	−0.520 44E−2	0.008 m	0.3421	1.112 21

* Measurement that failed the local test − for immediate identification of failed item.

The H_0 is not rejected (global test passes) at 96% confidence level. No more outlier(s) are suspected among the measurements. Repeat local tests by w-test for $\alpha_0 = 0.001$ and $\beta_0 = 0.20$, $\sqrt{\lambda_0} = 4.123$ (or $\lambda_0 = 17.0$), and $z_{1 - \alpha_0/2} = 3.291$.

The summary of the local test computations after removing the blunder is given in Table 11.4, while the resulting adjusted coordinates are given in Table 11.5. It can be seen from Table 11.4 that all the measurements satisfy the condition that $w_i \leq z_{0.9995} = 3.291$. Since all the measurements satisfy this acceptance criterion, there is no statistical reason to believe that there is still an outlier in the measurements at 95% confidence.

Table 11.4 Summary of local test computations with blunder removed.

i	v_i	σ_{ℓ_i} (rad and m)	r_i	$w_i = \left\|\dfrac{v_i}{\sigma_{\ell_i}\sqrt{r_i}}\right\|$
1	−0.464 18E−5	1.454 44E−5	0.7446	0.370 04
2	−0.109 20E−4	1.454 44E−5	0.7345	0.876 50
3	0.348 37E−5	1.454 44E−5	0.7460	0.277 45
4	0.120 78E−4	1.454 44E−5	0.7416	0.964 83
5	0.552 57E−5	1.454 44E−5	0.6989	0.454 66
6	−0.421 78E−5	1.454 44E−5	0.7274	0.340 18
7	0.638 71E−5	1.454 44E−5	0.7296	0.514 39
8	−0.769 50E−5	1.454 44E−5	0.6897	0.637 38
10	0.961 22E−4	0.009 m	0.3297	1.860 00
11	−0.115 40E−1	0.007 m	0.3982	2.612 55
12	0.112 18E−1	0.011 m	0.6099	1.305 81
13	0.741 73E−2	0.011 m	0.2342	1.393 33
14	0.193 78E−2	0.007 m	0.2112	0.602 34
15	0.873 59E−3	0.011 m	0.6603	0.977 35
16	−0.290 48E−2	0.004 m	0.1041	2.251 30
17	−0.150 21E−3	0.006 m	0.0823	0.087 29
18	−0.437 96E−2	0.006 m	0.2229	1.545 95
19	−0.791 75E−2	0.008 m	0.3349	1.710 11

Table 11.5 Adjusted coordinates after blunder removal.

Station	X (m)	Y (m)
A	**401 176.143**	5 501 862.458
B	**403 602.951**	5 505 663.535
C	406 443.729	5 504 112.713
D	406 650.303	5 501 161.305
E	403 625.704	5 503 700.730

Bolded coordinates are fixed.

c) Internal and external measures of reliability for the final adjustment of the network are based on the following:

For $\alpha_0 = 0.001$ and $\beta_0 = 0.20$, $\sqrt{\lambda_0} = 4.123$ $\lambda_0 = 17.0$.

The summary for the computations is given in Table 11.6, where $p_i = \dfrac{1}{\sigma_i^2}$

Table 11.6 Summary of final reliability computations.

$\hat{\ell}_i$		σ_{ℓ_i} (rad and m)	r_i	$\nabla_0\ell_i = \sigma_{\ell_i}\sqrt{\dfrac{\lambda_0}{r_i}}$ internal reliability	$\delta_{0i}^2 = \dfrac{1-r_i}{r_i}\lambda_0$ external reliability
1	0°00′00″	1.454 44E−5	0.7446	14.3″	5.8308
2	20°33′27.6″	1.454 44E−5	0.7345	14.4″	6.1445
3	34°18′41.7″	1.454 44E−5	0.7460	14.3″	5.7872
4	64°44′30.0″	1.454 44E−5	0.7416	14.4″	5.9240
5	0°00′00″	1.454 44E−5	0.6989	14.8″	7.3224
6	82°20′50.1″	1.454 44E−5	0.7274	14.5″	6.3696
7	130°40′50.2″	1.454 44E−5	0.7296	14.5″	6.3017
8	233°46′42.7″	1.454 44E−5	0.6897	14.9″	7.6486
10	4509.711 m	0.009 m	0.3297	0.065	34.5570
11	3062.624 m	0.007 m	0.3982	0.046	25.6925
12	5728.087 m	0.011 m	0.6099	0.058	10.8717
13	5518.873 m	0.011 m	0.2342	0.094	55.5828
14	3236.519 m	0.007 m	0.2112	0.063	63.4816
15	5436.573 m	0.011 m	0.6603	0.056	8.7459
16	1962.939 m	0.004 m	0.1041	0.051	146.3750
17	2958.628 m	0.006 m	0.0823	0.086	189.6739
18	2847.985 m	0.006 m	0.2229	0.052	59.2532
19	3949.296 m	0.008 m	0.3349	0.057	33.7558

Problems

11.1 The following residual vector $\hat{r}$ and estimated covariance matrix $C_{\hat{r}}$ were computed from a least squares adjustment using five independent observations with a standard deviation of $\sigma = 2$ mm and a degree of freedom $v = 3$:

$$\hat{r}^T = [4\ 2\ -3\ 6\ -5]\ (\text{mm})$$

$$C_{\hat{r}} = \begin{bmatrix} 3 & -1 & 2 & -1 & 2 \\ -1 & 2 & 1 & 1 & -2 \\ 2 & 1 & 2 & -2 & 1 \\ -1 & 1 & -2 & 3 & -1 \\ 2 & -2 & 1 & -1 & 3 \end{bmatrix}\ (\text{mm}^2)$$

Given $\alpha_0 = 0.01$, $\beta_0 = 0.20$

a) Conduct a global test (based on one-tailed test (Baarda's approach)) to decide if there exists outlier or not (you can use Table A.4).

b) If the test in (a) fails, conduct local (in-context) tests to locate the outlier (assuming the σ_0^2 is reliably known).

c) Conduct local (in-context) tests on the data (assuming the σ_0^2 is reliably known).

11.2 The following residual vector $\hat{r}$ and estimated covariance matrix $C_{\hat{r}}$ were computed from a least squares adjustment with five independent observations:

$$\hat{r}^T = [3.1 \ \ 5.1 \ \ -2.5 \ \ 7.9 \ \ -6.5] \ (mm)$$

$$\text{diag}(C_{\hat{r}}) = [12.1 \ \ 9.5 \ \ 15.3 \ \ 19.1 \ \ 14.3] \ (mm^2)$$

The a priori variance factor σ_0^2 assumed equal to 1 was estimated with the degrees of freedom of 3. Given $\alpha = 0.10$, perform in-context tests taking σ_0^2 as assumed (meaning that σ_0^2 is unreliable so that s_0^2 was used in estimating $C_{\hat{r}}$).

11.3 Continue from the adjustment results in question 1, and complete the following, assuming everything is the same as in question 1 except that the cofactor of the vector of residuals is given as follows:

$$Q_{\hat{r}} = \begin{bmatrix} 3 & -1 & 2 & -1 & 2 \\ -1 & 2 & 1 & 1 & -2 \\ 2 & 1 & 2 & -2 & 1 \\ -1 & 1 & -2 & 3 & -1 \\ 2 & -2 & 1 & -1 & 3 \end{bmatrix} \ (mm^2)$$

a) Marginally detectable error (MDE) is also commonly referred to as a measure of internal reliability. Give a mathematical expression for this measure with the variables well defined, and explain in your own words what you understand by "internal reliability."

b) Explain the important uses of local redundancy number and the global internal reliability measure.

c) Determine the global internal reliability measure for the network. From your result, is the network geometrically strong? Explain your answer.

d) Calculate the internal and external measures of reliability for the final adjustment of the network (at $\alpha_0 = 0.1\%$, $\beta_0 = 0.20$; Baarda's noncentrality parameter λ_0 at these probability values is 17).

11.4 a) Given the standard factor of unit weight obtained from the adjustment of a geodetic network as 1.40 with a degree of freedom of 10 and the a priori standard factor of unit weight as 1, conduct a statistical test (one tailed) to decide if the adjustment result is acceptable at a confidence level of 95%. Show all work including the formulas used and explain all steps.

      b) Explain Pope's approach to local test, including the purpose, choice of significance level, assumptions and uses of variance factor of unit weight, and the assumed statistical distributions for local tests (providing appropriate formulas and clearly explaining in your own words every term and symbol used).

12

Least Squares Adjustment of Conditional Models

OBJECTIVES

After studying this chapter, you should be able to:

1) Derive the steps involved in least squares method of adjusting conditional mathematical models in geomatics.
2) Formulate conditional model equations for different cases of survey system.
3) Perform adjustment of survey measurements (classical and GNSS) by conditional method of least squares adjustment.
4) Apply variance–covariance propagation laws to determine variance–covariance matrices of adjusted quantities and their functions.

Understanding Least Squares Estimation and Geomatics Data Analysis, First Edition.
John Olusegun Ogundare.
© 2019 John Wiley & Sons, Inc. Published 2019 by John Wiley & Sons, Inc.
Companion website: www.wiley.com/go/ogundare/Understanding-lse-and-gda

12.1   Introduction

As discussed in Section 4.3, mathematical equations involved in geomatics field are not always formulated only as parametric model equations ($\hat{\ell} = f(\hat{x})$, where $\hat{\ell}$ is a vector of observations and $\hat{x}$ is vector of adjusted parameters). Sometimes, one may not be immediately interested in using the field measurements to solve for any parameter (such as coordinates of network points); but one may be interested in only evaluating the measurements for their consistency with each other without having to solve for any parameters. For example, consider a case in which three sides of a triangle and its three interior angles are independently measured; it is expected that the measured interior angles should sum up to 180° with the interior angles and their corresponding measured sides of the triangle satisfying some sine and cosine rules. It is most likely that the measured interior angles will not sum up to 180° and the angles and the corresponding sides of the triangle will not satisfy the sine and cosine rules perfectly. This is a typical least squares adjustment problem that does not involve solving for any parameter (such as coordinates of points), but it involves formulating equations that connect observations together mathematically. The equations formulated in this form are referred to as conditional model equations. Conditional model, therefore, is a model containing only observations ($\hat{\ell}$) and no parameters and can be expressed mathematically as

$$f(\hat{\ell}) = 0 \qquad\qquad\qquad (12.1)$$

In another words, observations are the same as parameters in conditional models. Examples of conditional model equations are given in Section 12.2.

12.2   Conditional Model Equations

Conditional model equations are formulated in such a way that all measurements and imposed geometrical conditions are represented and accommodated uniquely. Once these have been satisfied, no further or unnecessary equations should be added. An equation will be considered unnecessary if any combination of already formed equations can reproduce that equation. Forming more equations than needed can result in the equations becoming impossible to solve. Since there are no parameters involved in conditional models, the number of condition equations is the same as the number of redundancies for the problem. This means that the *redundancy* or the *number of degrees of freedom* of conditional model equations is determined as the number of independent model equations formed. Forming unique equations is usually the most difficult aspect of conditional model least squares adjustment. The steps involved in the formulation of conditional model equations are:

1) Identify what constitutes the observations from the question. Note that the data given are identified as observations when standard deviations are associated with them or when the data are referred to as being measured or observed.
2) From the understanding of the type of model $(f(\hat{\ell}) = 0)$ to be formulated, the equations should include only the observations and some constant values (if necessary); there should not be any unknown parameters involved. Each equation must also be arranged so that it is set equal to zero; and it should be ensured that each equation is unique (based on sensible mathematical relationship occurring between the measurements) and cannot be formulated by combining (adding or subtracting or doing both on) any of the already formed equations.

A linear system of equations is called *linearly dependent system of equations* if one or more of the equations is redundant, i.e. if at least one of the equations is a linear combination of the others. A system is linearly dependent if one of the equations can be removed without affecting the overall solutions. If none of the equations can be obtained as a linear combination of the others, then the system is called *linearly independent system of equations*. The following examples are to illustrate how to form unique and independent conditional model equations, which exclude unnecessary equations that will make the problem unsolvable.

12.2.1 Examples of Model Equations

The following examples are given to illustrate how conditional model equations are formulated.

Example 12.1 Distance measurements were made between points A, B, and C as shown in Figure 12.1, where $\hat{\ell}_1$, $\hat{\ell}_2$, and $\hat{\ell}_3$ are the adjusted distances to be determined and the measured distances are 190.40, 203.16, and 393.65 m, respectively. Form the conditional model $f(\hat{\ell}) = 0$ for this problem.

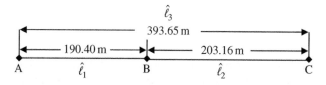

Figure 12.1 Baseline measurement.

Solution:

The conditional model equations should look like $f(\hat{\ell}) = 0$. Following the steps given in Section 12.2, the following equation is formed:

$$\hat{\ell}_3 - (\hat{\ell}_1 + \hat{\ell}_2) = 0 \tag{12.2}$$

where the hats on the measurements indicate that the measurements are to be adjusted; these are needed for the consistency of the equation, i.e. for the equation to have zero misclosure.

Example 12.2   In Figure 12.2, the three horizontal angles measured around the horizon are $\alpha = 22°12'13''$, $\beta = 59°56'15''$, $\gamma = 277°51'35''$. Form an appropriate equation $(f(\hat{\ell}) = 0)$ for adjusting the observations.

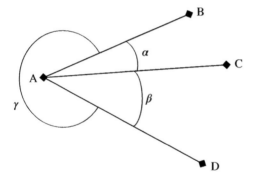

Figure 12.2 Horizontal angles around the horizon (not to scale).

Solution:

The conditional model equations should look like $f(\hat{\ell}) = 0$. Following the steps given in Example 12.1, the sensible mathematical relationship among the observations is that all angles in a circle must add up to 360. This is represented mathematically as

$$\hat{\alpha} + \hat{\beta} + \hat{\gamma} - 360° = 0 \tag{12.3}$$

Example 12.3   Given a closed leveling circuit in Figure 12.3 with measured elevation differences for four sections as Δh_1, Δh_2, Δh_3, Δh_4, give the conditional model for this problem.

Figure 12.3 Closed leveling circuit.

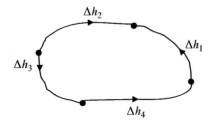

Solution:

In a closed leveling circuit, the sum of elevation differences (following the direction of leveling run indicated by arrows) must be equal to zero. The direction of the arrow for elevation difference Δh_1 will be used as a reference direction round the circuit (any direction can be used as a reference as long as consistency is maintained, any elevation difference in the same direction will be represented as positive, and any opposing direction will be represented as negative). The conditional model $f(\hat{\ell}) = 0$ is given as

$$\Delta \hat{h}_1 - \Delta \hat{h}_2 + \Delta \hat{h}_3 + \Delta \hat{h}_4 = 0 \tag{12.4}$$

Example 12.4 The angle and distance measurements for a simple triangulateration network in Figure 12.4 are as follows:

Angles : ℓ_1, ℓ_2, ℓ_3 Distances : ℓ_4, ℓ_5

Formulate the conditional model for the network, assuming the distance b is known and fixed.

Figure 12.4 Sample triangulateration network.

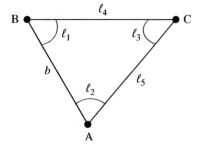

Solution (considering redundancy condition in step 2 in Section 12.2):

Using the sine laws on the triangle measurements and rearranging to look like $f(\hat{\ell}) = 0$ (noting that any other equation will be redundant),

$$\hat{\ell}_1 + \hat{\ell}_2 + \hat{\ell}_3 - 180 = 0 \tag{12.5}$$

$$b\sin\hat{\ell}_2 - \hat{\ell}_4 \sin\hat{\ell}_3 = 0 \tag{12.6}$$

$$b\sin\hat{\ell}_1 - \hat{\ell}_5 \sin\hat{\ell}_3 = 0 \tag{12.7}$$

Example 12.5 In the leveling network shown in Figure 12.5, the elevations h_A and h_B of points A and B, respectively, are to be determined. The benchmark elevations (in m) are shown in parentheses, and the field measurements are shown in the figure as dh_1, dh_2, dh_3, dh_4, and dh_5 with the arrows indicating the directions of leveling.

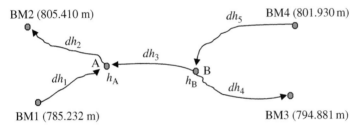

Figure 12.5 Leveling network.

Give the mathematical model ($f(\hat{\ell}) = 0$) for this problem, where $\hat{\ell}$ is a vector of adjusted observations.

Solution:

Take note of the following in formulating the equations:

- It can be seen from the leveling network in Figure 12.5 that there are only two unknowns (elevations of points A and B), needing only two measurements (elevation difference to A and to B) out of the five measurements to uniquely determine the two unknowns. The remaining three measurements are redundant, meaning there would be three conditional equations to be formulated.
- The sum of the elevation differences between any two known benchmarks must result in the difference between the elevations of the two benchmarks (following a particular arrow as positive value and the opposing arrow as negative value).

- Following the leveling route BM1-A-B-BM3 (dh$_3$ has an opposing arrow so that negative is assigned to it), the following is obtained:

$$d\hat{h}_1 - d\hat{h}_3 + d\hat{h}_4 - (794.881 - 785.232) = 0 \qquad (12.8)$$

- Following the leveling route BM4-B-A-BM2 (all the arrows are in the same direction), the following is obtained:

$$d\hat{h}_2 + d\hat{h}_3 + d\hat{h}_5 - (805.410 - 801.930) = 0 \qquad (12.9)$$

- Although all the measurements have been taken care of in Equations (12.8) and (12.9), there is still a need to ensure that the circuits BM2-A-BM1 and BM4-B-BM3 will close after the adjustment; if one of these circuits is satisfied together with the conditions stated in Equations (12.8) and (12.9), the other circuit will be automatically satisfied. The condition for fixing the circuit BM2-A-BM1 can be given as follows:

$$d\hat{h}_1 + d\hat{h}_2 - (805.410 - 785.232) = 0 \qquad (12.10)$$

or the condition for fixing the circuit BM4-B-BM3 as

$$d\hat{h}_5 + d\hat{h}_4 - (794.881 - 801.930) = 0 \qquad (12.11)$$

There are only three unique equations that accommodate all of the measurements. The number of unique condition equations can be determined approximately as the number of measurements minus the number of quantities not measured but whose values are being implicitly determined. In this problem, there are five measurements (dh$_1$, dh$_2$, dh$_3$, dh$_4$, dh$_5$) and two unknown quantities (h_A and h_B) not measured and whose values are being implicitly determined, giving five minus two condition equations (or three condition equations). The three simplified condition equations from Equations (12.8) to (12.10) can be given as follows:

$$d\hat{h}_1 - d\hat{h}_3 + d\hat{h}_4 - 9.649 = 0 \qquad (12.12)$$

$$d\hat{h}_2 + d\hat{h}_3 + d\hat{h}_5 - 3.480 = 0 \qquad (12.13)$$

$$d\hat{h}_1 + d\hat{h}_2 - 20.178 = 0 \qquad (12.14)$$

or from Equations (12.8), (12.9), and (12.11):

$$d\hat{h}_1 - d\hat{h}_3 + d\hat{h}_4 - 9.649 = 0 \qquad (12.15)$$

$$d\hat{h}_2 + d\hat{h}_3 + d\hat{h}_5 - 3.480 = 0 \qquad (12.16)$$

$$d\hat{h}_5 + d\hat{h}_4 + 7.049 = 0 \qquad (12.17)$$

As you can see above, only Equations (12.14) and (12.17) are different in the condition equations; the two equations should not appear together in order not to create problem when solving the equations. Condition equations must be unique (independent of each other). For example, it can be seen that Equation (12.12) plus Equation (12.13) minus Equation (12.14) gives Equation (12.17) exactly; hence there is no need of adding Equation (12.17) again to the three equations. Similarly, Equation (12.15) plus Equation (12.16) minus Equation (12.17) gives Equation (12.14) exactly. Considering Equations (12.8)–(12.11), Equation (12.11) is said to be redundant in the sense that it provides information that is already contained in the other equations.

The other way of formulating the conditional model equations for the problem in Figure 12.5 is as follows:

- Follow the leveling route BM1-A-BM2 to obtain the following equation:

$$\hat{dh}_1 + \hat{dh}_2 - (805.410 - 785.232) = 0 \tag{12.18}$$

- Follow the leveling route BM4-B-BM3 to obtain the following equation:

$$\hat{dh}_5 + \hat{dh}_4 - (794.881 - 801.930) = 0 \tag{12.19}$$

- Since there is only one more observation (dh_3) to be satisfied, any route that will accommodate this observation will be acceptable. Following route BM1-A-B-BM3 will give the following equation:

$$\hat{dh}_1 - \hat{dh}_3 + \hat{dh}_4 - (794.881 - 785.232) = 0 \tag{12.20}$$

Equations (12.18)–(12.20) are the linearly independent conditional equations that can be used instead of those given in Equations (12.12)–(12.17). Note that any of these three sets of equations will produce identical results in least squares adjustment of the equations.

Example 12.6   In a closed traverse shown in Figure 12.6, the following were measured: angles θ_1 and θ_2; distances d_1, d_2, and d_3; and bearings α_1 and α_F. If the coordinates of point 1 (x_1, y_1) and point 4 (x_4, y_4) are fixed, give all the necessary conditional model equations $(f(\hat{\ell}) = 0)$ for the measurements.

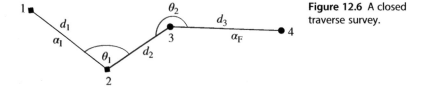

Figure 12.6 A closed traverse survey.

Solution:

All traverse adjustments have a maximum of three conditions to satisfy as follows:

1) The difference between the forward bearing α_F and the computed forward bearing α'_F based on the initial bearing α_I and the measured angles at the traverse points must be equal to zero. This is to check accuracy of angular measurements and can be expressed mathematically by

$$\hat{\alpha}_F - \left(\hat{\alpha}_I + \hat{\theta}_1 + \hat{\theta}_2\right) = 0 \tag{12.21}$$

2) The difference between change in eastings of the starting and the ending points of the traverse and the sum of changes in eastings of individual traverse legs must be equal to zero. This is to check both the errors in distances and the residual errors in angular measurements, along the easting direction. This can be expressed mathematically as follows:

$$(x_4 - x_1) - \hat{d}_1 \sin \hat{\alpha}_I - \hat{d}_2 \sin \left(\hat{\alpha}_I + \hat{\theta}_1 + 180\right) - \hat{d}_3 \sin \left(\hat{\alpha}_I + \hat{\theta}_1 + \hat{\theta}_2\right) = 0 \tag{12.22}$$

3) The difference between change in northings of the starting and the ending points of the traverse and the sum of changes in northings of individual traverse legs must be equal to zero. As in the easting case, this is also to check both the errors in distances and the residual errors in angular measurements, along the northing direction. This can also be expressed mathematically by

$$(y_4 - y_1) - \hat{d}_1 \cos \hat{\alpha}_I - \hat{d}_2 \cos \left(\hat{\alpha}_I + \hat{\theta}_1 + 180\right) - \hat{d}_3 \cos \left(\hat{\alpha}_I + \hat{\theta}_1 + \hat{\theta}_2\right) = 0 \tag{12.23}$$

12.3   Conditional Model Adjustment Formulation

The steps involved in conditional model adjustment are derived in this section. For example, consider the following conditional model:

$$f(\hat{\ell}) = 0 \ \ P \tag{12.24}$$

where f is the vector of mathematical model, $\hat{\ell}$ is the vector of adjusted observation, and P is the weight matrix of the original observation vector ℓ. The logical steps involved in obtaining the solution vector for this model can be given as follows:

- Linearizing the mathematical model if it is nonlinear.
- Formulating the variation function, which involves imposing the least squares criterion of the problem on the linearized model.
- Deriving the least squares normal equations from the variation function.
- Solving the least squares normal equations to obtain corrections to be applied to the approximate values of the unknown quantities.

In order to linearize the mathematical model in Equation (12.24), assume the following:

$$\hat{\ell} = \ell + v \tag{12.25}$$

where v is a vector of observation residuals to be applied to the original observation vector, ℓ. From Equations (12.24) to (12.25), the following can be obtained:

$$f(\ell + v) = 0 \tag{12.26}$$

By linearization by Taylor series expansion, Equation (12.26) in linearized form can be given as

$$f(\ell) + \frac{\partial f}{\partial \ell} v = 0 \tag{12.27}$$

The linearized Equation (12.27) can be expressed in simpler from as

$$w + Bv = 0 \tag{12.28}$$

where $B = \dfrac{\partial f}{\partial \ell}$ is commonly referred to as the second design matrix and $w = f(\ell)$ is a vector of misclosures. Equation (12.27) or Equation (12.28) is a linearized form of Equation (12.24).

The variation function can be formulated as follows. Since the weight matrix P is directly associated with the conditional model in Equation (12.24), the least squares criterion for this problem will be

$$v^T P v = \text{minimum} \tag{12.29}$$

Impose the least squares criterion on the linearized model in Equation (12.28) based on the Lagrangian approach to give the following variation function:

$$\varphi = v^T P v - 2k^T(Bv + w) = \text{minimum} \tag{12.30}$$

where k is a vector of correlates or a vector of Lagrangian multipliers. The least squares normal equation system can be derived by finding the partial derivatives of the variation function in Equation (12.30) with respect to the unknowns (i.e. v and k) and setting them to zero as follows:

$$\frac{\partial \varphi}{\partial v} = 2v^T P - 2k^T B = 0 \tag{12.31}$$

$$\frac{\partial \varphi}{\partial k^T} = -2(Bv + w) = 0 \tag{12.32}$$

The summary of the most expanded form of normal Equations (12.31) and (12.32) is:

$$v^T P - k^T B = 0 \tag{12.33}$$

$$Bv + w = 0 \tag{12.34}$$

Equations (12.33) and (12.34) can be solved simultaneously for the unknown quantities k and v as follows. From Equation (12.33), the following can be obtained:

$$v = P^{-1} B^T k \tag{12.35}$$

Put Equation (12.35) into Equation (12.34) to obtain

$$BP^{-1} B^T k + w = 0 \tag{12.36}$$

If $Q = P^{-1}$ and $M = BQB^T$, the following solution functions can be derived from Equations (12.36) and (12.35):

$$k = -M^{-1} w \tag{12.37}$$

$$v = QB^T k \tag{12.38}$$

From Equations (12.25) and (12.38), the following solution for the adjusted observation ($\hat{\ell}$) can be determined:

$$\hat{\ell} = \ell + v \tag{12.39}$$

Example 12.7   Linearize Equations (12.5)–(12.7) in Example 12.4, and present it in the form of Equation (12.28).

Solution:

For the Equations (12.5)–(12.7), the linearized model can be given as Equation (12.28) where vector of observations, $\ell = [\ell_1 \ \ell_2 \ \ell_3 \ \ell_4 \ \ell_5]^T$; the vector of the model equations, $f = [f_1 \; f_2 \; f_3]^T$; and the second design matrix (evaluated at ℓ) can be given as

$$
B = \frac{\partial f}{\partial \hat{\ell}}\bigg|_{\ell} =
\begin{bmatrix}
\dfrac{\partial f_1}{\partial \hat{\ell}_1} & \dfrac{\partial f_1}{\partial \hat{\ell}_2} & \dfrac{\partial f_1}{\partial \hat{\ell}_3} & \dfrac{\partial f_1}{\partial \hat{\ell}_4} & \dfrac{\partial f_1}{\partial \hat{\ell}_5} \\[2mm]
\dfrac{\partial f_2}{\partial \hat{\ell}_1} & \dfrac{\partial f_2}{\partial \hat{\ell}_2} & \dfrac{\partial f_2}{\partial \hat{\ell}_3} & \dfrac{\partial f_2}{\partial \hat{\ell}_4} & \dfrac{\partial f_2}{\partial \hat{\ell}_5} \\[2mm]
\dfrac{\partial f_3}{\partial \hat{\ell}_1} & \dfrac{\partial f_3}{\partial \hat{\ell}_2} & \dfrac{\partial f_3}{\partial \hat{\ell}_3} & \dfrac{\partial f_3}{\partial \hat{\ell}_4} & \dfrac{\partial f_3}{\partial \hat{\ell}_5}
\end{bmatrix}_{\ell}
$$

or

$$
B =
\begin{bmatrix}
1 & 1 & 1 & 1 & 1 \\
0 & b\cos\ell_2 & -\ell_4\cos\ell_3 & -\sin\ell_3 & 0 \\
b\cos\ell_1 & 0 & -\ell_5\cos\ell_3 & 0 & -\sin\ell_3
\end{bmatrix}
$$

The misclosure vector (w) evaluated at ℓ and the vector of residuals (v) can be given, respectively, as

$$
w =
\begin{bmatrix}
\ell_1 + \ell_2 + \ell_3 - 180 \\
b\sin\ell_2 - \ell_4\sin\ell_3 \\
b\sin\ell_1 - \ell_5\sin\ell_3
\end{bmatrix}, \quad
v =
\begin{bmatrix}
v_1 \\ v_2 \\ v_3 \\ v_4 \\ v_5
\end{bmatrix}
$$

The linearized model of Equations (12.5)–(12.7) can be given as

$$
\begin{bmatrix}
\ell_1 + \ell_2 + \ell_3 - 180 \\
b\sin\ell_2 - \ell_4\sin\ell_3 \\
b\sin\ell_1 - \ell_5\sin\ell_3
\end{bmatrix}
+
\begin{bmatrix}
1 & 1 & 1 & 1 & 1 \\
0 & b\cos\ell_2 & -\ell_4\cos\ell_3 & -\sin\ell_3 & 0 \\
b\cos\ell_1 & 0 & -\ell_5\cos\ell_3 & 0 & -\sin\ell_3
\end{bmatrix}
\begin{bmatrix}
v_1 \\ v_2 \\ v_3 \\ v_4 \\ v_5
\end{bmatrix}
= 0
$$

Example 12.8 Linearize Equations (12.21)–(12.23) in Example 12.6, and present it in the form of Equation (12.28).

Solution:

For the Equations (12.21)–(12.23), the linearized model can be given as Equation (12.28) where vector of observations, $\ell = [\alpha_I \ \ \alpha_F \ \ \theta_1 \ \ \theta_2 \ \ d_1 \ \ d_2 \ \ d_3]^T$; the vector of the model equations, $f = [f_1 \ \ f_2 \ \ f_3]^T$; and the second design matrix (evaluated at ℓ) can be given as

$$B = \frac{\partial f}{\partial \hat{\ell}}\bigg|_{\ell} = \begin{bmatrix} \dfrac{\partial f_1}{\partial \hat{\alpha}_I} & \dfrac{\partial f_1}{\partial \hat{\alpha}_F} & \dfrac{\partial f_1}{\partial \hat{\theta}_1} & \dfrac{\partial f_1}{\partial \hat{\theta}_2} & \dfrac{\partial f_1}{\partial \hat{d}_1} & \dfrac{\partial f_1}{\partial \hat{d}_2} & \dfrac{\partial f_1}{\partial \hat{d}_3} \\[2mm] \dfrac{\partial f_2}{\partial \hat{\alpha}_I} & \dfrac{\partial f_2}{\partial \hat{\alpha}_F} & \dfrac{\partial f_2}{\partial \hat{\theta}_1} & \dfrac{\partial f_2}{\partial \hat{\theta}_2} & \dfrac{\partial f_2}{\partial \hat{d}_1} & \dfrac{\partial f_2}{\partial \hat{d}_2} & \dfrac{\partial f_2}{\partial \hat{d}_3} \\[2mm] \dfrac{\partial f_3}{\partial \hat{\alpha}_I} & \dfrac{\partial f_3}{\partial \hat{\alpha}_F} & \dfrac{\partial f_3}{\partial \hat{\theta}_1} & \dfrac{\partial f_3}{\partial \hat{\theta}_2} & \dfrac{\partial f_3}{\partial \hat{d}_1} & \dfrac{\partial f_3}{\partial \hat{d}_2} & \dfrac{\partial f_3}{\partial \hat{d}_3} \end{bmatrix}_{\ell}$$

$$\frac{\partial f_1}{\partial \hat{\alpha}_I}\bigg|_{\ell} = -1; \quad \frac{\partial f_1}{\partial \hat{\alpha}_F}\bigg|_{\ell} = 1; \quad \frac{\partial f_1}{\partial \hat{\theta}_1}\bigg|_{\ell} = -1; \quad \frac{\partial f_1}{\partial \hat{\theta}_2}\bigg|_{\ell} = -1; \quad \frac{\partial f_1}{\partial \hat{d}_1}\bigg|_{\ell} = 0; \quad \frac{\partial f_1}{\partial \hat{d}_2}\bigg|_{\ell} = 0; \quad \frac{\partial f_1}{\partial \hat{d}_3}\bigg|_{\ell} = 0$$

$$\frac{\partial f_2}{\partial \hat{\alpha}_I}\bigg|_{\ell} = -d_1 \cos\left(\alpha_I\right) - d_2 \cos\left(\alpha_I + \theta_1 + 180\right) - d_3 \cos\left(\alpha_I + \theta_1 + \theta_2\right); \quad \frac{\partial f_2}{\partial \hat{\alpha}_F}\bigg|_{\ell} = 0$$

$$\frac{\partial f_2}{\partial \hat{\theta}_1}\bigg|_{\ell} = -d_2 \cos\left(\alpha_I + \theta_1 + 180\right) - d_3 \cos\left(\alpha_I + \theta_1 + \theta_2\right)$$

$$\frac{\partial f_2}{\partial \hat{\theta}_2}\bigg|_{\ell} = -d_3 \cos\left(\alpha_I + \theta_1 + \theta_2\right); \quad \frac{\partial f_2}{\partial \hat{d}_1}\bigg|_{\ell} = -\sin\alpha_I; \quad \frac{\partial f_2}{\partial \hat{d}_2}\bigg|_{\ell} = -\sin\left(\alpha_I + \theta_1 + 180\right)$$

$$\frac{\partial f_2}{\partial \hat{d}_3}\bigg|_{\ell} = -\sin\left(\alpha_I + \theta_1 + \theta_2\right)$$

$$\frac{\partial f_3}{\partial \hat{\alpha}_I}\bigg|_{\ell} = d_1 \sin\left(\alpha_I\right) + d_2 \sin\left(\alpha_I + \theta_1 + 180\right) + d_3 \sin\left(\alpha_I + \theta_1 + \theta_2\right); \quad \frac{\partial f_3}{\partial \hat{\alpha}_F}\bigg|_{\ell} = 0$$

$$\frac{\partial f_3}{\partial \hat{\theta}_1}\bigg|_{\ell} = d_2 \sin\left(\alpha_I + \theta_1 + 180\right) + d_3 \sin\left(\alpha_I + \theta_1 + \theta_2\right)$$

$$\frac{\partial f_3}{\partial \hat{\theta}_2}\bigg|_{\ell} = d_3 \sin\left(\alpha_I + \theta_1 + \theta_2\right); \quad \frac{\partial f_3}{\partial \hat{d}_1}\bigg|_{\ell} = -\cos\alpha_I; \quad \frac{\partial f_3}{\partial \hat{d}_2}\bigg|_{\ell} = -\cos\left(\alpha_I + \theta_1 + 180\right)$$

$$\frac{\partial f_3}{\partial \hat{d}_3}\bigg|_{\ell} = -\cos\left(\alpha_I + \theta_1 + \theta_2\right)$$

The misclosure vector (w) evaluated at ℓ and the vector of residuals (v) can be given, respectively, as

$$w = \begin{bmatrix} \alpha_F - (\alpha_I + \theta_1 + \theta_2) \\ (x_4 - x_1) - d_1 \sin\alpha_I - d_2 \sin(\alpha_I + \theta_1 + 180) - d_3 \sin(\alpha_I + \theta_1 + \theta_2) \\ (y_4 - y_1) - d_1 \cos\alpha_I - d_2 \cos(\alpha_I + \theta_1 + 180) - d_3 \cos(\alpha_I + \theta_1 + \theta_2) \end{bmatrix}, \quad v = \begin{bmatrix} v_1 \\ v_2 \\ v_3 \\ v_4 \\ v_5 \\ v_6 \\ v_7 \end{bmatrix}$$

The linearized model of Equations (12.5)–(12.7) can be given as

$$w + Bv = 0$$

where the quantities in the equation are as defined above.

12.3.1 Conditional Model Adjustment Steps

The summary of the steps for conditional least squares adjustment is given as follows:

1) Formulate the conditional model equations:

$$f\left(\hat{\ell}\right) = 0 \tag{12.40}$$

where

$$\hat{\ell} = \ell + v \tag{12.41}$$

2) Form the second design matrix (B) and the misclosure vector w:

$$B = \frac{\partial f}{\partial \ell} \text{ and } w = f(\ell) \tag{12.42}$$

3) Form the normal equations:

$$BQB^T k + w = 0 \tag{12.43}$$

or

$$Mk + w = 0 \tag{12.44}$$

where

$$M = BQB^T \tag{12.45}$$

$$Q = P^{-1} \tag{12.46}$$

or

$$Q = C_\ell \left(\text{assuming } \sigma_0^2 = 1 \right) \tag{12.47}$$

Q is the cofactor matrix, C_ℓ is the covariance matrix of observations, P is the weight matrix of observations, and k is a vector of correlates or vector of Lagrangian multipliers.

4) Solve for k in the normal equations in Equation (12.44):

$$k = -M^{-1}w \tag{12.48}$$

5) Compute the residual vector (v):

$$v = QB^T k \tag{12.49}$$

6) Compute the adjusted observations as

$$\hat{\ell} = \ell + v \tag{12.50}$$

The conditional model adjustment steps given above are illustrated numerically in the following examples.

Example 12.9 Three angles are measured in a plane triangle (referring to angles ℓ_1, ℓ_2, and ℓ_3 in Figure 12.4) as $30°45'20''$, $69°45'20''$, and $79°30'05''$; each measurement has the same standard deviation $\pm 5''$. Adjust the angles by the conditional method of least squares.

Solution:

Following the steps given in Section 12.3.1:

1) Form the conditional model: $f\left(\hat{\ell} \right) = 0$

$$\hat{\ell}_1 + \hat{\ell}_2 + \hat{\ell}_3 - 180 = 0$$

2) Form the second design matrix (B) from step 1 (with hats ignored):

$$B = \begin{bmatrix} \dfrac{\partial f}{\partial \ell_1} & \dfrac{\partial f}{\partial \ell_2} & \dfrac{\partial f}{\partial \ell_3} \end{bmatrix} \quad B = [1 \ 1 \ 1]$$

Form the misclosure vector w by substituting the observations into the conditional model:

$$w = \ell_1 + \ell_2 + \ell_3 - 180 = 45''$$

3) Form the normal equations.

Form cofactor matrix from the given standard deviations (since the same unit is used for the problem, leave the value in seconds): $Q = C_\ell$ (assuming $\sigma_0^2 = 1$)

$$Q = \begin{bmatrix} 5^2 & 0 & 0 \\ 0 & 5^2 & 0 \\ 0 & 0 & 5^2 \end{bmatrix}$$

Form the matrix $M = BQB^T$: $M = 75.0$.

4) Solve for k in the normal equations: $k = -M^{-1}w$

$$k = -75^{-1} \times 45 = -0.60$$

5) Compute the residual vector (v) from $v = QB^Tk$:

$$v = QB^Tk = \begin{bmatrix} -15 \\ -15 \\ -15 \end{bmatrix}$$

6) Compute the adjusted observations from $\hat{\ell} = \ell + v$:

$$\hat{\ell}_1 = 30°45'20'' - 15'' \quad \rightarrow \quad 30°45'05''$$
$$\hat{\ell}_2 = 69°45'20'' - 15'' \quad \rightarrow \quad 69°45'05''$$
$$\hat{\ell}_3 = 79°30'05'' - 15'' \quad \rightarrow \quad 79°29'55''$$

As a check on the condition: $\hat{\ell}_1 + \hat{\ell}_2 + \hat{\ell}_3 - 180 = 0$

Example 12.10   Figure 12.7 is a leveling network to be adjusted, with all of the height difference measurements necessary to carry out the adjustment given in Table 12.1. Assume all observations have the same standard deviation $\sigma_{\Delta h_{ij}} = 0.005$ m and points BMX and BMY are fixed points with known elevations of $H_{BMX} = 30.100$ m and $H_{BMY} = 32.331$ m. Perform conditional least squares adjustment on the leveling network in Figure 12.7 by computing the adjusted height differences.

Figure 12.7 Leveling network.

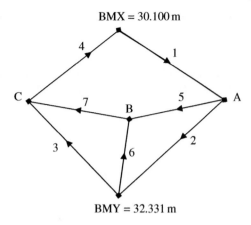

BMX = 30.100 m

BMY = 32.331 m

Table 12.1 Elevation differences.

No.	Δh_{ij} (m)
1	1.535
2	0.704
3	−0.376
4	−1.845
5	−0.205
6	−0.903
7	0.512

Solution:

The conditional model for the problem $\left(f\left(\hat{\ell}\right) = 0\right)$:

Given BMX and BMY as fixed points with $H_{BMX} = 30.100$ m and $H_{BMY} = 32.331$ m, the conditional model equations can be formulated (following the arrows on the network in Figure 12.7) as follows:

- From route BMX-A-B-C-BMX,

$$d\hat{h}_1 + d\hat{h}_5 + d\hat{h}_7 + d\hat{h}_4 = 0 \qquad (12.51)$$

- From route BMY-B-C-BMY,

$$d\hat{h}_6 + d\hat{h}_7 - d\hat{h}_3 = 0 \qquad (12.52)$$

- From route BMY-B-A-BMY,

$$d\hat{h}_6 - d\hat{h}_5 + d\hat{h}_2 = 0 \tag{12.53}$$

- Although all the measurements have been accommodated in Equations (12.51)–(12.53), there is still a condition that two points (BMX and BMY) must remain the same after adjustment or the elevation difference between them must be fixed. Any route connecting the two points should suffice. Considering route BMX-A-BMY, the following can be obtained:

$$d\hat{h}_1 + d\hat{h}_2 = \text{BMY} - \text{BMX}$$

or

$$d\hat{h}_1 + d\hat{h}_2 = 32.331 - 30.100$$

or

$$d\hat{h}_1 + d\hat{h}_2 - 2.231 = 0 \tag{12.54}$$

There are only four independent equations; Equation (12.54) is to ensure that the elevations of benchmarks BMX and BMY will remain the same after the adjustment. Forming another equation, for example, in the route BMY-C-BMX-A-BMY, will be dependent on the three Equations (12.51)–(12.53) already formed, making it dependent; so there is no need to form this additional equation. If dependent equation such as this is allowed, then it will be impossible to solve the equations.

$$\text{Given observations are}: \ell = \begin{bmatrix} dh_1 \\ dh_2 \\ dh_3 \\ dh_4 \\ dh_5 \\ dh_6 \\ dh_7 \end{bmatrix} \quad \text{or} \quad \ell = \begin{bmatrix} 1.535 \\ 0.704 \\ -0.376 \\ -1.845 \\ -0.205 \\ -0.903 \\ 0.512 \end{bmatrix}$$

Form the second design matrix (B) from Equations (12.51)–(12.54):

$$B = \begin{bmatrix} 1 & 0 & 0 & 1 & 1 & 0 & 1 \\ 0 & 0 & -1 & 0 & 0 & 1 & 1 \\ 0 & 1 & 0 & 0 & -1 & 1 & 0 \\ 1 & 1 & 0 & 0 & 0 & 0 & 0 \end{bmatrix}$$

Form misclosure vector (w) by substituting the measurements back into Equations (12.51)–(12.54) as follows:

$$w = \begin{bmatrix} -0.003 \\ -0.015 \\ 0.006 \\ 0.008 \end{bmatrix}$$

Form the cofactor matrix (Q) using the given standard deviations of the measurements:

$$Q_\ell = \text{diag}([2.5E-5 \ \ 2.5E-5 \ \ 2.5E-5 \ \ 2.5E-5 \ \ 2.5E-5 \ \ 2.5E-5 \ \ 2.5E-5])$$

Form the M-matrix:

$$M = BQB^T = \begin{bmatrix} 1.0E-4 & 2.5E-5 & -2.5E-5 & 2.5E-5 \\ 2.5E-5 & 7.5E-5 & 2.5E-5 & 0.0 \\ -2.5E-5 & 2.5E-5 & 7.5E-5 & 2.5E-5 \\ 2.5E-5 & 0.0 & 2.5E-5 & 5.0E-5 \end{bmatrix}$$

Solve normal equations for correlates (k):

$$k = -M^{-1}w = \begin{bmatrix} -97.1429 \\ -302.8571 \\ -211.429 \\ -5.7143 \end{bmatrix}$$

Solve for the residual vector (v):

$$v = QB^T k = \begin{bmatrix} -0.0026 \\ -0.0054 \\ -0.0076 \\ -0.0024 \\ 0.0029 \\ 0.0023 \\ 0.0051 \end{bmatrix}$$

The adjusted height differences are as follows: $\hat{\ell} = \ell + v$

$$\ell = \begin{bmatrix} 1.535 \\ 0.704 \\ -0.376 \\ -1.845 \\ -0.205 \\ -0.903 \\ 0.512 \end{bmatrix} + \begin{bmatrix} -0.0026 \\ -0.0054 \\ -0.0076 \\ -0.0024 \\ 0.0029 \\ 0.0023 \\ 0.0051 \end{bmatrix} \rightarrow \hat{\ell} = \begin{bmatrix} d\hat{h}_1 \\ d\hat{h}_2 \\ d\hat{h}_3 \\ d\hat{h}_4 \\ d\hat{h}_5 \\ d\hat{h}_6 \\ d\hat{h}_7 \end{bmatrix} = \begin{bmatrix} 1.5324 \\ 0.6986 \\ -0.3836 \\ -1.8474 \\ -0.2021 \\ -0.9007 \\ 0.5171 \end{bmatrix}$$

12.4 Stochastic Model of Conditional Adjustment

The stochastic model for conditional adjustments is derived as in the case of parametric model adjustment in Section 5.8. Recall from Section 1.7 and Chapter 5 the relationship between variance–covariance matrix (C_ℓ) of observation vector ℓ and cofactor matrix (Q_ℓ) of the observation vector as

$$C_\ell = \sigma_0^2 Q_\ell \tag{12.55}$$

where σ_0^2 is a priori variance factor of unit weight (which is always taken as the general population variance with a value of 1.00). Since the vector of observations ℓ and its variance–covariance matrix C_ℓ are used in computing the unknown quantities, such as the adjusted observations ($\hat{\ell}$) and the observation residuals (v), it is obvious that the errors in those observations will propagate into these quantities also. The concepts of variance–covariance propagation (Chapter 2) will be used as follows to derive the cofactor matrix of the adjusted observation ($Q_{\hat{\ell}}$) and the cofactor matrix of the observation residuals (Q_v).

12.4.1 Derivation of Cofactor Matrix of Adjusted Observations

The vector of least squares adjusted observations is given from Equations (12.42), (12.48), (12.49), and (12.50) as follows:

$$\hat{\ell} = \ell - Q_\ell B^T M^{-1} f(\ell) \tag{12.56}$$

Equation (12.56) is a form of $\hat{\ell} = f(\ell)$ with the usual propagated cofactor matrix (based on variance covariance propagation rules) given as

$$Q_{\hat{\ell}} = J Q_\ell J^T \tag{12.57}$$

where Q_ℓ is the cofactor matrix of the observations and J is the Jacobian matrix given as follows:

$$J = \frac{\partial \hat{\ell}}{\partial \ell} = I - Q_\ell B^T M^{-1} B \tag{12.58}$$

taking note that the partial derivative of the vector ℓ with respect to the vector ℓ gives an identity matrix I and the partial derivative of $f(\ell)$ with respect to ℓ gives the second design matrix B (based on Equation (12.42). Substituting Equation (12.58) into Equation (12.57) gives the following:

$$Q_{\hat{\ell}} = \left(I - Q_\ell B^T M^{-1} B\right) Q_\ell \left(I - Q_\ell B^T M^{-1} B\right)^T \tag{12.59}$$

By post-multiplying the first bracket by Q_ℓ and transposing the elements in the other bracket gives the following equation:

$$Q_{\hat{\ell}} = \left(I - Q_\ell B^T M^{-1} B\right) Q_\ell \left(I - B^T M^{-1} B Q_\ell\right)^T \tag{12.60}$$

Expanding Equation (12.60) gives the following:

$$Q_{\hat{\ell}} = Q_\ell - Q_\ell B^T M^{-1} B Q_\ell - Q_\ell B^T M^{-1} B Q_\ell + Q_\ell B^T M^{-1} \left(B^T Q_\ell B\right) M^{-1} B Q_\ell \tag{12.61}$$

Remembering that $M = (B^T Q_\ell B)$ and $(B^T Q_\ell B) M^{-1} = I$ reduces Equation (12.61) to the following:

$$Q_{\hat{\ell}} = Q_\ell - Q_\ell B^T M^{-1} B Q_\ell - Q_\ell B^T M^{-1} B Q_\ell + Q_\ell B^T M^{-1} B Q_\ell \tag{12.62}$$

The last three terms in Equation (12.62) are identical so that the last two terms cancel out, giving the reduced form as follows:

$$Q_{\hat{\ell}} = Q_\ell - Q_\ell B^T M^{-1} B Q_\ell \tag{12.63}$$

Equation (12.63) is the cofactor matrix of the adjusted observations $\hat{\ell}$.

12.4.2 Derivation of Cofactor Matrix of Observation Residuals

The vector of observation residuals is given from Equations (12.49), (12.48), and (12.42) as

$$v = -Q_\ell B^T M^{-1} f(\ell) \tag{12.64}$$

Equation (12.64) is a form of $v = f(\ell)$ with the usual propagated cofactor matrix (based on variance covariance propagation rules) given as

$$Q_v = J Q_\ell J^T \tag{12.65}$$

where Q_ℓ is the cofactor matrix of the observations and J is the Jacobian matrix given as follows:

$$J = \frac{\partial v}{\partial \ell} = -Q_\ell B^T M^{-1} B \tag{12.66}$$

Substituting Equation (12.66) into Equation (12.65) gives the following:

$$Q_v = \left(-Q_\ell B^T M^{-1} B\right) Q_\ell \left(-Q_\ell B^T M^{-1} B\right)^T \tag{12.67}$$

By applying the transpose to the elements in the second bracket in Equation (12.67) and letting the two negatives multiply out, the following is obtained:

$$Q_v = \left(Q_\ell B^T M^{-1} B\right) Q_\ell \left(B^T M^{-1} B Q_\ell\right) \tag{12.68}$$

Equation (12.68) in expanded form can be given as

$$Q_v = Q_\ell B^T M^{-1} \left(B Q_\ell B^T\right) M^{-1} B Q_\ell \tag{12.69}$$

Taking $M = (B^T Q_\ell B)$ and $(B^T Q_\ell B) M^{-1} = I$, Equation (12.69) reduces to the following:

$$Q_v = Q_\ell B^T M^{-1} B Q_\ell \tag{12.70}$$

Equation (12.70) is the cofactor matrix of the observation residuals (v).

12.4.3 Covariance Matrices of Adjusted Observations and Residuals

The variance–covariance matrices of the adjusted quantities can be given from Equations (12.63) and (12.70), respectively, as follows:

$$C_{\hat{\ell}} = s_0^2 Q_\ell - s_0^2 Q_\ell B^T M^{-1} B Q_\ell \tag{12.71}$$

$$C_v = s_0^2 Q_\ell B^T M^{-1} B Q_\ell \tag{12.72}$$

where $C_{\hat{\ell}}$ is the covariance matrix of the adjusted observations, C_v is the covariance matrix of the observation residuals, and s_0^2 is *a posteriori variance factor (APVF) of unit weight*, which is calculated after the least squares adjustment as follows:

$$s_0^2 = \frac{v^T P v}{r} \tag{12.73}$$

where v is the vector of residuals, P is the weight matrix of the observations, and r is the redundancy (or the degrees of freedom). The degrees of freedom r is the number of independent condition equations formulated. From

Equations (12.71) to (12.72), it can be seen that the covariance matrix of the adjusted observation can be given as

$$C_{\hat{\ell}} = \hat{C}_\ell - C_v \tag{12.74}$$

where $\hat{C}_\ell = s_0^2 Q_\ell$ is the scaled covariance matrix of the original observations ℓ, which should also not be confused with $C_\ell = \sigma_0^2 Q_\ell$ (with $\sigma_0^2 = 1$), which is the unscaled covariance matrix of the original observations, used at the beginning of the least squares adjustment. Equation (12.74) can be interpreted to mean that the covariance matrix of the adjusted observation is always less than the covariance of the original observation.

12.5 Assessment of Observations and Conditional Model

The misclosures in conditional model adjustment can be tested to check if they have normal probability density function following the idea of global and local tests in Chapter 11. This is done by testing the quadratic forms of the misclosures as follows:

i) If the covariance matrix (C_w) of the misclosure vector (w) is well known and the a priori variance factor of unit weight is well known, the quadratic form of the misclosures will be considered as having normal probability density function if the following is satisfied:

$$\chi^2_{\alpha/2, df\,=\,m} < w^T C_w^{-1} w < \chi^2_{1-\alpha/2, df\,=\,m} \tag{12.75}$$

where $\chi^2_{\alpha/2, df\,=\,m}$ is the lower-tailed chi-square density function with the number of degrees of freedom $df = m$ for m number of misclosures. This is similar to global test discussed in Chapter 11.

ii) If the scaled covariance matrix $(\hat{C}_w)$ of the misclosure vector (w) is used because it is not well known and the a priori variance factor of unit weight is unknown but the a posteriori variance factor of unit weight (s_0^2) is used, the quadratic form of the misclosures will be considered as having normal probability density function if the following is satisfied:

$$F_{\alpha/2, df_1\,=\,m, df_2\,=\,m-u} < \frac{w^T \hat{C}_w^{-1} w}{m} < F_{1-\alpha/2, df_1\,=\,m, df_2\,=\,m-u} \tag{12.76}$$

where $F_{\alpha/2, df_1\,=\,m, df_2\,=\,m-u}$ is the lower-tailed F (or Fisher) density function with the first degrees of freedom $df_1 = m$ (number of misclosures) and $df_2 = m - u$ with u as the number of unknown parameters in the least squares adjustment.

The individual misclosure can also be assessed to check if it has normal probability density function as follows:

a) If the covariance matrix (C_w) of the misclosure vector (w) is well known and the a priori variance factor of unit weight is well known, the standardized misclosure will be considered as having normal probability density function if the following is satisfied (meaning that the misclosure is not an outlier):

$$\left| \frac{w_i}{\sigma_{w_i}} \right| < z_{1-\alpha/2} \tag{12.77}$$

where σ_{w_i} is the standard deviation of the misclosure w_i.

b) If the scaled covariance matrix ($\hat{C}_w$) of the misclosure vector (w) is used because it is not well known and the a priori variance factor of unit weight is unknown but the a posteriori variance factor of unit weight (s_0^2) is used, the standardized misclosure will be considered as having normal probability density function if the following is satisfied (meaning that the misclosure is not an outlier):

$$\left| \frac{w_i}{\hat{\sigma}_{w_i}} \right| < t_{1-\alpha/2,m-u} \tag{12.78}$$

where $\hat{\sigma}_{w_i}$ is the scaled standard deviation of the misclosure w_i, $t_{1-\alpha/2,\,m-u}$ is the Student's t-density (assuming the observations used in computing w and s_0^2 are different), m is the number of misclosures, and u is the number of unknown parameters in the least squares adjustment. The $t_{1-\alpha/2,\,m-u}$ will be replaced with the τ density with the degrees of freedom $m-u$ if the observations used in computing w are also used in determining s_0^2.

12.6   Variance–Covariance Propagation for Derived Parameters from Conditional Adjustment

The relationship between the adjusted parameters ($\hat{x}$) and the adjusted observations ($\hat{\ell}$) from the conditional least squares adjustment can be given as follows:

$$\hat{x} = f\left(\hat{\ell}\right) \tag{12.79}$$

where the adjusted observations ($\hat{\ell}$) are used in deriving the unknown parameters ($\hat{x}$). By the laws of variance–covariance propagation, the covariance matrix ($C_{\hat{x}}$) of the adjusted parameters can be given as:

$$C_{\hat{x}} = J C_{\hat{\ell}} J^T \tag{12.80}$$

where J is the Jacobian matrix of Equation (12.79) with respect to all of the adjusted observations from the conditional least squares adjustment and $C_{\hat{l}}$ is the covariance matrix of the adjusted observations. The most important step in the computational process is to be able to formulate Equation (12.79), which is a set of equations used in determining the vector of adjusted parameters from the adjusted observations. The following examples are provided to illustrate the computational steps involved in deriving vectors of adjusted parameters and their covariance matrices based on results of least squares adjustment of conditional models.

Example 12.11   Continuing from Example 12.9, using the adjustment results for the problem, calculate the a posteriori variance factor of unit weight and the covariance matrix of the adjusted observations.

Solution:

From Example 12.9, the following were computed:

$$B = \begin{bmatrix} 1 & 1 & 1 \end{bmatrix} \quad Q_\ell = \begin{bmatrix} 5^2 & & \\ & 5^2 & \\ & & 5^2 \end{bmatrix} \quad M = 75.0$$

$$v = QB^T k = \begin{bmatrix} -15 \\ -15 \\ -15 \end{bmatrix} \quad P = Q_\ell^{-1}$$

The a posteriori variance factor of unit weight:

$$s_0^2 = \frac{v^T P v}{r} \quad s_0^2 = \frac{27}{1} \quad \rightarrow \quad 27.0$$

The cofactor matrix of the adjusted observations: $Q_{\hat{\ell}} = Q_\ell - Q_\ell B^T M^{-1} B Q_\ell$

$$Q_{\hat{\ell}} = \begin{bmatrix} 5^2 & & \\ & 5^2 & \\ & & 5^2 \end{bmatrix} - \begin{bmatrix} 5^2 & & \\ & 5^2 & \\ & & 5^2 \end{bmatrix} \begin{bmatrix} 1 \\ 1 \\ 1 \end{bmatrix} \frac{1}{75} \begin{bmatrix} 1 & 1 & 1 \end{bmatrix} \begin{bmatrix} 5^2 & & \\ & 5^2 & \\ & & 5^2 \end{bmatrix}$$

$$Q_{\hat{\ell}} = \begin{bmatrix} 16.667 & -8.333 & -8.333 \\ -8.333 & 16.667 & -8.333 \\ -8.333 & -8.333 & 16.667 \end{bmatrix}$$

Covariance matrix of the adjusted observations: $C_{\hat{\ell}} = s_0^2 Q_{\hat{\ell}}$ (for $s_0^2 = 27$)

$$C_{\hat{\ell}} = \begin{bmatrix} 450 & -225 & -225 \\ -225 & 450 & -225 \\ -225 & -225 & 450 \end{bmatrix}$$

Example 12.12 Determine the standard deviation of the adjusted elevation difference between benchmarks FH1 and FH2 in Figure 12.8 given the adjusted elevation differences and their corresponding standard deviations s_i in Table 12.2.

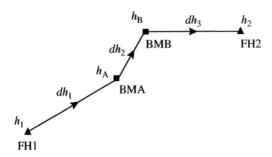

Figure 12.8 Sample differential leveling network.

Table 12.2 Adjusted elevation differences.

Leg i	BS	FS	Adjusted elevation difference dh_i (m)	Standard deviation (m)
1	FH1	BMA	−7.341	0.008
2	BMA	BMB	2.495	0.005
3	BMB	FH2	5.107	0.004

Solution:

The steps of the computation are given as follows:

1) Create the functional model between the adjusted parameter ($\hat{x}$) and the observations:

$$\hat{x} = d\hat{h}_1 + d\hat{h}_2 + d\hat{h}_3 \tag{12.81}$$

2) Determine the Jacobian matrix (J):

$$J = \frac{\partial \hat{x}}{\partial \hat{\ell}} = \left[\frac{\partial \hat{x}}{\partial d\hat{h}_1} \quad \frac{\partial \hat{x}}{\partial d\hat{h}_2} \quad \frac{\partial \hat{x}}{\partial d\hat{h}_3} \right] \tag{12.82}$$

$$J = \begin{bmatrix} 1 & 1 & 1 \end{bmatrix}$$

3) Create variance–covariance of the adjusted observations:

$$C_{\hat{\ell}} = \begin{bmatrix} s_1^2 & 0 & 0 \\ 0 & s_2^2 & 0 \\ 0 & 0 & s_3^2 \end{bmatrix} \tag{12.83}$$

or

$$C_{\hat{\ell}} = \begin{bmatrix} 0.008^2 & 0 & 0 \\ 0 & 0.005^2 & 0 \\ 0 & 0 & 0.004^2 \end{bmatrix} \tag{12.84}$$

4) Compute the variance–covariance matrix of the adjusted parameters $C_{\hat{x}}$:

$$C_{\hat{x}} = J C_{\hat{\ell}} J^T \tag{12.85}$$

$$C_{\hat{x}} = \begin{bmatrix} 1 & 1 & 1 \end{bmatrix} \begin{bmatrix} 0.008^2 & 0 & 0 \\ 0 & 0.005^2 & 0 \\ 0 & 0 & 0.004^2 \end{bmatrix} \begin{bmatrix} 1 \\ 1 \\ 1 \end{bmatrix}$$

$$C_{\hat{x}} = \begin{bmatrix} 0.000 \ 105 \end{bmatrix}$$

This is a one-by-one matrix with the only element being the variance of the adjusted parameter ($\hat{x}$).

5) The standard deviation of the adjusted parameter is as follows from step 4:

$$s_{\hat{x}} = \sqrt{0.000 \ 105} = 0.010 \, \text{m}$$

Example 12.13 Continuing from Example 12.10 and using the computed quantities in the problem, perform the following tasks with reference to Figure 12.7:

a) Compute the adjusted elevations of points A, B, and C.
b) Compute the a posteriori variance factor of unit weight.
c) Calculate the standard deviations of the adjusted elevations of points A, B, and C.

Solution:

Given the following from Example 12.10

$$B = \begin{bmatrix} 1 & 0 & 0 & 1 & 1 & 0 & 1 \\ 0 & 0 & -1 & 0 & 0 & 1 & 1 \\ 0 & 1 & 0 & 0 & -1 & 1 & 0 \\ 1 & 1 & 0 & 0 & 0 & 0 & 0 \end{bmatrix} \quad M = BQB^T = \begin{bmatrix} 1.0E-4 & 2.5E-5 & -2.5E-5 & 2.5E-5 \\ 2.5E-5 & 7.5E-5 & 2.5E-5 & 0.0 \\ -2.5E-5 & 2.5E-5 & 7.5E-5 & 2.5E-5 \\ 2.5E-5 & 0.0 & 2.5E-5 & 5.0E-5 \end{bmatrix}$$

the cofactor matrix (Q_ℓ) of observations is

$$Q_\ell = \text{diag}([\,2.5E-5 \quad 2.5E-5 \quad 2.5E-5 \quad 2.5E-5 \quad 2.5E-5 \quad 2.5E-5 \quad 2.5E-5\,])$$

$$\text{Adjusted observations}: \quad \hat{\ell} = \begin{bmatrix} \hat{dh}_1 \\ \hat{dh}_2 \\ \hat{dh}_3 \\ \hat{dh}_4 \\ \hat{dh}_5 \\ \hat{dh}_6 \\ \hat{dh}_7 \end{bmatrix} = \begin{bmatrix} 1.5324 \\ 0.6986 \\ -0.3836 \\ -1.8474 \\ -0.2021 \\ -0.9007 \\ 0.5171 \end{bmatrix}$$

a) The adjusted elevations $\hat{H}_A$, $\hat{H}_B$, and $\hat{H}_C$ of points A, B, and C, respectively, using the adjusted observations and any leveling route in the network, are computed as follows:

$$\hat{H}_A = BMX + \hat{dh}_1 \quad \rightarrow \quad 31.6324\,\text{m} \tag{12.86}$$

$$\hat{H}_B = BMY + \hat{dh}_6 \quad \rightarrow \quad 31.4303\,\text{m} \tag{12.87}$$

$$\hat{H}_C = BMY + \hat{dh}_3 \quad \rightarrow \quad 31.9474\,\text{m} \tag{12.88}$$

Equations (12.86)–(12.88) are in the form of $\hat{x} = f\left(\hat{\ell}\right)$.

b) The a posteriori variance factor of unit weight, s_0^2:

$$s_0^2 = \frac{v^T P v}{r} \quad s_0^2 = \frac{5.565\,714\,28}{4} \quad \rightarrow \quad 1.391\,43$$

where $P = Q_\ell^{-1}$.

c) The standard deviations of the adjusted elevations of points A, B, and C are calculated as follows. Remember that the adjusted observations are used in computing the adjusted elevations in question (a); the errors in the adjusted observations therefore propagate into the adjusted parameters through the use of the adjusted observations. In order to determine the covariance matrix of the adjusted elevations, the covariance matrix of the adjusted observations must be determined using the cofactor of the adjusted observations and the a posteriori variance factor of unit weight computed in question (b). Equation (12.74) is used as follows:

$$C_{\hat{l}} = \hat{C}_\ell - C_v \tag{12.89}$$

The scaled covariance matrix of the original observations is given as follows (for $s_0^2 = 1.391\,43$):

$$\hat{C}_\ell = s_0^2 Q_\ell = 1.391\,43 \times \text{diag}\begin{bmatrix} 0.005^2 & 0.005^2 & 0.005^2 & 0.005^2 & 0.005^2 & 0.005^2 & 0.005^2 \end{bmatrix}$$

or

$$\hat{C}_\ell = \text{diag}\begin{bmatrix} 3.48E-5 & 3.48E-5 & 3.48E-5 & 3.48E-5 & 3.48E-5 & 3.48E-5 & 3.48E-5 \end{bmatrix}$$

The covariance matrix of residuals is given as $C_v = s_0^2 Q_\ell B^T M^{-1} B Q_\ell$

$$C_v = \begin{bmatrix} 2.15E-5 & 1.33E-5 & -1.66E-6 & 1.66E-6 & 8.28E-6 & -4.97E-6 & 3.31E-6 \\ 1.33E-5 & 2.15E-5 & 1.66E-6 & -1.66E-6 & -8.28E-6 & 4.97E-6 & -3.31E-6 \\ -1.66E-6 & 1.66E-6 & 2.15E-5 & 1.33E-5 & -3.31E-6 & -4.97E-6 & -8.28E-6 \\ 1.66E-6 & -1.66E-6 & 1.33E-5 & 2.15E-5 & 3.31E-6 & 4.97E-6 & 8.28E-6 \\ 8.28E-6 & -8.28E-6 & -3.31E-6 & 3.31E-6 & 1.66E-5 & -9.94E-6 & 6.63E-6 \\ -4.97E-6 & 4.97E-6 & -4.97E-6 & 4.97E-6 & -9.94E-6 & 1.99E-5 & 9.94E-6 \\ 3.31E-6 & -3.31E-6 & -8.28E-6 & 8.28E-6 & 6.63E-6 & 9.94E-6 & 1.66E-5 \end{bmatrix}$$

From Equation (12.89), the covariance matrix of the adjusted observations is given as follows:

$$C_{\hat{l}} = \begin{bmatrix} 1.33E-5 & -1.33E-5 & 1.66E-6 & -1.66E-6 & -8.28E-6 & 4.97E-6 & -3.31E-6 \\ -1.33E-5 & 1.33E-5 & -1.66E-6 & 1.66E-6 & 8.28E-6 & -4.97E-6 & 3.31E-6 \\ 1.66E-6 & -1.66E-6 & 1.33E-5 & -1.33E-5 & 3.31E-6 & 4.97E-6 & 8.28E-6 \\ -1.66E-6 & 1.66E-6 & -1.33E-5 & 1.33E-5 & -3.31E-6 & -4.97E-6 & -8.28E-6 \\ -8.28E-6 & 8.28E-6 & 3.31E-6 & -3.31E-6 & 1.82E-5 & 9.94E-6 & -6.63E-6 \\ 4.97E-6 & -4.97E-6 & 4.97E-6 & -4.97E-6 & 9.94E-6 & 1.49E-5 & -9.94E-6 \\ -3.31E-6 & 3.31E-6 & 8.28E-6 & -8.28E-6 & -6.63E-6 & -9.94E-6 & 1.82E-5 \end{bmatrix}$$

The model Equations (12.86)–(12.88) constitute the functional model, $\hat{x} = f\left(\hat{\ell}\right)$, as follows:

$$\hat{H}_A = BMX + d\hat{h}_1 \tag{12.90}$$

$$\hat{H}_B = BMY + d\hat{h}_6 \tag{12.91}$$

$$\hat{H}_C = BMY + d\hat{h}_3 \tag{12.92}$$

Find the Jacobian matrix (J) of the Equations (12.90)–(12.92) with respect to all of the adjusted observations ($\hat{\ell}$) from the original adjustment (including those that are not used in Equations (12.90)–(12.92)) as follows:

$$\hat{\ell} = \begin{bmatrix} d\hat{h}_1 \\ d\hat{h}_2 \\ d\hat{h}_3 \\ d\hat{h}_4 \\ d\hat{h}_5 \\ d\hat{h}_6 \\ d\hat{h}_7 \end{bmatrix} \quad J = \frac{\partial \hat{x}}{\partial \hat{\ell}} = \begin{bmatrix} 1 & 0 & 0 & 0 & 0 & 0 & 0 \\ 0 & 0 & 0 & 0 & 0 & 1 & 0 \\ 0 & 0 & 1 & 0 & 0 & 0 & 0 \end{bmatrix}$$

From the laws of variance–covariance propagation:

$$C_{\hat{x}} = J C_{\hat{\ell}} J^T = \begin{bmatrix} 1.325\text{E}-5 & 4.969\text{E}-6 & 1.656\text{E}-6 \\ 4.969\text{E}-6 & 1.491\text{E}-5 & 4.969\text{E}-6 \\ 1.656\text{E}-6 & 4.969\text{E}-6 & 1.325\text{E}-5 \end{bmatrix} \tag{12.93}$$

The square roots of the diagonal elements of the matrix in Equation (12.93) are the standard deviations of the adjusted elevations of points A, B, and C, which are $\sigma_{\hat{H}_A} = 0.0036$ m, $\sigma_{\hat{H}_B} = 0.0039$ m, and $\sigma_{\hat{H}_C} = 0.0036$ m, respectively.

12.7 Simple GNSS Network Adjustment Example

A simple Global Navigation Satellite System (GNSS) triangular network is considered in this example to illustrate the application of conditional model adjustment. Consider a scenario where a control point C (x, y, z) is to be established from two fixed points A and B (referring to Figure 12.9) using GPS survey. The coordinates of the fixed points are given in three-dimensional GPS system as:

$$A(x = 402.351 \text{ m}, \quad y = -4\ 652\ 995.301 \text{ m}, \quad z = 4\ 349\ 760.778 \text{ m})$$

$$B(x = 8086.032 \text{ m}, \quad y = -4\ 642\ 712.847 \text{ m}, \quad z = 4\ 360\ 439.083 \text{ m})$$

The measured GPS baseline vectors are shown in Table 12.3, assuming the standard deviation of each observation is 0.010 m. Perform the following tasks:

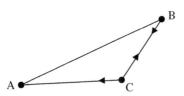

Figure 12.9 GPS baselines.

a) Compute the least squares adjusted coordinates of point C using the condition equations method.

Table 12.3 GPS baseline vectors.

From	To	dx (m)	dy (m)	dz (m)
C	A	−1116.452	−4596.161	−4355.906
C	B	6567.231	5686.293	6322.392
B	C	−6567.231	−5686.303	−6322.381

b) Compute the standard deviations of the vector of the adjusted observations.

Solution:

a) The least squares adjusted coordinates of point C using the condition equations method are determined as follows. The baseline vectors constitute the observations (ℓ) given as follows:

$$\ell = [dx_{CA}\ dx_{CB}\ dx_{BC}\ dy_{CA}\ dy_{CB}\ dy_{BC}\ dz_{CA}\ dz_{CB}\ dz_{BC}]^T$$

The condition equations are

$$d\hat{x}_{CA} + d\hat{x}_{AB} + d\hat{x}_{BC} = 0 \tag{12.94}$$

$$d\hat{y}_{CA} + d\hat{y}_{AB} + d\hat{y}_{BC} = 0 \tag{12.95}$$

$$d\hat{z}_{CA} + d\hat{z}_{AB} + d\hat{z}_{BC} = 0 \tag{12.96}$$

$$d\hat{x}_{CB} + d\hat{x}_{BC} = 0 \tag{12.97}$$

$$d\hat{y}_{CB} + d\hat{y}_{BC} = 0 \tag{12.98}$$

$$d\hat{z}_{CB} + d\hat{z}_{BC} = 0 \tag{12.99}$$

The B-matrix from Equations (12.94)–(12.99) (ignoring the hats):

$$B = \frac{\partial f}{\partial \ell} = \begin{bmatrix} 1 & 0 & 1 & 0 & 0 & 0 & 0 & 0 & 0 \\ 0 & 0 & 0 & 1 & 0 & 1 & 0 & 0 & 0 \\ 0 & 0 & 0 & 0 & 0 & 0 & 1 & 0 & 1 \\ 0 & 1 & 1 & 0 & 0 & 0 & 0 & 0 & 0 \\ 0 & 0 & 0 & 0 & 1 & 1 & 0 & 0 & 0 \\ 0 & 0 & 0 & 0 & 0 & 0 & 0 & 1 & 1 \end{bmatrix}$$

The misclosure vector (w) obtained by using the measurements in Equations (12.94)–(12.99) with the fixed vector of line AB as $d\hat{x}_{AB} = 7\,683.681$, $d\hat{y}_{AB} = 10\,282.454$, and $d\hat{z}_{AB} = 10\,678.305$ is as follows:

$$w = f(\ell) = \begin{bmatrix} -0.002 \\ -0.010 \\ 0.018 \\ 0.000 \\ -0.010 \\ 0.011 \end{bmatrix}$$

The cofactor matrix based on the variances of the baseline vectors:

$$Q = \text{diag}\left([1\ \ 1\ \ 1\ \ 1\ \ 1\ \ 1\ \ 1\ \ 1\ \ 1] \times (0.01)^2\right)$$

Form the M-matrix:

$$M = BQB^T = \begin{bmatrix}
2.0E-4 & 0.0 & 0.0 & 1.0E-4 & 0.0 & 0.0 \\
0.0 & 2.0E-4 & 0.0 & 0.0 & 1.0E-4 & 0.0 \\
0.0 & 0.0 & 2.0E-4 & 0.0 & 0.0 & 1.0E-4 \\
1.0E-4 & 0.0 & 0.0 & 2.0E-4 & 0.0 & 0.0 \\
0.0 & 1.0E-4 & 0.0 & 0.0 & 2.0E-4 & 0.0 \\
0.0 & 0.0 & 1.0E-4 & 0.0 & 0.0 & 2.0E-4
\end{bmatrix}$$

Solve for the correlates:

$$k = -M^{-1}w = \begin{bmatrix} 13.3334 \\ 33.334 \\ -83.3343 \\ -6.6666 \\ 33.334 \\ -13.3343 \end{bmatrix}$$

The residuals: $v = QB^T k$

$$v = [0.001\ \ -0.001\ \ 0.001\ \ 0.003\ \ 0.003\ \ 0.007\ \ -0.008\ \ -0.001\ \ -0.010]^T$$

The adjusted observations:

$$\hat{\ell} = \ell + v = \begin{bmatrix} -1116.452 \\ 6567.231 \\ -6567.231 \\ -4596.161 \\ 5686.293 \\ -5686.303 \\ -4355.906 \\ 6322.392 \\ -6322.381 \end{bmatrix} + \begin{bmatrix} 0.001 \\ -0.001 \\ 0.001 \\ 0.003 \\ 0.003 \\ 0.007 \\ -0.008 \\ -0.001 \\ -0.010 \end{bmatrix} = \begin{bmatrix} -1116.451 \\ 6567.230 \\ -6567.230 \\ -4596.158 \\ 5686.296 \\ -5686.296 \\ -4355.914 \\ 6322.391 \\ -6322.391 \end{bmatrix}$$

Use any route to determine the adjusted coordinates of C using the adjusted observations; using the baseline CA or CB, the following adjusted coordinates are obtained:

$$X = 1\ 518.802\,\text{m}$$

$$Y = -4\ 648\ 399.143\,\text{m}$$

$$Z = 4\ 354\ 116.692\,\text{m}$$

b) The standard deviations of the vector of the adjusted observations are computed as follows. First compute the a posteriori variance factor of unit weight:

$$s_0^2 = \frac{v^T P v}{r} \quad s_0^2 = \frac{2.35}{6} \quad \rightarrow \quad 0.392$$

Compute the scaled covariance of the observations:

$$\hat{C}_\ell = s_0^2 Q_\ell$$

$$\hat{C}_\ell = 0.392 \times \text{diag}\big([1\ 1\ 1\ 1\ 1\ 1\ 1\ 1\ 1]\times (0.01)^2\big)$$

Compute the covariance of the residuals:

$$C_v = s_0^2 Q_\ell B^T M^{-1} B Q_\ell = 0.392 \times$$

0.26E−4	−0.13E−4	0.13E−4	0	0	0	0	0	0
−0.13E−4	0.26E−4	0.13E−4	0	0	0	0	0	0
0.13E−4	0.13E−4	0.26E−4	0	0	0	0	0	0
0	0	0	0.26E−4	−0.13E−4	0.13E−4	0	0	0
0	0	0	−0.13E−4	0.26E−4	0.13E−4	0	0	0
0	0	0	0.13E−4	0.13E−4	0.26E−4	0	0	0
0	0	0	0	0	0	0.26E−4	−0.13E−4	0.13E−4
0	0	0	0	0	0	−0.13E−4	0.26E−4	0.13E−4
0	0	0	0	0	0	0.13E−4	0.13E−4	0.26E−4

Compute the covariance matrix of the adjusted observations:

$$C_{\hat{\ell}} = \hat{C}_\ell - C_v =$$

0.13E−4	0.13E−4	−0.13E−4	0	0	0	0	0	0
0.13E−4	0.13E−4	−0.13E−4	0	0	0	0	0	0
−0.13E−4	−0.13E−4	0.13E−4	0	0	0	0	0	0
0	0	0	0.13E−4	0.13E−4	−0.13E−4	0	0	0
0	0	0	0.13E−4	0.13E−4	−0.13E−4	0	0	0
0	0	0	−0.13E−4	−0.13E−4	0.13E−4	0	0	0
0	0	0	0	0	0	0.13E−4	0.13E−4	−0.13E−4
0	0	0	0	0	0	0.13E−4	0.13E−4	−0.13E−4
0	0	0	0	0	0	−0.13E−4	−0.13E−4	0.13E−4

All the adjusted observations in this case have the same standard deviation of $\sqrt{0.13\text{E}-4}$(or 0.004 m).

12.8 Simple Traverse Network Adjustment Example

A simple traverse network is used in this section to illustrate how to adjust a traverse network using conditional least squares method. The approach can be extended to a more complex traverse network with the steps given in this section applicable in the same way. For illustration, the two-dimensional fitted traverse network in Figure 12.10 with field measurements given in Table 12.4 is to be adjusted using conditional least squares adjustment method. The measurements consist of two angles, two bearings, and three distances with their standard deviations. The coordinates of the control points B and E (to be fixed in the adjustment) and the approximate coordinates of the unknown points C and D to be determined are given in Table 12.5. Compute the adjusted coordinates of points C and D and their standard deviations.

The conditional model equations $\left(f\left(\hat{\ell} \right) = 0 \right)$ are formulated as follows. As discussed in Example 12.6, three conditions are possible with this traverse problem:

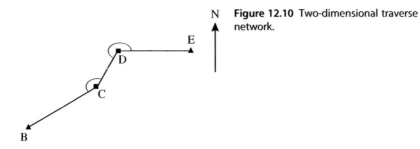

Figure 12.10 Two-dimensional traverse network.

Table 12.4 Field measurements.

Symbol	At	From	To	Measurement	Standard deviation
ℓ_1	C	B	D	149°59′45″	10″
ℓ_2	D	C	E	240°01′00″	10″
ℓ_3		B	C	59°59′15″	2″
ℓ_4		D	E	90°00′00″	2″
ℓ_5		B	C	199.880 m	3 mm + 2 ppm
ℓ_6		C	D	99.900 m	0.005 m
ℓ_7		D	E	177.000 m	0.005 m

Table 12.5 Coordinates of control points and approximate coordinates of unknown points.

Station	X (m)	Y (m)
B	1000.000	1000.000
E	1400.000	1186.500
C	1173	1100
D	1223	1186

1) Using the starting bearing of the first traverse leg and the measured angles at each setup point, compute the bearing of the end traverse leg and compare the computed value with the given bearing of that end traverse leg; the difference between them must be equal to zero. This is expressed mathematically as follows:

$$\hat{\ell}_3 + \hat{\ell}_1 + \hat{\ell}_2 - \hat{\ell}_4 - 360° = 0 \tag{12.100}$$

2) The sum of calculated changes in x coordinates from the beginning of the traverse to the end of traverse must be equal to the difference between the x coordinates of the first traverse point and the terminating traverse point. This is expressed mathematically as follows:

$$d\hat{X}_{BC} + d\hat{X}_{CD} + d\hat{X}_{DE} - (X_E - X_B) = 0 \tag{12.101}$$

which must be rewritten in terms of the observations, in order to be solvable, as follows:

$$\hat{\ell}_5 \sin\left(\hat{\ell}_3\right) + \hat{\ell}_6 \sin\left(\hat{\ell}_3 + 180 + \hat{\ell}_1\right) + \hat{\ell}_7 \sin\left(\hat{\ell}_3 + \hat{\ell}_1 + \hat{\ell}_2 - 360°\right) - (X_E - X_B) = 0 \tag{12.102}$$

3) Repeat step 2 for the changes in y coordinates, giving the following:

$$d\hat{Y}_{BC} + d\hat{Y}_{CD} + d\hat{Y}_{DE} - (Y_E - Y_B) = 0 \tag{12.103}$$

which also must be rewritten in terms of the observations, in order to be solvable, as follows:

$$\hat{\ell}_5 \cos\left(\hat{\ell}_3\right) + \hat{\ell}_6 \cos\left(\hat{\ell}_3 + 180 + \hat{\ell}_1\right) + \hat{\ell}_7 \cos\left(\hat{\ell}_3 + \hat{\ell}_1 + \hat{\ell}_2 - 360°\right) - (Y_E - Y_B) = 0 \tag{12.104}$$

Substitute the values in Table 12.4 into Equations (12.100) (expressed in radians), (12.102), and (12.104) to obtain the misclosure vector, given as follows:

$$w = f(\ell) = \begin{bmatrix} 0.0 \\ 0.004\ 181\ 484 \\ 0.008\ 226\ 479 \end{bmatrix}$$

The diagonal values of the cofactor matrix (Q) of the observations using the given standard deviations (with the standard deviations of angles converted to radians) are given as follows:

$$Q = \text{diag}[2.35\text{E}-9 \quad 2.35\text{E}-9 \quad 9.4\text{E}-11 \quad 9.4\text{E}-11 \quad 1.16\text{E}-5 \quad 2.5\text{E}-5 \quad 2.5\text{E}-5]$$

The second design (B) matrix, which is the Jacobian matrix of Equations (12.100), (12.102), and (12.104) with respect to the observations (with hats ignored), $\ell = [\ell_1 \ \ell_2 \ \ell_3 \ \ell_4 \ \ell_5 \ \ell_6 \ \ell_7]^T$, is obtained as follows. In the following partial derivatives, let Equation (12.100) be represented by w_1, Equation (12.102) by w_2, and Equation (12.104) by w_3.

$$\frac{\partial w_1}{\partial \ell_1} = 1.0; \quad \frac{\partial w_1}{\partial \ell_2} = 1.0; \quad \frac{\partial w_1}{\partial \ell_3} = 1.0; \quad \frac{\partial w_1}{\partial \ell_4} = -1.0; \quad \frac{\partial w_1}{\partial \ell_5} = 0.0; \quad \frac{\partial w_1}{\partial \ell_6} = 0.0; \quad \frac{\partial w_1}{\partial \ell_7} = 0.0$$

$$\frac{\partial w_2}{\partial \ell_1} = \ell_7 \cos(\ell_3 + \ell_1 + \ell_2) - \ell_6 \cos(\ell_3 + \ell_1) = 86.5305$$

$$\frac{\partial w_2}{\partial \ell_2} = \ell_7 \cos(\ell_3 + \ell_1 + \ell_2) = 0.0$$

$$\frac{\partial w_2}{\partial \ell_3} = \ell_7 \cos(\ell_3 + \ell_1 + \ell_2) - \ell_6 \cos(\ell_3 + \ell_1) + \ell_5 \cos(\ell_3) = 186.508$$

$$\frac{\partial w_2}{\partial \ell_4} = 0.0$$

$$\frac{\partial w_2}{\partial \ell_5} = \sin(\ell_3) = 0.866$$

$$\frac{\partial w_2}{\partial \ell_6} = -\sin(\ell_3 + \ell_1) = 0.500$$

$$\frac{\partial w_2}{\partial \ell_7} = \sin(\ell_3 + \ell_1 + \ell_2) = 1.000$$

$$\frac{\partial w_3}{\partial \ell_1} = -\ell_7 \sin(\ell_3 + \ell_1 + \ell_2) + \ell_6 \sin(\ell_3 + \ell_1) = -226.9248$$

$$\frac{\partial w_3}{\partial \ell_2} = -\ell_7 \sin(\ell_3 + \ell_1 + \ell_2) = -177.000$$

$$\frac{\partial w_3}{\partial \ell_3} = -\ell_7 \sin(\ell_3 + \ell_1 + \ell_2) + \ell_6 \sin(\ell_3 + \ell_1) - \ell_5 \sin(\ell_3) = -400.004$$

$$\frac{\partial w_3}{\partial \ell_4} = 0.0$$

$$\frac{\partial w_3}{\partial \ell_5} = \cos(\ell_3) = 0.500$$

$$\frac{\partial w_3}{\partial \ell_6} = -\cos(\ell_3 + \ell_1) = 0.866$$

$$\frac{\partial w_3}{\partial \ell_7} = \cos(\ell_3 + \ell_1 + \ell_2) = 0.000$$

The summary of the partial derivatives is given as second design matrix (B) as follows:

$$B = \begin{bmatrix} 1.0000 & 1.0000 & 1.0000 & -1.000 & 0.000 & 0.000 & 0.000 \\ 86.5305 & 0.0000 & 186.508 & 0.0000 & 0.866 & 0.500 & 1.000 \\ -226.9248 & -1.7700 & -400.004 & 0.000 & 0.500 & 0.866 & 0.000 \end{bmatrix}$$

$$M = BQB^T = \begin{bmatrix} 4.888\ 9E-9 & 2.209\ 2E-7 & -9.870\ 1E-7 \\ 2.209\ 2E-7 & 6.078\ 09E-5 & -3.733\ 9E-5 \\ -9.870\ 1E-7 & -3.733\ 9E-5 & 2.313\ 64E-4 \end{bmatrix}$$

The computed vector of correlates:

$$k = -M^{-1}w = \begin{bmatrix} -47\ 449.87 \\ -47.2 \\ -245.60 \end{bmatrix}$$

The residual vector:

$$v = QB^T k = \begin{bmatrix} 9.866\ 370\ 35E-6 \\ -9.352\ 682\ 83E-6 \\ 3.947\ 433\ 059E-6 \\ 4.461\ 120\ 585E-6 \\ -1.892\ 605\ 3E-3 \\ -5.907\ 991E-3 \\ -1.180\ 110E-3 \end{bmatrix}$$

The adjusted observations:

$$\hat{\ell} = \ell + v = \begin{bmatrix} 2.617\ 931\ 022\ 309\ 7 \\ 4.189\ 071\ 740\ 312\ 2 \\ 1.046\ 983\ 332\ 473\ 2 \\ 1.570\ 800\ 787\ 915\ 5 \\ 199.878\ 11 \\ 99.894\ 09 \\ 176.998\ 82 \end{bmatrix}$$

The *a posteriori* variance factor of unit weight:

$$s_0^2 = \frac{v^T Q^{-1} v}{df} = \frac{2.217\ 787}{3} = 0.7393$$

From Equation (12.71), the covariance matrix of the adjusted observations can be given as

$$C_{\hat{\ell}} = s_0^2 Q - s_0^2 Q B^T M^{-1} B Q$$

$$C_{\hat{\ell}} = \begin{bmatrix} 0.639 & -0.558 & -0.088 & -0.007 & -6.620 & 16.700 & -41.50 \\ -0.558 & 0.609 & 0.023 & 0.074 & 7.840 & -15.50 & 44.90 \\ -0.088 & 0.023 & 0.056 & -0.009 & 0.190 & 7.290 & -6.58 \\ -0.007 & 0.074 & -0.009 & 0.059 & 1.41 & 8.49 & -3.21 \\ -6.620 & 7.840 & 0.190 & 1.41 & 6637.0 & -3880.0 & -3270.0 \\ 16.700 & -15.50 & 7.290 & 8.49 & -3880.0 & 6820.0 & -2860.0 \\ -41.50 & 44.90 & -6.58 & -3.21 & -3270.0 & -2860.0 & 9080.0 \end{bmatrix} \times E-9$$

The adjusted coordinates of points C and D can be calculated using the adjusted observations as follows:

$$\hat{X}_C = 1000.000 + \hat{\ell}_5 \sin\left(\hat{\ell}_3\right) = 1173.0781 \text{ m} \tag{12.105}$$

$$\hat{Y}_C = 1000.000 + \hat{\ell}_5 \cos\left(\hat{\ell}_3\right) = 1099.9761 \text{ m} \tag{12.106}$$

$$\hat{X}_D = 1000.000 + \hat{\ell}_5 \sin\left(\hat{\ell}_3\right) + \hat{\ell}_6 \sin\left(\hat{\ell}_3 + 180 + \hat{\ell}_1\right) = 1223.0012 \text{ m} \tag{12.107}$$

$$\hat{Y}_D = 1000.000 + \hat{\ell}_5 \cos\left(\hat{\ell}_3\right) + \hat{\ell}_6 \cos\left(\hat{\ell}_3 + 180 + \hat{\ell}_1\right) = 1186.5008 \text{ m} \tag{12.108}$$

$$\hat{X}_E = 1000.000 + \hat{\ell}_5 \sin\left(\hat{\ell}_3\right) + \hat{\ell}_6 \sin\left(\hat{\ell}_3 + 180 + \hat{\ell}_1\right) + \hat{\ell}_7 \sin\left(\hat{\ell}_3 + \hat{\ell}_1 + \hat{\ell}_2\right)$$
$$= 1400.000 \text{ m} \tag{12.109}$$

$$\hat{Y}_E = 1000.000 + \hat{\ell}_5 \cos\left(\hat{\ell}_3\right) + \hat{\ell}_6 \cos\left(\hat{\ell}_3 + 180 + \hat{\ell}_1\right) + \hat{\ell}_7 \cos\left(\hat{\ell}_3 + \hat{\ell}_1 + \hat{\ell}_2\right)$$
$$= 1186.500 \text{ m} \tag{12.110}$$

It can be seen from Equations (12.109) and (12.110) that the adjusted coordinates of the control point E match the given coordinate values for the control points; this is to check that the adjustment is done properly. The variance–covariance propagation laws can now be applied on Equations (12.105)–(12.108) to determine the variance–covariance matrix ($C_{\hat{x}}$) of the adjusted parameters $\hat{X}_C$, $\hat{Y}_C$, $\hat{X}_D$, and $\hat{Y}_D$, as follows:

$$C_{\hat{x}} = G C_{\hat{\ell}} G^T \tag{12.111}$$

where G is the Jacobian matrix (the partial derivatives) of the Equations (12.105)–(12.108) with respect to the adjusted observations $\hat{\ell} = \begin{bmatrix} \hat{\ell}_1 & \hat{\ell}_2 & \hat{\ell}_3 & \hat{\ell}_4 & \hat{\ell}_5 & \hat{\ell}_6 & \hat{\ell}_7 \end{bmatrix}^T$. The partial derivatives are given as follows:

$$\frac{\partial \hat{X}_C}{\partial \hat{\ell}_1} = 0.0; \quad \frac{\partial \hat{X}_C}{\partial \hat{\ell}_2} = 0.0; \quad \frac{\partial \hat{X}_C}{\partial \hat{\ell}_3} = \hat{\ell}_5 \cos\left(\hat{\ell}_3\right) = 99.976\ 13; \quad \frac{\partial \hat{X}_C}{\partial \hat{\ell}_4} = 0.0;$$

$$\frac{\partial \hat{X}_C}{\partial \hat{\ell}_5} = \sin\left(\hat{\ell}_3\right) = 0.865\ 918; \quad \frac{\partial \hat{X}_C}{\partial \hat{\ell}_6} = 0.0; \quad \frac{\partial \hat{X}_C}{\partial \hat{\ell}_7} = 0.0$$

$$\frac{\partial \hat{Y}_C}{\partial \hat{\ell}_1} = 0.0; \quad \frac{\partial \hat{Y}_C}{\partial \hat{\ell}_2} = 0.0; \quad \frac{\partial \hat{Y}_C}{\partial \hat{\ell}_3} = -\hat{\ell}_5 \sin\left(\hat{\ell}_3\right) = -173.078\ 11; \quad \frac{\partial \hat{Y}_C}{\partial \hat{\ell}_4} = 0.0;$$

$$\frac{\partial \hat{Y}_C}{\partial \hat{\ell}_5} = \cos\left(\hat{\ell}_3\right) = 0.500\ 19; \quad \frac{\partial \hat{Y}_C}{\partial \hat{\ell}_6} = 0.0; \quad \frac{\partial \hat{Y}_C}{\partial \hat{\ell}_7} = 0.0$$

$$\frac{\partial \hat{X}_D}{\partial \hat{\ell}_1} = -\hat{\ell}_6 \cos\left(\hat{\ell}_1 + \hat{\ell}_3\right) = 86.524\ 66; \quad \frac{\partial \hat{X}_D}{\partial \hat{\ell}_2} = 0.0;$$

$$\frac{\partial \hat{X}_D}{\partial \hat{\ell}_3} = \hat{\ell}_5 \cos\left(\hat{\ell}_3\right) - \hat{\ell}_6 \cos\left(\hat{\ell}_1 + \hat{\ell}_3\right) = 186.500\ 8;$$

$$\frac{\partial \hat{X}_D}{\partial \hat{\ell}_4} = 0.0; \quad \frac{\partial \hat{X}_D}{\partial \hat{\ell}_5} = \sin\left(\hat{\ell}_3\right) = 0.865\ 92;$$

$$\frac{\partial \hat{X}_D}{\partial \hat{\ell}_6} = -\sin\left(\hat{\ell}_1 + \hat{\ell}_3\right) = 0.499\ 76; \quad \frac{\partial \hat{X}_D}{\partial \hat{\ell}_7} = 0.0;$$

$$\frac{\partial \hat{Y}_D}{\partial \hat{\ell}_1} = \hat{\ell}_6 \sin\left(\hat{\ell}_1 + \hat{\ell}_3\right) = 49.923\ 07; \quad \frac{\partial \hat{Y}_D}{\partial \hat{\ell}_2} = 0.0;$$

$$\frac{\partial \hat{Y}_D}{\partial \hat{\ell}_3} = -\hat{\ell}_5 \sin\left(\hat{\ell}_3\right) + \hat{\ell}_6 \sin\left(\hat{\ell}_1 + \hat{\ell}_3\right) = -223.001;$$

$$\frac{\partial \hat{Y}_D}{\partial \hat{\ell}_4} = 0.0; \quad \frac{\partial \hat{Y}_D}{\partial \hat{\ell}_5} = \cos\left(\hat{\ell}_3\right) = 0.500\ 19;$$

$$\frac{\partial \hat{Y}_D}{\partial \hat{\ell}_6} = -\cos\left(\hat{\ell}_1 + \hat{\ell}_3\right) = 0.866\ 16; \quad \frac{\partial \hat{Y}_D}{\partial \hat{\ell}_7} = 0.0;$$

The G matrix is given as follows:

$$G = \begin{bmatrix} 0.0 & 0.0 & 99.976\ 13 & 0.0 & 0.865\ 918 & 0.0 & 0.0 \\ 0.0 & 0.0 & -173.078\ 11 & 0.0 & 0.500\ 19 & 0.0 & 0.0 \\ 86.524\ 66 & 0.0 & 186.500\ 8 & 0.0 & 0.865\ 92 & 0.499\ 76 & 0.0 \\ -49.923\ 07 & 0.0 & -223.001 & 0.0 & 0.500\ 19 & 0.866\ 16 & 0.0 \end{bmatrix}$$

Table 12.6 Adjusted coordinates with their standard deviations.

Station	X (m)	Standard deviation (m)	Y (m)	Standard deviation (m)
C	1173.0781	0.0024	1099.9761	0.0018
D	1223.0012	0.0030	1186.5008	0.0014

Substituting G and $C_{\hat{l}}$ into Equation (12.111) gives

$$
C_{\hat{x}} = \begin{bmatrix}
5.565 & 1.894 & 3.491 & 0.056 \\
1.894 & 3.292 & 0.498 & 0.402 \\
3.491 & 0.498 & 9.081 & 0.569 \\
0.056 & 0.402 & 0.569 & 1.841
\end{bmatrix} \times \mathrm{E}-6
$$

The adjusted coordinates ($\hat{X}_C$, $\hat{Y}_C$, $\hat{X}_D$, and $\hat{Y}_D$) calculated from Equations (12.105)–(12.108) and their standard deviations $C_{\hat{x}}$ are given in Table 12.6.

The results in Table 12.6 are consistent with the results determined for the same traverse adjustment problem based on parametric least squares method given in Chapter 6.

Problems

12.1 Given the linearized conditional model equations as:

$$m + Br = 0 \quad C_\lambda$$

where $B = \dfrac{\partial f}{\partial \lambda}$ is the second design matrix; $m = f(\lambda)$ a vector of misclosures; r is a vector of residuals; and C_λ is variance–covariance matrix of the observation vector λ.
   a) Derive the least squares normal equation. Show your derivation steps and define every other variable used.
   b) Derive the least squares solution from the normal equations.

12.2 Five height differences have been observed between four stations as shown in the following table and Figure P12.2. The heights of stations P1 and P3 (fixed and errorless) are 276.359 and 268.308 m above the datum, respectively. The standard deviation of an observed height difference can be determined using 0.005 m $\sqrt{k}$ (where k is the length of the level route in km).

Field Measurements

Stations			
From	To	Observed height differences (m)	Approx. distance (km)
P4	P1	61.478	10
P1	P2	16.994	15
P2	P3	−25.051	9
P3	P4	−53.437	18
P4	P2	78.465	20

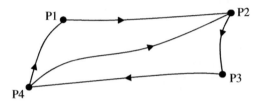

Figure P12.2

a) Using the conditional least squares adjustment procedure, compute the adjusted height differences for the leveling network in the figure.

b) Compute the *a posteriori* variance–covariance factor s_0^2.

12.3 Figure P12.3 shows a leveling network to be adjusted; the height difference measurements Δh_{ij} and their corresponding standard deviations are provided. Point P1 is a fixed point with known elevation of $H_1 = 107.500$ m. Answer the following questions.

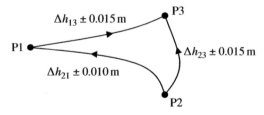

Figure P12.3

a) Compute what would be the covariance matrices of the vector of residuals and the vector of the adjusted observations, given the *a posteriori* variance–covariance factor of unit weight as $s_0^2 = 0.7273$.

b) If the adjusted elevations of points P2 and P3 will be calculated, respectively, as

$$\hat{H}_2 = 107.500 - \Delta\hat{h}_{21}$$
$$\hat{H}_3 = 107.500 + \Delta\hat{h}_{13}$$

where $\Delta\hat{h}_{21}$ is adjusted elevation difference from points P2 to P1 and $\Delta\hat{h}_{13}$ is the adjusted elevation difference from points P1 to P3. Compute the standard deviations of the adjusted elevations of points P2 and P3.

12.4 At a given station, a transit was used to measure four angles, closing the horizon. The value for each angle was found several times, and a standard deviation (σ) for each angle was found, as shown in the table below.

Field measurements

Angle	Observed mean value	σ
AOB	34°15′31″	±03″
BOC	23°49′55″	±10″
COD	190°50′54″	±05″
DOA	111°03′49″	±02″

After the least squares adjustment of the conditional model, the *a posteriori* variance–covariance factor is $s_0^2 = 0.586\,96$.
a) Give the conditional model and the M matrix.
b) Compute the covariance matrix of the residual vector C_v.
c) Compute the variance–covariance matrix of the adjusted observations $C_{\hat{\ell}}$.

13

Least Squares Adjustment of General Models

OBJECTIVES

After studying this chapter, you should be able to:

1) Derive and formulate the steps involved in the least squares method of adjusting the general mathematical models in geomatics.
2) Perform least squares adjustment of general mathematical models.
3) Apply the variance–covariance propagation law to determine the variance–covariance matrices of the adjusted quantities.

Understanding Least Squares Estimation and Geomatics Data Analysis, First Edition.
John Olusegun Ogundare.
© 2019 John Wiley & Sons, Inc. Published 2019 by John Wiley & Sons, Inc.
Companion website: www.wiley.com/go/ogundare/Understanding-lse-and-gda

4) Derive the least squares steps for solving the general models with constraints.
5) Perform least squares adjustment with weight constraints on the parameters.

13.1 Introduction

General models are considered in this chapter as mathematical equations in which the observations (or measurements) and parameters (unknown quantities) are intertwined. In these types of equations, it is usually difficult to separate the observations and the parameters in such a way as to make one a subject of the other as in the case of parametric models. It is, however, possible to rearrange parametric equations to look like general model equations, but real general model equations cannot be rearranged to form parametric model equations. Symbolically, a general model can be expressed mathematically as

$$f(\hat{x}, \hat{\ell}) = 0 \qquad (13.1)$$

where $\hat{x}$ is a vector of adjusted parameters and $\hat{\ell}$ is a vector of adjusted observations. From Equation (13.1), it can be understood that both the parameters and the observations are formulated in the equations in such a way that they are both on the left-hand side of the equality sign with zeros on the right-hand side. If the general model equations are not too convoluted, however, it may be possible to rearrange the equations to produce parametric model equations. In this case, the parametric model can be considered as a special form of general model. Another special form of general model may have parameters as functions of observations, where parameters are calculated directly from the observations.

The determination of degrees of freedom or *redundancy of general model equations* is similar to that of parametric adjustment. In parametric adjustment the number of degrees of freedom is the number of observations (which is the same as the number of parametric equations) minus the number of parameters; in general model adjustment, it is the number of general model equations minus the number of unknown parameters. This chapter discusses general model equation formulation and the adjustment procedure with computational examples.

13.2 General Model Equation Formulation

In formulating general model equations in the form given in Equation (13.1), the following should be considered:

1) Identify from the problem what constitutes the parameters (x) and what constitutes the observations (ℓ), and use suitable symbols to represent each of them; without this, one will not be able to formulate the equations.

2) Arrange the parameters (x) and the observations (ℓ) in the equations such that each equation is equated to zero. Each of the already identified parameters and observations should be represented, for now, in form of symbol, deferring the use of their numerical values till when the equations are to be solved.

3) In this book, hats are only attached to parameters and observations in order to identify them as the quantities to be adjusted; hats, of course, can be ignored if their use will be inconvenient.

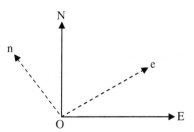

Figure 13.1 Relationship between two specified coordinate systems.

Typical general model equations are formulated as follows. For example, consider a transformation problem with regard to two different types of two-dimensional coordinate systems illustrated in Figure 13.1: the local photo coordinate (e, n) *system* and the global photogrammetric model coordinate (E, N) *system*. In order to determine a transformation system for computing the coordinates of any given point in one coordinate system given the coordinates of that point in the other system, the coordinates of two points (1 and 2) were measured in the two systems as given in Table 13.1.

This is a typical two-parameter transformation problem relating local photo coordinates (e, n) *system* with global photogrammetric model coordinates (E, N) *system*, which can be expressed by the following equations:

$$\hat{E}_i - \hat{a}\hat{e}_i + \hat{b}\hat{n}_i = 0 \tag{13.2}$$

$$\hat{N}_i - \hat{b}\hat{e}_i - \hat{a}\hat{n}_i = 0 \tag{13.3}$$

where $i = 1, 2$, for points 1 and 2, respectively; the coordinates of points 1 and 2 (in both the local and global systems) are considered as measurements, and the quantities a and b are the unknown parameters to be determined.

Table 13.1 Coordinates of points 1 and 2.

Point	Local system (photo)		Global system (model)	
i	e_i (cm)	n_i (cm)	E_i (cm)	N_i (cm)
1	0.0	1.0	−2.1	1.1
2	1.0	0.0	1.0	2.0

Equations (13.2)–(13.3) constitute the general model for this problem. In completing the equation formulation, the following steps can be taken:

1) Since the four coordinates under the local system and the corresponding ones under the global system (Table 13.1) are stated in the question as measurements, they are considered as observations in this problem, giving a total of eight observations. These observations are represented in symbolic form for now, as follows:

$$\ell = [e_1 \; n_1 \; e_2 \; n_2 \; E_1 \; N_1 \; E_2 \; N_2]^T$$

where e_i and n_i represent the easting and northing (in the local system) for point i ($i = 1, 2$) and E_i and N_i are the easting and the northing (in the global system) for point i ($i = 1, 2$). The unknown parameters (x) in Equations (13.2) and (13.3) are specified as a and b.

2) Looking at Equations (13.2) and (13.3), one can see that the parameters (x) and the observations (ℓ) are in the equations with their arrangement conforming to the expected format for general model in Equation (13.1). Since each point is associated with two equations (Equations (13.2) and (13.3)), the complete general model equations may be given as

$$\hat{E}_1 - \hat{a}\hat{e}_1 + \hat{b}\hat{n}_1 = 0 \tag{13.4}$$

$$\hat{N}_1 - \hat{b}\hat{e}_1 - \hat{a}\hat{n}_1 = 0 \tag{13.5}$$

$$\hat{E}_2 - \hat{a}\hat{e}_2 + \hat{b}\hat{n}_2 = 0 \tag{13.6}$$

$$\hat{N}_2 - \hat{b}\hat{e}_2 - \hat{a}\hat{n}_2 = 0 \tag{13.7}$$

where $(\hat{e}_1, \hat{n}_1)$ and $(\hat{e}_2, \hat{n}_2)$ represent the adjusted photo coordinates of points 1 and 2, respectively, $(\hat{E}_1, \hat{N}_1)$ and $(\hat{E}_2, \hat{N}_2)$ represent the adjusted model coordinates of points 1 and 2 respectively, and $(\hat{a}, \hat{b})$ represent the adjusted parameters (unknowns). It can be seen in the model equations above that the adjusted parameters ($\hat{x}$) and the adjusted observations ($\hat{\ell}$) are mixed and formed on the left-hand side of the equal sign with all of the equations looking like the general model in Equation (13.1). This type of order must be maintained when formulating any general model equations.

Example 13.1 In the absence of lens distortion, the direct linear transformation (collinearity condition equations) between an object point (X, Y, Z) in space and its corresponding image space coordinates (x, y) can be expressed as follows:

$$x = \frac{L_1 X + L_2 Y + L_3 Z + L_4}{L_9 X + L_{10} Y + L_{11} Z + 1} \tag{13.8}$$

$$y = \frac{L_5 X + L_6 Y + L_7 Z + L_8}{L_9 X + L_{10} Y + L_{11} Z + 1} \tag{13.9}$$

where L_1 to L_{11} are the transformation parameters, (X, Y, Z) are the ground coordinates of the object point, and (x, y) are the corresponding image (or photo) coordinates of the point. Answer the following.

i) Arrange Equations (13.8) and (13.9) in general model form $(f(\hat{x}, \hat{\ell}) = 0)$, identifying what could be considered as the unknown parameters (x) and the observables (ℓ).

Solution:

General model form of Equations (13.8) and (13.9) can be given as follows:

$$x - \frac{L_1 X + L_2 Y + L_3 Z + L_4}{L_9 X + L_{10} Y + L_{11} Z + 1} = 0 \tag{13.10}$$

$$y - \frac{L_5 X + L_6 Y + L_7 Z + L_8}{L_9 X + L_{10} Y + L_{11} Z + 1} = 0 \tag{13.11}$$

The unknown parameters: $x^T = [L_1 \ L_2 \ L_3 \ L_4 \ L_5 \ L_6 \ L_7 \ L_8 \ L_9 \ L_{10} \ L_{11}]$

The observables: (X, Y, Z) per ground point and the corresponding (x, y) image coordinates of the same point.

ii) Determine the minimum number of observables needed in order to be able to determine the values of the unknown parameters (x). How many general model equations will be possible with the minimum number of observations?

Solution:

To solve for the 11 transformation parameters, there must be at least 6 registration marks in the image whose ground coordinates are known. Since each registration mark will form 2 equations, there will be a minimum number of 12 general model equations formed in total.

13.3 Linearization of General Model

In preparation for least squares adjustment of general models, the models must first be linearized. The concepts of linearization by Taylor series expansion can be applied to general model equations as follows. Given the following general mathematical model,

$$f(\hat{x}, \hat{\ell}) = 0 \quad C_\ell \tag{13.12}$$

where f is the vector of general mathematical model equations, $\hat{x}$ is a vector of adjusted parameters, $\hat{\ell}$ is a vector of adjusted observations, and C_ℓ is the covariance matrix of the observations ℓ. Assume the following:

$$\hat{\ell} = \ell + v \quad \text{and} \quad \hat{x} = x^0 + \delta \tag{13.13}$$

where v is a vector of observation residuals, x^0 is a vector of approximate values of the parameters, and δ is the vector of corrections to the approximate parameters. Using Equation (13.13) in Equation (13.12) gives the following:

$$f\left(x^0 + \delta, \ell + v\right) = 0 \tag{13.14}$$

The Taylor series expansion (up to the first order) of the model in Equation (13.14) can be given as

$$f\left(x^0, \ell\right) + \frac{\partial f}{\partial x}(\delta) + \frac{\partial f}{\partial \ell}(v) = 0 \tag{13.15}$$

where Equation (13.15) is the approximate linear form of the function given in Equation (13.12), the first term (the original function evaluated at the given initial numerical values of the parameters (x^0) and the original observation (ℓ)) is the zero-order term of Taylor series, $\frac{\partial f}{\partial x}(\delta)$ and $\frac{\partial f}{\partial \ell}(v)$ represent the first-order terms of the Taylor series, $\frac{\partial f}{\partial x}$ is a Jacobian matrix containing the partial derivatives with respect to the parameters, and $\frac{\partial f}{\partial \ell}$ is a partial derivative with respect to the observation. Equation (13.15) can be expressed in simpler form as follows:

$$w + A\delta + Bv = 0 \tag{13.16}$$

where

$$A = \frac{\partial f}{\partial x} \text{ is the first design matrix} \tag{13.17}$$

$$B = \frac{\partial f}{\partial \ell} \text{ is the second design matrix} \tag{13.18}$$

and

$$w = f\left(x^0, \ell\right) \text{ is a vector of misclosures} \tag{13.19}$$

Equation (13.15) or Equation (13.16) is the linearized general model, which is to be adjusted by the method of least squares.

Example 13.2 Linearize general model Equations (13.4)–(13.7) and present it in the form of Equation (13.16).

Solution:

For Equations (13.4)–(13.7), the linearized model can be given as Equation (13.16) where parameter vector is $x = \begin{bmatrix} a \\ b \end{bmatrix}$; given the vector of approximate parameters $x^0 = \begin{bmatrix} a^0 \\ b^0 \end{bmatrix}$, the first design matrix (evaluated at x^0) can be given as

$$A = \left. \frac{\partial f}{\partial \hat{x}} \right|_{x^0} = \begin{bmatrix} \dfrac{\partial f_1}{\partial \hat{a}} & \dfrac{\partial f_1}{\partial \hat{b}} \\ \dfrac{\partial f_2}{\partial \hat{a}} & \dfrac{\partial f_2}{\partial \hat{b}} \\ \dfrac{\partial f_3}{\partial \hat{a}} & \dfrac{\partial f_3}{\partial \hat{b}} \\ \dfrac{\partial f_4}{\partial \hat{a}} & \dfrac{\partial f_4}{\partial \hat{b}} \end{bmatrix}_{x^0} \qquad A = \begin{bmatrix} -e_1 & n_1 \\ -n_1 & -e_1 \\ -e_2 & n_2 \\ -n_2 & -e_2 \end{bmatrix}$$

For the given vector of observations, $\ell = \begin{bmatrix} e_1 & n_1 & e_2 & n_2 & E_1 & N_1 & E_2 & N_2 \end{bmatrix}^T$, the second design matrix (B) can be given as

$$B = \left. \frac{\partial f}{\partial \hat{\ell}} \right|_{\ell} = \begin{bmatrix} \dfrac{\partial f_1}{\partial \hat{e}_1} & \dfrac{\partial f_1}{\partial \hat{n}_1} & \dfrac{\partial f_1}{\partial \hat{e}_2} & \dfrac{\partial f_1}{\partial \hat{n}_2} & \dfrac{\partial f_1}{\partial \hat{E}_1} & \dfrac{\partial f_1}{\partial \hat{N}_1} & \dfrac{\partial f_1}{\partial \hat{E}_2} & \dfrac{\partial f_1}{\partial \hat{N}_2} \\ \dfrac{\partial f_2}{\partial \hat{e}_1} & \dfrac{\partial f_2}{\partial \hat{n}_1} & \dfrac{\partial f_2}{\partial \hat{e}_2} & \dfrac{\partial f_2}{\partial \hat{n}_2} & \dfrac{\partial f_2}{\partial \hat{E}_1} & \dfrac{\partial f_2}{\partial \hat{N}_1} & \dfrac{\partial f_2}{\partial \hat{E}_2} & \dfrac{\partial f_2}{\partial \hat{N}_2} \\ \dfrac{\partial f_3}{\partial \hat{e}_1} & \dfrac{\partial f_3}{\partial \hat{n}_1} & \dfrac{\partial f_3}{\partial \hat{e}_2} & \dfrac{\partial f_3}{\partial \hat{n}_2} & \dfrac{\partial f_3}{\partial \hat{E}_1} & \dfrac{\partial f_3}{\partial \hat{N}_1} & \dfrac{\partial f_3}{\partial \hat{E}_2} & \dfrac{\partial f_3}{\partial \hat{N}_2} \\ \dfrac{\partial f_4}{\partial \hat{e}_1} & \dfrac{\partial f_4}{\partial \hat{n}_1} & \dfrac{\partial f_3}{\partial \hat{e}_2} & \dfrac{\partial f_4}{\partial \hat{n}_2} & \dfrac{\partial f_4}{\partial \hat{E}_1} & \dfrac{\partial f_4}{\partial \hat{N}_1} & \dfrac{\partial f_4}{\partial \hat{E}_2} & \dfrac{\partial f_4}{\partial \hat{N}_2} \end{bmatrix}_{\ell}$$

or

$$B = \left. \frac{\partial f}{\partial \hat{\ell}} \right|_{\ell} = \begin{bmatrix} -a^0 & b^0 & 0 & 0 & 1 & 0 & 0 & 0 \\ -b^0 & -a^0 & 0 & 0 & 0 & 1 & 0 & 0 \\ 0 & 0 & -a^0 & b^0 & 0 & 0 & 1 & 0 \\ 0 & 0 & -b^0 & -a^0 & 0 & 0 & 0 & 1 \end{bmatrix}_{\ell}$$

where the vector of the model equations is $f = \begin{bmatrix} f_1 & f_2 & f_3 & f_4 \end{bmatrix}^T$. The misclosure vector (w) evaluated at x^0, the vector of residuals (v), and the vector of corrections to the approximate parameters (δ) can be given, respectively, as

$$
w = \begin{bmatrix} E_1 - a^0 e_1 + b^0 n_1 \\ N_1 - a^0 n_1 - b^0 e_1 \\ E_2 - a^0 e_2 + b^0 n_2 \\ N_2 - a^0 n_2 - b^0 e_2 \end{bmatrix}, \quad v = \begin{bmatrix} v_1 \\ v_2 \\ v_3 \\ v_4 \\ v_5 \\ v_6 \\ v_7 \\ v_8 \end{bmatrix}, \quad \delta = \begin{bmatrix} \delta a \\ \delta b \end{bmatrix}
$$

The linearized model of Equations (13.4)–(13.7) (evaluated at x^0 and ℓ) can be given as

$$
\begin{bmatrix} -e_1 & n_1 \\ -n_1 & -e_1 \\ -e_2 & n_2 \\ -n_2 & -e_2 \end{bmatrix} \begin{bmatrix} \delta a \\ \delta b \end{bmatrix} + \begin{bmatrix} -a^0 & b^0 & 0 & 0 & 1 & 0 & 0 & 0 \\ -b^0 & -a^0 & 0 & 0 & 0 & 1 & 0 & 0 \\ 0 & 0 & -a^0 & b^0 & 0 & 0 & 1 & 0 \\ 0 & 0 & -b^0 & -a^0 & 0 & 0 & 0 & 1 \end{bmatrix}_\ell \begin{bmatrix} v_1 \\ v_2 \\ v_3 \\ v_4 \\ v_5 \\ v_6 \\ v_7 \\ v_8 \end{bmatrix}
$$

$$
+ \begin{bmatrix} E_1 - a^0 e_1 + b^0 n_1 \\ N_1 - a^0 n_1 - b^0 e_1 \\ E_2 - a^0 e_2 + b^0 n_2 \\ N_2 - a^0 n_2 - b^0 e_2 \end{bmatrix} = 0
$$

13.4 Variation Function for Linearized General Model

Continuing from Section 13.3, it can be seen from Equation (13.12) that the covariance matrix C_ℓ of the observations is directly associated with the general model. Based on this model, the least squares criterion can be given as

$$
v^T C_\ell^{-1} v = \text{minimum} \tag{13.20}
$$

where the weight matrix is taken as C_ℓ^{-1}, with $\sigma_0^2 = 1$. Imposing the least squares criterion on the linearized general model Equation (13.16) using the Lagrangian approach discussed in Chapter 5 gives the following variation function:

$$\phi = v^T C_\ell^{-1} v - 2k^T (A\delta + Bv + w) = \text{minimum} \tag{13.21}$$

where k is a vector of correlates (or a vector of Lagrangian multipliers). Equation (13.21) is the variation function for the general model equations.

13.5 Normal Equation System and the Least Squares Solution

The least squares normal equation system and its least squares solution are derived as follows. The normal equations are derived from the variation function given in Equation (13.21) by finding the partial derivatives of the function with respect to the unknowns (i.e. v, δ, k) and setting the derivatives to zero:

$$\frac{\partial \phi}{\partial v} = 2v^T C_\ell^{-1} - 2k^T B = 0 \tag{13.22}$$

$$\frac{\partial \phi}{\partial \delta} = -2k^T A = 0 \tag{13.23}$$

$$\frac{\partial \phi}{\partial k^T} = -2(A\delta + Bv + w) = 0 \tag{13.24}$$

Equations (13.22)–(13.24) can be rewritten to give the most expanded form of the least squares normal equation system as

$$C_\ell^{-1} v - B^T k = 0 \tag{13.25}$$

$$-A^T k = 0 \tag{13.26}$$

$$A\delta + Bv + w = 0 \tag{13.27}$$

Equations (13.25)–(13.27) can be presented in matrix form as follows:

$$\begin{bmatrix} C_\ell^{-1} & 0 & B^T \\ 0 & 0 & A^T \\ B & A & 0 \end{bmatrix} \begin{bmatrix} v \\ \delta \\ -k \end{bmatrix} + \begin{bmatrix} 0 \\ 0 \\ w \end{bmatrix} = \begin{bmatrix} 0 \\ 0 \\ 0 \end{bmatrix} \tag{13.28}$$

Equations (13.25)–(13.27) or Equation (13.28) can be solved as follows. From Equation (13.25),

$$v = C_\ell B^T k \tag{13.29}$$

Since $C_\ell = Q_\ell$ (for $\sigma_0^2 = 1$), then Equation (13.29) can be rewritten as

$$v = Q_\ell B^T k \tag{13.30}$$

Substituting Equation (13.30) into Equation (13.27) gives

$$A\delta + BQ_\ell B^T k + w = 0 \tag{13.31}$$

or

$$k = -\left(BQ_\ell B^T\right)^{-1}(A\delta + w) \tag{13.32}$$

Substituting Equation (13.32) into Equation (13.26) gives

$$A^T\left(BQ_\ell B^T\right)^{-1}(A\delta + w) = 0 \tag{13.33}$$

or

$$A^T\left(BQ_\ell B^T\right)^{-1}A\delta + A^T\left(BQ_\ell B^T\right)^{-1}w = 0 \tag{13.34}$$

If $M = BQ_\ell B^T$, then Equation (13.34) can be rewritten and solved as follows:

$$A^T M^{-1} A\delta + A^T M^{-1} w = 0 \tag{13.35}$$

or

$$\delta = -\left(A^T M^{-1} A\right)^{-1} A^T M^{-1} w \tag{13.36}$$

After solving for the vector of corrections (δ) to the approximate parameters (x^0) from Equation (13.36), the solution can be used to solve for the vector of correlates (k) from Equation (13.32) and then also for v in Equation (13.30).

13.6 Steps for General Model Adjustment

On the basis of the discussions in Sections 13.2–13.5, the steps for performing an adjustment of general model equations can be given as follows:

1) Form the general mathematical model: $f(\hat{x}, \hat{\ell}) = 0$.
2) Form the first and second design matrices A and B, respectively: $A = \dfrac{\partial f}{\partial x}$ and $B = \dfrac{\partial f}{\partial \ell}$.
3) Form the cofactor matrix (Q_ℓ) from the variances of the observations, and the misclosure vector w: $Q_\ell = C_\ell$ (for $\sigma_0^2 = 1$) and $w = f(x^0, \ell)$.
4) Form the elements of the most reduced normal equations: $M = BQ_\ell B^T$, $N = A^T M^{-1} A$, and $u = A^T M^{-1} w$.
5) Solve for the vector of corrections to the approximate parameters (x^0): $\delta = -N^{-1}u$.
6) Solve for the vector of correlates: $k = -M^{-1}(A\delta + w)$.

7) Solve for the vector of residuals: $v = Q_\ell B^T k$.
8) Compute the adjusted parameters ($\hat{x}$) and the adjusted observations ($\hat{\ell}$): $\hat{x} = x^0 + \delta$ and $\hat{\ell} = \ell + v$.

13.7 General Model Adjustment Examples

The following examples are given to illustrate how to perform least squares adjustment of general models. The examples are based on the steps given in Section 13.6.

13.7.1 Coordinate Transformations

One of the important applications of general model adjustment is in coordinate transformation problems, where transformation parameters based on common points are used to convert the coordinates of new points from one system into corresponding coordinates in another system. There are many ways of carrying out the transformations: *similarity, affine, projective,* and *orthogonal* transformations. In *similarity transformation,* the same scale factor is used in all directions, thereby preserving the angles, but lengths of lines and positions of points may change; rotation of network is allowed. *Affine transformation* allows angular distortions but preserves the parallelism of lines; lines that are originally parallel will remain parallel after transformations, but lengths and angles are not preserved. The *projective transformation* allows both angular and length distortions to occur during the transformation, but does not preserve parallel lines. The *orthogonal transformation* is a similarity transformation in which the scale factor is unity, the angles and distances in the network are preserved, but the positions of points may change.

13.7.1.1 Two-dimensional Similarity Transformation Example

Assume the main control surveying network in an open-pit mine consists of five points (10, 11, 12, 13, and 14) as shown in Figure 13.2. The original (global) positions of these points, which are in the Universal Transverse Mercator (UTM) projection (u, v) coordinate system, are given in Table 13.2. Due to mining activities, the control marker 13 was destroyed. The approximate location of the point was then staked by the surveyor, and a traverse was conducted using a selected local (X, Y) coordinate system, and the local coordinates of all the five points were determined. There is, therefore, a need to transform the local coordinates of the marker 13 into global (UTM) coordinates in order to determine how far away the staked point is from its original location. The general model equations for the transformation from local (X, Y) system to global (u, v) are given as follows:

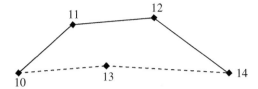

Figure 13.2 Control surveying network (not to scale).

$$u - aX - bY - c_1 = 0 \tag{13.37}$$
$$v + bX - aY - c_2 = 0 \tag{13.38}$$

The transformation parameters a, b, c_1, and c_2 in Equations (13.37) and (13.38) are the unknown parameters to be estimated. Taking the two sets of rectangular coordinates (global and local) as observed quantities assumed made with the same precision, perform general model least squares adjustment to determine the adjusted values of the unknown parameters a, b, c_1, and c_2. The following initial values of the parameters can be assumed: $a^0 = 1.0$, $b^0 = 1.0$, $c_1^0 = 514\,317.0$, and $c_2^0 = 5\,120\,373.0$.

From the question, the following observation and parameter vectors are stated:

Observation vector:

$$\ell^T = \begin{bmatrix} u_{10} & v_{10} & u_{11} & v_{11} & u_{14} & v_{14} & u_{12} & v_{12} & X_{10} & Y_{10} & X_{11} & Y_{11} & X_{12} & Y_{12} & X_{14} & Y_{14} \end{bmatrix}$$

Parameter vector:

$$x^T = \begin{bmatrix} a & b & c_1 & c_2 \end{bmatrix}$$

Note that the variables (or observables) u_i, v_i, X_i, and Y_i are used above to represent the observations given in Table 13.2; their numerical values are not used at this stage of the adjustment. The general model equations $(f(\hat{x}, \hat{\ell}) = 0)$ are formed for each point that has observations associated with it as follows:

$$\hat{u}_{10} - \hat{a}\hat{X}_{10} - \hat{b}\hat{Y}_{10} - \hat{c}_1 = 0 \tag{13.39}$$
$$\hat{v}_{10} + \hat{b}\hat{X}_{10} - \hat{a}\hat{Y}_{10} - \hat{c}_2 = 0 \tag{13.40}$$
$$\hat{u}_{11} - \hat{a}\hat{X}_{11} - \hat{b}\hat{Y}_{11} - \hat{c}_1 = 0 \tag{13.41}$$
$$\hat{v}_{11} + \hat{b}\hat{X}_{11} - \hat{a}\hat{Y}_{11} - \hat{c}_2 = 0 \tag{13.42}$$
$$\hat{u}_{12} - \hat{a}\hat{X}_{12} - \hat{b}\hat{Y}_{12} - \hat{c}_1 = 0 \tag{13.43}$$
$$\hat{v}_{12} + \hat{b}\hat{X}_{12} - \hat{a}\hat{Y}_{12} - \hat{c}_2 = 0 \tag{13.44}$$

Table 13.2 Coordinates of control points.

| Point | Global coordinates (UTM) | | Local coordinates (model) | | Comments |
	u	v	X	Y	
10	514 316.937	5 120 373.171	1 503.101	1 503.101	
11	516 004.947	5 121 890.360	2 659.130	3 457.329	
14	519 033.863	5 119 651.091	6 221.443	2 228.538	
12	517 142.532	5 121 693.517	3 803.857	3 610.606	
13	—	—	3 828.952	2 211.451	Destroyed

$$\hat{u}_{14} - \hat{a}\hat{X}_{14} - \hat{b}\hat{Y}_{14} - \hat{c}_1 = 0 \tag{13.45}$$

$$\hat{v}_{14} + \hat{b}\hat{X}_{14} - \hat{a}\hat{Y}_{14} - \hat{c}_2 = 0 \tag{13.46}$$

The first design matrix, A:

$$A = \begin{bmatrix} -X_{10} & -Y_{10} & -1 & 0 \\ -Y_{10} & X_{10} & 0 & -1 \\ -X_{11} & -Y_{11} & -1 & 0 \\ -Y_{11} & X_{11} & 0 & -1 \\ -X_{14} & -Y_{14} & -1 & 0 \\ -Y_{14} & X_{14} & 0 & -1 \\ -X_{12} & -Y_{12} & -1 & 0 \\ -Y_{12} & X_{12} & 0 & -1 \end{bmatrix} \quad A = \begin{bmatrix} -1503.101 & -1503.101 & -1.0 & 0 \\ -1503.101 & 1503.101 & 0 & -1.0 \\ -2659.130 & -3457.329 & -1.0 & 0 \\ -3457.329 & 2659.130 & 0 & -1.0 \\ -3803.857 & -3610.606 & -1.0 & 0 \\ -3610.606 & 3803.857 & 0 & -1.0 \\ -6221.443 & -2228.538 & -1.0 & 0 \\ -2228.538 & 6221.443 & 0 & -1.0 \end{bmatrix}$$

The second design matrix, B:

$$B = \begin{bmatrix} 1 & 0 & 0 & 0 & 0 & 0 & 0 & 0 & -1 & -1 & 0 & 0 & 0 & 0 & 0 & 0 \\ 0 & 1 & 0 & 0 & 0 & 0 & 0 & 0 & 1 & -1 & 0 & 0 & 0 & 0 & 0 & 0 \\ 0 & 0 & 1 & 0 & 0 & 0 & 0 & 0 & 0 & 0 & -1 & -1 & 0 & 0 & 0 & 0 \\ 0 & 0 & 0 & 1 & 0 & 0 & 0 & 0 & 0 & 0 & 1 & -1 & 0 & 0 & 0 & 0 \\ 0 & 0 & 0 & 0 & 0 & 0 & 1 & 0 & 0 & 0 & 0 & 0 & -1 & -1 & 0 & 0 \\ 0 & 0 & 0 & 0 & 0 & 0 & 0 & 1 & 0 & 0 & 0 & 0 & 1 & -1 & 0 & 0 \\ 0 & 0 & 0 & 0 & 1 & 0 & 0 & 0 & 0 & 0 & 0 & 0 & 0 & 0 & -1 & -1 \\ 0 & 0 & 0 & 0 & 0 & 1 & 0 & 0 & 0 & 0 & 0 & 0 & 0 & 0 & 1 & -1 \end{bmatrix}$$

$$Q = \text{diag}[1\ 1\ 1\ 1\ 1\ 1\ 1\ 1\ 1\ 1\ 1\ 1\ 1\ 1\ 1\ 1]$$

$$M = BB^T = \text{diag}[3\ 3\ 3\ 3\ 3\ 3\ 3\ 3]$$

$$N = A^T M^{-1} A = \begin{bmatrix} 3.157\ 38E+7 & 0.0 & 4.729\ 18E+3 & 3.599\ 86E+3 \\ 0.0 & 3.157\ 38E+7 & 3.599\ 86E+3 & -4.729\ 18E+3 \\ 4.729\ 18E+3 & 3.599\ 86E+3 & 1.333\ 333 & 0 \\ 3.599\ 86E+3 & -4.729\ 18E+3 & 0 & 1.333\ 333 \end{bmatrix}$$

Misclosure vector w:

$$w = f\left(x^0, \ell\right) = \begin{bmatrix} -3.006\ 265E+3 \\ 0.171\ 00 \\ -4.428\ 512E+3 \\ 7.191\ 610E+2 \\ -4.588\ 931E+3 \\ 1.513\ 768E+3 \\ -3.733\ 118E+3 \\ 3.270\ 996E+3 \end{bmatrix} \qquad u = A^T M^{-1} w = \begin{bmatrix} 1.391\ 131\ 18E+7 \\ 2.424\ 627\ 73E+7 \\ 5.252\ 275\ 33E+3 \\ -1.834\ 698\ 7E+3 \end{bmatrix}$$

The solution vector:

$$\delta = -N^{-1} u = \begin{bmatrix} -0.046\ 37 \\ -0.700\ 345 \\ -1\ 883.875\ 57 \\ -982.820\ 504 \end{bmatrix}$$

The adjusted parameters:

$$\hat{x} = x^0 + \delta = \begin{bmatrix} \hat{a} \\ \hat{b} \\ \hat{c}_1 \\ \hat{c}_2 \end{bmatrix} = \begin{bmatrix} 0.953\ 63 \\ 0.299\ 65 \\ 512\ 433.124\ 4 \\ 5\ 119\ 390.179\ 5 \end{bmatrix}$$

Using the adjusted parameters and the X and Y local coordinates of point 13 in Equations (13.37) and (13.38) gives the global coordinates of point 13 as $u = 516$ $747.194\ 9$ m and $v = 5\ 120\ 351.718\ 6$ m. The MATLAB code for the general model least squares adjustment is given in Table 13.3.

Table 13.3 MATLAB code for the general model least squares adjustment.

```
>>% Define the variables using syms
>> syms u10 v10 u11 v11 u14 v14 u12 v12 X10 Y10 X11 Y11 X12 Y12
X14 Y14
>> syms a b c1 c2
>>% Form the 8 General model Equations (13.39)-(13.46)
>> L1 =u10-a*X10 -b*Y10 - c1;
>> L2=v10 + b*X10 -a*Y10-c2;
>> L3=u11-a*X11 - b*Y11 -c1;
>> L4=v11 +b*X11-a*Y11 - c2;
>> L5=u12 - a*X12 -b*Y12 - c1;
>> L6= v12+b*X12 -a*Y12 - c2;
>> L7=u14 - a*X14 -b*Y14-c1;
>> L8=v14 +b*X14 -a*Y14 - c2;
>>% form Jacobian of the equations wrt parameters a, b, c1, c2
as J1 and wrt observations as J2
>> J1=jacobian([L1;L2;L3;L4;L5;L6;L7;L8], [a b c1 c2]);
>> J2=jacobian([L1;L2;L3;L4;L5;L6;L7;L8], [u10 v10 u11 v11
u14 v14 u12 v12 X10 Y10 X11 Y11 X12 Y12 X14 Y14]);
>>%assign values to parameters and observations
>>
u10=514316.937;v10=5120373.171;u11=516004.947;
v11=5121890.360;u14=519033.863;v14=5119651.091;
u12=517142.532;v12=5121693.517;
>>
X10=1503.101;Y10=1503.101;X11=2659.130;Y11=3457.329;
X14=6221.443;Y14=2228.538;X12=3803.857;Y12=3610.606;
>> a=1.0;b=1.0;c1=514317;c2=5120373;
>>% evaluate Jacobians as A and B matrices and solve for
corrections d to parameters
>> A=eval(J1);
>> B=eval(J2);
>> M=B*B';
>> N=A'*inv(M)*A;
>> w=eval([L1;L2;L3;L4;L5;L6;L7;L8]);
>> u=A'*inv(M)*w;
>> d=-inv(N)*u
d =
    1.0e+03 *
  -0.000046371315225
  -0.000700345145864
  -1.883875574860158
  -0.982820504349043
>>
```

13.7.2 Parabolic Vertical Transition Curve Example

Consider Figure 13.3 showing a vertical section along the centerline of a parabolic vertical transition curve having height y_0 and length L with the curve joining two gradients $g_1 = +4.55\%$ and $g_2 = -3.05\%$. The beginning of vertical curve (BVC) is at a chainage point $1 + 303.790$ m that is currently inaccessible. Shown in Table 13.4 are distances (x_i) along the centerline from the start of the transition curve and their corresponding heights (y_i). The standard deviations of distances chained (x_i) and the corresponding heights (y_i) are ± 0.005 m.

The equation of the parabolic curve can be given as

$$y = y_0 + g_1 x + \frac{(g_2 - g_1)x^2}{2L} \tag{13.47}$$

Perform the least squares adjustment of the general models, and determine the unknown parameters (y_0, L). Let the approximate values of the parameters be $\left[y_0^0 = 45 \quad L^0 = 220 \right]^T$.

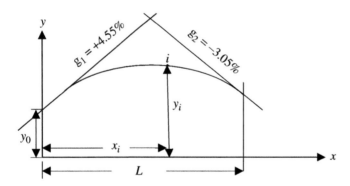

Figure 13.3 Vertical section along the centerline of a parabolic vertical transition curve.

Table 13.4 Measurements along a vertical transition curve.

Point i	Chainage (m)	Distance (x_i) (m)	Height (y_i) (m)
1	$1 + 340.000$	36.210	45.301
2	$1 + 400.000$	96.210	46.659
3	$1 + 460.000$	156.210	46.773
4	$1 + 520.000$	216.210	45.643

The general model equations $f(\ell,x) = 0$ (with hats ignored) based on the given four points and the Equation (13.47) can be given as follows:

$$2Ly_1 - (g_2 - g_1)x_1^2 - 2Lg_1x_1 - 2Ly_0 = 0 \qquad (13.48)$$

$$2Ly_2 - (g_2 - g_1)x_2^2 - 2Lg_1x_2 - 2Ly_0 = 0 \qquad (13.49)$$

$$2Ly_3 - (g_2 - g_1)x_3^2 - 2Lg_1x_3 - 2Ly_0 = 0 \qquad (13.50)$$

$$2Ly_4 - (g_2 - g_1)x_4^2 - 2Lg_1x_4 - 2Ly_0 = 0 \qquad (13.51)$$

The parameters (x) and the observations (ℓ) are listed as follows:

$$x = \begin{bmatrix} y_0 \\ L \end{bmatrix} \quad \ell = \begin{bmatrix} y_1 \\ x_1 \\ y_2 \\ x_2 \\ y_3 \\ x_3 \\ y_4 \\ x_4 \end{bmatrix} \quad [g_1 = 0.0455 \quad g_2 = -0.0305] \text{ Let } \begin{bmatrix} y_0^0 = 45 & L^0 = 220 \end{bmatrix}^T$$

The first and second design matrices A and B, respectively, are as follows:

$$A = \frac{\partial f}{\partial x} = \begin{bmatrix} -2L & 2y_1 - 2g_1x_1 - 2y_0 \\ -2L & 2y_2 - 2g_1x_2 - 2y_0 \\ -2L & 2y_3 - 2g_1x_3 - 2y_0 \\ -2L & 2y_4 - 2g_1x_4 - 2y_0 \end{bmatrix} \quad A = \begin{bmatrix} -440.00 & -2.6931 \\ -440.00 & -5.4371 \\ -440.00 & -10.6691 \\ -440.00 & -18.3891 \end{bmatrix}$$

$$B = \frac{\partial f}{\partial \ell} = \begin{bmatrix} 2L & -2(g_2-g_1)x_1 - 2Lg_1 & & & \\ & & 2L & -2(g_2-g_1)x_2 - 2Lg_1 & & \\ & & & & 2L & -2(g_2-g_1)x_3 - 2Lg_1 & \\ & & & & & & 2L & -2(g_2-g_1)x_4 - 2Lg_1 \end{bmatrix}$$

$$B = \begin{bmatrix} 440 & -14.5161 & & \\ & & 440 & -5.3961 & \\ & & & & 440 & 3.7239 & \\ & & & & & & 440 & 12.8439 \end{bmatrix}$$

The misclosure vector, w:

$$w = f\left(\ell, x^0\right) \quad w = \begin{bmatrix} -492.8357 \\ -492.6805 \\ -492.6853 \\ -492.8501 \end{bmatrix}$$

$$Q = \text{diag}\begin{bmatrix} 0.005^2 & 0.005^2 & 0.005^2 & 0.005^2 & 0.005^2 & 0.005^2 & 0.005^2 & 0.005^2 \end{bmatrix}$$

$$M = BQB^T = \begin{bmatrix} 4.8453 & & & \\ & 4.8407 & & \\ & & 4.8403 & \\ & & & 4.8441 \end{bmatrix}$$

$$N = A^T M^{-1} A = \begin{bmatrix} 1.5991E+5 & 3.38E+3 \\ 3.38E+3 & 1.00E+2 \end{bmatrix}$$

$$N^{-1} = \begin{bmatrix} 2.137E-5 & -7.155E-4 \\ -7.155E-4 & 3.3861E-2 \end{bmatrix}$$

$$u = A^T M^{-1} w = \begin{bmatrix} 179\ 089.719\ 2 \\ 3\ 784.229\ 3 \end{bmatrix}$$

The solution vector (δ) and the adjusted parameters ($\hat{x}$):

$$\delta = -N^{-1}u = \begin{bmatrix} -1.1198 \\ -0.0037 \end{bmatrix} \quad \hat{x} = x^0 + \delta = \begin{bmatrix} 43.8802 \\ 219.9963 \end{bmatrix} \text{m}$$

$$A\delta + w = \begin{bmatrix} -0.0970 \\ 0.0682 \\ 0.0826 \\ -0.0539 \end{bmatrix} \quad k = -M^{-1}(A\delta + w) = \begin{bmatrix} 0.0200 \\ -0.0141 \\ -0.0171 \\ 0.0111 \end{bmatrix}$$

The residual vector (v) and the a posteriori variance factor (APVF) of unit weight (s_0^2):

$$v = QB^T k = \begin{bmatrix} 2.203E-4 \\ -7.30E-6 \\ -1.55E-4 \\ 1.90E-6 \\ -1.877E-4 \\ -1.60E-6 \\ 1.225E-4 \\ 3.60E-6 \end{bmatrix} \text{m} \quad s_0^2 = \frac{v^T P v}{df} = \frac{0.0049}{2} = 0.0025$$

The MATLAB code for solving the parabolic vertical transition curve is given in Table 13.5.

Table 13.5 MATLAB code for solving parabolic vertical transition curve.

```
>>% Use syms to define the parameters and observations as
variables
>> syms y0 L
>> syms y1 x1 y2 x2 y3 x3 y4 x4 g1 g2
>>% form the 4 general model Equations (13.48)-(13.51)
>> L1=2*L*y1-(g2-g1)*x1^2-2*L*g1*x1-2*L*y0;
>> L1=2*L*y2-(g2-g1)*x2^2-2*L*g1*x2-2*L*y0;
>> L1=2*L*y3-(g2-g1)*x3^2-2*L*g1*x3-2*L*y0;
>> L1=2*L*y1-(g2-g1)*x1^2-2*L*g1*x1-2*L*y0;
>> L2=2*L*y2-(g2-g1)*x2^2-2*L*g1*x2-2*L*y0;
>> L3=2*L*y3-(g2-g1)*x3^2-2*L*g1*x3-2*L*y0;
>> L4=2*L*y4-(g2-g1)*x4^2-2*L*g1*x4-2*L*y0;
>>% determine Jacobians wrt parameters and wrt
observations as J and J2
>> J=jacobian([L1;L2;L3;L4],[y0 L]);
>> J2=jacobian([L1;L2;L3;L4],[y1 x1 y2 x2 y3 x3 y4 x4]);
>>% assign values to parameters and observations and
evaluate Jacobians as A and B
>> g1=0.0455;g2=-0.0305;y0=45;L=220;
>>
x1=36.210;y1=45.301;x2=96.210;y2=46.659;x3=156.210;
y3=46.773;x4=216.210;y4=45.643;
>> A=eval(J);
>> B=eval(J2);
>>% determine misclosure w, cofactor of observations Q
>> w=eval([L1;L2;L3;L4]);
>> Q=diag([0.005^2 0.005^2 0.005^2 0.005^2 0.005^2 0.005^2
0.005^2 0.005^2]);
>>% solve for corrections d to parameters
>> M=B*Q*B';
>> N=A'*inv(M)*A;
>> format LONG
>> Ninv=inv(N);
>> u=A'*inv(M)*w;
>> format SHORT
>> d=-Ninv*u
d =
   -1.1198
   -0.0037
>>
```

13.8 Stochastic Properties of General Model Adjustment

The stochastic properties of general model adjustment are derived in a similar way as those for parametric model adjustment in Section 5.8. The usual relationship between variance–covariance matrix (C_ℓ) of observation vector ℓ and cofactor matrix (Q_ℓ) of the observation vector can be given as

$$C_\ell = \sigma_0^2 Q_\ell \tag{13.52}$$

where σ_0^2 is a priori variance factor of unit weight (which is always taken as the general population variance with a value of 1). Since a vector of observations ℓ and its variance–covariance matrix C_ℓ are used in computing unknown quantities, such as the adjusted parameters $(\hat{x})$, the adjusted observations $(\hat{\ell})$, and the observation residuals (v), it is obvious that errors in those observations will propagate into these quantities also. The concepts of variance–covariance propagation (Chapter 2) will be used as follows to derive the cofactor matrix of the adjusted parameters $(Q_{\hat{x}})$, the cofactor matrix of the adjusted observation $(Q_{\hat{\ell}})$, and the cofactor matrix of the observation residuals (Q_v).

13.8.1 Derivation of Cofactor Matrix of Adjusted Parameters

The vector of least squares adjusted parameters is given from Equations (13.13), (13.36), and (13.19) as

$$\hat{x} = x^0 - \left(A^T M^{-1} A\right)^{-1} A^T M^{-1} f\left(\ell, x^0\right) \tag{13.53}$$

Equation (13.53) is a form of $\hat{x} = G(\ell, x^0)$ with the usual propagated cofactor matrix (based on variance–covariance propagation rules) given as

$$Q_{\hat{x}} = J Q_\ell J^T \tag{13.54}$$

where Q_ℓ is the cofactor matrix of the observations and J is the Jacobian matrix given as follows:

$$J = \frac{\partial \hat{x}}{\partial \ell} = -\left(A^T M^{-1} A\right)^{-1} A^T M^{-1} B \tag{13.55}$$

taking note that the partial derivative of $f(\ell, x^0)$ in Equation (13.53) with respect to ℓ is B as shown in Equation (13.18) and x^0 is a vector of constant approximate values. Substituting Equation (13.55) into Equation (13.54) gives the following:

$$Q_{\hat{x}} = \left(A^T M^{-1} A\right)^{-1} A^T M^{-1} \left(B Q_\ell B^T\right) M^{-1} A \left(A^T M^{-1} A\right)^{-1} \tag{13.56}$$

Since by definition $M = (BQ_\ell B^T)$, then $M^{-1}(BQ_\ell B^T) = I$ (an identity matrix with all the principal diagonal elements as one and all the remaining elements as zero), so that Equation (13.56) can be reduced to the following:

$$Q_{\hat{x}} = \left(A^T M^{-1} A\right)^{-1} \left(A^T M^{-1} A\right) \left(A^T M^{-1} A\right)^{-1} \tag{13.57}$$

Similarly, $(A^T M^{-1} A)(A^T M^{-1} A)^{-1} = I$, so that Equation (13.57) can further be reduced to the following:

$$Q_{\hat{x}} = \left(A^T M^{-1} A\right)^{-1} \tag{13.58}$$

Equation (13.58) is the cofactor matrix of the adjusted parameters ($\hat{x}$).

13.8.2 Derivation of Cofactor Matrices of Adjusted Observations and Residuals

The vector of least squares adjusted observations is given from Equations (13.13), (13.30), and (13.32) as

$$\hat{\ell} = \ell - Q_\ell B^T \left(BQ_\ell B^T\right)^{-1} (A\delta + w) \tag{13.59}$$

Substituting Equations (13.36) and (13.19) into Equation (13.59) gives the following:

$$\hat{\ell} = \ell + Q_\ell B^T M^{-1} A \left(A^T M^{-1} A\right)^{-1} A^T M^{-1} f\left(\ell, x^0\right) - Q_\ell B^T M^{-1} f\left(\ell, x^0\right) \tag{13.60}$$

where $M = (BQ_\ell B^T)$. Equation (13.60) is a form of $\hat{\ell} = G(\ell, x^0)$ with the usual propagated cofactor matrix of adjusted observations (based on variance–covariance propagation rules) given as

$$Q_{\hat{\ell}} = JQ_\ell J^T \tag{13.61}$$

where Q_ℓ is the cofactor matrix of the observations and J is the Jacobian matrix given as follows:

$$J = \frac{\partial \hat{\ell}}{\partial \ell} = I + Q_\ell B^T M^{-1} A \left(A^T M^{-1} A\right)^{-1} A^T M^{-1} B - Q_\ell B^T M^{-1} B \tag{13.62}$$

with the partial derivative of $f(\ell, x^0)$ in Equation (13.60) with respect to ℓ being equal to the second design matrix B according to Equation (13.18). Substituting Equation (13.62) into Equation (13.61) gives the following:

$$Q_{\hat{\ell}} = Q_\ell - Q_\ell B^T M^{-1} B Q_\ell + Q_\ell B^T M^{-1} A N^{-1} A^T M^{-1} B Q_\ell \tag{13.63}$$

where $N = A^T M^{-1} A$. Equation (13.63) is the cofactor matrix of the adjusted observations.

From the relationships among the covariance matrix of the adjusted observations, the original observations, and the observation residuals, for parametric model adjustment (Equation (5.119)) and conditional model adjustment (Equation (12.74)), the cofactor matrix of residuals can be given as follows:

$$Q_v = Q_\ell - Q_{\hat{\ell}} \tag{13.64}$$

Relating Equation (13.63) with Equation (13.64), the cofactor matrix of residuals (v) can be deduced as

$$Q_v = Q_\ell B^T M^{-1} B Q_\ell - Q_\ell B^T M^{-1} A N^{-1} A^T M^{-1} B Q_\ell \tag{13.65}$$

13.8.3 Covariance Matrices of Adjusted Quantities

The variance–covariance matrices of the adjusted quantities can be given from Equations (13.58), (13.63), and (13.65), respectively, as follows:

$$C_{\hat{x}} = s_0^2 \left(A^T M^{-1} A\right)^{-1} \tag{13.66}$$

$$C_{\hat{\ell}} = s_0^2 Q_\ell - s_0^2 Q_\ell B^T M^{-1} B Q_\ell + s_0^2 Q_\ell B^T M^{-1} A N^{-1} A^T M^{-1} B Q_\ell \tag{13.67}$$

$$C_v = s_0^2 Q_\ell B^T M^{-1} B Q_\ell - s_0^2 Q_\ell B^T M^{-1} A N^{-1} A^T M^{-1} B Q_\ell \tag{13.68}$$

where $C_{\hat{x}}$ is the covariance matrix of the adjusted parameters, $C_{\hat{\ell}}$ is the covariance matrix of the adjusted observations, C_v is the covariance matrix of the observation residuals, and s_0^2 is *APVF of unit weight*, which is calculated after the least squares adjustment as follows:

$$s_0^2 = \frac{v^T P v}{r - u} \tag{13.69}$$

where r is the number of general model equations formed, u is the number of parameters in the equations, and P is the weight matrix of observations formed as $P = Q_\ell^{-1}$. If Equation (13.30) is transposed and post-multiplied by P, the following will be obtained:

$$v^T P = k^T B Q_\ell P \tag{13.70}$$

Since $Q_\ell P = I$ (an identity matrix), then Equation (13.70) can be reduced to the following:

$$v^T P = k^T B \tag{13.71}$$

Post-multiply Equation (13.71) by v to give the following:

$$v^T P v = k^T B v \tag{13.72}$$

From Equation (13.27),

$$B v = -(A\delta + w) \tag{13.73}$$

Substitute Bv in Equation (13.73) into Equation (13.72):

$$v^T P v = -k^T (A\delta + w) \tag{13.74}$$

From Equation (13.26), it can be seen that $-k^T A \delta$ term in Equation (13.74) can be set to zero, giving

$$v^T P v = -k^T w \tag{13.75}$$

From Equations (13.69) and (13.75), the following can be obtained:

$$s_0^2 = \frac{-k^T w}{r - u} \tag{13.76}$$

Either Equation (13.69) or Equation (13.76) can be used to calculate the APVF of unit weight s_0^2 in general model adjustment; but Equation (13.76) will require less computational steps.

13.8.4 Summary of Stochastic Properties of General Model Adjustment

The summary of stochastic properties of general model adjustment as derived in Sections 13.8.1–13.8.3 is as follows:

1) Cofactor matrix of the adjusted parameters ($\hat{x}$):

$$Q_{\hat{x}} = \left(A^T M^{-1} A\right)^{-1} \tag{13.77}$$

where $M = (B Q_\ell B^T)$, A is the first design matrix, B is the second design matrix, and Q_ℓ is the cofactor matrix of observations based on the variances of the observations.

2) Cofactor matrix of the adjusted observation ($\hat{\ell}$):

$$Q_{\hat{\ell}} = Q_\ell - Q_\ell B^T M^{-1} B Q_\ell + Q_\ell B^T M^{-1} A N^{-1} A^T M^{-1} B Q_\ell \tag{13.78}$$

3) Cofactor matrix of the observation residuals (v):

$$Q_v = Q_\ell - Q_{\hat{\ell}} \tag{13.79}$$

or

$$Q_v = Q_\ell B^T M^{-1} B Q_\ell - Q_\ell B^T M^{-1} A N^{-1} A^T M^{-1} B Q_\ell \tag{13.80}$$

4) Covariance matrices of the adjusted parameters, the adjusted observations, and the observation residuals are, respectively, as follows:

$$C_{\hat{x}} = s_0^2 Q_{\hat{x}} \tag{13.81}$$

$$C_{\hat{\ell}} = s_0^2 Q_{\hat{\ell}} \tag{13.82}$$

$$C_v = s_0^2 Q_v \tag{13.83}$$

where s_0^2 is the APVF of unit weight, which can be expressed as

$$s_0^2 = \frac{v^T P v}{r - u}$$

(13.84)

or

$$s_0^2 = \frac{-k^T w}{r - u}$$

(13.85)

13.9 Horizontal Circular Curve Example

The surveyor, using Real-time kinematic global positioning system (RTK GPS) surveys, measured the UTM coordinates of three points that belong to a circular curve shown in Figure 13.4. The measurements are given in Table 13.6 with all coordinates measured with accuracy of ±0.01 m and the radius of the curve as 28.88 m ± 0.01 m. The adjusted coordinates of the center of the circular curve, the adjusted measurements, and the standard deviations of the coordinates of the center of the circular curve are to be determined using the general model adjustment method as follows.

The general mathematical model (circular curve) equations are given in form of $f(x, \ell) = 0$ as follows:

$$(x_1 - x_0)^2 + (y_1 - y_0)^2 - r^2 = 0 \quad \text{for point 1}$$

(13.86)

Figure 13.4 Circular curve of a road.

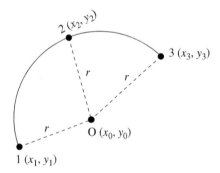

Table 13.6 Measurements.

Point	x (m)	y (m)
1	491 573.24	5 454 923.11
2	491 587.97	5 454 956.30
3	491 617.65	5 454 954.27

$$(x_2 - x_0)^2 + (y_2 - y_0)^2 - r^2 = 0 \quad \text{for point 2} \tag{13.87}$$

$$(x_3 - x_0)^2 + (y_3 - y_0)^2 - r^2 = 0 \quad \text{for point 3} \tag{13.88}$$

where (x_0, y_0) and r are the coordinates and the radius of the center of the circular curve, respectively, and the vector of unknown parameters to be determined is

$$x = \begin{bmatrix} x_0 \\ y_0 \end{bmatrix} \tag{13.89}$$

The field observations given in Table 13.6 with the given radius of the curve taken as a measurement are listed as follows:

$$\ell = \begin{bmatrix} x_1 = 491\,573.24 \\ y_1 = 5\,454\,923.11 \\ x_2 = 491\,587.97 \\ y_2 = 5\,454\,956.30 \\ x_3 = 491\,617.65 \\ y_3 = 5\,454\,954.27 \\ r = 28.88 \end{bmatrix} \text{m} \tag{13.90}$$

Remember that in any model adjustment (including general model adjustment), any quantities whose standard deviations are given are considered as observations and are treated as such in the adjustment. The coordinates of the three points can be used to determine the approximate coordinates of the center point of the circle using the principle of geometry that states that the perpendicular bisectors of two chords will meet at the center. Let the perpendicular bisectors of chords 1-2 and 2-3 be L1 and L2, respectively, with the computed gradients of 1-2 and 2-3 as 2.253 22 and −0.068 396 2, respectively. The coordinates of the midpoints of 1-2 and 2-3 are, respectively, (x = 491 580.61, y = 5 454 939.71) and (x = 491 602.81, y = 5 454 955.29). Since L1 and L2 are perpendicular to 1-2 and 1-3, respectively, their gradients can be determined from the gradients of 1-2 and 2-3 as −1/2.253 22 (or −0.443 81) and 1/0.068 396 2 (or 14.620 7), respectively. Using the coordinates of the midpoints of 1-2 and 2-3 and the gradients of the perpendicular bisectors L1 and L2, the following two equations for the bisectors can be formulated:

$$\text{For L1}: \quad y_0 - 5\,454\,939.71 = -0.443\,81(x_0 - 491\,580.61) \tag{13.91}$$

$$\text{For L2}: \quad y_0 - 5\,454\,955.29 = 14.620\,7(x_0 - 491\,602.81) \tag{13.92}$$

with x_0 and y_0 representing the coordinates of the center of the given circle. Solving for x_0 and y_0 in Equations (13.91) and (13.92) gives x_0 = 491 601 m and y_0 = 5 454 931 m. The vector of approximate values of the unknown parameters is then given as

$$x^0 = \begin{bmatrix} 491\ 601 \\ 5\ 454\ 931 \end{bmatrix} \tag{13.93}$$

The misclosure vector (w) is obtained by substituting field observations in Equation (13.90) and the approximate values of parameters in Equation (13.93) into Equations (13.86)–(13.88), giving

$$w = f(x^0, \ell) = \begin{bmatrix} -1.1847 \\ -24.1835 \\ -15.3390 \end{bmatrix}$$

Determine the first design matrix A (partial derivatives of the model Equations (13.86)–(13.88) with respect to the unknown parameters x) as follows:

$$A = \frac{\partial f}{\partial x} = \begin{bmatrix} -2(x_1 - x_0) & -2(y_1 - y_0) \\ -2(x_2 - x_0) & -2(y_2 - y_0) \\ -2(x_3 - x_0) & -2(y_3 - y_0) \end{bmatrix}$$

Substituting the approximate values of the parameters and the values of the observations into A-matrix gives the following:

$$A = \begin{bmatrix} 55.520 & 15.780 \\ 26.060 & -50.600 \\ -33.300 & -46.540 \end{bmatrix}$$

Determine the second design matrix, B (partial derivatives of the model Equations (13.86)–(13.88) with respect to the observations (ℓ) listed in Equation (13.90)):

$$B = \frac{\partial f}{\partial \ell} = \begin{bmatrix} 2(x_1 - x_0) & 2(y_1 - y_0) & 0 & 0 & 0 & 0 & -2r \\ 0 & 0 & 2(x_2 - x_0) & 2(y_2 - y_0) & 0 & 0 & -2r \\ 0 & 0 & 0 & 0 & 2(x_3 - x_0) & 2(y_3 - y_0) & -2r \end{bmatrix}$$

Substituting the approximate values of the parameters and the values of the observations into B-matrix gives the following:

$$B = \begin{bmatrix} -55.520 & -15.780 & 0 & 0 & 0 & 0 & -57.760 \\ 0 & 0 & -26.060 & 50.600 & 0 & 0 & -57.760 \\ 0 & 0 & 0 & 0 & 33.300 & 46.540 & -57.760 \end{bmatrix}$$

The standard deviation of each observation is 0.01 m; the cofactor matrix of the observations is diagonal as follows:

$$Q_\ell = \text{diag}\begin{bmatrix} 0.01^2 & 0.01^2 & 0.01^2 & 0.01^2 & 0.01^2 & 0.01^2 & 0.01^2 \end{bmatrix}$$

$$M = BQ_\ell B^T = \begin{bmatrix} 0.666\ 77 & 0.333\ 62 & 0.333\ 62 \\ 0.333\ 62 & 0.657\ 57 & 0.333\ 62 \\ 0.333\ 62 & 0.333\ 62 & 0.661\ 11 \end{bmatrix}$$

$$N = A^T M^{-1} A = \begin{bmatrix} 12\ 993.18 & 6\ 297.57 \\ 6\ 297.57 & 10\ 077.55 \end{bmatrix}$$

$$N^{-1} = \left(A^T M^{-1} A\right)^{-1} = \begin{bmatrix} 0.110\ 40\text{E}-3 & -0.069\ 00\text{E}-3 \\ -0.069\ 00\text{E}-3 & 0.142\ 344\text{E}-3 \end{bmatrix}$$

$$u = A^T M^{-1} w = \begin{bmatrix} 914.732\ 1 \\ 3\ 316.237\ 3 \end{bmatrix} \qquad \delta = -N^{-1} u = \begin{bmatrix} 0.127\ 8 \\ -0.408\ 9 \end{bmatrix}$$

$$\hat{x} = x^0 + \delta = \begin{bmatrix} 491\ 601.128 \\ 5\ 454\ 930.591 \end{bmatrix}$$

The residuals and the adjusted measurements can be determined as follows. The vector of correlates:

$$k = -M^{-1}(A\delta + w) = \begin{bmatrix} 0.677\ 090\ 5 \\ -0.480\ 963\ 2 \\ 0.752\ 497\ 4 \end{bmatrix}$$

The residual vector (v) and the adjusted observations ($\hat{\ell}$):

$$v = Q_\ell B^T k = \begin{bmatrix} -0.003\ 76 \\ -0.001\ 07 \\ 0.001\ 25 \\ -0.002\ 43 \\ 0.002\ 51 \\ 0.003\ 50 \\ -0.005\ 48 \end{bmatrix} \qquad \hat{\ell} = \ell + v = \begin{bmatrix} 491\ 573.236 \\ 5\ 454\ 923.109 \\ 491\ 587.971 \\ 5\ 454\ 956.298 \\ 491\ 617.653 \\ 5\ 454\ 954.274 \\ 28.875 \end{bmatrix}$$

The covariance matrix of the adjusted parameters is computed as follows. Compute the APVF of unit weight (s_0^2):

$$s_0^2 = \frac{v P^T v}{r - u} = \frac{-k^T w}{r - u} \qquad s_0^2 = \frac{0.713\ 33}{3 - 2} = 0.713\ 33$$

where r (the number of model equations) is 3 and u (the number of unknown parameters) is 2, giving the degrees of freedom (or redundancy) as 3–2 (or 1). The covariance matrix of the adjusted parameters is given as

$$C_{\hat{x}} = s_0^2 \left(A^T M^{-1} A \right)^{-1} = \begin{bmatrix} 7.875\text{E}-5 & -4.921\ 4\text{E}-5 \\ -4.921\ 4\text{E}-5 & 1.015\ 39\text{E}-4 \end{bmatrix}$$

and the standard deviations of the adjusted parameters are as follows:

$$S_{x_0} = \sqrt{7.875\text{E}-5} = 0.009 \text{ m}$$
$$S_{y_0} = \sqrt{1.015\ 39\text{E}-4} = 0.010 \text{ m}$$

The MATLAB code for the general model adjustment to determine the coordinates of the center of the circular curve is given in Table 13.7.

Table 13.7 MATLAB code for calculating adjusted coordinates of the center of circular curve.

```
>> syms x0 y0
>> syms x1 y1 x2 y2 x3 y3 r
>> L1=(x1-x0)^2+(y1-y0)^2-r^2;
>> L2=(x2-x0)^2+(y2-y0)^2-r^2;
>> L3=(x3-x0)^2+(y3-y0)^2-r^2;
>> J1=jacobian([L1;L2;L3],[x0 y0]);
>> J2=jacobian([L1;L2;L3],[x1 y1 x2 y2 x3 y3 r]);
>> x0=491601;y0=5454931;
>> xap=[491601;5454931];
>>
x1=491573.24;y1=5454923.11;x2=491587.97;y2=5454956.30;
x3=491617.65;y3=5454954.27;r=28.88;
>> L=[x1;y1;x2;y2;x3;y3;r];
>> format long
>> w=eval([L1;L2;L3]);
>> A=eval(J1);
>> B=eval(J2);
>> QL=diag(0.01^2*[1 1 1 1 1 1 1]);
>> M=B*QL*B';
>> Minv=inv(M);
>> N=A'*Minv*A;
>> Ninv=inv(N);
>> u=A'*Minv*w;
% Corrections to approximate parameters,d
>> d=-Ninv*u
d =
   0.127804117905152
  -0.408937749640790
>> xadj=xap+d;
```

Table 13.7 (Continued)

```
>> k=-Minv* (A*d+w) ;
>> v=QL*B'*k;
>> Ladj=L+v;
>> ktw=-k'*w
ktw =
    0.713331916905039
% Variance factor of unit weight s02
>> s02=ktw/1
s02 =
    0.713331916905039
>> Cx=s02*Ninv;
>> sx=sqrt (Cx(1,1)) ;
>>
```

Example 13.3 In order to determine if the shape of the upper region of an arch (shown in Figure 13.5) is parabolic or not, the coordinates (x, y) of six points on the arch, chosen on both sides of a vertical line of symmetry of the arch, were measured from an aerial photograph of the arch. The measurements are given in Table 13.8. The following general parabolic equation is to be fitted

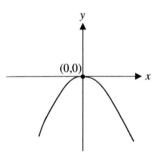

Figure 13.5 Arch frame.

Table 13.8 Photogrammetric measurements of an arch.

Point	x (cm)	y (cm)
1	−1.0	−0.32
2	1.0	−0.32
3	−2.0	−1.42
4	2.0	−1.42
5	−3.0	−4.0
6	3.0	−4.0

to the measurements in order to determine the unknown parameters a and b:
$y = ax^2 + bx$.

Perform the least squares adjustment of the general model, and determine the unknown parameters a and b. Assume equal weight for each measurement and the approximate values of the parameters, $x^0 = \begin{bmatrix} a^0 = -0.5 & b^0 = -0.2 \end{bmatrix}^T$.

Solution:

$$f(\ell, x) = 0 \quad y = ax^2 + bx$$

$$y_1 - ax_1^2 - bx_1 = 0 \tag{13.94}$$

$$y_2 - ax_2^2 - bx_2 = 0 \tag{13.95}$$

$$y_3 - ax_3^2 - bx_3 = 0 \tag{13.96}$$

$$y_4 - ax_4^2 - bx_4 = 0 \tag{13.97}$$

$$y_5 - ax_5^2 - bx_5 = 0 \tag{13.98}$$

$$y_6 - ax_6^2 - bx_6 = 0 \tag{13.99}$$

The observation vector: $\ell = \begin{bmatrix} x_1 & y_1 & x_2 & y_2 & x_3 & y_3 & x_4 & y_4 & x_5 & y_5 & x_6 & y_6 \end{bmatrix}^T$
Given approximate parameters, $x^0 = \begin{bmatrix} a^0 = -0.5 & b^0 = -0.2 \end{bmatrix}^T$
The first design matrix, A:

$$A = \frac{\partial f}{\partial x} = \begin{bmatrix} -x_1^2 & -x_1 \\ -x_2^2 & -x_2 \\ -x_3^2 & -x_3 \\ -x_4^2 & -x_4 \\ -x_5^2 & -x_5 \\ -x_6^2 & -x_6 \end{bmatrix} \quad A = \begin{bmatrix} -1 & 1 \\ -1 & -1 \\ -4 & 2 \\ -4 & -2 \\ -9 & 3 \\ -9 & -3 \end{bmatrix}$$

The second design matrix, B:

$$B = \frac{\partial f}{\partial \ell} = \begin{bmatrix} -b - 2ax_1 & 1 \\ & & -b - 2ax_2 & 1 \\ & & & & -b - 2ax_3 & 1 \\ & & & & & & \ddots \\ & & & & & & & & -b - 2ax_6 & 1 \end{bmatrix}$$

$$B = \begin{bmatrix} -0.8 & 1 & & & & \\ & 1.2 & 1 & & & \\ & & -1.8 & 1 & & \\ & & & 2.2 & 1 & \\ & & & & -2.8 & 1 \\ & & & & & 3.2 & 1 \end{bmatrix}$$

The misclosure vector (w):

$$w = f\left(\ell, x^0\right) = w = \begin{bmatrix} -0.02 \\ 0.38 \\ 0.18 \\ 0.98 \\ -0.10 \\ 1.1 \end{bmatrix}$$

$$Q = \mathrm{diag}\begin{bmatrix} 1^2 & 1^2 & 1^2 & 1^2 & 1^2 & 1^2 & \cdots & 1^2 \end{bmatrix}$$

$$M = BQB^T = \begin{bmatrix} 1.64 & & & & & \\ & 2.44 & & & & \\ & & 4.24 & & & \\ & & & 5.84 & & \\ & & & & 8.84 & \\ & & & & & 11.24 \end{bmatrix}$$

$$N = A^T M^{-1} A = \begin{bmatrix} 23.902\ 204\ 7 & -1.369\ 012\ 88 \\ -1.369\ 012\ 88 & 4.466\ 731\ 2 \end{bmatrix}$$

$$N^{-1} = \begin{bmatrix} 4.258\ 47\mathrm{E}-2 & 1.305\ 183\mathrm{E}-3 \\ 1.305\ 183\mathrm{E}-3 & 0.227\ 877\ 6 \end{bmatrix}$$

$$u = A^T M^{-1} W = \begin{bmatrix} -1.763\ 560 \\ -0.746\ 175 \end{bmatrix} \qquad \delta = -N^{-1} u = \begin{bmatrix} 0.084\ 84 \\ 0.193\ 05 \end{bmatrix}$$

$$\text{Adjusted parameters}: \begin{bmatrix} \hat{a} \\ \hat{b} \end{bmatrix} = x^0 + \delta = \begin{bmatrix} -0.415 \\ -0.007 \end{bmatrix} \mathrm{m}$$

13.10 Adjustment of General Model with Weight Constraints

A general model with weight constraint assumes that there was a previous adjustment of the network in which the coordinates of the network points were determined with the covariance matrix (C_x) of the adjustment provided. The derivation of the steps for the adjustment of general model with weight constraints is similar to that of parametric model adjustment given in Chapter 5 and of the general model adjustment in Sections 13.3–13.5. The steps are derived as follows. Given the following mathematical model (with weight constraint on the parameters),

$$f(\hat{x}, \hat{\ell}) = 0 \quad C_\ell \quad C_x \tag{13.100}$$

where f, $\hat{x}$, and $\hat{\ell}$ represent the mathematical model, adjusted parameter vector, and adjusted observation vector, respectively, C_ℓ is the covariance matrix of the unadjusted observations, and C_x is the covariance matrix of the unadjusted parameters. The steps for the adjustment of the general model in Equation (13.100) with imposed weight constraint based on C_x require the formulation of the following:

- Variation function.
- Normal equation system and solution.
- Stochastic models of adjusted quantities.

These steps are derived in the following sections with some numerical examples given later to explain the least squares adjustment steps involved when weight (C_x) constraints are imposed on the parameters.

13.10.1 Variation Function for General Model with Weight Constraints

The variation function for general model with weight constraint expressed in Equation (13.100) can be formulated by first linearizing the model by Taylor series expansion as follows. As usual, let the following be assumed:

$$\hat{x} = x^0 + \delta \quad \text{and} \quad \hat{\ell} = \ell + v \tag{13.101}$$

where x^0 is a vector of approximate values of unknown parameters, δ is a vector of corrections to the approximate values of the unknown parameters, ℓ is a vector of observations, and v is a vector of observation residuals. Substituting Equation (13.101) into Equation (13.100) gives the following:

$$f(x^0 + \delta, \ell + v) = 0 \tag{13.102}$$

By linearizing Equation (13.102) by Taylor series expansion, the following is obtained:

$$f\left(x^0, \ell\right) + \frac{\partial f}{\partial x}\delta + \frac{\partial f}{\partial \ell}v = 0 \tag{13.103}$$

The linearized Equation (13.103) can be simplified as follows:

$$w + A\delta + Bv = 0 \tag{13.104}$$

where $A = \dfrac{\partial f}{\partial x}$ and $B = \dfrac{\partial f}{\partial \ell}$ are the first and second design matrices, respectively, and $w = f(x^0, \ell)$ is a vector of misclosures of the model. Equation (13.103) or Equation (13.104) is the linearized form of the model Equation (13.102). Since the covariance matrices of observations (C_ℓ) and of the parameters (C_x) are directly associated with the model Equation (13.100), the least squares criterion for the problem can be given with regard to the covariance matrices as follows:

$$v^T C_\ell^{-1} v + \delta^T C_x^{-1} \delta = \text{minimum} \tag{13.105}$$

where C_ℓ^{-1} and C_x^{-1} are the weight matrices of the observations and the parameters, respectively, assuming the a priori variance factor, $\sigma_0^2 = 1$. By imposing the least squares criterion on the linearized model (13.104), the following variation function is obtained based on the Lagrangian approach:

$$\varphi = v^T C_\ell^{-1} v + \delta^T C_x^{-1} \delta - 2k^T \left(w + A\delta + Bv\right) = \text{minimum} \tag{13.106}$$

where k is a vector of Lagrangian multipliers and assuming $\sigma_0^2 = 1$.

13.10.2 Normal Equation System and Solution

The most extended form of the least squares normal equation system for the general model with weight constraint can be derived from the variation function given in Equation (13.106). This is done as follows by finding the partial derivatives of the function with respect to the unknown quantities in the function (i.e. v, k, and δ) and setting the partial derivatives to zero:

$$\frac{\partial \varphi}{\partial v} = 2v^T C_\ell^{-1} - 2k^T B = 0 \tag{13.107}$$

$$\frac{\partial \varphi}{\partial k^T} = -2\left(w + A\delta + Bv\right) = 0 \tag{13.108}$$

$$\frac{\partial \varphi}{\partial \delta} = 2\delta^T C_x^{-1} - 2k^T A = 0 \tag{13.109}$$

Equations (13.107)–(13.109) can be rewritten as follows. Divide Equation (13.107) by 2 and transpose all the elements to obtain the following:

$$C_\ell^{-1} v - B^T k = 0 \tag{13.110}$$

Divide Equation (13.108) by 2 to give

$$-(w + A\delta + Bv) = 0 \tag{13.111}$$

and divide Equation (13.109) by 2 and transpose all the elements to obtain the following:

$$C_x^{-1} \delta - A^T k = 0 \tag{13.112}$$

Equations (13.110)–(13.112) constitute the most expanded form of the least squares normal equation system, which can be given in matrix form as follows:

$$\begin{bmatrix} C_\ell^{-1} & -B^T & 0 \\ -B & 0 & -A \\ 0 & -A^T & C_x^{-1} \end{bmatrix} \begin{bmatrix} v \\ k \\ \delta \end{bmatrix} + \begin{bmatrix} 0 \\ -w \\ 0 \end{bmatrix} = \begin{bmatrix} 0 \\ 0 \\ 0 \end{bmatrix} \tag{13.113}$$

The vector of corrections to the unknown parameters can be determined from the matrix in Equation (13.113) or by appropriate solution of Equations (13.110)–(13.112) as follows:

$$\delta = -\left(C_x^{-1} + A^T M^{-1} A\right)^{-1} A^T M^{-1} w \tag{13.114}$$

where $M = BC_\ell B^T$ and $Q_\ell = C_\ell$ (for $\sigma_0^2 = 1$). The solution in Equation (13.114) can be rewritten as follows:

$$\delta = -\left(P_x + A^T M^{-1} A\right)^{-1} A^T M^{-1} w \tag{13.115}$$

where $P_x = C_x^{-1}$ and $P = C_\ell^{-1}$, assuming the a priori variance factor, $\sigma_0^2 = 1$. From Equation (13.110) and taking $Q_\ell = C_\ell$,

$$v = Q_\ell B^T k \tag{13.116a}$$

and from Equations (13.110) and (13.111),

$$k = -\left(B Q_\ell B^T\right)^{-1} (A\delta + w) \tag{13.116b}$$

The adjusted parameters are determined in the usual way as

$$\hat{x} = x^0 + \delta \tag{13.116c}$$

13.10.3 Stochastic Models of Adjusted Quantities

The covariance matrix of the adjusted parameters can be deduced from the general understanding from the previous derivation of stochastic properties in Section 13.8 that the inverse of the coefficient matrix of the parameters is the cofactor matrix of the parameters. This gives the cofactor matrix of the weight-constrained adjustment of the parameters as

$$Q_{\hat{x}} = \left(P_x + A^T M^{-1} A\right)^{-1} \tag{13.117}$$

and the covariance matrix of the adjusted parameters as

$$C_{\hat{x}} = s_0^2 \left(P_x + A^T M^{-1} A\right)^{-1} \tag{13.118}$$

where the APVF of unit weight is given as

$$s_0^2 = \frac{v^T P v + \delta^T P_x \delta}{r - (u - u_w)} \tag{13.119}$$

where r is the number of model equations, u is the total number of parameters to be determined in the adjustment (including those whose covariance matrix is given), u_w is the number of parameters whose covariance matrix is known, P_x is the given weight matrix of parameters, P is the weight matrix of observations, v is the residual vector, and δ is the vector of corrections to the approximate parameters.

Example 13.4 The local horizontal coordinates of points A, B, and C in the network in Figure 13.6 were initially determined by RTK GPS techniques. The computed coordinates and their corresponding variance–covariance matrix are given in Table 13.9 and the following matrix, respectively.

Figure 13.6 Traverse survey network (not to scale).

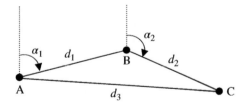

Table 13.9 Adjusted coordinates from previous survey.

Point	X (easting) (m)	Y (northing) (m)
A	1000.000	1000.000
B	1531.985	1178.140
C	1898.206	955.940

Variance–covariance matrix of the adjusted coordinates:

$$
C_{\hat{x}} =
\begin{bmatrix}
9.0E-5 & -5.6E-5 & 0 & 0 & 0 & 0 \\
-5.6E-5 & 9.4E-5 & 0 & 0 & 0 & 0 \\
0 & 0 & 3.3E-4 & 1.3E-4 & 0 & 0 \\
0 & 0 & 1.3E-4 & 1.6E-3 & 0 & 0 \\
0 & 0 & 0 & 0 & 9.8E-4 & -9.6E-5 \\
0 & 0 & 0 & 0 & -9.6E-5 & 9.4E-4
\end{bmatrix}
$$

At a later date, a traverse survey was run from point A and closed on C. The measurements include two bearings (α_1, α_2) and three distances (d_1, d_2, d_3) as shown in Figure 13.6. The measurements are stochastically independent (i.e. the measurement covariance matrix has zero off-diagonal elements), and their values with the corresponding standard deviations (s) are as follows:

$\alpha_1 = 71°29'13''$, $\quad s_{\alpha_1} = 10''$

$\alpha_2 = 121°14'50''$, $\quad s_{\alpha_2} = 10''$

$d_1 = 561.025\,\text{m}$, $\quad s_{d_1} = 0.005\,\text{m}$

$d_2 = 428.356\,\text{m}$, $\quad s_{d_2} = 0.005\,\text{m}$

$d_3 = 899.290\,\text{m}$, $\quad s_{d_3} = 0.010\,\text{m}$

Calculate the least squares adjusted parameters with weight constraint on the parameters.

Solution:

Note that this problem can also be solved by using parametric least squares adjustment method since the model can also be presented as parametric model. The complete general model equations ($f(\hat{x}, \ell) = 0$) of the problem can be given as follows:

$$
\operatorname{atan}\left(\frac{\hat{x}_B - \hat{x}_A}{\hat{y}_B - \hat{y}_A}\right) - \hat{\alpha}_1 = 0 \tag{13.120}
$$

$$
\operatorname{atan}\left(\frac{\hat{x}_C - \hat{x}_B}{\hat{y}_C - \hat{y}_B}\right) + 180 - \hat{\alpha}_2 = 0 \tag{13.121}
$$

$$
\sqrt{(\hat{x}_B - \hat{x}_A)^2 + (\hat{y}_B - \hat{y}_A)^2} - d_1 = 0 \tag{13.122}
$$

$$
\sqrt{(\hat{x}_C - \hat{x}_B)^2 + (\hat{y}_C - \hat{y}_B)^2} - d_2 = 0 \tag{13.123}
$$

$$
\sqrt{(\hat{x}_C - \hat{x}_A)^2 + (\hat{y}_C - \hat{y}_A)^2} - d_3 = 0 \tag{13.124}
$$

Form the first design matrix, A, using the given coordinates as approximate coordinates, and assume the following in the partial derivatives for the A-matrix:

$$d^2_{AB} = (x_B - x_A)^2 + (y_B - y_A)^2$$
$$d^2_{BC} = (x_C - x_B)^2 + (y_C - y_B)^2$$
$$d^2_{AC} = (x_C - x_A)^2 + (y_C - y_A)^2$$

Form the following A-matrix as the Jacobian of Equations (13.120)–(13.124) (ignoring the hats) with respect to all of the six coordinates, $x = [x_A \; y_A \; x_B y_B x_C y_C]^T$, since they all have standard deviations associated with them:

$$A = \frac{\partial f}{\partial x} = \begin{bmatrix} \dfrac{-(y_B - y_A)}{d^2_{AB}} & \dfrac{(x_B - x_A)}{d^2_{AB}} & \dfrac{(y_B - y_A)}{d^2_{AB}} & \dfrac{-(x_B - x_A)}{d^2_{AB}} & 0 & 0 \\[2mm] 0 & 0 & \dfrac{-(y_C - y_B)}{d^2_{BC}} & \dfrac{(x_C - x_B)}{d^2_{BC}} & \dfrac{(y_C - y_B)}{d^2_{BC}} & \dfrac{-(x_C - x_B)}{d^2_{BC}} \\[2mm] \dfrac{-(x_B - x_A)}{d_{AB}} & \dfrac{-(y_B - y_A)}{d_{AB}} & \dfrac{(x_B - x_A)}{d_{AB}} & \dfrac{(y_B - y_A)}{d_{AB}} & 0 & 0 \\[2mm] 0 & 0 & \dfrac{-(x_C - x_B)}{d_{BC}} & \dfrac{-(y_C - y_B)}{d_{BC}} & \dfrac{(x_C - x_B)}{d_{BC}} & \dfrac{(y_C - y_B)}{d_{BC}} \\[2mm] \dfrac{-(x_C - x_A)}{d_{AC}} & \dfrac{-(y_C - y_A)}{d_{AC}} & 0 & 0 & \dfrac{(x_C - x_A)}{d_{AC}} & \dfrac{(y_C - y_A)}{d_{AC}} \end{bmatrix}$$

Substituting the approximate coordinates (the original coordinates from the previous survey) into the A-matrix gives the following:

$$A = \begin{bmatrix} -5.659\,88E-4 & 1.690\,23E-3 & 5.659\,88E-4 & -1.690\,23E-3 & 0 & 0 \\ 0 & 0 & 1.210\,96E-3 & 1.995\,86E-3 & -1.210\,96E-3 & -1.995\,86E-3 \\ -0.948\,25 & -0.317\,53 & 0.948\,25 & 0.317\,53 & 0 & 0 \\ 0 & 0 & -0.854\,94 & 0.518\,73 & 0.854\,94 & -0.518\,73 \\ -0.998\,80 & 0.048\,99 & 0 & 0 & 0.998\,80 & -0.048\,99 \end{bmatrix}$$

Find the partial derivatives of Equations (13.120)–(13.124) (ignoring the hats for brevity) with respect to the observation vector, $\ell = [\alpha_1 \; \alpha_2 \; d_1 \; d_2 \; d_3]^T$, giving the following second design matrix, B:

$$B = \frac{\partial f}{\partial \ell} = \begin{bmatrix} -1 & 0 & 0 & 0 & 0 \\ 0 & -1 & 0 & 0 & 0 \\ 0 & 0 & -1 & 0 & 0 \\ 0 & 0 & 0 & -1 & 0 \\ 0 & 0 & 0 & 0 & -1 \end{bmatrix}$$

The given coordinates are used as approximate coordinates:

$$
x = \begin{bmatrix} x_A \\ y_A \\ x_B \\ y_B \\ x_C \\ y_C \end{bmatrix} \rightarrow x^0 = \begin{bmatrix} 1000.000 \\ 1000.000 \\ 1531.985 \\ 1178.140 \\ 1898.206 \\ 955.940 \end{bmatrix}
$$

The cofactor matrix and other derived matrices:

$$
Q_\ell = \begin{bmatrix} 2.350\ 44E-9 & 0 & 0 & 0 & 0 \\ 0 & 2.350\ 44E-9 & 0 & 0 & 0 \\ 0 & 0 & 2.5E-5 & 0 & 0 \\ 0 & 0 & 0 & 2.5E-5 & 0 \\ 0 & 0 & 0 & 0 & 1.0E-4 \end{bmatrix}
$$

$$
P = \begin{bmatrix} 4.254\ 525E8 & 0 & 0 & 0 & 0 \\ 0 & 4.254\ 525E8 & 0 & 0 & 0 \\ 0 & 0 & 4.0E4 & 0 & 0 \\ 0 & 0 & 0 & 4.0E4 & 0 \\ 0 & 0 & 0 & 0 & 1.0E4 \end{bmatrix}
$$

Use the covariance matrix of the adjusted parameters from the previous survey to construct the weight matrix of parameters (assuming the a priori variance factor of unit weight can be assumed to be $\sigma_0^2 = 1$):

$$
P_{\hat{x}} = C_{\hat{x}}^{-1} = \begin{bmatrix} 17\ 655.897\ 8 & 10\ 518.407\ 2 & 0 & 0 & 0 & 0 \\ 10\ 518.407\ 2 & 16\ 904.583 & 0 & 0 & 0 & 0 \\ 0 & 0 & 3\ 130.502\ 84 & -254.353\ 4 & 0 & 0 \\ 0 & 0 & -254.353\ 4 & 645.666\ 2 & 0 & 0 \\ 0 & 0 & 0 & 0 & 1\ 030.719\ 8 & 105.265\ 0 \\ 0 & 0 & 0 & 0 & 105.265\ 0 & 1\ 074.580\ 3 \end{bmatrix}
$$

$$
M = BQ_\ell B^T = \begin{bmatrix} 2.350\ 44E-9 & 0 & 0 & 0 & 0 \\ 0 & 2.350\ 44E-9 & 0 & 0 & 0 \\ 0 & 0 & 2.5E-5 & 0 & 0 \\ 0 & 0 & 0 & 2.5E-5 & 0 \\ 0 & 0 & 0 & 0 & 1.0E-4 \end{bmatrix}
$$

$$M^{-1} = \begin{bmatrix} 4.254\ 525E8 & 0 & 0 & 0 & 0 \\ 0 & 4.254\ 525E8 & 0 & 0 & 0 \\ 0 & 0 & 4.0E4 & 0 & 0 \\ 0 & 0 & 0 & 4.0E4 & 0 \\ 0 & 0 & 0 & 0 & 1.0E4 \end{bmatrix}$$

$$A^T M^{-1} A = \begin{bmatrix} 46\ 079.285 & 11\ 147.512 & -36\ 103.289 & -11\ 636.868 & -9\ 975.995 & 489.356 \\ 11\ 147.512 & 5\ 272.466 & -11\ 636.868 & -5\ 248.461 & 489.356 & -24.005 \\ -36\ 103.289 & -11\ 636.868 & 65\ 964.164 & -5\ 074.026 & -29\ 860.875 & 16\ 710.894 \\ -11\ 636.868 & -5\ 248.461 & -5\ 074.026 & 17\ 706.246 & 16\ 710.894 & -12\ 457.785 \\ -9\ 975.995 & 489.356 & -29\ 860.875 & 16\ 710.894 & 39\ 836.871 & -17\ 200.249 \\ 489.356 & -24.005 & 16\ 710.894 & -12\ 457.785 & -17\ 200.249 & 12\ 481.790 \end{bmatrix}$$

$$(P_x + A^T M^{-1} A)^{-1} = \begin{bmatrix} 6.822E-5 & -4.1527E-5 & 5.4144E-5 & -8.3806E-6 & 5.9126E-5 & -2.4216E-6 \\ -4.1527E-5 & 8.1798E-5 & -2.6409E-5 & 3.8752E-5 & -3.6099E-5 & 2.4288E-5 \\ 5.4144E-5 & -2.6409E-5 & 7.8876E-5 & -5.6526E-5 & 6.5696E-5 & -6.8332E-5 \\ -8.3806E-6 & 3.8752E-5 & -5.6526E-5 & 2.4929E-4 & -4.3634E-5 & 2.4412E-4 \\ 5.9126E-5 & -3.6099E-5 & 6.5696E-5 & -4.3634E-5 & 1.1346E-4 & 1.9797E-5 \\ -2.4216E-6 & 2.4288E-5 & -6.8332E-5 & 2.4412E-4 & 1.9797E-5 & 4.0743E-4 \end{bmatrix}$$

The misclosure vector, w:

$$w = \begin{bmatrix} -9.310E-6 \\ -8.115E-6 \\ -6.373E-3 \\ 2.099E-3 \\ -4.004E-3 \end{bmatrix}$$

The first two elements in the above misclosure vector (w) are in radians; remember that whenever angle or directional measurements and linear measurements such as distances are used together in an adjustment, the angle or directional measurements must be expressed in radians in the weight matrices and the misclosure vectors for consistency sake:

$$u = A^T M^{-1} w = \begin{bmatrix} 283.960\ 45 \\ 72.286\ 56 \\ -299.371\ 10 \\ -50.059\ 31 \\ 15.410\ 65 \\ -22.227\ 25 \end{bmatrix}$$

The vector of corrections (δ) to the approximate coordinates:

$$\delta = -\left(P_x + A^T M^{-1} A\right)^{-1}\left(A^T M^{-1} w\right) = \begin{bmatrix} -0.001\ 545 \\ 0.001\ 009 \\ 0.004\ 786 \\ 0.001\ 234\ 1 \\ 0.001\ 994\ 8 \\ -0.000\ 553 \end{bmatrix}$$

The adjusted parameters (with weight constraint) (in m):

$$\hat{x} = x^0 + \delta = \begin{bmatrix} 1\ 000.000 \\ 1\ 000.000 \\ 1\ 531.985 \\ 1\ 178.140 \\ 1\ 898.206 \\ 955.940 \end{bmatrix} + \begin{bmatrix} -0.001\ 545 \\ 0.001\ 009 \\ 0.004\ 786 \\ 0.001\ 234\ 1 \\ 0.001\ 994\ 8 \\ -0.000\ 553 \end{bmatrix} = \begin{bmatrix} 999.998\ 5 \\ 1\ 000.001\ 0 \\ 1\ 531.989\ 8 \\ 1\ 178.141\ 2 \\ 1\ 898.208\ 0 \\ 955.939\ 4 \end{bmatrix}$$

Example 13.5 Adjust the leveling network in Figure 13.7 with the leveling notes given in Table 13.10 using the general model method of least squares adjustment. In the adjustment, assume the elevations of control stations FH1 and FH2 are known as 100.000 and 99.729 m, respectively, and consider the elevation of each control station as having the same standard deviation of ±0.004 m.

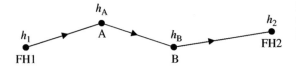

Figure 13.7 Leveling network with arrows indicating leveling directions.

Table 13.10 Leveling field notes.

Leg i	BS	FS	dh_i (m)	Standard dev. s_i (m)
1	FH1	A	−7.640	0.007
2	A	B	2.370	0.006
3	B	FH2	5.030	0.005

a) Determine the least squares adjusted elevations of the station points.

Solution:

Form the general model equations $(f(\hat{x}, \ell) = 0)$ as follows:

$$d\hat{h}_1 - \hat{h}_A + \hat{h}_1 = 0$$
$$d\hat{h}_2 - \hat{h}_B + \hat{h}_A = 0$$
$$d\hat{h}_3 - \hat{h}_2 + \hat{h}_B = 0$$

Let the approximate heights of points be

$$\begin{bmatrix} h_1^0 \\ h_A^0 \\ h_B^0 \\ h_2^0 \end{bmatrix} = \begin{bmatrix} 100.000 \\ 92.659 \\ 95.154 \\ 99.729 \end{bmatrix}$$

The first design matrix (A):

$$A = \begin{bmatrix} 1 & -1 & 0 & 0 \\ 0 & 1 & -1 & 0 \\ 0 & 0 & 1 & -1 \end{bmatrix}$$

The second design matrix (B):

$$B = \begin{bmatrix} 1 & 0 & 0 \\ 0 & 1 & 0 \\ 0 & 0 & 1 \end{bmatrix}$$

The misclosure vector (w):

$$w = \begin{bmatrix} -7.640 - 92.659 + 100 \\ 2.370 - 95.154 + 92.659 \\ 5.030 - 99.729 + 95.154 \end{bmatrix} \quad \text{or} \quad w = \begin{bmatrix} -0.299 \\ -0.125 \\ +0.455 \end{bmatrix}$$

$$Q_\ell = \begin{bmatrix} (0.007)^2 & & \\ & (0.006)^2 & \\ & & (0.005)^2 \end{bmatrix}$$

$$P = Q_\ell^{-1} = \begin{bmatrix} 20\ 408.163 & & \\ & 27\ 777.8 & \\ & & 40\ 000.0 \end{bmatrix}$$

$$M = BQ_\ell B^T = \begin{bmatrix} 4.9\text{E}-5 & & \\ & 3.6\text{E}-5 & \\ & & 2.5\text{E}-5 \end{bmatrix}$$

$$M^{-1} = \begin{bmatrix} 20\ 408.163 & & \\ & 27\ 777.8 & \\ & & 40\ 000.0 \end{bmatrix}$$

$$A^T M^{-1} A = \begin{bmatrix} 20\ 408.163\ 26 & -20\ 408.163\ 26 & 0.0 & 0.0 \\ -20\ 408.163\ 26 & 48\ 185.941\ 04 & -27\ 777.777\ 8 & 0.0 \\ 0.0 & -27\ 777.777\ 8 & 67\ 777.777\ 8 & -40\ 000 \\ 0.0 & 0.0 & -40\ 000 & 40\ 000 \end{bmatrix}$$

$$u = A^T M^{-1} w = \begin{bmatrix} -6\ 102.040\ 82 \\ 2\ 629.818\ 59 \\ 21\ 672.222\ 2 \\ -18\ 200.0 \end{bmatrix}$$

The solution vector is given as follows. Similar to parametric model adjustment with weight constraints, those points with no standard deviations will have weights of zeros, and those points with standard deviations will have the inverse of their variances as weights, as given below:

$$P_x = \begin{bmatrix} \dfrac{1}{(0.004)^2} & & & \\ & 0.0 & & \\ & & 0.0 & \\ & & & \dfrac{1}{(0.004)^2} \end{bmatrix} = \begin{bmatrix} 62\ 500 & & & \\ & 0.0 & & \\ & & 0.0 & \\ & & & 62\ 500 \end{bmatrix}$$

$$\left(P_x + A^T M^{-1} A\right)^{-1} = \begin{bmatrix} 1.419\ 72\text{E}-5 & 8.676\ 06\text{E}-6 & 4.619\ 72\text{E}-6 & 1.802\ 82\text{E}-6 \\ 8.676\ 06\text{E}-6 & 3.524\ 6\text{E}-5 & 1.876\ 7\text{E}-5 & 7.323\ 9\text{E}-6 \\ 4.619\ 72\text{E}-6 & 1.876\ 7\text{E}-5 & 2.916\ 19\text{E}-5 & 1.138\ 0\text{E}-5 \\ 1.802\ 82\text{E}-6 & 7.323\ 9\text{E}-6 & 1.138\ 0\text{E}-5 & 1.419\ 7\text{E}-5 \end{bmatrix}$$

The vector of corrections (δ) to the approximate parameters:

$$\delta = -\left(P_x + A^T M^{-1} A\right)^{-1}\left(A^T M^{-1} w\right) = \begin{bmatrix} -3.493\ 0\text{E}-3 \\ -0.313\ 19 \\ -0.446\ 05 \\ 3.493\ 0\text{E}-3 \end{bmatrix}$$

The adjusted parameters (with weight constraint):

$$\hat{x} = x^0 + \delta \;\longrightarrow\; \begin{bmatrix} 100.000 \\ 92.659 \\ 95.154 \\ 99.729 \end{bmatrix} + \begin{bmatrix} -3.493\ 0E-3 \\ -0.313\ 19 \\ -0.446\ 05 \\ 3.493\ 0E-3 \end{bmatrix} \;\longrightarrow\; \begin{bmatrix} 99.997 \\ 92.346 \\ 94.708 \\ 99.732 \end{bmatrix} \text{m}$$

b) Calculate the adjusted observations and the APVF of unit weight.

Solution:

The vector of residuals (v):

$$A\delta + w = \begin{bmatrix} 0.010\ 697\ 2 \\ 0.007\ 859\ 1 \\ 0.005\ 457\ 7 \end{bmatrix} \quad \text{with } k = -M^{-1}(A\delta + w) = \begin{bmatrix} -218.309\ 859 \\ -218.309\ 859 \\ -218.309\ 859 \end{bmatrix}$$

$$\text{and } v = Q_\ell B^T k = \begin{bmatrix} -0.010\ 69 \\ -0.007\ 859 \\ -0.005\ 46 \end{bmatrix}$$

The adjusted observations ($\hat{\ell}$):

$$\hat{\ell} = \ell + v = \begin{bmatrix} -7.640 \\ 2.370 \\ 5.030 \end{bmatrix} + \begin{bmatrix} -0.010\ 69 \\ -0.007\ 859 \\ -0.005\ 46 \end{bmatrix} \;\longrightarrow\; \begin{bmatrix} -7.650\ 7 \\ 2.362\ 1 \\ 5.024\ 5 \end{bmatrix}$$

The APVF of unit weight can be determined from Equation (13.119):

$$s_0^2 = \frac{v^T P v + \delta^T P_x \delta}{r - (u - u_w)}$$

Since the weights of unknown points A and B are set to zero, then $u = 4$, $u_w = 2$ (for the elevations of the remaining two points considered as pseudo-observations), and $r = 3$. Given the following weight matrices for the parameters and the observations,

$$P_x = \begin{bmatrix} 62\ 500 & & & \\ & 0.0 & & \\ & & 0.0 & \\ & & & 62\ 500 \end{bmatrix} \qquad P = \begin{bmatrix} 20\ 408.163 & & \\ & 27\ 777.8 & \\ & & 40\ 000.0 \end{bmatrix}$$

$$s_0^2 = \frac{5.240\ 294\ 01 + 1.525\ 131\ 125}{3 - (4 - 2)} = 6.765\ 425$$

This is a large number, indicating that there are possible outliers in measurements or the precisions of the observations and the pseudo-observations were not well scaled.

Example 13.6 In Example 13.4, the parametric model equations were re-modified to look like the general model equations. Formulate different set of general model equations ($f(\hat{x}, \ell) = 0$) for the problem using forward traverse computation approach.

Solution:

General model equations ($f(\hat{x}, \ell) = 0$) based on forward traverse computation approach:

Unknown parameters: $x = \begin{bmatrix} x_A & y_A & x_B & y_B & x_C & y_C \end{bmatrix}^T$
Observations: $\ell = \begin{bmatrix} \alpha_1 & \alpha_2 & d_1 & d_2 & d_3 \end{bmatrix}^T$

By traversing from A to B, the following general model equations are obtained:

$$\hat{x}_A + \hat{d}_1 \sin \hat{\alpha}_1 - \hat{x}_B = 0 \tag{13.125}$$

$$\hat{y}_A + \hat{d}_1 \cos \hat{\alpha}_1 - \hat{y}_B = 0 \tag{13.126}$$

By traversing from B to C, the following general model equations are obtained:

$$\hat{x}_B + \hat{d}_2 \sin \hat{\alpha}_2 - \hat{x}_C = 0 \tag{13.127}$$

$$\hat{y}_B + \hat{d}_2 \cos \hat{\alpha}_2 - \hat{y}_C = 0 \tag{13.128}$$

Form an equation for the remaining observation (d_3) as follows:

$$\sqrt{(\hat{x}_C - \hat{x}_A)^2 + (\hat{y}_C - \hat{y}_A)^2} - \hat{d}_3 = 0 \tag{13.129}$$

Equations (13.125)–(13.129) constitute a set of different general model equations compared with those given in Example 13.4.

Example 13.7 Given the following mathematical models with weight constraints on some parameters,

$$f_1\left(\hat{\ell}_1, \hat{x}_1\right) = 0 \quad C_{\ell_1} \quad C_{x_1} \tag{13.130}$$

$$f_2\left(\hat{\ell}_2, \hat{x}_1, \hat{x}_2\right) = 0 \quad C_{\ell_2} \quad C_{x_2} \tag{13.131}$$

where f_i, $\hat{x}_i$, $\hat{\ell}_i$, and C_{ℓ_i} (for $i = 1, 2$) represent the mathematical models, the adjusted parameter vectors, the adjusted observation vectors, and the covariance matrices of the observation vectors. Formulate the variation function.

Solution:

Equation (13.130) can be considered as representing a survey network with the observations (ℓ_1) taken at an epoch t_1 and the parameters (x_1) with given covariance matrix (C_{x_1}) relating to that network; Equation (13.131) can be considered as representing the same network (x_1) with different observables (ℓ_2) measured at another epoch t_2 with different instruments (e.g. distance observable measured with an EDM whose scale factor (x_2) is known with given covariance matrix (C_{x_2})). Assume the following:

$$\hat{\ell}_1 = \ell_1 + v_1 \quad \text{and} \quad \hat{x}_1 = x_1^0 + \delta_1 \tag{13.132}$$

$$\hat{\ell}_2 = \ell_2 + v_2 \quad \text{and} \quad \hat{x}_2 = x_2^0 + \delta_2 \tag{13.133}$$

where ℓ_1, ℓ_2 are vectors of original observations; v_1, v_2 are vectors of corrections to the original observations; x_1^0, x_2^0 are vectors of approximate values of the parameters; and δ_1, δ_2 are the vectors of corrections to the approximate values of parameters. Equations (13.130) and (13.131) can be rewritten as follows:

$$f_1\left(\ell_1 + v_1, x_1^0 + \delta_1\right) = 0 \tag{13.134}$$

$$f_2\left(\ell_2 + v_2, x_1^0 + \delta_1, x_2^0 + \delta_2\right) = 0 \tag{13.135}$$

Linearize Equation (13.134) by Taylor series expansion to obtain the following linearized model:

$$\frac{\partial f_1}{\partial x_1}\delta_1 + \frac{\partial f_1}{\partial \ell_1}v_1 + f_1\left(\ell_1, x_1^0\right) = 0 \tag{13.136}$$

or in simpler form as

$$A_1\delta_1 + B_1v_1 + w_1 = 0 \tag{13.137}$$

where $A_1 = \dfrac{\partial f_1}{\partial x_1}$ is the first design matrix, $B_1 = \dfrac{\partial f_1}{\partial \ell_1}$ is the second design matrix, and $w_1 = f_1\left(\ell_1, x_1^0\right)$ is the misclosure vector. Similarly, linearize Equation (13.135) by Taylor series expansion to obtain the following linearized model:

$$\frac{\partial f_2}{\partial x_1}\delta_1 + \frac{\partial f_2}{\partial x_2}\delta_2 + \frac{\partial f_2}{\partial \ell_2}v_2 + f_2\left(\ell_2, x_1^0, x_2^0\right) = 0 \tag{13.138}$$

or in simpler form as

$$A_2\delta_1 + A_3\delta_2 + B_2v_2 + w_2 = 0 \tag{13.139}$$

where $A_2 = \dfrac{\partial f_2}{\partial x_1}$ and $A_3 = \dfrac{\partial f_2}{\partial x_2}$ are the first design matrices for the model, $B_2 = \dfrac{\partial f_2}{\partial \ell_2}$ is the second design matrix for the model, and $w_2 = f_2\left(\ell_2, x_1^0, x_2^0\right)$ is the misclosure vector. Since the given covariance matrices are directly associated with the given models, the least squares criterion for the problem can be given as follows:

$$v_1^T C_{\ell_1}^{-1} v_1 + v_2^T C_{\ell_2}^{-1} v_2 + \delta_1^T C_{x_1}^{-1} \delta_1 + \delta_2^T C_{x_2}^{-1} \delta_2 = \text{minimum} \tag{13.140}$$

Assuming the weight matrices can be expressed as $P_{x_1} = C_{x_1}^{-1}$, $P_{x_2} = C_{x_2}^{-1}$, $P_1 = C_{\ell_1}^{-1}$, and $P_2 = C_{\ell_2}^{-1}$ with the a priori variance factor, $\sigma_0^2 = 1$, and by imposing least squares criterion on the linearized models (13.137) and (13.139), the following variation function is obtained using Lagrangian approach:

$$\varphi = v_1^T C_{\ell_1}^{-1} v_1 + v_2^T C_{\ell_2}^{-1} v_2 + \delta_1^T C_{x_1}^{-1} \delta_1 + \delta_2^T C_{x_2}^{-1} \delta_2 - 2k_1^T \left(A_1 \delta_1 + B_1 v_1 + w_1\right)$$

$$- 2k_2^T \left(A_2 \delta_1 + A_3 \delta_2 + B_2 v_2 + w_2\right) = \text{minimum}$$

$$\tag{13.141}$$

where k_1 and k_2 are the vectors of Lagrangian multipliers.

Problems

13.1 Given the following mathematical models

$$f_1(\lambda_1, x_1, x_2) = 0 \quad C_{\lambda_1} \quad C_{x_1}$$

$$f_2(\lambda_2, x_1, x_3) = 0 \quad C_{\lambda_2} \quad C_{x_3}$$

where f_1 and f_2 are vectors of mathematical models, x_1, x_2, and x_3 are vectors of unknown parameters, λ_1 and λ_2 are vectors of observations, and C_{λ_1}, C_{λ_2}, C_{x_1}, and C_{x_3} are covariance matrices, answer the following.
a) Formulate the variation function (showing the least squares criterion and defining every unknown symbol).
b) Derive the most extended form of the least squares normal equation system.

13.2 The equation of plane can be given as

$$ax + by + cz + 1 = 0$$

where a, b, and c are the unknown constants to be determined and the (x, y, z) coordinates of the four corners of a building measured from a stereo pair of photographs are given in the following table.

Measured coordinates

Point	x (cm)	y (cm)	z (cm)
1	1.1	−1.0	0.9
2	−2.0	2.0	1.0
3	2.0	−2.0	1.0
4	−1.1	1.0	0.9

Assuming these four points must lie in a plane and taking the approximate values of the parameters as $[a^0 = 1, b^0 = 1, c^0 = -1]^T$, answer the following.

a) Give the general model equations for this problem.
b) Form the first design matrix from the general model equations, providing the numerical values of the matrix elements.
c) Determine the misclosure vector for this problem.
d) What is the size (row by column) of the second design matrix for this problem?
e) Determine the degrees of freedom for the adjustment, showing clearly how you arrive at your answer.

13.3 A leveling survey was run as shown in Figure P13.3. The height differences between stations and the corresponding lengths of the leveling routes (in km) (routes 1, 2, 3, and 4) are shown in the following table. The elevation of point A is to be fixed at 100.000 m above the datum, and the standard deviation for each height difference is to be determined using 0.005 m $\sqrt{k}$ (where k is the length of the level route in km). Perform the following tasks using least squares adjustment of general models.

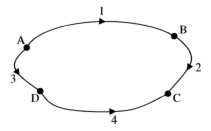

Figure P13.3

Field measurements

Line	Height differences (m)	Length (km)
1	9.138	3
2	9.500	5
3	4.310	4
4	14.293	2

   a) Determine the least squares adjusted elevations (using general models) of the station points B, C, and D. (Your answer must include the model equations, design matrices, misclosure vector, cofactor matrix of measurements, $(A^T M^{-1} A)^{-1}$, $A^T M^{-1} w$, and the solution vector.)

   b) Calculate the standard deviations of the adjusted elevations of points B, C, and D.

13.4 The following table contains the reduced calibration data (distance measurements) over an EDM baseline in Maryland for an EDMI with manufacturer's stated accuracy of 10 mm ± 10 ppm. The published data and the reduced measured data are as follows.

From sta. (i)	To sta. (j)	Published dist. (m) horizontal (p_{ij})	Std. dev. (mm)	Reduced measured dist. (m) horizontal (m_{ij})	Std. dev. (m)
1	2	149.9929	1.0	149.9899	0.0101
1	3	449.990	1.0	449.9916	0.0112
1	4	1649.9959	2.0	1649.9600	0.0193

Compute the following:

   a) Adjusted unknown parameters (scale factor S and system constant C) using the general model $f(x, \ell) = p - Sm - C = 0$ where p is a vector of published distances, m is a vector of measured distances, S is the scale factor, and C is the system (instrument/reflector) constant. Assume the approximate values of $S^0 = 1.000$ and $C^0 = 0.000$.

   b) Standard deviations of the adjusted parameters.

13.5 In order to determine the horizontal distance (D) and the height (H) of a total station from a setup point P to a benchmark Q, three zenith angles (z_i) were measured to various graduations (r_i) of a vertical leveling rod at point Q. The zenith angles and the rod graduation readings, which are made with precisions ±3″ and ±0.5 mm, respectively, are given in the following table and Figure P13.5. Given the general model equation relating the measurements and parameters as

$$\tan z = \frac{D}{(r-H)}$$

perform general model least squares adjustment to determine the adjusted parameters and their standard deviations.

Zenith angles and rod readings

Zenith (z)	Rod reading (r) (m)
87°53′40″	2.954
89°20′25″	1.818
90°43′23″	0.732

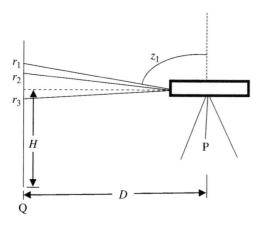

Figure P13.5

14

Datum Problem and Free Network Adjustment

OBJECTIVES

After studying this chapter, you should be able to:

1) Describe datum problems and solution approaches.
2) Formulate free network adjustment constraint equations.
3) Perform inner constraint and external constraint network adjustments.
4) Transform adjusted quantities from one minimal constraint datum to another.

14.1 Introduction

The observations (such as angles, distances, etc.) in classical survey define *internal network geometry*, while coordinates and elevations of points constitute the *external network geometry*. Those parameters that are fixed in order to solve a network adjustment problem constitute the datum of the adjustment model. If the coordinates of the network points are fixed in defining the datum, then the network is said to be *externally constrained*. If coordinate parameters, however, are not involved (as in conditional model adjustment), there will be no datum problem. The minimal number of coordinate parameters that must be fixed so

Understanding Least Squares Estimation and Geomatics Data Analysis, First Edition.
John Olusegun Ogundare.
© 2019 John Wiley & Sons, Inc. Published 2019 by John Wiley & Sons, Inc.
Companion website: www.wiley.com/go/ogundare/Understanding-lse-and-gda

that the matrix of coefficients of normal equations in least squares adjustment (*N*) becomes nonsingular is called *rank deficiency (or datum deficiency)*. *Datum deficiency* can be considered as the difference between the number of unknown parameters and the rank of the matrix *N*. Refer to Chapter 4 for further discussions on datum deficiency.

Rank deficiency is the condition of a matrix (such as *N*-matrix) that will make the matrix impossible to solve, e.g. the matrix cannot be inverted. This will usually be the case if there is datum deficiency; note that grammatically, deficiency means "insufficiency or lack." The amount by which the information for solving a problem is insufficient or is lacking is called datum deficiency. For example, the maximum deficiency in two-dimensional network survey is four (if no two coordinates, no one bearing, and no one distance for scale, are fixed); if two coordinates of the starting point are given for the survey, there will remain two datum defects (due to missing one bearing and one distance for scale). It should be mentioned that measured distances in a network provide the necessary scale for a network. No matter the number of distances measured (by the same instrument), all the distance measurements will still be counted as constituting one scale for the network. This is to say that distance measurements (no matter the number) in a network do not overconstrain a network adjustment. If every parameter needed for fixing the datum is provided, the datum deficiency will be zero, and the corresponding rank deficiency will also be zero. This is to say that the amount of datum deficiency is the same as the amount of rank deficiency of the corresponding matrix.

14.2   Minimal Datum Constraint Types

Classical solution to overcoming rank deficiencies or datum defects is to adequately define network datum through the addition of absolute or weight constraints on the unknown coordinate parameters. Network adjustment that incorporates a minimal amount of information necessary to define a reference coordinate system, to obtain a unique coordinate solution, is called *minimal constraint (minimal datum constraint)* or *free network* adjustment (Leick 1982). Such a network is considered free in that its *geometrical size and shape* are determined while it remains essentially independent of the reference coordinate system. In minimal constraint system, the choice of datum parameters to fix is usually a problem. When a network point is fixed as part of a datum definition, the network is said to be *externally constrained*; if minimal constraints conditions are still satisfied, the associated adjustment will be *external constraint minimal constraint adjustment*. Arbitrarily fixing network points in minimal constraint adjustments, however, usually leads to different coordinates of points and their associated precisions, but without limiting or affecting the freedom of the network to translate, rotate, or change in size. The term "free adjustment" is sometimes used to describe an estimation based on minimal constraints. In other words, a minimal constrained network adjustment is a *free network adjustment.*

Solving a network adjustment problem means finding the inverse associated with the adjustment model. The type of generalized inverse important to surveyor is *pseudo-inverse* that provides unique solution of the adjusted parameters ($\hat{x}$). Since pseudo-inverse is used in least squares adjustment, the sum of the weighted squares of residuals is always a minimum (i.e. $v^T P v$ = minimum). A form of minimal constraint adjustment known as *inner constraint* adjustment also provides similar result as pseudo-inverse. A network is inner (or internally) constrained if, instead of fixing one of the network points, the center of gravity (or the centroid) of the network is fixed. Inner constraints for least squares adjustment are specified as follows:

- There will be no change in the coordinates of the centroid after adjustment, e.g. for 2-D case, translations from the centroid $dx_G = 0$ and $dy_G = 0$. This means there will be no translation of the centroid between epochs.
- The average bearing from the centroid (or center of gravity) to each of the other points will remain unchanged after adjustment, meaning that there will be no differential change in rotation.
- The average distance from the centroid (or center of gravity) to each of the other points will remain unchanged after adjustment, meaning that there will be no differential change in scale.

Generally, inner constraints ensure that the initial coordinate values assigned to each of the network points at the start of the iterative least squares solution define the datum.

14.3 Free Network Adjustment Model

In free network adjustment, two types of models are created: the model relating observations to parameters (parametric model) and the model that will constraint the parameters to allow the solution of the parametric model (constraint model). These two models can be expressed mathematically as follows:

$$\hat{\ell} = f_1(\hat{x}) \quad C_\ell \tag{14.1}$$

$$f_2(\hat{x}) = 0 \tag{14.2}$$

where f_1 and f_2 are the parametric model and constraint model, respectively, $\hat{x}$ is the vector of the adjusted parameters, and C_ℓ is the covariance matrix of the observations. As usual, the models can be linearized by first assuming the following:

$$\ell + v = f_1\left(x^0 + \delta\right) \tag{14.3}$$

$$f_2\left(x^0 + \delta\right) = 0 \tag{14.4}$$

where v is a vector of residuals, x^0 is a vector of approximate values of the unknown parameters, and δ is the vector of corrections to the approximate values of the parameters. By linearizing Equation (14.3) by Taylor series expansion,

$$v = f_1\left(x^0\right) + \frac{\partial f_1}{\partial x}\delta - \ell \tag{14.5}$$

Equation (14.5) can be given in simpler form as

$$A\delta - v + w = 0 \tag{14.6}$$

where $A = \dfrac{\partial f_1}{\partial x}$ is the first design matrix, $w = f_1(x^0) - \ell$ is a vector of misclosures, and ℓ is a vector of observations. Similarly, linearize Equation (14.4) by Taylor series expansion as follows:

$$f_2\left(x^0\right) + \frac{\partial f_2}{\partial x}\delta = 0 \tag{14.7}$$

Equation (14.7) can be given in simpler form as follows:

$$w_c + G^T\delta = 0 \tag{14.8}$$

where $G^T = \dfrac{\partial f_2}{\partial x}$ and $w_c = f_2(x^0)$ is a vector of misclosures for the constraint equations. Since the covariance matrix of observations (C_ℓ) is associated with the model Equation (14.1), the least squares criterion for this problem can be given as

$$v^T C_\ell^{-1} v = \text{minimum} \tag{14.9}$$

By imposing the least squares criterion on Equations (14.6) and (14.8), the following variation function is obtained:

$$\varphi = v^T C_\ell^{-1} v - 2k_1^T \left(G^T\delta + w_c\right) - 2k_2^T \left(A\delta - v + w\right) = \text{minimum} \tag{14.10}$$

The least squares normal equation can be derived from the variation function in Equation (14.10) by finding the partial derivatives of the function with respect to the unknown quantities (i.e. v, δ, k_1, k_2) as follows:

$$\frac{\partial \varphi}{\partial v} = 2v^T C_\ell^{-1} + 2k_2^T = 0 \tag{14.11}$$

$$\frac{\partial \varphi}{\partial \delta} = -2k_1^T G^T - 2k_2^T A = 0 \tag{14.12}$$

$$\frac{\partial \varphi}{\partial k_1^T} = -2\left(G^T \delta + w_c\right) = 0 \tag{14.13}$$

$$\frac{\partial \varphi}{\partial k_2^T} = -2(A\delta - v + w) = 0 \tag{14.14}$$

Equations (14.11)–(14.14) can be rewritten in the most expanded form of the least squares normal equation system as

$$v^T C_\ell^{-1} + k_2^T = 0 \tag{14.15}$$

$$k_1^T G^T + k_2^T A = 0 \tag{14.16}$$

$$G^T \delta + w_c = 0 \tag{14.17}$$

$$A\delta - v + w = 0 \tag{14.18}$$

Transpose Equation (14.15) and rearrange

$$k_2 = -C_\ell^{-1} v \tag{14.19}$$

Transpose Equation (14.16) and substitute Equation (14.19) into it to give the following equation:

$$Gk_1 - A^T C_\ell^{-1} v = 0 \tag{14.20}$$

From Equation (14.18)

$$v = A\delta + w \tag{14.21}$$

Substituting Equation (14.21) into Equation (14.20) gives the following:

$$Gk_1 - A^T C_\ell^{-1} A\delta - A^T C_\ell^{-1} w = 0 \tag{14.22}$$

which can be rearranged to give

$$A^T C_\ell^{-1} A\delta - Gk_1 + A^T C_\ell^{-1} w = 0 \tag{14.23}$$

Equations (14.17) and (14.23), which are left to be solved, can be expressed in matrix form as follows:

$$\begin{bmatrix} N & G \\ G^T & 0 \end{bmatrix} \begin{bmatrix} \delta \\ -k_1 \end{bmatrix} + \begin{bmatrix} u \\ w_c \end{bmatrix} = 0 \tag{14.24}$$

where $N = A^T C_\ell^{-1} A$ and $u = A^T C_\ell^{-1} w$. The solution vector cannot be obtained directly, but the pseudosolution can be expressed from Equation (14.24) as follows, assuming the vector $w_c = 0$ (see the derivation in Frankich 2006):

$$\delta = -Q_{11} u \tag{14.25}$$

where

$$Q_{11} = \left(N + GG^T\right)^{-1} N \left(N + GG^T\right)^{-1} \tag{14.26}$$

is the pseudo-inverse representing the cofactor matrix of the adjusted parameters. Equation (14.25) can now be used as the solution to the two models given in Equations (14.1) and (14.2). The formulation of the parametric model (Equation (14.1)) is straightforward, but formulating the constraint model needs special attention.

14.4 Constraint Model for Free Network Adjustment

The constraint model equations are most of the time based on the datum definition as discussed in Chapter 4. For free two-dimensional network adjustment (inner constraint case), the maximum number of datum defects possible will be four; this will result in formulating four constraint equations to define the datum. The constraint model will consist of the following four general constraint equations, for the case of two-dimensional network:

1) Equations that will define the origin (or translations from this origin) of the coordinate system. The usually chosen equations for *inner constraint adjustment* impose the positional constraints or translational constraints on the centroid (with coordinates X_G, Y_G) of the network by specifying that changes in those coordinates of the center of mass of the network remain zero after the adjustment, i.e. $\delta X_G = 0$ and $\delta Y_G = 0$. The center of mass of the network can be determined as

$$X_G = \frac{\sum_{i=1}^{n} X_i}{n} \quad Y_G = \frac{\sum_{i=1}^{n} Y_i}{n} \tag{14.27}$$

where (X_i, Y_i) (for $i = 1, 2, ..., n$) are the approximate coordinates of the network points. In order to satisfy the constraint condition that $\delta X_G = 0$ and $\delta Y_G = 0$, the total differentials of equations in Equation (14.27) are done with respect to the coordinates of the network points to be used for defining the datum (this could be a subset of the network points or the whole network points) and the total differentials set equal to zero. The following two constraint Equations (14.28) and (14.29) are then obtained:

$$\sum_{i=1}^{n} \delta x_i = 0 \quad \text{(for network } x\text{–translation)} \tag{14.28}$$

$$\sum_{i=1}^{n} \delta y_i = 0 \quad \text{(for network } y\text{–translation)} \tag{14.29}$$

2) Equations that will define the orientations (or rotations) of the coordinate system. The equation imposes rotational constraint by specifying that the average bearing from the centroid to each of the network datum point must not change after the adjustment. This will require that the sum of all the changes in bearings from all the datum network points to the centroid must be equal to zero. This is done by calculating bearings from the centroid to the datum network points (using the tangent function and the coordinates of the centroid and the corresponding points), finding their total differentials, and setting their sum to zero. The following constraint equation is then obtained:

$$\sum_{i=1}^{n} (Y_i \delta x_i - X_i \delta y_i) = 0 \quad \text{(for network rotation)} \tag{14.30}$$

3) Equations that will define the scale (or provide the idea of distance) of the coordinate system.

The equation imposes scalar constraint by specifying that the average distance from the centroid to each of the network datum point must not change after the adjustment. This requires that the sum of changes in distances from the centroid to the datum network points must be equal to zero. In this case, the distance equations from the datum network points to the centroid are formed, and the sum of total differentials of those distances are set equal to zero. The following constraint equation is then obtained:

$$\sum_{i=1}^{n} (X_i \delta x_i + Y_i \delta y_i) = 0 \quad \text{(for network scale)} \tag{14.31}$$

Equations (14.28)–(14.31) are the constraints, mathematically expressing that the network is invariant in its shape with respect to small differential translation, rotation, and scale change. This means that the adjustment preserves the shape of the network defined by the observations. Equations (14.28)–(14.31) can be expanded and presented as G-matrix as follows for horizontal network where there are no fixed point, no azimuth, and no scale (with the number of datum deficiencies equal to four):

$$
G^T \delta =
\begin{bmatrix}
1 & 0 & 1 & 0 & \cdots & 1 & 0 \\
0 & 1 & 0 & 1 & \cdots & 0 & 1 \\
Y_1 & -X_1 & Y_2 & -X_2 & \cdots & Y_n & -X_n \\
X_1 & Y_1 & X_2 & Y_2 & \cdots & X_n & Y_n
\end{bmatrix}
\begin{bmatrix}
\delta x_1 \\
\delta y_1 \\
\delta x_2 \\
\delta y_2 \\
\vdots \\
\delta x_n \\
\delta y_n
\end{bmatrix}
= 0
\tag{14.32}
$$

where

$$
G^T =
\begin{bmatrix}
1 & 0 & 1 & 0 & \cdots & 1 & 0 \\
0 & 1 & 0 & 1 & \cdots & 0 & 1 \\
Y_1 & -X_1 & Y_2 & -X_2 & \cdots & Y_n & -X_n \\
X_1 & Y_1 & X_2 & Y_2 & \cdots & X_n & Y_n
\end{bmatrix}
\begin{matrix}
x-\text{translation} \\
y-\text{translation} \\
\text{Orientation} \\
\text{Scale}
\end{matrix}
\tag{14.33}
$$

$$
\delta = [\delta x_1\ \delta y_1\ \delta x_2\ \delta y_2\ \cdots\ \delta x_n\ \delta y_n]^T
\tag{14.34}
$$

and n is the number of points defining the datum, which could be all of the points or a subset of points in the network being adjusted. If n is a subset of points in the network, then the corresponding columns in matrix G^T will be zeros. The coordinates used in the G^T-matrix in Equation (14.33) can be reduced to the center of mass of the datum points as shown in Equation (14.35) to reduce rounding error:

$$
x_i = X_i - X_G \quad y_i = Y_i - Y_G \quad (\text{for } i = 1, 2, \ldots, n).
\tag{14.35}
$$

However, it can be shown that the solutions obtained when using the original coordinates $(X_i,\ Y_i)$ or the reduced coordinates $(x_i,\ y_i)$ will be identical.

Equation (14.32) can be extended to three dimensions to give the following set of seven inner constraint equations:

$$
G^T \delta =
\begin{bmatrix}
1 & 0 & 0 & : & 1 & 0 & 0 & : & \cdots & 0 \\
0 & 1 & 0 & : & & & & : & \cdots & 0 \\
0 & 0 & 1 & : & 0 & 0 & 1 & : & \cdots & 1 \\
0 & Z_1 & -Y_1 & : & 0 & Z_2 & -Y_2 & : & \cdots & -Y_n \\
-Z_1 & 0 & X_1 & : & -Z_2 & 0 & X_2 & : & \cdots & X_n \\
Y_1 & -X_1 & 0 & : & Y_2 & -X_2 & 0 & : & \cdots & 0 \\
X_1 & Y_1 & Z_1 & : & X_2 & Y_2 & Z_2 & : & \cdots & Z_n
\end{bmatrix}
\begin{bmatrix}
\delta x_1 \\
\delta y_1 \\
\delta z_1 \\
\delta x_2 \\
\delta y_2 \\
\delta z_2 \\
\vdots \\
\delta z_n
\end{bmatrix}
= 0
\tag{14.36}
$$

The constraint equations for the one-dimensional (leveling) case can be given as follows:

$$G^T \delta = [1 \quad 1 \quad \cdots \quad 1] \begin{bmatrix} \delta z_1 \\ \delta z_2 \\ \vdots \\ \delta z_n \end{bmatrix} = 0 \tag{14.37}$$

In free network adjustment, the following properties are also satisfied:

$$|\delta| = \delta^T \delta = \text{minimum} \tag{14.38}$$

and

$$\text{trace}(Q_{\hat{x}}) = \text{minimum} \tag{14.39}$$

where δ is the vector of corrections to the approximate values of the unknown parameters (as given in Equation (14.34)) and $Q_{\hat{x}}$ is cofactor matrix of the adjusted parameters. The resulting estimated parameter vector $\hat{x}$ from the adjustment based on the above constraints is referred to as the *best linear unbiased estimates* (BLUE) for the unknown parameter x.

14.5 Summary of Free Network Adjustment Procedure

The summary of the steps for formulating the solution to a network problem that have constraint equations imposed on it is given as follows. Note that the initial coordinate values used at the start of the least squares solution essentially define the datum for the inner constraint adjustment; errors or changes in those approximate coordinates will alter the adjustment results. The general steps for the solution are as follows:

1) Form the parametric model

$$\hat{\ell} = f_1(\hat{x}) \tag{14.40}$$

and derive the first design matrix A and misclosure vector (w) from Equation (14.40) and form the covariance matrix (C_ℓ) of the observations (ℓ).
2) Form the constraint model

$$f_2(\hat{x}) = 0 \tag{14.41}$$

or, alternatively, construct the constraint matrix G^T as discussed in Section 14.4 according to the type of datum defects of the given network. The G^T must be formulated so that the appropriate row(s) will be removed depending on the type of datum elements contained in the measurements. For example, if distances (or any distance) are measured in a two-dimensional network, the last row in Equation (14.33) will be discarded; if both azimuths and distances (or an azimuth and a distance) are measured in the network, the last two rows of the equation will be discarded; etc.

3) Solve for the vector of corrections (δ) to the approximate values of the unknown parameters (x^0), expressed as

$$\delta = -Q_{11}u \tag{14.42}$$

where the pseudo-inverse Q_{11} is given as

$$Q_{11} = \left(N + GG^T\right)^{-1} N \left(N + GG^T\right)^{-1} \tag{14.43}$$

with $N = A^T PA$, $u = A^T Pw$, and $P = C_\ell^{-1}$. The solution vector in Equation (14.42) can be modified if nuisance parameters (δ_2) are involved; in this case, one would want to eliminate the effect of the nuisance parameters before the inversion is done. The solution vector can be modified as follows. Assume the vector of corrections to the approximate coordinate parameters is δ_1, δ_2 is a vector of corrections to approximate unknown nuisance parameters, A_1 is the first design matrix based on the coordinate parameters, A_2 is the first design matrix based on the nuisance parameters, and P is the weight matrix of the observations. The following can be derived from Equations (14.42) and (14.43):

i) The following pseudo-inverse (with nuisance parameters eliminated) is modified from Equation (14.43):

$$Q_{11}^* = \left(N^* + GG^T\right)^{-1} N^* \left(N^* + GG^T\right)^{-1} \tag{14.44}$$

where

$$N^* = A_1^T P^* A_1 \tag{14.45}$$

$$P^* = P\left[I - A_2\left(A_2^T PA_2\right)^{-1} A_2^T P\right] \tag{14.46}$$

ii) The modified solution vector is given from Equation (14.42) as

$$\delta_1 = -Q_{11}^* u_1^* \tag{14.47}$$

where

$$u_1^* = A_1^T P^* w \tag{14.48}$$

iii) The nuisance parameters are calculated as

$$\delta_2 = -\left(A_2^T PA_2\right)^{-1} \left(A_2^T PA_1 \delta_1 + u_2\right) \tag{14.49}$$

where

$$u_2 = A_2^T P w \tag{14.50}$$

iv) The residual vector of observations can be given as

$$v = A_1\delta_1 + A_2\delta_2 + w \tag{14.51}$$

where w is the misclosure vector.

v) The covariance matrix of the solution vector can be given as

$$C_{\hat{x}} = s_0^2 Q_{11}^* \tag{14.52}$$

where

$$s_0^2 = \frac{v^T P v}{n + d - u} \tag{14.53}$$

with n as the number of observations, d as the number of datum deficiencies (number of parameters to fix in order to define the datum), and u as the number of parameters in the network.

The following example is given to illustrate the computational steps involved in a free network adjustment.

Example 14.1 The horizontal coordinates of points A, B, and C in the network in Figure 14.1 are to be determined. In order to do this, a traverse survey was run from point A and closed on C. The measurements include two bearings (α_1, α_2) and three distances (d_1, d_2, d_3) as shown in Figure 14.1.

Figure 14.1 Simple traverse survey (not to scale).

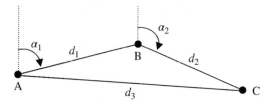

The measurements are stochastically independent (meaning the covariance matrix has zero off-diagonal elements), and their values together with the corresponding standard deviations are as follows:

$$\alpha_1 = 71°29'13'', \quad s_{\alpha_1} = 10''$$
$$\alpha_2 = 121°14'50'', \quad s_{\alpha_2} = 10''$$
$$d_1 = 561.025 \text{ m}, \quad s_{d_1} = 0.005 \text{ m}$$
$$d_2 = 428.356 \text{ m}, \quad s_{d_2} = 0.005 \text{ m}$$
$$d_3 = 899.290 \text{ m}, \quad s_{d_3} = 0.010 \text{ m}$$

The approximate coordinates of the network points are given in Table 14.1.

Table 14.1 Approximate coordinates of points.

Point	X (easting) (m)	Y (northing) (m)
A	1000.000	1000.000
B	1531.985	1178.140
C	1898.206	955.940

Perform the free network adjustment (fixing the center of gravity of the network): Determine the adjusted coordinates of the network points and their cofactor matrix and the a posteriori variance factor of unit weight.

Solution:

Given the observation vector, $\ell = \begin{bmatrix} \alpha_1 & \alpha_2 & d_1 & d_2 & d_3 \end{bmatrix}^T$, the complete parametric model equations ($\hat{\ell} = f(\hat{x})$) of the problem can be given as follows:

$$\hat{\alpha}_1 = a \tan \left(\frac{\hat{x}_B - \hat{x}_A}{\hat{y}_B - \hat{y}_A} \right) \tag{14.54}$$

$$\hat{\alpha}_2 = a \tan \left(\frac{\hat{x}_C - \hat{x}_B}{\hat{y}_C - \hat{y}_B} \right) + 180^\circ \tag{14.55}$$

$$\hat{d}_1 = \sqrt{\left(\hat{x}_B - \hat{x}_A \right)^2 + \left(\hat{y}_B - \hat{y}_A \right)^2} \tag{14.56}$$

$$\hat{d}_2 = \sqrt{\left(\hat{x}_C - \hat{x}_B \right)^2 + \left(\hat{y}_C - \hat{y}_B \right)^2} \tag{14.57}$$

$$\hat{d}_3 = \sqrt{\left(\hat{x}_C - \hat{x}_A \right)^2 + \left(\hat{y}_C - \hat{y}_A \right)^2} \tag{14.58}$$

$$\text{Approximate values of coordinates of points } x^0 = \begin{bmatrix} 1000.000 \\ 1000.000 \\ 1531.985 \\ 1178.140 \\ 1898.206 \\ 955.940 \end{bmatrix}.$$

Form the first design matrix, A, by finding the partial derivatives of Equations (14.54)–(14.58) (ignoring the hats for simplicity) with respect to all of the six coordinates, $x = [x_A \ y_A \ x_B \ y_B \ x_C \ y_C]^T$, since none of them is fixed:

$$A = \frac{\partial f}{\partial x} = \begin{bmatrix} \dfrac{-(y_B - y_A)}{d_{AB}^2} & \dfrac{(x_B - x_A)}{d_{AB}^2} & \dfrac{(y_B - y_A)}{d_{AB}^2} & \dfrac{-(x_B - x_A)}{d_{AB}^2} & 0 & 0 \\[2ex] 0 & 0 & \dfrac{-(y_C - y_B)}{d_{BC}^2} & \dfrac{(x_C - x_B)}{d_{BC}^2} & \dfrac{(y_C - y_B)}{d_{BC}^2} & \dfrac{-(x_C - x_B)}{d_{BC}^2} \\[2ex] \dfrac{-(x_B - x_A)}{d_{AB}} & \dfrac{-(y_B - y_A)}{d_{AB}} & \dfrac{(x_B - x_A)}{d_{AB}} & \dfrac{(y_B - y_A)}{d_{AB}} & 0 & 0 \\[2ex] 0 & 0 & \dfrac{-(x_C - x_B)}{d_{BC}} & \dfrac{-(y_C - y_B)}{d_{BC}} & \dfrac{(x_C - x_B)}{d_{BC}} & \dfrac{(y_C - y_B)}{d_{BC}} \\[2ex] \dfrac{-(x_C - x_A)}{d_{AC}} & \dfrac{-(y_C - y_A)}{d_{AC}} & 0 & 0 & \dfrac{(x_C - x_A)}{d_{AC}} & \dfrac{(y_C - y_A)}{d_{AC}} \end{bmatrix}$$

$$(14.59)$$

where

$$d_{AB} = \sqrt{(x_B - x_A)^2 + (y_B - y_A)^2}$$
$$d_{BC} = \sqrt{(x_C - x_B)^2 + (y_C - y_B)^2} \qquad (14.60)$$
$$d_{AC} = \sqrt{(x_C - x_A)^2 + (y_C - y_A)^2}$$

Substituting the approximate coordinates into Equations (14.59) and (14.60) gives the following first design matrix, A:

$$A = \begin{bmatrix} -5.659\,88E{-}4 & 1.690\,23E{-}3 & 5.659\,88E{-}4 & -1.690\,23E{-}3 & 0.0 & 0.0 \\ 0.0 & 0.0 & 1.210\,96E{-}3 & 1.995\,86E{-}3 & -1.210\,96E{-}3 & -1.995\,86E{-}3 \\ -0.948\,25 & -0.317\,53 & 0.948\,25 & 0.317\,530 & 0.0 & 0.0 \\ 0.0 & 0.0 & -0.854\,94 & 0.518\,725 & 0.854\,941 & -0.518\,725 \\ -0.998\,80 & 0.048\,99 & 0.0 & 0.0 & 0.998\,799 & -0.048\,994 \end{bmatrix}$$

The cofactor matrix of observations (Q_ℓ) and the weight matrix (P):

$$Q_\ell = \begin{bmatrix} 2.350\,44E{-}9 & 0 & 0 & 0 & 0 \\ 0 & 2.350\,44E{-}9 & 0 & 0 & 0 \\ 0 & 0 & 2.5E{-}5 & 0 & 0 \\ 0 & 0 & 0 & 2.5E{-}5 & 0 \\ 0 & 0 & 0 & 0 & 1.0E{-}4 \end{bmatrix}$$

$$P = \begin{bmatrix} 4.254\ 525E8 & 0 & 0 & 0 & 0 \\ 0 & 4.254\ 525E8 & 0 & 0 & 0 \\ 0 & 0 & 40\ 000 & 0 & 0 \\ 0 & 0 & 0 & 40\ 000 & 0 \\ 0 & 0 & 0 & 0 & 10\ 000 \end{bmatrix}$$

The misclosure vector (w):

$$\ell = \begin{bmatrix} 1.247\ 682\ 6 \\ 2.116\ 163\ 24 \\ 561.025 \\ 428.356 \\ 899.290 \end{bmatrix} \quad f(x^0) = \begin{bmatrix} 1.247\ 673\ 2 \\ 2.116\ 155\ 72 \\ 561.018\ 63 \\ 428.358\ 1 \\ 899.286\ 0 \end{bmatrix}$$

$$w = f(x^0) - \ell = \begin{bmatrix} -9.310E - 6 \\ -8.115E - 6 \\ -6.373E - 3 \\ 2.099E - 3 \\ -4.004E - 3 \end{bmatrix}$$

Form the G-matrix from Equation (14.33) (for x–y translations only). The orientation of the network is defined by the measured bearings, and the scale is defined by the measured distances. The only undefined quantities will be the two translations; the G-matrix is thus formed to take care of the translations as follows:

$$G^T = \begin{bmatrix} 1 & 0 & 1 & 0 & 1 & 0 \\ 0 & 1 & 0 & 1 & 0 & 1 \end{bmatrix}$$

$$A^T Pw = \begin{bmatrix} 283.960\ 448 \\ 72.286\ 56 \\ -299.371\ 096 \\ -50.059\ 311\ 7 \\ 15.410\ 647\ 6 \\ -22.227\ 248 \end{bmatrix}$$

Compute the corrections to the approximate coordinates (δ):

$$\delta = -\left(A^TPA + GG^T\right)^{-1}A^TPA\left(A^TPA + GG^T\right)^{-1}A^TPw = \begin{bmatrix} -0.003\ 8 \\ 0.002\ 4 \\ 0.003\ 8 \\ -0.000\ 2 \\ 0.000\ 05 \\ -0.002\ 3 \end{bmatrix}$$

The adjusted coordinates by free network method (inner constraint) are as follows:

$$\hat{x} = x^0 + \delta \rightarrow \begin{bmatrix} 1\ 000.000 \\ 1\ 000.000 \\ 1\ 531.985 \\ 1\ 178.140 \\ 1\ 898.206 \\ 955.940 \end{bmatrix} + \begin{bmatrix} -0.003\ 8 \\ 0.002\ 4 \\ 0.003\ 8 \\ -0.000\ 2 \\ 0.000\ 05 \\ -0.002\ 3 \end{bmatrix} \rightarrow \begin{bmatrix} 999.996\ 2 \\ 1\ 000.002\ 5 \\ 1\ 531.988\ 8 \\ 1\ 178.140\ 0 \\ 1\ 898.206\ 1 \\ 955.937\ 7 \end{bmatrix}$$

Determine the cofactor matrix of the adjusted coordinates as follows:

$$Q_{\hat{x}} = \left(A^TPA + GG^T\right)^{-1}A^TPA\left(A^TPA + GG^T\right)^{-1}$$

$$= \begin{bmatrix} 1.983\ 8\text{E}-5 & -3.843\ 4\text{E}-5 & -8.649\ 6\text{E}-6 & 2.158\ 28\text{E}-5 & -1.118\ 86\text{E}-5 & 1.685\ 16\text{E}-5 \\ -3.843\ 4\text{E}-5 & 2.963\ 38\text{E}-4 & 6.979\ 57\text{E}-5 & -6.758\ 9\text{E}-5 & -3.136\ 1\text{E}-5 & -2.287\ 5\text{E}-4 \\ -8.649\ 6\text{E}-6 & 6.979\ 57\text{E}-5 & 2.560\ 49\text{E}-5 & -6.318\ 0\text{E}-6 & -1.695\ 5\text{E}-5 & -6.347\ 8\text{E}-5 \\ 2.158\ 28\text{E}-5 & -6.758\ 9\text{E}-5 & -6.318\ 0\text{E}-6 & 5.397\ 58\text{E}-5 & -1.526\ 48\text{E}-5 & 1.361\ 36\text{E}-5 \\ -1.118\ 86\text{E}-5 & -3.136\ 1\text{E}-5 & -1.695\ 5\text{E}-5 & -1.526\ 48\text{E}-5 & 2.814\ 39\text{E}-5 & 4.662\ 61\text{E}-5 \\ 1.685\ 16\text{E}-5 & -2.287\ 5\text{E}-4 & -6.347\ 8\text{E}-5 & 1.361\ 36\text{E}-5 & 4.662\ 61\text{E}-5 & 2.151\ 35\text{E}-4 \end{bmatrix}$$

Form the residual vector:

$$v = A\delta + w = \begin{bmatrix} -4.7092\text{E}-6 \\ 4.7092\text{E}-6 \\ -2.2873\text{E}-5 \\ -2.1591\text{E}-5 \\ 9.8218\text{E}-4 \end{bmatrix}$$

$$s_0^2 = \frac{v^TPv}{n+d-u} = \frac{0.000\ 32}{5+2-6} = 0.000\ 32$$

The MATLAB code for the free network adjustment is given in Table 14.2.

Table 14.2 MATLAB code for free network adjustment.

```
>>% use syms to define the parameters
>> syms xA yA xB yB xC yC
>>% form the 5 parametric Equations (14.54) - (14.58)
>> a1=atan((xB-xA)/(yB-yA));
>> a2=atan((xC-xB)/(yC-yB))+pi;
>> d1=sqrt((xB-xA)^2+(yB-yA)^2);
>> d2=sqrt((xC-xB)^2+(yC-yB)^2);
>> d3=sqrt((xC-xA)^2+(yC-yA)^2);
>>% form the Jacobian J wrt parameters
>> J=jacobian([a1;a2;d1;d2;d3],[xA yA xB yB xC yC]);
>>% assign values to parameters and evaluate J as A matrix
>> x0=[1000.000;1000.000;1531.985;1178.140;1898.206; 955.940];
>>
xA=1000.000;yA=1000.000;xB=1531.985;yB=1178.140;
xC=1898.206;yC=955.940;
>> A=eval(J);
>>% form the weight matrix of observations P, observation
vector L, computed observations L0 and misclosures w
>> P=inv(diag([(10/206265)^2 (10/206265)^2 0.005^2
0.005^2 0.010^2]));
>>
L=[pi*(71+29/60+13/3600)/180;pi*(121+14/60+50/3600)/
180;561.025;428.356;899.290];
>> L0=eval([a1;a2;d1;d2;d3]);
>> w=L0-L;
>>% form the G-matrix for one-dimensional case and determine
corrections d and adjusted parameters xa
>> G=[1 0;0 1;1 0;0 1;1 0;0 1];
>> u=A'*P*w;
>> N1=A'*P*A+G*G';
>> d=-inv(N1)*A'*P*A*inv(N1)*u;
>> xa=x0+d;
>>% determine cofactor matrix Qx of adjusted parameters,
residuals and aposteriori variance s02
>> Qx=inv(N1)*A'*P*A*inv(N1);
>> v=A*d+w;
>> s02=v'*P*v/1;
>>
```

14.6 Datum Transformation

The following outcomes of an adjustment are invariant regardless of the choice of minimal constraints. They are invariant in the sense that they are not dependent on which of the network points is fixed as datum; every minimal constraint adjustment of the same set of measurements will produce the same results irrespective of choice of datum point with regard to the following datum invariant quantities:

- Vector of residuals (v).
- Vector of adjusted observations ($\hat{\ell}$).
- Covariance matrix of residuals (C_v) and the covariance matrix of the adjusted observations ($C_{\hat{\ell}}$).
- The product $v^T P v$, where P is the weight matrix of observations.
- The a posteriori variance factor of unit weight, s_0^2.

The corrections to approximate values of unknown parameters (δ) and the cofactor matrix of adjusted parameters ($Q_{\hat{x}}$), however, will vary with the choice of a datum. In order to preserve the network geometry from one datum to another, the so-called Helmert's transformation or similarity transformation is commonly used. *Helmert's* or *similarity transformation* translates, rotates, and scales a given network differentially into the reference coordinate system defined by the approximate coordinate values of the network stations. The scale factor used in this transformation is the same in all directions; if that scale factor is one (unity), the transformation is referred to as *orthogonal transformation*. The similarity transformation is also referred to as *S-transformation*, which can be used to transform given (δ_i, Q_i) of a certain set of constraints i into (δ_j, Q_j) of a different set of constraints j. These S-transformations are given as follows (Van Mierlo 1981):

$$\delta_j = S_j \delta_i \tag{14.61}$$

$$Q_j = S_j Q_i S_j^T \tag{14.62}$$

where

$$S_j = I - G \left(B_j^T G \right)^{-1} B_j^T \tag{14.63}$$

S is the transformation matrix (having the same dimension as the normal equation matrix); $B_j = E_j G$; I is the usual identity matrix (a matrix with all of the main diagonal elements equal to one and all of the off-diagonal elements equal to zero) with the matrix size equal to the size of the matrix product being subtracted from it; E_j is a special case of identity matrix in which the diagonal elements are only equal to unity for those corresponding to the coordinates of the datum points. When all points of a control network constitute the datum, like in a free network (inner constraint) adjustment, the matrix E becomes an identity matrix. The following example is given to illustrate the computational steps involved in performing S-transformation from inner constraint adjustment to external constraint adjustment.

Example 14.2 Continuing from Example 14.1, use S-transformation to transform your results (adjusted parameters and the cofactor of the adjusted parameters) in the question into minimum constraint adjustment results in which point A is fixed.

Solution:

This problem is about performing S-transformation (from inner constraint to external constraint, keeping point A fixed). Form the E-matrix with the elements corresponding to point A having values of 1 as follows:

$$E = \begin{bmatrix} 1 & 0 & 0 & 0 & 0 & 0 \\ 0 & 1 & 0 & 0 & 0 & 0 \\ 0 & 0 & 0 & 0 & 0 & 0 \\ 0 & 0 & 0 & 0 & 0 & 0 \\ 0 & 0 & 0 & 0 & 0 & 0 \\ 0 & 0 & 0 & 0 & 0 & 0 \end{bmatrix}$$

$$G^T EG = \begin{bmatrix} 1 & 0 \\ 0 & 1 \end{bmatrix}$$

$$G(G^T EG)^{-1} G^T E^T = \begin{bmatrix} 1 & 0 & 0 & 0 & 0 & 0 \\ 0 & 1 & 0 & 0 & 0 & 0 \\ 1 & 0 & 0 & 0 & 0 & 0 \\ 0 & 1 & 0 & 0 & 0 & 0 \\ 1 & 0 & 0 & 0 & 0 & 0 \\ 0 & 1 & 0 & 0 & 0 & 0 \end{bmatrix}$$

Form the similarity transformation (S) matrix:

$$S = I - G(G^T EG)^{-1} G^T E^T = \begin{bmatrix} 0 & 0 & 0 & 0 & 0 & 0 \\ 0 & 0 & 0 & 0 & 0 & 0 \\ -1 & 0 & 1 & 0 & 0 & 0 \\ 0 & -1 & 0 & 1 & 0 & 0 \\ -1 & 0 & 0 & 0 & 1 & 0 \\ 0 & -1 & 0 & 0 & 0 & 1 \end{bmatrix}$$

where I is the usual identity matrix (with all of the main diagonal elements equal to one and all of the off-diagonal elements equal to zero) with the matrix size of 6×6. Use the vector of corrections (δ) determined in Example 14.1 to obtain a new vector of transformed corrections (δ_T) as follows:

$$\delta_T = S\delta = \begin{bmatrix} 0.0 \\ 0.0 \\ 0.0076 \\ -0.0027 \\ 0.0039 \\ -0.0047 \end{bmatrix}$$

Determine the adjusted coordinates based on the transformation:

$$\hat{x} = x^0 + \delta_T = \begin{bmatrix} 1000.000 \\ 1000.000 \\ 1531.985 \\ 1178.140 \\ 1898.206 \\ 955.940 \end{bmatrix} + \begin{bmatrix} 0.0 \\ 0.0 \\ 0.0076 \\ -0.0027 \\ 0.0039 \\ -0.0047 \end{bmatrix} \rightarrow \begin{bmatrix} 1000.000 \\ 1000.000 \\ 1531.9926 \\ 1178.1373 \\ 1898.2099 \\ 955.9353 \end{bmatrix}$$

Transform the cofactor matrix of the adjusted parameters $Q_{\hat{x}}$:

$$Q_T = SQ_{\hat{x}}S^T = \begin{bmatrix} 0.0 & 0.0 & 0.0 & 0.0 & 0.0 & 0.0 \\ 0.0 & 0.0 & 0.0 & 0.0 & 0.0 & 0.0 \\ 0.0 & 0.0 & 6.2742E-5 & -1.3613E-4 & 2.2721E-5 & -1.8856E-4 \\ 0.0 & 0.0 & -1.3613E-4 & 4.8549E-4 & -4.3921E-5 & 6.0629E-4 \\ 0.0 & 0.0 & 2.2721E-5 & -4.3921E-5 & 7.0360E-5 & 2.2701E-5 \\ 0.0 & 0.0 & -1.8856E-4 & 6.0629E-4 & 2.2701E-5 & 9.6897E-4 \end{bmatrix}$$

The MATLAB code for performing the S-transformation from inner constraint to external constraint is given in Table 14.3.

Table 14.3 MATLAB code for S-transformation (continuing from Table 14.2).

```
>>% Continuing from Table 14.2, form
>> E=zeros(6,6);
>> E(1,1)=1.0;E(2,2)=1.0;
>> R=G*inv(G'*E*G)*G'*E';
>> I=eye(6);
>>% form the S equation
>> S=I-R;
>>% determine the transformed correction vector d from
Table 14.2 and adjusted parameters xa2
>> dT=S*d
dT =
                        0
                        0
    0.007596151433752
   -0.002686042224871
    0.003875411798696
   -0.004730638967096
>> xa2=x0+dT
xa2 =
   1.0e+03 *
   1.000000000000000
   1.000000000000000
   1.531992596151434
   1.178137313957775
   1.898209875411798
   0.955935269361033
>>% determine the transformed cofactor matrix Qx of adjusted
parameters
>> QT=S*Qx*S';
>>
```

Example 14.3 Continuing from Example 14.1, perform the minimal constraint (fixing point A) adjustment on the network, and compare the adjusted coordinates of points B and C with the transformed values obtained in Example 14.2 for the two points.

Solution:

Find partial derivatives of Equations (14.54)–(14.58) in Example 14.1 with respect to X_B, Y_B, X_C, Y_C:

$$A = \begin{bmatrix} 5.659\ 88E-4 & -1.690\ 23E-3 & 0.0 & 0.0 \\ 1.210\ 96E-3 & 1.995\ 86E-3 & -1.210\ 96E-3 & -1.995\ 86E-3 \\ 0.948\ 248 & 0.317\ 530 & 0.0 & 0.0 \\ -0.854\ 941 & 0.518\ 725 & 0.854\ 941 & -0.518\ 725 \\ 0.0 & 0.0 & 0.998\ 799 & -0.048\ 994 \end{bmatrix}$$

$$A^T PA = \begin{bmatrix} 65\ 964.164\ 3 & -5\ 074.025\ 7 & -29\ 860.875\ 1 & 16\ 710.893\ 5 \\ -5\ 074.025\ 7 & 17\ 706.246\ 5 & 16710.893\ 5 & -12\ 457.785\ 3 \\ -29\ 860.875\ 1 & 16\ 710.893\ 5 & 39\ 836.870\ 5 & -17\ 200.249\ 3 \\ 16\ 710.893\ 5 & -12\ 457.785\ 3 & -17\ 200.249\ 3 & 12\ 481.789\ 9 \end{bmatrix}$$

Using the same P and w as in the question in Example 14.1,

$$A^T Pw = \begin{bmatrix} -319.926\ 44 \\ -37.592\ 086 \\ 35.961\ 45 \\ -34.695\ 72 \end{bmatrix}$$

$$\delta = -\left(A^T PA\right)^{-1} A^T Pw = \begin{bmatrix} 0.0076 \\ -0.0027 \\ 0.0039 \\ -0.0047 \end{bmatrix}$$

The adjusted coordinates by minimal constraint method (external constraint):

$$\hat{x} = x^0 + \delta = \begin{bmatrix} 1531.895 \\ 1178.140 \\ 1898.206 \\ 955.940 \end{bmatrix} + \begin{bmatrix} 0.0076 \\ -0.0027 \\ 0.0039 \\ -0.0047 \end{bmatrix} \rightarrow \begin{bmatrix} 1531.9926 \\ 1178.1373 \\ 1898.2099 \\ 955.9353 \end{bmatrix}$$

The cofactor matrix from externally constrained adjustment (point A fixed):

$$Q_{\hat{x}} = \left(A^T PA\right)^{-1} = \begin{bmatrix} 6.2742E-5 & -1.3613E-4 & 2.2721E-5 & -1.8856E-4 \\ -1.3613E-4 & 4.8549E-4 & -4.3921E-5 & 6.0629E-4 \\ 2.2721E-5 & -4.3921E-5 & 7.0359E-5 & 2.2701E-5 \\ -1.8856E-4 & 6.0629E-4 & 2.2701E-5 & 9.6897E-4 \end{bmatrix}$$

By comparing the adjusted coordinates of points B and C from Examples 14.2 and 14.3, it can be seen that the two sets of coordinates are identical; similarly, their corresponding cofactor matrices are identical. The MATLAB code for the minimal constraint adjustment (by fixing point A) is given in Table 14.4.

Table 14.4 MATLAB code for minimal constraint adjustment by fixing point A (continuing from Examples 14.1 and 14.2).

```
>>% Continuing from Table 4.2, the A1 matrix can be extracted
>> A1=[A(1,3) A(1,4) A(1,5) A(1,6);A(2,3) A(2,4) A(2,5) A
(2,6);A(3,3) A(3,4) A(3,5) A(3,6);A(4,3) A(4,4) A(4,5) A
(4,6);A(5,3) A(5,4) A(5,5) A(5,6)];
>>% determine the solution d1, adjusted parameters xa2 and
cofactor matrix of adjusted parameters Qx
>> Nn=A1'*P*A1;
>> u=A1'*P*w;
>> d1=-inv(Nn)*u
d1 =
    0.007596151433738
   -0.002686042224871
    0.003875411798685
   -0.004730638967097
>> x02=[1531.985;1178.140;1898.206;955.940];
>> xa2=x02+d1
xa2 =
    1.0e+03 *
    1.531992596151434
    1.178137313957775
    1.898209875411798
    0.955935269361033
>> Qx=inv(Nn)
Qx =
    1.0e-03 *
    0.062742177566871   -0.136130848974333
    0.022721046095414   -0.188559421785849
   -0.136130848974333    0.485492808985260
   -0.043920688112238    0.606290171652727
    0.022721046095414   -0.043920688112238
    0.070359272462566    0.022701354960008
   -0.188559421785849    0.606290171652727
    0.022701354960008    0.968971462168621
>>
```

14.6.1   Iterative Weighted Similarity Transformation

The S-transformations expressed by Equations (14.61)–(14.63) can be modified for use in the process of deformation analysis based on what is known as iterative weighted similarity transformation (IWST) (Ogundare 2016). The process starts by formulating the G-matrix according to Section 14.4 using the epoch network (in the case of two-epoch analysis) with higher number of datum defects. The transformation equations are then rewritten for a deformation network of n points as follows:

$$\hat{d}^{(k+1)} = S_k \hat{d}^{(k)} \tag{14.64}$$

$$Q_{\hat{d}^{(k+1)}} = S_k Q_{\hat{d}^{(k)}} S_k^T \tag{14.65}$$

where

$$S_k = I - G\left(B_k^T G\right)^{-1} B_k^T \tag{14.66}$$

$$B_k = E_k G \tag{14.67}$$

$$E_k = \begin{bmatrix} \frac{1}{\left(d\hat{x}_1^2 + d\hat{y}_1^2\right)^{1/2} + c} & & & & & & \\ & \frac{1}{\left(d\hat{x}_1^2 + d\hat{y}_1^2\right)^{1/2} + c} & & & & & \\ & & \ddots & & & & \\ & & & \frac{1}{\left(d\hat{x}_i^2 + d\hat{y}_i^2\right)^{1/2} + c} & & & \\ & & & & \frac{1}{\left(d\hat{x}_i^2 + d\hat{y}_i^2\right)^{1/2} + c} & & \\ & & & & & \ddots & \\ & & & & & & \frac{1}{\left(d\hat{x}_n^2 + d\hat{y}_n^2\right)^{1/2} + c} & \\ & & & & & & & \frac{1}{\left(d\hat{x}_n^2 + d\hat{y}_n^2\right)^{1/2} + c} \end{bmatrix}^k \tag{14.68}$$

$\hat{d}_i^{(k)} = \begin{bmatrix} d\hat{x}_i \\ d\hat{y}_i \end{bmatrix}^{(k)}$ is the ith component of the vector of the adjusted displacements $\hat{d}^{(k)}$ at the kth iteration; $\hat{d}^{(k+1)}$ is a vector of adjusted displacements at the $k+1$th iteration; $Q_{\hat{d}^{(k)}}$ and $Q_{\hat{d}^{(k+1)}}$ are the cofactor matrices of the adjusted displacements at the kth and $k+1$th iterations, respectively; I is the usual identity matrix (a matrix with all of the main diagonal elements equal to one); and c is a small constant chosen to avoid the denominators becoming zero during the iterations. The iteration is considered to have converged when the maximum value of the absolute differences between $\hat{a}^{(k)}$ and $\hat{a}^{(k+1)}$ is less than a prese-lected tolerance value, which can be chosen arbitrarily.

Problems

14.1 Figure P14.1 is a leveling network with the height difference measure-ments Δh_{ij} and their corresponding standard deviations (Std.) given in the following table. This network is to be adjusted assuming the approx-imate elevations of the network points are P1 = 107.500 m, P2 = 106.000 m, and P3 = 102.500 m.

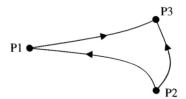

Figure P14.1

No.	i	j	Δh_{ij} (m)	Std. (m)
1	P2	P1	1.340	0.010
2	P1	P3	-5.000	0.015
3	P2	P3	-3.680	0.015

Answer the following questions.
a) Perform the free network adjustment (with all of the station points used in defining the center of gravity to be fixed), and determine the adjusted coordinates and the cofactor matrix of all the points.
b) Use S-transformation to transform your results (adjusted parameters and the cofactor of the adjusted parameters) in Question (a) into min-imum constraint adjustment results with point P1 fixed as the datum.

14.2 Continue from Question 14.1 by using the data given in order to adjust the network by the minimal constraint least squares adjustment method (by fixing point P1). Determine the adjusted elevations and the cofactor matrix of the unknown points (assuming point P1 is fixed). Compare the coordinates of point P2 obtained here with the values obtained in Question 14.1b.

14.3 The leveling network in Figure P14.3 was measured in a leveling survey. The five height differences with their corresponding standard deviations observed between the four stations and the approximate elevations of the stations are given in the following tables.

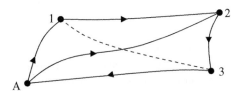

Figure P14.3

Field measurements

Stations		Observed height	Standard
From	To	differences (m)	deviation (m)
A	1	61.478	0.005
1	2	16.994	0.010
2	3	−25.051	0.005
3	A	−53.437	0.010
A	2	78.465	0.010

Elevations of points

Point	Elevation (m)
A	214.880
1	276.358
2	293.352
3	268.301

a) Perform the free network adjustment (with all of the station points used in defining the center of gravity to be fixed), and determine the adjusted coordinates and the cofactor matrix of all the points.

b) Use S-transformation to transform your results (adjusted parameters and the cofactor of the adjusted parameters) in Question (a) into minimum constraint adjustment results with point A used to determine the datum.

14.4   Continue from Question 14.3 by using the data given in order to adjust the network by the minimal constraint least squares adjustment method (by fixing point A). Determine the adjusted elevations and the cofactor matrix of the unknown points (assuming point A is fixed). Compare the coordinates of point 1 obtained here with the values obtained in Question 14.3b.

14.5   A horizontal control network of five points has been measured by classical methods of triangulation and trilateration. The measured quantities are 8 total station directions and 10 distances. The network is to be adjusted by inner constraint method. Answer the following.
a) How many datum defects are there in the original network, explaining what constitutes the defects?
b) What is the size (row by column) of the *G*-matrix for the inner constraint adjustment? Explain how you arrived at the numbers specified for row and column in the matrix.

14.6   A small two-dimensional network of five points (A, 1, 2, 3, B) is to be adjusted by the parametric method of least squares. All the nine distances were measured in the field. In addition, four total station directions (to stations 1, 2, 3, B) were measured from point A, and three total station directions (to stations 3, A, 1) were measured from point 2. Answer the following.
a) If the network is to be adjusted without fixing any of the network points, how many datum defects are contained in the network? List the types of parameters needed to just fix the datum problem.
b) How many unknown parameters does the network contain if a minimum constrained adjustment is to be performed? Give details of how you obtain your answer.
c) How many parameters are there if the network is fully constrained with fixed stations A and B?
d) How many degrees of freedom are there in the network if it is adjusted as a free network? Give details of how you arrive at your answer.

14.7   A two-dimensional triangular survey network consisting of points P, Q, and R had all its three angles and three distances independently measured using total station equipment. The goal is to determine the coordinates P (X_P, Y_P), Q (X_Q, Y_Q), and R (X_R, Y_R) of the network points. Answer the following, assuming the coordinates of all the network points are unknown and are to be determined by the method of least squares (free network) adjustment.
a) Give the size (row by column) of the first design matrix (*A*) for this problem if no coordinates of the network points are to be constrained.

b) How many datum defects are in the network, explaining what constitutes the defects?
c) Give (symbolically) the *G*-matrix (or inner constraints matrix) for adjusting this network by inner constraints (or free network) method, clearly giving every step of your answer (assume that all the coordinates of the network points are to be used to define the datum).
d) Determine the degrees of freedom for the adjustment of this network by inner constraints (or free network) method, clearly giving every step of your answer.
e) Determine the degrees of freedom for the adjustment of this network by minimal external constraints method, clearly giving every step of your answer.

15

Introduction to Dynamic Mode Filtering and Prediction

OBJECTIVES

After studying this chapter, you should be able to:

1) Distinguish between filtering and prediction.
2) Construct simple filtering equations (dynamic and measurement equations).
3) Solve simple filtering problems.
4) Discuss the relationship between filtering and sequential least squares adjustment.

15.1   Introduction

A *filter* is an *optimal recursive* data processing *algorithm* for estimating the *state vector* (or *state variables*) of a *dynamic system* given a series of *data* that contains random errors. The main elements involved in the definition of a filter are optimality, recursiveness, algorithm, state, dynamic system, and data. A filter is said to be *optimal* because it produces the estimates that would be

Understanding Least Squares Estimation and Geomatics Data Analysis, First Edition.
John Olusegun Ogundare.
© 2019 John Wiley & Sons, Inc. Published 2019 by John Wiley & Sons, Inc.
Companion website: www.wiley.com/go/ogundare/Understanding-lse-and-gda

better (in terms of precision) than those estimates that any of the input data would have produced if processed independently. The filter has an ability to simultaneously process all available data (measurements and predicted values), regardless of their precision unlike in the least squares adjustment in which the optimality is based on its ability to minimize errors in measurements only (predicted values not included). The term "optimality" as used here includes all of the following:

a) Smaller errors are associated with the updated state vectors than what would have been obtained from individual data used in the filtering.
b) Predicted state vector and measurements for the current time are processed simultaneously for best possible result.
c) Random errors are minimized in the outcome of filtering.

A filter is *recursive* since the filtering process is repeated all over each time with only the input (new measurements and predicted state) changing. In this case, the filter does not require all the previous measurements and all the previous states to be kept in memory and reprocessed every time a new measurement is available; recursive solution allows sequential, rather than batch, processing of the measurements. A filter as an algorithm follows a set of rules in solving for the state vector (or state variables). A *state* of an object is a set of instantaneous quantities that fully describe any of the object's location, velocity, attitude, or condition at a given time. This is described by the *state vector* or the *state variables*. For example, in Global Navigation Satellite System (GNSS) satellite orbit determination, some of the state variables involved are the position (X, Y, Z coordinates) and the velocity components (three components in the coordinate axes' directions). The *dynamic system* involved in a filter can be considered as a mathematical model that describes how one object's *state* transforms into the next state over the course of time. The *data* involved are the current *predicted state vector* and the *current measurements* (with their associated random errors and corresponding covariance matrices) relating to the instantaneous location of the moving object. Predicted state vector is obtained through prediction.

There are many filtering techniques; a particular one known as Kalman filter is perhaps the best known of the techniques. *Kalman filter* is simply an optimal recursive data processing algorithm; it is an important data fusion algorithm in use today. However, in order to use it in the present context, the associated mathematical models must be linear. If, however, the processes involved (such as in GNSS and inertial navigation systems (INS) are nonlinear, a method of linearizing the processes about some known reference process will be needed in order to use the filter. It should be pointed out that the least squares adjustment method is a special case of a Kalman filtering where special conditions have been implicitly imposed on the behavior of the parameters in time. Some of the common applications of Kalman filter in geomatics are as follows:

1) Kalman filter is now being used in a wide variety of applications both in GNSS kinematic and static positioning and also in satellite orbit determination. GNSS broadcast ephemerides are determined based on predictions made using the knowledge of the dynamics of the GNSS satellites motion; the precise ephemerides are obtained later as filtered positions of the satellites when the satellite positions independently measured from the ground-based control stations are used to correct the broadcast ephemerides. The advantage of using Kalman filtering over a single-point, least squares fix, is that the equation of motion can smooth the GNSS noise, thereby improving the performance. In GNSS kinematic positioning, where the receiver is moving, the state variables (parameters changing with time) are usually the clock phase errors, the tropospheric delays, the ionospheric delay, and the satellite positions.

2) Kalman filter has become the main integration method of choice in GNSS-based navigation systems (at sea, on land, or in space). Kalman filter provides improved overall navigation performance in integrating GNSS with an INS. Over short period of time, the GNSS output is noisy and the INS output is smooth; however, over a long period of time, the GNSS stays accurate, while the INS drifts off. Kalman filter uses statistical models of both the GNSS and INS systems (taking advantage of their different error characteristics) to provide optimal output.

3) In hydrography, apart from navigation purpose, Kalman filter is used in deriving accurate bathymetric maps, and also in providing accurate offshore positions, by integrating INS and GNSS. The INS can provide real-time, accurate horizontal and vertical positioning; the velocity can drift, needing aid from external source. As long as the real-world location (coordinates) of the starting point of the INS is known, INS can be used to generate a displacement from this start point.

4) Kalman filtering is applied in photogrammetry when integrating data from high-accuracy inertial measurement units (IMU), precision GNSS positioning techniques, light detection and ranging (LIDAR) system, and camera systems with multiple computer systems (with advanced processing system) on board a mapping aircraft. The integration process makes it possible to practically or nearly eliminate ground control and the exterior orientation parameters in aerial triangulation and to remove the need for setting stereo models (using survey or ground points) for capture of an elevation model and planimetric details.

5) Used in radar systems in tracking targets, such as aircraft and missiles.

6) Currently, Kalman filters are at work in every smartphone and many computer games. For example, when using a smartphone (integrated with GNSS receiver) for navigation through a short tunnel, the navigation system temporarily continues to show movement forward in the tunnel even when the GNSS receiver is not getting any new position from the GNSS satellites; this is due to Kalman filter in the system producing the estimates of the system's next state.

Further reading on the application of Kalman filter in integrated navigation systems can be found in Groves (2013).

15.1.1 Prediction, Filtering, and Smoothing

Prediction and filtering are closely related. *Prediction* refers to using the best estimates of a state vector at the current time and at the current location of a moving object to estimate the state vector of the object at a future time. It is a process in which the time of interest occurs after the last available measurement. In summary, prediction as used in this book includes all of the following:

a) Best estimates of the current state vector are used in predicting the next state vector; not any estimate of the current state vector should be used in predicting the next state vector, but the best filtered values.
b) Predicted state vector relates to a different location and a time that is different from the current time; it will be incorrect to declare that the approximate estimates of the state vector at the current time are predicted from the current location of the dynamic system.
c) The system involved must be dynamic or constantly in motion.

In simple terms, a mathematical filter is a set of rules that can be applied repeatedly on measurements and predictions, which are simultaneously processed in order to determine the "best" possible average state (coordinates and velocity components) of an object in motion. In a domestic sense, a filter is a device for removing impurities from drinking water. If one likens survey measurements to drinking water and the measurement noise to impurities, we can consider a mathematical filter as a tool for removing noise from survey measurements. Based on this, it can be said that *mathematical filtering* is the process of estimating state variables at the present time with the effects of noise removed. If the estimates made refer to a future time, the process is referred to as *prediction*; if, however, the estimates refer to a time in the past using all observed data including those of the present by working backward from the present state to update the previous estimates, the process is called *interpolation* or *smoothing*. In filtering, the time at which an estimate is desired coincides with the last measurement point, while in smoothing process the time of interest falls within the span of all the available measurement data. In smoothing steps, all past measurements are reprocessed after the last measurement has been made and the filtering step has been completed; this makes smoothing a non-real-time data processing algorithm.

The concepts of *prediction, filtering,* and *smoothing* are closely related and are best illustrated using a case of a moving vehicle whose state variables at some time t are of interest. For example, in determining the orbit of a satellite, the state variables (such as the X, Y, and Z coordinates and the velocity components in X-, Y-, and Z-directions) may be measured with a great deal of corruption by noise using radar or similar system. A filter is to exploit these radar measurements and the knowledge of the dynamics of the satellite (which govern the motion of the satellite in its orbit) in order to remove the effects of the noise and to get good estimates of the state variables of the satellite. In summary the following simplification can be made with regard to filtering as used in dynamic systems:

a) The approximate estimates of the current state vector are predicted from the previous location of the dynamic system.
b) Measurements at the current time are used to update the approximate estimates of the state vector at the current time.
c) Accuracy of the updated state vector is improved over the accuracy of the approximate estimates.
d) All the past measurements are indirectly involved since they are carried forward through prediction process.

15.2 Static Mode Filter

This section is to illustrate the static mode scenario of a filter for the purpose of laying the foundation for the understanding of dynamic filter, which is the subject of this chapter. A *static system* can be considered as a model describing the state of an object that is not varying over time (i.e. the components and structures of the object are at rest under the action of external forces of equilibrium). A static system is similar to a dynamic system except that a dynamic system records only the new state and has no memory of the past. In the following sections, an algorithm for real-time calculation of running averages discussed in Section 10.4.1 and the sequential least squares adjustment model discussed in Section 10.5. are used to illustrate how a static system relates to a dynamic system.

15.2.1 Real-time Moving Averages as Static Mode Filter

Consider the algorithm for real-time calculation of running averages in which previous measurements will not have to be used when new measurement (having equal weight as the previous measurements) becomes available later

(Section 10.4.1). Let the old measurements up to n times be given as $\ell_1, \ell_2, \ell_3, \ldots,$ ℓ_n with its average as the predicted value $\bar{x}_n$ given from Equation (10.38) as

$$\bar{x}_n = \frac{1}{n}\sum_{i=1}^{n}\ell_i \tag{15.1}$$

The sequentially adjusted value $(\hat{x}_{n+1})$ as new observation ℓ_{n+1} becomes available can be likened to a filtered value, which can be given from Equations (10.38) to (10.42) as

$$\hat{x}_{n+1} = \frac{1}{n+1}\sum_{i=1}^{n+1}\ell_i = \frac{1}{n+1}\left(n\bar{x}_n + \ell_{n+1}\right) = \bar{x}_n + \frac{1}{n+1}\left(\ell_{n+1} - \bar{x}_n\right) \tag{15.2}$$

where

$$\frac{1}{n+1}\left(n\bar{x}_n + \ell_{n+1}\right) = \frac{1}{n+1}\left(n\bar{x}_n + \bar{x}_n + \ell_{n+1} - \bar{x}_n\right) = \bar{x}_n + \frac{1}{n+1}\left(\ell_{n+1} - \bar{x}_n\right) \tag{15.3}$$

Equation (15.2) can be simplified as

$$\hat{x}_{n+1} = \bar{x}_n + K\left(\ell_{n+1} - \bar{x}_n\right) \tag{15.4}$$

where K is a form of Kalman gain that can be given from Equation (15.2) as

$$K = \frac{1}{n+1} \tag{15.5}$$

If Equation (15.4) is considered as a filtering problem, the predicted value $(\bar{x}_n)$ determined using the linear model in Equation (15.1) and the new measurement (ℓ_{n+1}) will be considered as input to the filter. The output of the filter is $\hat{x}_{n+1}$, which will be as close to the predicted value $(\bar{x}_n)$ as possible and is determined from Equation (15.4) by scaling the difference between the predicted value $(\bar{x}_n)$ and the measurement (ℓ_{n+1}) with the factor K and adding it to the predicted value. From the propagated variance for the sequentially adjusted value from Equation (10.53), the following variance of the filtered value in Equation (15.4) can be given:

$$\sigma_{\hat{x}_{n+1}}^2 = \sigma_{\hat{x}_n}^2 - K\sigma_{\hat{x}_n}^2 \tag{15.6}$$

Equations (15.4) and (15.6) constitute the "filtering form" in static mode for real-time calculation of running averages and their variances. In terms of filtering, it can be said that the mean squared error $\sigma_{\hat{x}_n}^2$ is minimized by a factor K in Equation (15.6); in dynamic mode, K will be referred to as Kalman gain. Since Equations (15.4) and (15.6) relate only to static mode of adjustment, they are referred to as the least squares adjustment solution of *sequential* model equations. In this case, the least squares sequential adjustment method (which is a static mode case) is considered as a special case of filtering in which some special conditions have been implicitly imposed on the behavior of the parameters in time.

15.2.2   Sequential Least Adjustment as Static Mode Filter

Sequential least squares adjustment in Sections 10.4 and 10.5 will be compared with the static mode "filtering" discussed in Section 15.3. The sequential least squares adjusted vector of updated corrections (δ) to approximate parameters is given from Equations (10.101) and (10.104) as

$$\delta = \delta' + \Delta\delta \tag{15.7}$$

with

$$\Delta\delta = -N_1^{-1}A_2^T M(A_2\delta' + w_2) \tag{15.8}$$

$$A_2\delta + w_2 - v_2 = 0 \tag{15.9}$$

$$w_2 = f_2(x^0) - \ell_2 \tag{15.10}$$

$$M = [P_2^{-1} + A_2 N_1^{-1} A_2^T] \tag{15.11}$$

$$N_1 = A_1^T P_1 A_1 \tag{15.12}$$

where

δ' is a vector of previously adjusted corrections based on the first set of observations (ℓ_1)

$\Delta\delta$ is a set of updates to be applied to previous corrections, due to the effects of new set of observations (ℓ_2)

$f_2(x^0)$ is a function of approximate parameter vector x^0 for estimating the approximate measurements corresponding with the second batch of measurements ℓ_2

v_2 and w_2 are the residual and misclosure vectors, respectively, for the second batch of measurements (ℓ_2)

A_1 and A_2 are the Jacobian matrices of the parametric equations of the first and second batches of measurements with respect to the unknown parameters (state vector), respectively

P_1 and P_2 are, respectively, the weight matrices of the first and second batches of measurements

In the corresponding cofactor matrix of the vector of updated corrections (δ) to the approximate parameters given from Equations (10.105) and (10.106) are

$$Q_\delta = N_1^{-1} - N_1^{-1}A_2^T M A_2 N_1^{-1} \tag{15.13}$$

or

$$Q_\delta = Q_{\delta'} + \Delta Q_\delta \tag{15.14}$$

where $Q_{\delta'} = N_1^{-1}$ is the cofactor of the vector of corrections to the approximate parameters based on the first batch of observations (ℓ_1); $\Delta Q_\delta = -N_1^{-1} A_2^T M A_2 N_1^{-1}$ is the correction matrix to be applied to the previous cofactor matrix, due to the new observation vector (ℓ_2); and other variables are as defined in Section 10.5. By substituting Equations (15.8) and (15.10) into Equation (15.7), the following is obtained:

$$\delta = \delta' + N_1^{-1} A_2^T M \left[\ell_2 - \left(A_2\delta' + f_2(x^0)\right)\right] \tag{15.15}$$

From Equations (15.9) and (15.10), the following is derived:

$$A_2\delta + f_2(x^0) = \ell_2 + v_2 \tag{15.16}$$

where $\hat{\ell}_2 = \ell_2 + v_2$ is a vector of the adjusted new observations and δ is the vector of updated corrections. Since δ' (a vector of previously adjusted corrections based on the first set of observations (ℓ_1)) is used in Equation (15.15), the term $(A_2\delta' + f_2(x^0))$ can be considered as the "predicted observations" $(\bar{\ell}_2)$ using only the corrections based on the previous observations (ℓ_1). Equation (15.15) can then be rewritten in "filtering form" as

$$\delta = \delta' + K\left[\ell_2 - \bar{\ell}_2\right] \tag{15.17}$$

where $\bar{\ell}_2 = (A_2\delta' + f_2(x^0))$ and $K = N_1^{-1} A_2^T M$. From Equations (15.13) and (15.14), it can be seen that

$$Q_\delta = Q_{\delta'} - KA_2 Q_{\delta'} \tag{15.18}$$

It can be seen that Equations (15.17) and (15.4) are similar and Equations (15.18) and (15.6) are also similar. These equations can similarly be compared with filtering in the dynamic mode.

15.3 Dynamic Mode Filter

Dynamic system is a cause-response system. It has a memory of the state of the system as a function of time, which is then used in predicting the next state of the system in the future time. Considering a dynamic system such as a moving vehicle, the unknown variables (coordinates and the velocity) will form the elements of the "state vector," which are time dependent. These time-dependent vectors may be predicted for at any instant of time by means of the *dynamic models* or *prediction equations*. The values predicted from the dynamic models can be updated by observations containing information on some components of the state vector. This step-by-step procedure of "predicting and updating" is known as *Kalman filtering*. Kalman filtering can be considered as a form of least squares that allows parameters to vary with time.

Prediction and filtering are closely related. The mathematical form of the filter consists of two independent models: the *dynamic* model (for prediction) and *measurement* model (for updating the prediction). In order to use the Kalman filter for estimating the state vector of a process given only a sequence of noisy observations, the process must first be modeled according to the framework of the Kalman filter. This means specifying the transition matrix (H) from time t_{k-1} to t_k, the design matrix for the measurement model (A_k), the cofactor matrix for the dynamic model noise (Q_w), and the cofactor matrix of the measurement noise (Q_ℓ). In this chapter, the cofactor matrix will be used to mean the same thing as covariance matrix; this is based on the assumption that the a priori variance factor is 1 and is well known. The dynamic model or prediction equation in Kalman filter can be expressed as follows (cf. Faragher 2012):

$$\bar{x}_k = H\hat{x}_{k-1} + B_{k-1}u_{k-1} + w_{k-1} \tag{15.19}$$

and the measurement model or filtering equation (which must be linear in the case of simple Kalman filtering) can be given as

$$\ell_k = A_k\hat{x}_k + e_k \tag{15.20}$$

where

$\bar{x}_k$ is the predicted state vector for the next time t_k

$\hat{x}_k$ is the filtered state vector at time t_k

$\hat{x}_{k-1}$ is the filtered state vector at time t_{k-1}

ℓ_k is the observation vector (the actual measurements of state variables at the current time) at time t_k

H is the transition matrix from time t_{k-1} to t_k, which is the Jacobian of the dynamic model with respect to the current state vector

w_{k-1} is the noise vector representing the dynamic model at time t_{k-1} (referred to as white noise since it is not correlated and most especially not correlated with past values of the noise)

u_{k-1} is the vector of control inputs of the dynamic system, such as steering angle change, throttle push, breaking force effect, etc.

B_{k-1} is the control input matrix for applying the effect of each control input parameter in the vector u_{k-1} on the state vector

A_k is the design matrix for the measurement model, which is the Jacobian of the measurement model with respect to the current state vector

e_k is the noise vector of the measurement model (assumed uncorrelated with any other variables) at time t_k.

The term $B_{k-1}u_{k-1}$ in Equation (15.19) represents an additional information that are not related to the state itself but are affecting the system, such as when the train operator pushes on the throttle of the train, causing the train to accelerate. This effect must be added as correction to the prediction; however, the term can be omitted for a very simple system if no external influences are exerted on the dynamic system. The other external influences on the prediction, which are unknown random quantities, are treated as model noises (w_{k-1}) with cofactor matrix $Q_{w_{k-1}}$. This covariance matrix constitutes the uncertainty due to the process noise terms, which include the $B_{k-1}u_{k-1}$ term (the noisy control inputs). The mean values of w_{k-1} and e_k are assumed to be zero and are uncorrelated between epochs with any other variables; also, no correlation exists between w_{k-1} and e_k. In this case, $Q_{w_{k-1}}$ can be given as

$$Q_{w_{k-1}} = E\left(w_{k-1}w_{k-1}^T\right) \tag{15.21}$$

and the covariance matrix (Q_{ℓ_k}) for the measurement model noises can be given as

$$Q_{\ell_k} = E\left(e_k e_k^T\right) \tag{15.22}$$

where $E(\cdot)$ means the expected value. The prediction and filtering equations of the linear Kalman filter can be summarized as follows. Since w_{k-1} is the model random error with its mean value zero, it can be set equal to zero in using Equation (15.19) for prediction as follows:

$$\bar{x}_k = H\hat{x}_{k-1} + B_{k-1}u_{k-1} \tag{15.23}$$

where $\bar{x}_k$ is the predicted state vector for the next epoch t_k using the data from the current epoch t_{k-1} and $\hat{x}_{k-1}$ is the filtered estimate of the state vector at the current time t_{k-1}. The following cofactor matrix of predicted state vector is obtained by applying the variance–covariance propagation laws to Equation (15.19):

$$Q_{\bar{x}_k} = HQ_{\hat{x}_{k-1}}H^T + Q_{w_{k-1}} \tag{15.24}$$

where $Q_{\bar{x}_k}$ is the cofactor matrix of predicted state vector at t_k and $Q_{\hat{x}_{k-1}}$ is the cofactor matrix of the filtered state vector at t_{k-1}. By variance–covariance laws, the cofactor matrix of the model noises ($Q_{w_{k-1}}$) can be given from Equation (15.23) as

$$Q_{w_{k-1}} = B_{k-1}Q_{u_{k-1}}B_{k-1}^T + Q_{n_{k-1}} \tag{15.25}$$

where the first term of the equation represents the process cofactor matrix component associated with the vector of known noisy control inputs (u_{k-1}), $Q_{u_{k-1}}$ is the cofactor matrix of u_{k-1}, and $Q_{n_{k-1}}$ is the cofactor representing new uncertainty after prediction, which are due to untracked system noises. Using the prediction outcomes in Equations (15.23) and (15.24) and the Equation (15.20), the *filtering equations* can be given as follows:

$$\hat{x}_k = \bar{x}_k + K_k\left(\ell_k - A_k\bar{x}_k\right) \tag{15.26}$$

where $\hat{x}_k$ is the filtered state vector at the time (t_k) when the measurements (ℓ_k) were taken and K_k is the Kalman gain formulated for the epoch t_k. The cofactor ($Q_{\hat{x}_k}$) of the filtered state vector is given as

$$Q_{\hat{x}_k} = (I - K_k A_k) Q_{\bar{x}_k}^T \tag{15.27}$$

where the Kalman gain is expressed by

$$K_k = Q_{\bar{x}_k} A_k^T \left(A_k Q_{\bar{x}_k} A_k^T + Q_{\ell_k} \right)^{-1} \tag{15.28}$$

Equations (15.23)–(15.28) are the simple Kalman filter equations and are applied at every epoch for which observations are available. As it can be seen in Equation (15.26), the measurements (ℓ_k) at the current time step k are used to refine the prediction ($\bar{x}_k$) in order to arrive at a new more accurate state estimate ($\hat{x}_k$). Generally, it can be said that Kalman filter relates to a dynamic case, while the method of least squares *sequential adjustment* relates to a static case.

15.3.1 Summary of Kalman Filtering Process

The main steps of simple Kalman filter procedure can be summarized as follows:

1) Perform prediction by using previously adjusted state vector $\hat{x}_{k-1}$ to predict the state vector of the next kth point:

$$\bar{x}_k = H\hat{x}_{k-1} + B_{k-1}u_{k-1} \tag{15.29}$$

and the predicted cofactor matrix:

$$Q_{\bar{x}_k} = HQ_{\hat{x}_{k-1}}H^T + Q_{w_{k-1}} \tag{15.30}$$

where

$$Q_{w_{k-1}} = E\left(w_{k-1}w_{k-1}^T\right) \tag{15.31}$$

If the process error (w_{k-1}) is associated only with the noisy control inputs in Equation (15.29), the propagated covariance matrix can be given from Equation (15.25) as

$$Q_{w_{k-1}} = B_{k-1}Q_{u_{k-1}}B_{k-1}^T \tag{15.32}$$

2) Perform correction to the prediction in the following four steps:
   a) Formulate the measurement model:

$$\ell_k = A_k\hat{x}_k + e_k \tag{15.33}$$

and form the measurement cofactor, Q_{ℓ_k}, using the variances of the measurements.

b) Compute the Kalman gain:

$$K_k = Q_{\bar{x}_k} A_k^T \left(A_k Q_{\bar{x}_k} A_k^T + Q_{\ell_k} \right)^{-1} \tag{15.34}$$

c) Update the predicted state vector ($\bar{x}_k$) to obtain the filtered state vector ($\hat{x}_k$):

$$\hat{x}_k = \bar{x}_k + K_k (\ell_k - A_k \bar{x}_k) \tag{15.35}$$

d) Update the covariance matrix ($Q_{\bar{x}_k}$) of the predicted state vector to obtain the covariance matrix ($Q_{\hat{x}_k}$) of the filtered state vector:

$$Q_{\hat{x}_k} = (I - K_k A_k) Q_{\bar{x}_k}^T \tag{15.36}$$

3) Use the corrected quantities in the next prediction as follows:

$$\bar{x}_{k+1} = H\hat{x}_k + B_k u_k \tag{15.37}$$

$$Q_{\bar{x}_{k+1}} = H Q_{\hat{x}_k} H^T + Q_{w_k} \tag{15.38}$$

$$Q_{w_k} = E\left(w_k w_k^T \right) \tag{15.39}$$

The simple Kalman filtering process discussed in this section is illustrated in Figure 15.1.

As can be seen in Figure 15.1, for each cycle of the simple Kalman filter equations, the measurement matrix A_k (assumed stationary over time) and the measurements covariance Q_{ℓ_k} must be known a priori or computed based on the measurements and a partial prior knowledge. For each cycle of the dynamic model, the state transition matrix H (also assumed stationary over time) and

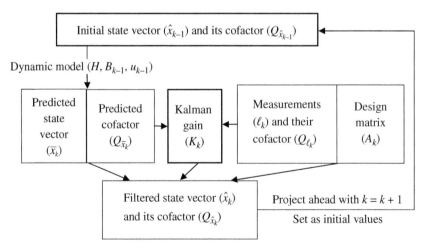

Figure 15.1 Summary of Kalman filtering process.

the dynamic model covariance matrix Q_{w_k} must be known. Simple Kalman filter is recursive and takes some time to settle down in practice. Note that in simple Kalman filtering, the measurements models must be linear; usually, distance and angle measurements are nonlinear and cannot be used in simple Kalman filtering. Extended Kalman filtering process or nonlinear Kalman filter model (which is beyond the scope of this book) should be used to process distance and angle measurements directly. The following examples will illustrate how the Kalman filtering is applied. In the examples, the Kalman filter processes after certain recursion are presented in order to simply illustrate the numerical procedure of computations for prediction and filtering. The numbers are arbitrarily chosen and are not representing any particular navigation system.

15.4 Kalman Filtering Examples

The examples given in this section are to illustrate the steps for formulating Kalman filtering problems and for solving those problems.

Example 15.1   Consider a hypothetical train wagon that is capable of running smoothly on a perfectly frictionless long straight rail. Assume the initial state of the wagon is such that it was stationary at position 0 and, when it is given a tap, it runs at a constant speed with a random acceleration of 0.1 m s^{-2}. Answer the following filtering problem:

a) Derive the Kalman filter dynamic and observation models for the current state of the wagon and what its velocity will be if the position of the wagon is measured at a later time Δt.
b) Determine the optimal position and velocity of the wagon if the position of the wagon was measured imprecisely as 50.0 m with a standard deviation of 1 m at $\Delta t = 30$ seconds.

Solution (a):

Derivation of Kalman filter.

Given the filtered state vector at the current time $k - 1$ as $\hat{x}_{k-1} = [\hat{p}_{k-1} \ \hat{v}_{k-1}]^T$ and the predicted state vector for time k as $\bar{x}_k = [\bar{p}_k \ \bar{v}_k]^T$, where $\hat{p}_{k-1}$ is the filtered position, $\hat{v}_{k-1}$ is the filtered velocity, $\bar{p}_k$ is the predicted position, and $\bar{v}_k$ is the predicted velocity, the following dynamic (prediction) model can be formulated. Based on the general laws of physics, the prediction equations can be formulated as follows:

$$\bar{p}_k = \hat{p}_{k-1} + (\Delta t)\hat{v}_{k-1} + \frac{1}{2}(\Delta t^2)a_{k-1} + w_{p_{k-1}} \tag{15.40}$$

where Δt is time step, a_{k-1} is the acceleration of the system at time $k-1$, and $w_{p_{k-1}}$ is the position noise. Similarly, the predicted velocity ($\bar{v}_k$) for time k can be given as

$$\bar{v}_k = \hat{v}_{k-1} + (\Delta t)a_{k-1} + w_{v_{k-1}} \qquad (15.41)$$

where $w_{v_{k-1}}$ is the velocity noise at time $k-1$. The transition matrix (H) can be derived as the Jacobian of Equations (15.40) and (15.41) with respect to the currently filtered state vector, as follows:

$$H = \begin{bmatrix} \dfrac{\partial \bar{p}_k}{\partial \hat{p}_{k-1}} & \dfrac{\partial \bar{p}_k}{\partial \hat{v}_{k-1}} \\[2mm] \dfrac{\partial \bar{v}_k}{\partial \hat{p}_{k-1}} & \dfrac{\partial \bar{v}_k}{\partial \hat{v}_{k-1}} \end{bmatrix} \quad \text{or } H = \begin{bmatrix} 1 & \Delta t \\ 0 & 1 \end{bmatrix} \qquad (15.42)$$

Equations (15.40) and (15.41) can be expressed in the following matrix form:

$$\bar{x}_k = \begin{bmatrix} 1 & \Delta t \\ 0 & 1 \end{bmatrix} \hat{x}_{k-1} + \begin{bmatrix} \dfrac{\Delta t^2}{2} \\ \Delta t \end{bmatrix} a_{k-1} + w_{k-1} \qquad (15.43)$$

where $w_{k-1} = [w_{p_{k-1}} \ \ w_{v_{k-1}}]^T$ is a vector of the system noise consisting of position and velocity noises. Comparing with Equation (15.19), $B_{k-1} = \left[\dfrac{\Delta t^2}{2} \ \ \Delta t\right]^T$ and $u_{k-1} = [a_{k-1}]$. Since position at the given time (t_k) is the quantity measured directly, the measurement equation can be formulated for position only as follows:

$$\hat{\ell}_{p_k} = \hat{p}_k \qquad (15.44)$$

The design matrix A_k can be derived as the Jacobian of Equation (15.44) with respect to the filtered state vector $\hat{x}_k = [\hat{p}_k \ \ \hat{v}_k]^T$ as

$$A_k = \begin{bmatrix} \dfrac{\partial \hat{\ell}_{p_k}}{\partial \hat{p}_k} & \dfrac{\partial \hat{\ell}_{p_k}}{\partial \hat{v}_k} \end{bmatrix} \quad \text{or } A_k = \begin{bmatrix} 1 & 0 \end{bmatrix} \qquad (15.45)$$

Equation (15.45) can be expressed in the following matrix form:

$$\hat{\ell}_k = \begin{bmatrix} 1 & 0 \end{bmatrix} \begin{bmatrix} \hat{p}_k \\ \hat{v}_k \end{bmatrix} + e_k \qquad (15.46)$$

where $e_k = [e_{p_k}]$ is a vector of measurement noise due to measurement scheme with e_{p_k} as the noise in measuring position; the noise in measuring the velocity is zero since it is not measured in this problem. Equation (15.43) is the dynamic model and Equation (15.46) is the observation model.

Solution (b):

Determination of the optimal position and velocity of the wagon.

The position of the wagon was imprecisely measured as 50.0 m with a standard deviation of 1 m at $\Delta t = 30$ seconds. The position and velocity of the wagon are described by the linear state vector given by Equation (15.43).

Given $\Delta t = 30$ seconds. and the position measurement as $50.0\,m \pm 1\,m$, the prediction Equation (15.43) becomes

$$\bar{x}_k = \begin{bmatrix} 1 & 30 \\ 0 & 1 \end{bmatrix} \hat{x}_{k-1} + \begin{bmatrix} 450 \\ 30 \end{bmatrix} a_{k-1} \tag{15.47}$$

or

$$\bar{x}_k = H\hat{x}_{k-1} + B_{k-1}u_{k-1} \tag{15.48}$$

where

$$H = \begin{bmatrix} 1 & 30 \\ 0 & 1 \end{bmatrix}; \ B_{k-1} = \begin{bmatrix} 450 \\ 30 \end{bmatrix}$$

and for random acceleration (the acceleration is not constant) $u_{k-1} = a_{k-1} = 0.0\,m\,s^{-2}$ and w_{k-1} is a noise with the mean value of zero. The position of the vehicle was only measured so that the measurement model will reflect only the measurements as given below and based on Equation (15.46), $A_k = \begin{bmatrix} 1 & 0 \end{bmatrix}$. Since the initial state of the wagon is known with perfect precision, the filter can be initialized as

$$\hat{x}_{k-1} = \begin{bmatrix} 0 \\ 0 \end{bmatrix}$$

It can also be specified to the filter that the exact position is known perfectly at the given time by giving the position a zero covariance matrix, as follows:

$$Q_{\hat{x}_{k-1}} = \begin{bmatrix} 0 & 0 \\ 0 & 0 \end{bmatrix}$$

If the initial position and velocity are not known perfectly, the covariance matrix should be initialized with a suitably large number on its diagonal. With this, the filter will prefer the information from the measurements over the information already in the model. The filtering process can be summarized as follows:

1) Compute the predicted state vector and its covariance matrix.
   From Equation (15.47), the predicted state vectors are (substituting values into the equation and setting the noise to zero, and $a_{k-1} = 0.0$)

$$\bar{x}_k = \begin{bmatrix} 0.0 \\ 0.0 \end{bmatrix} + \begin{bmatrix} 0.0 \\ 0.0 \end{bmatrix} \quad \rightarrow \quad \begin{bmatrix} 0.0 \\ 0.0 \end{bmatrix}$$

The predicted covariance matrix for the predicted state vector:

$$Q_{\bar{x}_k} = HQ_{\hat{x}_{k-1}}H^T + Q_{w_{k-1}}$$

where $Q_{w_{k-1}}$ is the process noise covariance matrix associated with noisy control inputs $u_{k-1} = [a_{k-1}]$. The matrix can be propagated as shown in Equation (15.32) based on the laws of variance–covariance propagation as follows (taking note that the random acceleration is given as $\sigma_a = 0.1 \text{ m s}^{-2}$):

$$Q_{w_{k-1}} = B_{k-1}[\sigma_a^2]B_{k-1}^T \tag{15.49}$$

or

$$Q_{w_{k-1}} = \sigma_a^2 \begin{bmatrix} 450 \\ 30 \end{bmatrix} [450 \quad 30] \quad \rightarrow \quad Q_{w_{k-1}} = (0.1)^2 \begin{bmatrix} 202\,500 & 13\,500 \\ 13\,500 & 900 \end{bmatrix}$$

$$Q_{w_{k-1}} = \begin{bmatrix} 2025.00 & 135.00 \\ 135.00 & 9.00 \end{bmatrix}$$

Given:

$$Q_{\hat{x}_{k-1}} = \begin{bmatrix} 0.0 & 0.0 \\ 0.0 & 0.0 \end{bmatrix} \quad HQ_{\hat{x}_{k-1}}H^T = \begin{bmatrix} 0.0 & 0.0 \\ 0.0 & 0.0 \end{bmatrix}$$

$$Q_{\bar{x}_k} = HQ_{\hat{x}_{k-1}}H^T + Q_{w_{k-1}} = \begin{bmatrix} 2025.00 & 135.00 \\ 135.00 & 9.00 \end{bmatrix} \tag{15.50}$$

The covariance matrix of measurement (for single measurement) can be given as

$$Q_{\ell_k} = \left[\sigma_{e_k}^2\right] \quad \text{or} \quad Q_{\ell_k} = \left[(1)^2\right]$$

2) Compute the Kalman gain:

$$K_k = Q_{\bar{x}_k}A_k^T\left(A_k Q_{\bar{x}_k}A_k^T + Q_{\ell_k}\right)^{-1} \text{ with } A_k = [1 \quad 0]$$

$$K_k = \begin{bmatrix} 0.999\,506 \\ 0.066\,634 \end{bmatrix}$$

3) Compute the filtered state vector:

$$\hat{x}_k = \bar{x}_k + K_k(\ell_k - A_k\bar{x}_k)$$

$$\hat{x}_k = \begin{bmatrix} 0.0 \\ 0.0 \end{bmatrix} + \begin{bmatrix} 0.999\ 506 \\ 0.066\ 634 \end{bmatrix}(50.0 - 0.0)$$

$$\hat{x}_k = \begin{bmatrix} 49.975 \\ 3.332 \end{bmatrix}$$

4) Compute the covariance matrix of the filtered state vector:

$$Q_{\hat{x}_k} = (I - K_k A_k)Q_{\bar{x}_k}^T$$

$$Q_{\hat{x}_k} = \begin{bmatrix} 0.999\ 506 & 0.666\ 34 \\ 0.666\ 34 & 0.004\ 44 \end{bmatrix}$$

Example 15.2   A vehicle with a radio antenna mounted on its roof is traveling along a perfectly frictionless long straight road with a constant acceleration of $0.3\ \text{m s}^{-2} \pm 0.06\ \text{m s}^{-2}$; and its current position and velocity along the straight road are measured with a radar system as $61.0\ \text{m} \pm 2.5\ \text{m}$ and $4.1\ \text{m s}^{-1} \pm 0.5$ m s^{-1}, respectively. Answer the following.

a)   If the state of the vehicle is measured at every 10-second interval, what are the optimal values of the position and velocity of the vehicle at the kth recursion? Assume the values of the state vector $\hat{x}_{k-1} = [\hat{p}_{k-1}\ \ \hat{v}_{k-1}]^T$ (where $\hat{p}_{k-1}$ and $\hat{v}_{k-1}$ are the filtered position and the velocity component at time $k - 1$, respectively) and its covariance matrix $(Q_{\hat{x}_{k-1}})$ from the $(k - 1)$th recursion are given as follows:

$$\hat{x}_{k-1} = \begin{bmatrix} 16.7 \\ 4.3 \end{bmatrix} \begin{matrix} \text{m} \\ \text{m s}^{-1} \end{matrix}$$

$$Q_{\hat{x}_{k-1}} = \begin{bmatrix} 4.50 & 0.90 \\ 0.90 & 0.18 \end{bmatrix}$$

Solution:

The filtering equations used in Example 15.1, with slight modifications, are still applicable in this problem. Given $\Delta t = 10$ seconds, Equation (15.43) becomes

$$\bar{x}_k = \begin{bmatrix} 1 & 10 \\ 0 & 1 \end{bmatrix}\hat{x}_{k-1} + \begin{bmatrix} 50 \\ 10 \end{bmatrix}a_{k-1} \qquad (15.51)$$

or

$$\bar{x}_k = H\hat{x}_{k-1} + B_{k-1}u_{k-1} \tag{15.52}$$

where

$$H = \begin{bmatrix} 1 & 10 \\ 0 & 1 \end{bmatrix}; \quad B_{k-1} = \begin{bmatrix} 50 \\ 10 \end{bmatrix};$$

$a_{k-1} = 0.3 \text{ m s}^{-2}$ and w_{k-1} is the dynamic model noise with the mean value of zero. The measured position and the velocity of the vehicle can be modeled as

$$\hat{\ell}_{p_k} = \hat{p}_k \tag{15.53}$$

$$\hat{\ell}_{v_k} = \hat{v}_k \tag{15.54}$$

The first design matrix (A_k) of Equations (15.53) and (15.54) with respect to the current state vector $\hat{x}_k = [\hat{p}_k \quad \hat{v}_k]^T$ can be given as

$$A_k = \begin{bmatrix} \dfrac{\partial \hat{\ell}_{p_k}}{\partial \hat{p}_k} & \dfrac{\partial \hat{\ell}_{p_k}}{\partial \hat{v}_k} \\ \dfrac{\partial \hat{\ell}_{v_k}}{\partial \hat{p}_k} & \dfrac{\partial \hat{\ell}_{v_k}}{\partial \hat{v}_k} \end{bmatrix} \quad \text{or } A_k = \begin{bmatrix} 1 & 0 \\ 0 & 1 \end{bmatrix} \tag{15.55}$$

The outcomes of the filtering process are as follows. From Equation (15.51), the predicted state vectors (substituting values into the equation and setting the noise to zero) can be given as

$$\bar{x}_k = \begin{bmatrix} 59.70 \\ 4.30 \end{bmatrix} + \begin{bmatrix} 15.00 \\ 3.00 \end{bmatrix} \rightarrow \begin{bmatrix} 74.70 \\ 7.30 \end{bmatrix}$$

and the corresponding covariance matrix of the predicted state vector can be given as

$$Q_{\bar{x}_k} = HQ_{\hat{x}_{k-1}}H^T + Q_{w_{k-1}} \tag{15.56}$$

where $Q_{w_{k-1}}$ is the propagated covariance matrix for the system noisy control inputs based on the laws of variance–covariance propagation from Equation (15.32) (with $\sigma_a = 0.06 \text{ m s}^{-2}$ as the random acceleration), given as follows:

$$Q_{w_{k-1}} = B_{k-1}[\sigma_a^2]B_{k-1}^T \tag{15.57}$$

or

$$Q_{w_{k-1}} = \sigma_a^2 \begin{bmatrix} 50 \\ 10 \end{bmatrix} [50 \ 10] \rightarrow Q_{w_{k-1}} = (0.06)^2 \begin{bmatrix} 2500 & 500 \\ 500 & 100 \end{bmatrix}$$

$$Q_{w_{k-1}} = \begin{bmatrix} 9.00 & 1.80 \\ 1.80 & 0.36 \end{bmatrix}$$

Given:

$$Q_{\hat{x}_{k-1}} = \begin{bmatrix} 4.50 & 0.90 \\ 0.90 & 0.18 \end{bmatrix} \quad HQ_{\hat{x}_{k-1}}H^T = \begin{bmatrix} 40.50 & 2.70 \\ 2.70 & 0.18 \end{bmatrix}$$

$$Q_{\bar{x}_k} = HQ_{\hat{x}_{k-1}}H^T + Q_{w_{k-1}} = \begin{bmatrix} 49.50 & 4.50 \\ 4.50 & 0.54 \end{bmatrix}$$

Given the cofactor matrix of measurement (for single measurement) as follows:

$$Q_{\ell_k} = \begin{bmatrix} \sigma_{\ell_{p_k}}^2 & 0 \\ 0 & \sigma_{\ell_{v_k}}^2 \end{bmatrix} \quad \text{or} \quad Q_{\ell_k} = \begin{bmatrix} 2.5^2 & 0.0 \\ 0.0 & 0.5^2 \end{bmatrix}$$

The Kalman gain can be computed from

$$K_k = Q_{\bar{x}_k} A_k^T \left(A_k Q_{\bar{x}_k} A_k^T + Q_{\ell_k} \right)^{-1}$$

$$K_k = \begin{bmatrix} 0.7925 & 1.1821 \\ 0.0473 & 0.4142 \end{bmatrix}$$

The filtered state vector can be computed as follows:

$$\hat{x}_k = \bar{x}_k + K_k(\ell_k - A_k\bar{x}_k)$$

$$\hat{x}_k = \begin{bmatrix} 74.70 \\ 7.30 \end{bmatrix} + \begin{bmatrix} 0.7925 & 1.1821 \\ 0.0473 & 0.4142 \end{bmatrix} \left(\begin{bmatrix} 61.00 \\ 4.100 \end{bmatrix} - \begin{bmatrix} 74.70 \\ 7.30 \end{bmatrix} \right)$$

$$\hat{x}_k = \begin{bmatrix} 60.0604 \\ 5.3268 \end{bmatrix}$$

The cofactor matrix of the filtered state vector is computed as

$$Q_{\hat{x}_k} = (I - K_k A_k)Q_{\bar{x}_k}^T$$

$$Q_{\hat{x}_k} = \begin{bmatrix} 4.9530 & 0.2955 \\ 0.2955 & 0.1036 \end{bmatrix}$$

The MATLAB code for the solution given above is given in Table 15.1.

Table 15.1 MATLAB code for completing Question 15.2a.

```
>> %input initial state vector xk1 and its cofactor
>> xk1=[16.7;4.3];
>> Qxk1=[4.50 0.90;0.90 0.18];
>> %input measurement Lk and its cofactor QL
>> Lk=[61.0;4.1];
>> QL=[2.5^2 0.0; 0.0 0.5^2];
>> %set transition matrix H
>> H=eye(2); H(1,2)=10;
>> %form the B, u and Bu and Qww
>> B=[50;10];
>> u=0.3;Qu=0.06^2; Qw=B*Qu*B'
>> %predict next state and cofactor
>> xkPred=H*xk1+B*u
xkPred =
   74.7000
    7.3000
>> QxkPred=H*Qxk1*H'+Qw
QxkPred =
   49.5000    4.5000
    4.5000    0.5400
>> %form the design matrix Ak and calculate Kaman gain Kk
>> Ak=eye(2);
>> Kk=QxkPred*Ak'*inv(Ak*QxkPred*Ak'+QL)
Kk =
    0.7925    1.1821
    0.0473    0.4142
>> %Calculate new filtered state vector xkfiltered and its
cofactor matrix Qxkfiltered
>> xkfiltered=xkPred+Kk*(Lk-Ak*xkPred);
>> QxkFiltered=QxkPred'-Kk*Ak*QxkPred'
QxkFiltered =
    4.9530    0.2955
    0.2955    0.1036
```

b) Continuing from the answers to Question 15.2a and using the state vector measurements and their corresponding time steps (at 10 seconds intervals) in Table 15.2, provide a summary (numerically and graphically) of the values of the filtered state variables and the corresponding standard deviations at the given time steps (assuming uniform standard deviations of 2.5 m and 0.5 m s^{-1} for the position and velocity component measurements, respectively).

Table 15.2 State vector measurements at different time steps after the given initial state.

Time step (seconds)	Position measurement (m) (ℓ_{p_k})	Velocity component (m s^{-1}) (ℓ_{v_k})
10	61.0	4.1
20	103.5	4.0
30	141.8	4.5
40	185.2	4.9
50	230.1	4.4
60	276.2	4.5

Solution:

Continuing from solution to Question 15.2a with the next time step as $\Delta t = 10$ seconds, the next prediction can be given as

$$\bar{x}_{k+1} = \begin{bmatrix} 1 & 10 \\ 0 & 1 \end{bmatrix} \hat{x}_k + \begin{bmatrix} 50 \\ 10 \end{bmatrix} a_k \tag{15.58}$$

or

$$\bar{x}_{k+1} = H\hat{x}_k + B_k u_k \tag{15.59}$$

where H is as in Question 15.2a, $B_k = B_{k-1}$, $u_k = a_k = 0.3$ m s^{-2}, and w_k is the dynamic model noise with the mean value of zero. The measured position and the velocity of the vehicle are modeled as above, giving the same design matrix $A_k = A_{k+1}$ as

$$A_{k+1} = \begin{bmatrix} 1 & 0 \\ 0 & 1 \end{bmatrix}$$

The outcomes of the filtering process are as follows. From Equation (15.58), the predicted state vectors (substituting values into the equation and setting the noise to zero) can be given as

$$\bar{x}_{k+1} = \begin{bmatrix} 113.3284 \\ 5.3268 \end{bmatrix} + \begin{bmatrix} 15.00 \\ 3.00 \end{bmatrix} \rightarrow \begin{bmatrix} 128.3284 \\ 8.3268 \end{bmatrix}$$

and the corresponding covariance matrix of the predicted state vector as

$$Q_{\bar{x}_{k+1}} = HQ_{\hat{x}_k}H^T + Q_{w_k} \tag{15.60}$$

where

$$Q_{w_k} = B_k\left[\sigma_a^2\right]B_k^T \tag{15.61}$$

or

$$Q_{\bar{x}_{k+1}} = \begin{bmatrix} 30.2230 & 3.1315 \\ 3.1315 & 0.4636 \end{bmatrix}$$

With the cofactor matrix of measurement (for single measurement) as before (i.e. $Q_{\ell_{k+1}} = Q_{\ell_k}$), the Kalman gain can be computed from

$$K_{k+1} = Q_{\bar{x}_{k+1}} A_{k+1}^T \left(A_{k+1} Q_{\bar{x}_{k+1}} A_{k+1}^T + Q_{\ell_{k+1}} \right)^{-1}$$

$$K_{k+1} = \begin{bmatrix} 0.7250 & 1.2066 \\ 0.0483 & 0.4379 \end{bmatrix}$$

The filtered state vector can be computed as follows:

$$\hat{x}_{k+1} = \bar{x}_{k+1} + K_{k+1} \left(\ell_{k+1} - A_{k+1} \bar{x}_{k+1} \right)$$

$$\hat{x}_{k+1} = \begin{bmatrix} 128.3284 \\ 8.3268 \end{bmatrix} + \begin{bmatrix} 0.7250 & 1.2066 \\ 0.0483 & 0.4379 \end{bmatrix} \left(\begin{bmatrix} 103.500 \\ 4.000 \end{bmatrix} - \begin{bmatrix} 128.3284 \\ 8.3268 \end{bmatrix} \right)$$

$$\hat{x}_{k+1} = \begin{bmatrix} 105.1070 \\ 5.2355 \end{bmatrix}$$

$$(15.62)$$

The covariance matrix of the filtered state vector is computed as

$$Q_{\hat{x}_{k+1}} = (I - K_{k+1} A_{k+1}) Q_{\bar{x}_{k+1}}^T$$

$$Q_{\hat{x}_{k+1}} = \begin{bmatrix} 4.5315 & 0.3016 \\ 0.3016 & 0.1095 \end{bmatrix} \qquad (15.63)$$

The filtered values are summarized in Tables 15.4 and 15.5 and Figures 15.2 and 15.3; they are obtained by recursively changing from the solution to Question 15.2a the values of $\hat{x}_{k-1}$, $Q_{\hat{x}_{k-1}}$, and $\ell_k = \begin{bmatrix} \ell_{p_k} & \ell_{v_k} \end{bmatrix}^T$ as shown in the MATLAB code in Table 15.3.

Table 15.3 MATLAB code for completing Question 15.2 part (b).

```
>> % only initial state xk1, cofactor Qxk1, input
measurements Lk and cofactor QL need to change for the
remaining iterations as follows
>> %input initial state vector xk1 and its cofactor
>> xk1=[60.0604;5.3268];
>> Qxk1=[4.953 0.2955;0.2955 0.1036];
>> %input measurement Lk and its cofactor
>> Lk=[103.5;4.0];
```

Table 15.3 (Continued)

```
>> %predict next state and cofactor
>> xkPred=H*xk1+B*u
xkPred =
  128.3284
    8.3268
>> QxkPred=H*Qxk1*H'+Qw
QxkPred =
   30.2230    3.1315
    3.1315    0.4636
>> Kk=QxkPred*Ak'*inv(Ak*QxkPred*Ak'+QL)
Kk =
    0.7250    1.2066
    0.0483    0.4379
>> xkfiltered=xkPred+Kk*(Lk-Ak*xkPred)
xkfiltered =
  105.1060
    5.2339
>> QxkFiltered=QxkPred'-Kk*Ak*QxkPred'
QxkFiltered =
    4.5315    0.3016
    0.3016    0.1095
```

In Figures 15.2 and 15.3, it can be seen that the measured and the filtered state values are almost identical since the standard deviations of the measurements are better than those of the predictions as shown in Tables 15.4 and 15.5; the filtered values are located between the predicted and measured quantities as illustrated in Figures 15.2 and 15.3. It can also be seen from Tables 15.4 and 15.5 that the standard deviations of the predicted and filtered state vector values stabilize with time.

Table 15.4 Summary of predicted and filtered position vectors.

Time step (seconds)	Predicted position (m)	Standard deviation (m)	Filtered position (m)	Standard deviation (m)
10	74.700	7.036	60.060	2.226
20	128.328	5.498	105.106	2.129
30	172.445	5.524	145.694	2.126
40	211.748	5.525	188.605	2.126
50	257.706	5.523	232.810	2.126
60	300.949	5.523	278.361	2.126

Table 15.5 Summary of predicted and filtered velocity vectors.

Time step (seconds)	Predicted velocity (m s⁻¹)	Standard deviation (m s⁻¹)	Filtered velocity (m s⁻¹)	Standard deviation (m s⁻¹)
0			4.300	
10	7.300	0.292	5.327	0.322
20	8.327	0.681	5.233	0.331
30	8.234	0.685	5.105	0.329
40	8.105	0.684	5.410	0.329
50	8.414	0.684	5.314	0.329
60	8.314	0.684	5.443	0.329

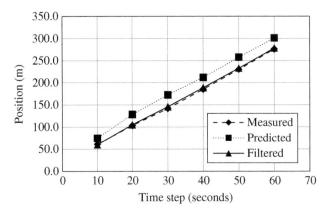

Figure 15.2 Measured, predicted, and filtered positions compared.

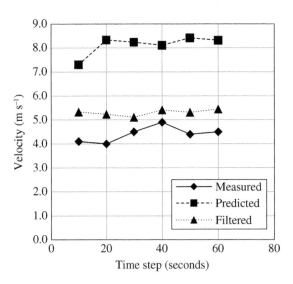

Figure 15.3 Measured, predicted, and filtered velocity components compared.

The Kalman gains for all the time steps are given as follows:

$$K_k = \begin{bmatrix} 0.7925 & 1.1821 \\ 0.0473 & 0.4142 \end{bmatrix} \tag{15.64}$$

$$K_{k+1} = \begin{bmatrix} 0.7250 & 1.2066 \\ 0.0483 & 0.4379 \end{bmatrix} \tag{15.65}$$

$$K_{k+2} = \begin{bmatrix} 0.7230 & 1.2307 \\ 0.0492 & 0.4338 \end{bmatrix} \tag{15.66}$$

$$K_{k+3} = \begin{bmatrix} 0.7233 & 1.2296 \\ 0.0492 & 0.4335 \end{bmatrix} \tag{15.67}$$

$$K_{k+4} = \begin{bmatrix} 0.7232 & 1.2296 \\ 0.0492 & 0.4335 \end{bmatrix} \tag{15.68}$$

$$K_{k+5} = \begin{bmatrix} 0.7232 & 1.2296 \\ 0.0492 & 0.4335 \end{bmatrix} \tag{15.69}$$

It can be seen from Equations (15.68)–(15.69) that the Kalman gains tend to stabilize with time.

Example 15.3 Given the values of the state vector $\hat{x}_{k-1} = \begin{bmatrix} \hat{p}_{x_{k-1}} & \hat{p}_{y_{k-1}} & \hat{v}_{x_{k-1}} & \hat{v}_{y_{k-1}} \end{bmatrix}$ and its covariance matrix $(Q_{\hat{x}_{k-1}})$ resulting from the $(i-1)$th recursion for a ship traveling on a high sea, respectively, as

$$\hat{x}_{k-1} = \begin{bmatrix} 15\,969.933 \text{ m} \\ 25\,030.638 \text{ m} \\ 2.922\,14 \text{ m s}^{-1} \\ 2.005\,28 \text{ m s}^{-1} \end{bmatrix} = \begin{bmatrix} \text{Easting} \\ \text{Northing} \\ \text{Velocity eastward} \\ \text{Velocity northward} \end{bmatrix}$$

$$Q_{\hat{x}_{k-1}} = \begin{bmatrix} 29.020\,576 & 11.740\,694 & 0.092\,973 & 0.029\,312 \\ 11.740\,694 & 20.661\,862 & 0.029\,312 & 0.072\,305 \\ 0.092\,973 & 0.029\,312 & 0.000\,655 & 0.000\,111 \\ 0.029\,312 & 0.072\,305 & 0.000\,111 & 0.000\,576 \end{bmatrix}$$

where $(\hat{p}_{x_{k-1}}, \hat{p}_{y_{k-1}})$ are the filtered two-dimensional position of the system at time $k-1$ and $(\hat{v}_{x_{k-1}}, \hat{v}_{y_{k-1}})$ are the filtered velocity components at time $k-1$.

If the standard deviation of ship's random acceleration (σ) for fix interval (Δt) of 60 seconds is 0.0002 m s^{-2}, covariance matrix of each position fix is

$$Q_\ell = \begin{bmatrix} 91.6 & 42.7 & 0.0 & 0.0 \\ 42.7 & 91.6 & 0.0 & 0.0 \\ 0.0 & 0.0 & 0.0008 & 0.0001 \\ 0.0 & 0.0 & 0.0001 & 0.0007 \end{bmatrix}$$

and the result of the position fix at the kth point (i.e. the two-dimensional position and velocity according to Kalman filtering) is

$$\ell_k = \begin{bmatrix} 16\ 145.292\ \text{m} \\ 25\ 158.442\ \text{m} \\ 2.922\ 14\ \text{m s}^{-1} \\ 2.005\ 28\ \text{m s}^{-1} \end{bmatrix}$$

compute the ship's filtered state vector (two-dimensional position and velocity) at t_k and the predicted state vector at t_{k+1}.

Solution:

The previous filtered state vector $\hat{x}_{k-1}$ can be used to predict the state vector of the next kth point as follows:

$$\bar{x}_k = H\hat{x}_{k-1} + B_{k-1}u_{k-1} \tag{15.70}$$

and the predicted cofactor matrix of the prediction:

$$Q_{\bar{x}_k} = HQ_{\hat{x}_{k-1}}H^T + Q_{w_{k-1}} \tag{15.71}$$

The prediction equations for the position components at time k can be formulated for two-dimensional (x, y) case from Equation (15.40) as follows:

$$\bar{p}_{x_k} = \hat{p}_{x_{k-1}} + (\Delta t)\hat{v}_{x_{k-1}} + \frac{1}{2}(\Delta t^2)a_{x_{k-1}} + w_{p_{x_{k-1}}} \tag{15.72}$$

$$\bar{p}_{y_k} = \hat{p}_{y_{k-1}} + (\Delta t)\hat{v}_{y_{k-1}} + \frac{1}{2}(\Delta t^2)a_{y_{k-1}} + w_{p_{y_{k-1}}} \tag{15.73}$$

where $(\bar{p}_{x_k}, \bar{p}_{y_k})$ are the predicted position components, Δt is time step, $a_{k-1} = \begin{bmatrix} a_{x_{k-1}} & a_{y_{k-1}} \end{bmatrix}^T$ is a vector of the acceleration components of the system at time $k-1$, and $w_{p_{k-1}} = \begin{bmatrix} w_{p_{x_{k-1}}} & w_{p_{y_{k-1}}} \end{bmatrix}^T$ is the vector of position noises. Similarly, the prediction equations for the velocity components at time k can be formulated for two-dimensional (x, y) case from Equation (15.41) as follows:

$$\bar{v}_{x_k} = \hat{v}_{x_{k-1}} + (\Delta t)a_{x_{k-1}} + w_{v_{x_{k-1}}} \tag{15.74}$$

$$\bar{v}_{y_k} = \hat{v}_{y_{k-1}} + (\Delta t)a_{y_{k-1}} + w_{v_{y_{k-1}}} \tag{15.75}$$

where $(\bar{v}_{x_k}, \bar{v}_{y_k})$ are the predicted velocity components and $w_{v_{k-1}} = \left[w_{v_{x_{k-1}}} \ w_{v_{y_{k-1}}} \right]^T$ is the vector of velocity noises. From Equations (15.72) to (15.75), the prediction model can be summarized as Equation (15.70), where the transition matrix (H) can be derived as the Jacobian of Equations (15.72)–(15.75) with respect to the current filtered state vector, as follows:

$$
H = \begin{bmatrix} \dfrac{\partial \bar{p}_{x_k}}{\partial \hat{p}_{x_{k-1}}} & \dfrac{\partial \bar{p}_{x_k}}{\partial \hat{p}_{y_{k-1}}} & \dfrac{\partial \bar{p}_{x_k}}{\partial \hat{v}_{x_{k-1}}} & \dfrac{\partial \bar{p}_{x_k}}{\partial \hat{v}_{y_{k-1}}} \\[2mm] \dfrac{\partial \bar{p}_{y_k}}{\partial \hat{p}_{x_{k-1}}} & \dfrac{\partial \bar{p}_{y_k}}{\partial \hat{p}_{y_{k-1}}} & \dfrac{\partial \bar{p}_{y_k}}{\partial \hat{v}_{x_{k-1}}} & \dfrac{\partial \bar{p}_{y_k}}{\partial \hat{v}_{y_{k-1}}} \\[2mm] \dfrac{\partial \bar{v}_{x_k}}{\partial \hat{p}_{x_{k-1}}} & \dfrac{\partial \bar{v}_{x_k}}{\partial \hat{p}_{y_{k-1}}} & \dfrac{\partial \bar{v}_{x_k}}{\partial \hat{v}_{x_{k-1}}} & \dfrac{\partial \bar{v}_{x_k}}{\partial \hat{v}_{y_{k-1}}} \\[2mm] \dfrac{\partial \bar{v}_{y_k}}{\partial \hat{p}_{x_{k-1}}} & \dfrac{\partial \bar{v}_{y_k}}{\partial \hat{p}_{y_{k-1}}} & \dfrac{\partial \bar{v}_{y_k}}{\partial \hat{v}_{x_{k-1}}} & \dfrac{\partial \bar{v}_{y_k}}{\partial \hat{v}_{y_{k-1}}} \end{bmatrix} \quad \text{or} \quad H = \begin{bmatrix} 1 & 0 & \Delta t & 0 \\ 0 & 1 & 0 & \Delta t \\ 0 & 0 & 1 & 0 \\ 0 & 0 & 0 & 1 \end{bmatrix} \tag{15.76}
$$

$$
B_{k-1} = \left[\frac{\Delta t^2}{2} \ \frac{\Delta t^2}{2} \ \Delta t \ \Delta t \right]^T \tag{15.77}
$$

and

$$
u_{k-1} = a_{k-1} = \left[a_{x_{k-1}} \ a_{y_{k-1}} \right]^T \tag{15.78}
$$

The predicted state vector at $\Delta t = 60$, given that $u_k = \begin{bmatrix} 0 & 0 \end{bmatrix}^T$, can be given from Equation (15.70) as

$$
\bar{x}_k = \begin{bmatrix} 1 & 0 & 60 & 0 \\ 0 & 1 & 0 & 60 \\ 0 & 0 & 1 & 0 \\ 0 & 0 & 0 & 1 \end{bmatrix} \begin{bmatrix} 15\,969.933 \\ 25\,030.638 \\ 2.922\,14 \\ 2.005\,28 \end{bmatrix} + \begin{bmatrix} 0 \\ 0 \\ 0 \\ 0 \end{bmatrix} = \begin{bmatrix} 16\,145.262 \text{ m} \\ 25\,150.955 \text{ m} \\ 2.922\,14 \text{ m s}^{-1} \\ 2.005\,28 \text{ m s}^{-1} \end{bmatrix}
$$

The covariance matrix (Q_w) of the dynamic model noise can be predicted based on the known system noisy control inputs using Equation (15.32) (with $\sigma_a = 0.0002$ m s^{-2} as the random acceleration), given as follows:

$$
Q_{w_{k-1}} = B_{k-1} \left[\sigma_a^2 \right] B_{k-1}^T \tag{15.79}
$$

or

$$
Q_{w_{k-1}} = \sigma_a^2 \begin{bmatrix} \dfrac{\Delta t^2}{2} \\[2mm] \dfrac{\Delta t^2}{2} \\[2mm] \Delta t \\[2mm] \Delta t \end{bmatrix} \begin{bmatrix} \dfrac{\Delta t^2}{2} & \dfrac{\Delta t^2}{2} & \Delta t & \Delta t \end{bmatrix}
$$

For $\Delta t = 60$ seconds

$$Q_{w_{k-1}} = (0.000\ 2)^2 \begin{bmatrix} 1\ 800 \\ 1\ 800 \\ 60 \\ 60 \end{bmatrix} [1\ 800\ \ 1\ 800\ \ 60\ \ 60]$$

$$Q_{w_{k-1}} = \begin{bmatrix} 0.129\ 6 & 0.129\ 6 & 0.004\ 32 & 0.004\ 32 \\ 0.129\ 6 & 0.129\ 6 & 0.004\ 32 & 0.004\ 32 \\ 0.004\ 32 & 0.004\ 32 & 0.000\ 144 & 0.000\ 144 \\ 0.004\ 32 & 0.004\ 32 & 0.000\ 144 & 0.000\ 144 \end{bmatrix}$$

Given that

$$Q_{\hat{x}_{k-1}} = \begin{bmatrix} 29.020\ 576 & 11.740\ 694 & 0.092\ 973 & 0.029\ 312 \\ 11.740\ 694 & 20.661\ 862 & 0.029\ 312 & 0.072\ 305 \\ 0.092\ 973 & 0.029\ 312 & 0.000\ 655 & 0.000\ 111 \\ 0.029\ 312 & 0.072\ 305 & 0.000\ 111 & 0.000\ 576 \end{bmatrix}$$

$$Q_{\bar{x}_k} = HQ_{\hat{x}_{k-1}}H^T + Q_{w_{k-1}} \quad Q_{w_{k-1}} = B_{k-1}[\sigma_a^2]B_{k-1}^T$$

the covariance matrix of the predicted state vector can be given as follows:

$$Q_{\bar{x}_k} = \begin{bmatrix} 42.664\ 94 & 15.787\ 33 & 0.136\ 59 & 0.040\ 29 \\ 15.787\ 33 & 31.541\ 66 & 0.040\ 29 & 0.111\ 19 \\ 0.136\ 59 & 0.040\ 29 & 0.000\ 80 & 0.000\ 26 \\ 0.040\ 29 & 0.111\ 19 & 0.000\ 26 & 0.000\ 72 \end{bmatrix}$$

The measurement model can be formulated as $\ell_k = A_k \hat{x}_k + e_k$

For direct measurement of the state vector, the measurement model can be given as

$$\hat{\ell}_{p_{x_k}} = \hat{p}_{x_k} \tag{15.80}$$

$$\hat{\ell}_{p_{y_k}} = \hat{p}_{y_k} \tag{15.81}$$

$$\hat{\ell}_{v_{x_k}} = \hat{v}_{x_k} \tag{15.82}$$

$$\hat{\ell}_{v_{y_k}} = \hat{v}_{y_k} \tag{15.83}$$

The design matrix A_k can be derived as the Jacobian of Equations (15.80)–(15.83) with respect to the filtered state vector $\hat{x}_k = \begin{bmatrix} \hat{p}_{x_k} & \hat{p}_{y_k} & \hat{v}_{x_k} & \hat{v}_{y_k} \end{bmatrix}$ at time k as

$$A_k = \begin{bmatrix} \dfrac{\partial \hat{\ell}_{p_{x_k}}}{\partial \hat{p}_{x_k}} & \dfrac{\partial \hat{\ell}_{p_{x_k}}}{\partial \hat{p}_{y_k}} & \dfrac{\partial \hat{\ell}_{p_{x_k}}}{\partial \hat{v}_{x_k}} & \dfrac{\partial \hat{\ell}_{p_{x_k}}}{\partial \hat{v}_{y_k}} \\[2ex] \dfrac{\partial \hat{\ell}_{p_{y_k}}}{\partial \hat{p}_{x_k}} & \dfrac{\partial \hat{\ell}_{p_{y_k}}}{\partial \hat{p}_{y_k}} & \dfrac{\partial \hat{\ell}_{p_{y_k}}}{\partial \hat{v}_{x_k}} & \dfrac{\partial \hat{\ell}_{p_{y_k}}}{\partial \hat{v}_{y_k}} \\[2ex] \dfrac{\partial \hat{\ell}_{v_{x_k}}}{\partial \hat{p}_{x_k}} & \dfrac{\partial \hat{\ell}_{v_{x_k}}}{\partial \hat{p}_{y_k}} & \dfrac{\partial \hat{\ell}_{v_{x_k}}}{\partial \hat{v}_{x_k}} & \dfrac{\partial \hat{\ell}_{v_{x_k}}}{\partial \hat{v}_{y_k}} \\[2ex] \dfrac{\partial \hat{\ell}_{v_{y_k}}}{\partial \hat{p}_{x_k}} & \dfrac{\partial \hat{\ell}_{v_{y_k}}}{\partial \hat{p}_{y_k}} & \dfrac{\partial \hat{\ell}_{v_{y_k}}}{\partial \hat{v}_{x_k}} & \dfrac{\partial \hat{\ell}_{v_{y_k}}}{\partial \hat{v}_{y_k}} \end{bmatrix} \quad \text{or} \quad A_k = \begin{bmatrix} 1 & 0 & 0 & 0 \\ 0 & 1 & 0 & 0 \\ 0 & 0 & 1 & 0 \\ 0 & 0 & 0 & 1 \end{bmatrix} \tag{15.84}$$

and the random noise e_k is set to zero. The cofactor of observations (Q_{ℓ_k}) is given as

$$Q_{\ell_k} = \begin{bmatrix} 91.6 & 42.7 & 0.0 & 0.0 \\ 42.7 & 91.6 & 0.0 & 0.0 \\ 0.0 & 0.0 & 0.0008 & 0.0001 \\ 0.0 & 0.0 & 0.0001 & 0.0007 \end{bmatrix}$$

The Kalman gain matrix can be computed using the equation $K_k = Q_{\bar{x}_k} A_k^T \left(A_k Q_{\bar{x}_k} A_k^T + Q_{\ell_k} \right)^{-1}$.

Substituting all of the given quantities into the Kalman gain equation gives

$$K_k = \begin{bmatrix} 0.262\ 76 & 0.008\ 19 & 0.000\ 63 & -0.001\ 26 \\ 0.008\ 19 & 0.194\ 94 & -0.000\ 18 & 0.000\ 53 \\ 0.000\ 63 & -0.000\ 18 & 0.435\ 47 & 0.055\ 77 \\ -0.001\ 26 & 0.000\ 53 & 0.055\ 77 & 0.454\ 89 \end{bmatrix}$$

The predicted state vector is computed using the equation $\hat{x}_k = \bar{x}_k + K_k(\ell_k - A_k\bar{x}_k)$.

The filtered estimates of the state vector are calculated as

$$\hat{x}_k = \begin{bmatrix} \text{Easting} \\ \text{Northing} \\ \text{Velocity eastward} \\ \text{Velocity northward} \end{bmatrix} = \begin{bmatrix} 16\ 145.094\ 0 \text{ m} \\ 25\ 152.414\ 6 \text{ m} \\ 2.920\ 9 \text{ m s}^{-1} \\ 2.009\ 3 \text{ m s}^{-1} \end{bmatrix}$$

The covariance matrix of the filtered state vector is calculated using the equation

$$Q_{\hat{x}_k} = (I - K_k A_k) Q_{\bar{x}_k}^T$$

The computed covariance matrix of state vector is

$$Q_{\hat{x}_k} = \begin{bmatrix} 23.068\ 7 & 9.074\ 0 & 0.050\ 3 & 0.011\ 3 \\ 9.074\ 0 & 18.206\ 2 & 0.010\ 91 & 0.043\ 5 \\ 0.050\ 3 & 0.010\ 91 & 0.000\ 36 & 0.000\ 09 \\ 0.011\ 3 & 0.043\ 5 & 0.000\ 09 & 0.000\ 32 \end{bmatrix}$$

Prediction for state vector at point $k + 1$ is done by using the dynamic model equation (assuming $u_{k+1} = [0\ \ 0]^T$):

$$\bar{x}_{k+1} = H\hat{x}_k$$

$$\bar{x}_{k+1} = \begin{bmatrix} \text{Easting} \\ \text{Northing} \\ \text{Velocity eastward} \\ \text{Velocity northward} \end{bmatrix} = \begin{bmatrix} 16\ 320.345\ \text{m} \\ 25\ 272.971\ \text{m} \\ 2.920\ 9\ \text{m s}^{-1} \\ 2.009\ 3\ \text{m s}^{-1} \end{bmatrix}$$

The MATLAB code for solving the Kalman filtering problem in Example 15.3 is given in Table 15.6.

Table 15.6 MATLAB code for solving Kalman filtering problem in Example 15.3.

```
>>%input the initial state vector xk1 and its cofactor
matrix Qxk1
>> xk1=[15969.933;25030.638;2.92214;2.00528];
>> Qxk1=[29.020576 11.740694 0.092973 0.029312;11.740694
20.661862 0.029312 0.072305;0.092973 0.029312 0.000655
0.000111;0.029312 0.072305 0.000111 0.000576];
>>% input the cofactor matrix of measurements QL and the
measurements
>> QL=[91.6 42.7 0 0; 42.7 91.6 0 0; 0 0 0.0008 0.0001; 0 0
0.0001 0.0007];
>> Lk=[16145.292;25158.442;2.92214;2.00528];
>>% set transition matrix H initially to identity and input
the remaining elements
>> H=eye(4);
>> H(1,3)=60;H(2,4)=60;
>> % compute w vector for Qw matrix computation
```

Table 15.6 (Continued)

```
>> w=[60^2/2;60^2/2;60;60];
>> format long
>>% predict the next state vector xkpred and its cofactor
matrix Qxkpred
>> xkpred=H*xk1;
>> Qw=0.0002^2*w*w';
>> Qxkpred=H*Qxk1*H'+Qw;
>>% form the design matrix Ak and calculate Kalman Gain Kk
>> Ak=eye(4);
>> Kk=Qxkpred*Ak'*inv(Ak*Qxkpred*Ak'+QL);
>>% Calculate new filtered state vector xkfiltered and its
cofactor matrix Qxkfiltered
>> xkfiltered=xkpred+Kk*(Lk-Ak*xkpred)
xkfiltered =
    1.0e+04 *
    1.614509403119584
    2.515241461181566
    0.000292085052733
    0.000200927123248
>> Qxkfiltered=Qxkpred'-Kk*Ak*Qxkpred'
Qxkfiltered =
    23.068687973626911    9.073970079370358
    0.050290694777670     0.011280617681275
    9.073970079370357    18.206199133270403
    0.010911032988134     0.043514134462516
    0.050290694777670     0.010911032988134
    0.000355023352460     0.000090101680067
    0.011280617681275     0.043514134462516
    0.000090101680067     0.000323995822716
>> % use transition matrix H and current filtered state to
predict for the next state xk1pred
>> xk1pred=H*xkfiltered
xk1pred =
    1.0e+04 *
    1.632034506283553
    2.527297088576444
    0.000292085052733
    0.000200927123248
```

Example 15.4   A projectile was launched at a slope angle $\alpha = 35°$ with a velocity of 40.0 m s^{-1} (as shown in Figure 15.4). After the launch, the projectile accelerates under gravity (9.8 m s^{-2}) with a standard deviation of ± 0.01 m s^{-2}. The height (z) of the projectile was measured imprecisely as 17.5 m ± 0.5 m at a time interval $\Delta t = 1$ second after launch. If the state of the projectile was measured every one second, determine the (h, z) position and the velocity components of the projectile at the interval $\Delta t = 1$ second after launch using the Kalman filtering approach.

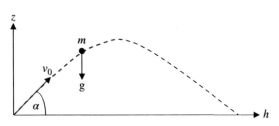

Figure 15.4 Trajectory of a projectile.

Solution:

From the knowledge of the dynamics of the projectile:

Force = mass × acceleration

Acceleration of projectile (a) = acceleration due to gravity (g)

In this problem the position and the velocity (the state variables) of the projectile are being tracked in the air. Considering this problem as a two-dimensional case, let (h_k, z_k) be the position of the object at step k and (v_{hk}, v_{zk}) the components of its speed in the coordinates directions at step k. The equations relating the different components can be given as follows. At the launch of the projectile, the velocities in the h (horizontal) and z (vertical) directions can be given as

$$v_{h0} = v_0 \cos\alpha \tag{15.85}$$

$$v_{z0} = v_0 \sin\alpha \tag{15.86}$$

where v_{h0} and v_{z0} are the components of velocity at the launch of the projectile, v_0 is the initial velocity of launch, and α is the angle above the horizontal with which the projectile is thrown. Given the filtered state vector at the current time $k-1$ as $\hat{x}_{k-1} = \begin{bmatrix} \hat{h}_{k-1} & \hat{z}_{k-1} & \hat{v}_{h_{k-1}} & \hat{v}_{z_{k-1}} \end{bmatrix}^T$ and the predicted state vector for time k as $\bar{x}_k = \begin{bmatrix} \bar{h}_k & \bar{z}_k & \bar{v}_{h_k} & \bar{v}_{z_k} \end{bmatrix}^T$, where the coordinates $(\hat{h}_{k-1}, \hat{z}_{k-1})$ represent the filtered position at current time $k-1$, $(\hat{v}_{h_{k-1}}, \hat{v}_{z_{k-1}})$ are the filtered velocity components at time $k-1$, $(\bar{h}_k, \bar{z}_k)$ represent the predicted position for time

k, and $(\bar{v}_{h_k}, \bar{v}_{z_k})$ are the predicted velocity components at time k, the following dynamic (prediction) model can be formulated. Based on the general laws of physics, the prediction equations can be formulated for the projectile as

$$\bar{h}_k = \hat{h}_{k-1} + \hat{v}_{h_{k-1}} \Delta t \tag{15.87}$$

$$\bar{z}_k = \hat{z}_{k-1} + \hat{v}_{z_{k-1}} \Delta t - \frac{1}{2} g \Delta t^2 \tag{15.88}$$

$$\bar{v}_{h_k} = \hat{v}_{h_{k-1}} \tag{15.89}$$

$$\bar{v}_{z_k} = \hat{v}_{z_{k-1}} - g \Delta t \tag{15.90}$$

where Equations (15.87) and (15.88) represent the model for the prediction of horizontal and vertical coordinates of the projectile, respectively, and Equations (15.89) and (15.90) are for the prediction of horizontal and vertical components of the velocity of the projectile. The transition matrix H can be determined as the Jacobian of Equations (15.87)–(15.90) with respect to the state vector $\hat{x}_{k-1} = \begin{bmatrix} \hat{h}_{k-1} & \hat{z}_{k-1} & \hat{v}_{h_{k-1}} & \hat{v}_{z_{k-1}} \end{bmatrix}^T$ as follows:

$$H = \begin{bmatrix} \dfrac{\partial \bar{h}_k}{\partial \hat{h}_{k-1}} & \dfrac{\partial \bar{h}_k}{\partial \hat{z}_{k-1}} & \dfrac{\partial \bar{h}_k}{\partial \hat{v}_{h_{k-1}}} & \dfrac{\partial \bar{h}_k}{\partial \hat{v}_{z_{k-1}}} \\[2mm] \dfrac{\partial \bar{z}_k}{\partial \hat{h}_{k-1}} & \dfrac{\partial \bar{z}_k}{\partial \hat{z}_{k-1}} & \dfrac{\partial \bar{z}_k}{\partial \hat{v}_{h_{k-1}}} & \dfrac{\partial \bar{z}_k}{\partial \hat{v}_{z_{k-1}}} \\[2mm] \dfrac{\partial \bar{v}_{h_k}}{\partial \hat{h}_{k-1}} & \dfrac{\partial \bar{v}_{h_k}}{\partial \hat{z}_{k-1}} & \dfrac{\partial \bar{v}_{h_k}}{\partial \hat{v}_{h_{k-1}}} & \dfrac{\partial \bar{v}_{h_k}}{\partial \hat{v}_{z_{k-1}}} \\[2mm] \dfrac{\partial \bar{v}_{z_{k-1}}}{\partial \hat{h}_{k-1}} & \dfrac{\partial \bar{v}_{z_{k-1}}}{\partial \hat{z}_{k-1}} & \dfrac{\partial \bar{v}_{z_{k-1}}}{\partial \hat{v}_{h_{k-1}}} & \dfrac{\partial \bar{v}_{z_{k-1}}}{\partial \hat{v}_{z_{k-1}}} \end{bmatrix} \quad \text{or} \quad H = \begin{bmatrix} 1 & 0 & \Delta t & 0 \\ 0 & 1 & 0 & \Delta t \\ 0 & 0 & 1 & 0 \\ 0 & 0 & 0 & 1 \end{bmatrix} \tag{15.91}$$

Equations (15.87)–(15.90) can be written in matrix form as follows:

$$\hat{x}_k = H\hat{x}_{k-1} + B_{k-1} u_{k-1} \tag{15.92}$$

where

$$B_{k-1} = \begin{bmatrix} 0.0 & -\dfrac{\Delta t^2}{2} & 0.0 & -\Delta t \end{bmatrix}^T ; \; u_{k-1} = g \tag{15.93}$$

Given the time step as $\Delta t = 1$ seconds; initial velocity, $v_0 = 40 \text{ m s}^{-1}$; slope angle, $\alpha = 35°$; acceleration, $g = 9.8 \text{ m s}^{-2}$; standard deviation of acceleration, $\sigma_g = \pm 0.01 \text{ m s}^{-2}$; and measured height about the ground, $z_k = 17.5 \pm 0.5 \text{ m}$, the velocity components from Equations (15.85) to (15.86) are

$$v_{h0} = 40 \cos 35° \rightarrow 32.8 \text{ m s}^{-1}$$
$$v_{z0} = 40 \sin 35° \rightarrow 22.9 \text{ m s}^{-1}$$

From Equations (15.87) to (15.90), the following prediction model (similar to Equation (15.92)) is formed:

$$\bar{x}_k = \begin{bmatrix} 1 & 0 & 1 & 0 \\ 0 & 1 & 0 & 1 \\ 0 & 0 & 1 & 0 \\ 0 & 0 & 0 & 1 \end{bmatrix} \hat{x}_{k-1} + \begin{bmatrix} 0.0 \\ -0.5 \\ 0.0 \\ -1.0 \end{bmatrix} g_{k-1} \tag{15.94}$$

where acceleration $u_{k-1} = g_{k-1} = 9.8 \text{ m s}^{-2}$, $B_{k-1} = [0.0 \ -0.5 \ 0.0 \ -1.0]^T$, and w_{k-1} is the model noise with the mean value of zero. The height (z) of the projectile was only measured so that the measurement model will reflect only this measurement as follows:

$$\hat{\ell}_{z_k} = \hat{z}_k \tag{15.95}$$

The design matrix A_k can be derived as the Jacobian of Equation (15.95) with respect to the filtered state vector $\hat{x}_k = \begin{bmatrix} \hat{h}_k & \hat{z}_k & \hat{v}_{h_k} & \hat{v}_{z_k} \end{bmatrix}^T$ at time k as

$$A_k = \begin{bmatrix} \dfrac{\partial \hat{\ell}_{z_k}}{\partial \hat{h}_k} & \dfrac{\partial \hat{\ell}_{z_k}}{\partial \hat{z}_k} & \dfrac{\partial \hat{\ell}_{z_k}}{\partial \hat{v}_{h_k}} & \dfrac{\partial \hat{\ell}_{z_k}}{\partial \hat{v}_{z_k}} \end{bmatrix} \quad \text{or} \quad A_k = [0 \ 1 \ 0 \ 0] \tag{15.96}$$

Since the initial state vector is known with perfect precision, the filter can be initialized (with the first two elements for position and the last two elements for the corresponding velocity components) as

$$\hat{x}_{k-1} = \begin{bmatrix} 0.0 \\ 0.0 \\ 32.8 \\ 22.9 \end{bmatrix}$$

It can also be specified to the filter that the exact position is known by giving the position a zero covariance matrix as follows:

$$Q_{\hat{x}_{k-1}} = \begin{bmatrix} 0 & 0 & 0 & 0 \\ 0 & 0 & 0 & 0 \\ 0 & 0 & 0 & 0 \\ 0 & 0 & 0 & 0 \end{bmatrix}$$

From Equation (15.94), the predicted state vectors are (substituting values into the equation and setting the noise (w_{k-1}) to zero)

$$\bar{x}_k = \begin{bmatrix} 32.77 \\ 22.94 \\ 32.77 \\ 22.94 \end{bmatrix} + \begin{bmatrix} 0.0 \\ -4.90 \\ 0.0 \\ -9.80 \end{bmatrix} \rightarrow \begin{bmatrix} 32.77 \\ 18.04 \\ 32.77 \\ 13.14 \end{bmatrix}$$

with the cofactor matrix of the predicted state vector as

$$Q_{\bar{x}_k} = H Q_{\hat{x}_{k-1}} H^T + Q_{w_{k-1}}$$

where $Q_{w_{k-1}}$ is the propagated cofactor matrix (for the system noisy control inputs) based on the laws of variance–covariance propagation. This cofactor can be propagated from Equation (15.32) as follows, where σ_g is the random error in the acceleration:

$$Q_{w_{k-1}} = B_{k-1} \left[\sigma_g^2 \right] B_{k-1}^T$$

or

$$Q_{w_{k-1}} = \sigma_g^2 \begin{bmatrix} 0 & 0 & 0 & 0 \\ 0 & \dfrac{\Delta t^4}{4} & 0 & \dfrac{\Delta t^3}{2} \\ 0 & 0 & 0 & 0 \\ 0 & \dfrac{\Delta t^3}{2} & 0 & \Delta t^2 \end{bmatrix} \quad a \quad (0.01)^2 \begin{bmatrix} 0 & 0 & 0 & 0 \\ 0 & \dfrac{1}{4} & 0 & \dfrac{1}{2} \\ 0 & 0 & 0 & 0 \\ 0 & \dfrac{1}{2} & 0 & 1 \end{bmatrix} \tag{15.97}$$

$$Q_{w_{k-1}} = \begin{bmatrix} 0 & 0 & 0 & 0 \\ 0 & 0.000\ 025 & 0 & 0.000\ 05 \\ 0 & 0 & 0 & 0 \\ 0 & 0.000\ 05 & 0 & 0.000\ 10 \end{bmatrix}$$

Given:

$$Q_{\hat{x}_{k-1}} = \begin{bmatrix} 0 & 0 & 0 & 0 \\ 0 & 0 & 0 & 0 \\ 0 & 0 & 0 & 0 \\ 0 & 0 & 0 & 0 \end{bmatrix} \quad H Q_{\hat{x}_{k-1}} H^T = \begin{bmatrix} 0 & 0 & 0 & 0 \\ 0 & 0 & 0 & 0 \\ 0 & 0 & 0 & 0 \\ 0 & 0 & 0 & 0 \end{bmatrix}$$

$$Q_{\bar{x}_k} = H Q_{\hat{x}_{k-1}} H^T + Q_{w_{k-1}} = \begin{bmatrix} 0 & 0 & 0 & 0 \\ 0 & 0.000\ 025 & 0 & 0.000\ 05 \\ 0 & 0 & 0 & 0 \\ 0 & 0.000\ 05 & 0 & 0.000\ 10 \end{bmatrix}$$

Given the covariance matrix of measurement (for single z-coordinate measurement) as follows:

$$Q_{\ell_k} = \left[\sigma_z^2 \right] \quad \text{or} \quad Q_{\ell_k} = \left[(0.5)^2 \right]$$

The Kalman gain is computed as follows:

$$K_k = Q_{\bar{x}_k} A_k^T \left(A_k Q_{\bar{x}_k} A_k^T + Q_{\ell_k}\right)^{-1} \quad \text{with} \quad A_k = \begin{bmatrix} 0 & 1 & 0 & 0 \end{bmatrix}$$

$$K_k = \begin{bmatrix} 0.0 \\ 0.0001 \\ 0.0 \\ 0.0002 \end{bmatrix}$$

The filtered state vector is computed as follows:

$$\hat{x}_k = \bar{x}_k + K_k(\ell_k - A_k \bar{x}_k)$$

$$\hat{x}_k = \begin{bmatrix} 32.77 \\ 18.04 \\ 32.77 \\ 13.14 \end{bmatrix} + \begin{bmatrix} 0.0 \\ 0.0001 \\ 0.0 \\ 0.0002 \end{bmatrix} (17.5 - 18.043)$$

$$\hat{x}_k = \begin{bmatrix} 32.77 \\ 18.04 \\ 32.77 \\ 13.14 \end{bmatrix}$$

The propagated covariance matrix of the filtered state vector is computed as

$$Q_{\hat{x}_k} = (I - K_k A_k) Q_{\bar{x}_k}^T$$

$$Q_{\hat{x}_k} = \begin{bmatrix} 0 & 0 & 0 & 0 \\ 0 & 0.000\,025 & 0 & 0.000\,05 \\ 0 & 0 & 0 & 0 \\ 0 & 0.000\,05 & 0 & 0.000\,10 \end{bmatrix}$$

If the height of the projectile for the next time step is given as $z_{k+1} = 25.0 \pm 0.5$ m, and using the filtered state values above, the next filtered state values can be calculated as follows. The predicted state vector and its covariance matrix are given as

$$\bar{x}_{k+1} = \begin{bmatrix} 65.60 \\ 31.10 \\ 32.80 \\ 13.10 \end{bmatrix} + \begin{bmatrix} 0.0 \\ -4.90 \\ 0.0 \\ -9.80 \end{bmatrix} \rightarrow \begin{bmatrix} 65.60 \\ 26.20 \\ 32.80 \\ 3.30 \end{bmatrix}$$

$$Q_{\bar{x}_{k+1}} = HQ_{\hat{x}_k}H^T + Q_{w_k} = \begin{bmatrix} 0 & 0 & 0 & 0 \\ 0 & 0.000\ 025 & 0 & 0.000\ 20 \\ 0 & 0 & 0 & 0 \\ 0 & 0.000\ 20 & 0 & 0.000\ 20 \end{bmatrix}$$

The computed Kalman gain:

$$K_{k+1} = \begin{bmatrix} 0.0 \\ 0.0010 \\ 0.0 \\ 0.0008 \end{bmatrix}$$

The state vector and its covariance matrix are given, respectively, as follows:

$$\hat{x}_{k+1} = \begin{bmatrix} 65.60 \\ 26.20 \\ 32.80 \\ 3.30 \end{bmatrix}$$

$$Q_{\hat{x}_{k+1}} = \begin{bmatrix} 0 & 0 & 0 & 0 \\ 0 & 0.000\ 025 & 0 & 0.000\ 20 \\ 0 & 0 & 0 & 0 \\ 0 & 0.000\ 20 & 0 & 0.000\ 20 \end{bmatrix}$$

15.5 Kalman Filter and the Least Squares Method

Filtering is a form of least squares adjustment if the prediction step in filtering is considered as a step just for providing fairly good a priori estimates of the unknown parameters. In this case, a filter will be seen as the most general form of the adjustment process in which filtered parameter estimates are the weighted combinations of predicted quantities and the measurement data. Specifically, filtering will be reduced to sequential least squares adjustment if weights are attributed only to measurements (but not to predicted quantities) and the state vectors (i.e. the unknown parameters) are not dependent on time. In filtering, however, the number of observations can be less than the number of unknown parameters, and an optimal result will still be obtained; in order to solve this type of problem using least squares parameter estimation, special precautions are required to reduce the number of parameters to be determined or to constrain their solution (Blais 2010). A filter satisfies two important criteria: It determines the state vectors so that their expected average values are to be equal to their true values, and it provides values with the smallest possible error variances. The classical least squares methods only guarantee estimated values with the smallest possible error variances; the least squares adjusted parameters may not be equal to their true values.

15.5.1 Filtering and Sequential Least Squares Adjustment: Similarities and Differences

The following simplification can be made to show the main similarities between the sequential least squares and the mathematical filtering:

a) Approximate estimates of the current state vector are determined only from the past data in both cases.
b) Measurements involved in both cases relate to the state at the current time of the measurements.
c) Approximate estimates of the current state vector are updated in both cases.

The main difference between the sequential least squares adjustment and the mathematical filtering can simply be stated that state vectors in sequential are static, while those in filtering are dynamic. It can be stated generally that Kalman filter is powerful in the estimation of a "network in motion" since it uses a dynamic model in addition to the measurement model, unlike the least squares method that uses the measurement model only. From Section 15.3, the following Kalman filter equations are recalled:

Filtered state vector at time t_k :

$$\hat{x}_k = \bar{x}_k + K_k(\ell_k - A_k\bar{x}_k) \tag{15.98}$$

Cofactor of filtered state vector at t_k :

$$Q_{\hat{x}_k} = (I - K_kA_k)Q_{\bar{x}_k}^T \tag{15.99}$$

Kalman gain matrix:

$$K_k = Q_{\bar{x}_k}A_k^T\left(A_kQ_{\bar{x}_k}A_k^T + Q_\ell\right)^{-1} \tag{15.100}$$

From Sections 10.5 and 15.5.1, the following sequential least squares adjustment equations are recalled:

Vector of updated corrections to parameters : $\delta = \delta' + K\left(\ell_2 - \bar{\ell}_2\right)$

$$\tag{15.101}$$

with $\bar{\ell}_2$ as a vector of "predicted observations" and ℓ_2 a vector of new observations. For a linear model, one can set $f_2(x^0)$ to zero in $\bar{\ell}_2 = A_2\delta' + f_2(x^0)$ so that $\bar{\ell}_2 = A_2\delta'$ and the adjusted quantities will become the adjusted parameters (with $x^0 = 0$). In this case, Equation (15.101) will become

Vector of updated parameters : $\delta = \delta' + K(\ell_2 - A_2\delta')$ $\tag{15.102}$

Updated cofactor matrix of updated parameters: $Q_\delta = Q_{\delta'} - KA_2Q_{\delta'}$

$$(15.103)$$

or

Updated cofactor matrix of updated parameters: $Q_\delta = (I - KA_2)Q_{\delta'}$

$$(15.104)$$

"Kalman gain" type:

$$K = N_1^{-1}A_2^T M \qquad (15.105)$$

From Equation (10.93), it is defined that

$$M = \left(A_2 N_1^{-1}A_2^T + P_2^{-1}\right)^{-1} \qquad (15.106)$$

Note also that $P_2^{-1} = Q_{\ell_2}$ is the cofactor matrix of the new set of observations for sequential adjustment. Substituting Equation (15.106) into Equation (15.105) gives the following for sequential least squares adjustment: "Kalman gain" type:

$$K = N_1^{-1}A_2^T\left(A_2 N_1^{-1}A_2^T + Q_{\ell_2}\right)^{-1} \qquad (15.107)$$

With regard to filtering, it can be seen from Equations (15.98) to (15.100) that the formulations involve only the cofactor matrices, while in the case of sequential least squares adjustment, it can be seen from Equations (15.101) to (15.106) that the inverses of the N_1 and P_2 matrices are involved. In this case, the inverses of N_1 and P_2 matrices must exist.

The summary of Kalman filtering equations and sequential least squares equations is given in Table 15.7.

Table 15.7 Similarity between Kalman filter equations and sequential adjustment equations.

Filter	Sequential
Filtered state vector equation (15.98): $\hat{x}_k = \bar{x}_k + K_k(\ell_k - A_k\bar{x}_k)$	Updated parameter equation (15.102): $\delta = \delta' + K(\ell_2 - A_2\delta')$
$\hat{x}_k$: filtered state vector	δ: updated parameters
$\bar{x}_k$: predicted state vector	δ': adjusted parameters from the first set of observations (ℓ_1)
$K_k = Q_{\bar{x}_k}A_k^T\left(A_k Q_{\bar{x}_k}A_k^T + Q_\ell\right)^{-1}$ (Equation (15.100))	$K = N_1^{-1}A_2^T\left(A_2 N_1^{-1}A_2^T + Q_{\ell_2}\right)^{-1}$ (Equation (15.107))
ℓ_k: new measurements at current time for the current new state	ℓ_2: new measurements at current time for the same old state
A_k: first design matrix based on new observation model	A_2: first design matrix based on new observation model
$Q_{\hat{x}_k} = (I - K_kA_k)Q_{\bar{x}_k}^T$ (Equation (15.99))	$Q_\delta = (I - KA_2)Q_{\delta'}$ (Equation (15.104))

Problems

15.1 A mapping aircraft is moving due east. The position and velocity (due east) at time $t_0 = 0$ second are easting $= 3\,878\,948.9$ m and velocity eastward $= 100$ m s^{-1}. Some of the matrices for the filtering are given as follows:

$$Q_{\hat{x}_{k-1}} = \begin{bmatrix} 0.16 & 0.10 \\ 0.10 & 0.25 \end{bmatrix}; \quad H = \begin{bmatrix} 1 & \Delta t \\ 0 & 1 \end{bmatrix}; \quad Q_w = \sigma_a^2 \begin{bmatrix} \dfrac{\Delta t^2}{2} \\ \Delta t \end{bmatrix} \begin{bmatrix} \dfrac{\Delta t^2}{2} & \Delta t \end{bmatrix}$$

Using constant velocity model, the standard deviation of the aircraft's acceleration is 0.015 m s^{-2}. The position from GNSS at $t_1 = 60$ seconds is $3\,884\,945.5$ with standard deviation of 0.5 m; at the same time $t_1 = 60$ seconds, the inertial measurement units (IMU) onboard the aircraft shows the velocity of the aircraft eastward as 98.5 m s^{-1} with standard deviation of 0.05 m s^{-1}. Answer the following.

a) Give the initial state vector for this problem.
b) Give the measurement vector at t_1 for this problem.
c) Give the measurement equations (in parametric equation form), the first design matrix, and the covariance matrix of the measurements.
d) Determine the predicted state vector at t_1 and its propagated covariance matrix for this problem.
e) Compute the gain matrix.
f) Compute the aircraft filtered state vector and the covariance matrix at t_1.
g) Compute the aircraft predicted state vector and its covariance matrix at $t_2 = 120$ seconds (next interval).

15.2 A vehicle with a radio antenna mounted on its roof is traveling along a perfectly frictionless long straight road with a constant acceleration of 0.3 m s^{-2} $\pm$ 0.06 m s^{-2}, and its current position is measured with a radio ranging system as 61.0 m $\pm$ 3 m. If the state of the vehicle is measured at every 10-second interval, what are the optimal values of the position and velocity of the vehicle at the kth recursion? Assume the values of the state vector $\hat{x}_{k-1} = [\hat{p}_{k-1} \ \ \hat{v}_{k-1}]^T$ (where $\hat{p}_{k-1}$ and $\hat{v}_{k-1}$ are the filtered position and the velocity component at time $k - 1$, respectively) and its covariance matrix $(Q_{\hat{x}_{k-1}})$ from the $(k - 1)$th recursion are given as follows:

$$\hat{x}_{k-1} = \begin{bmatrix} 16.7 \\ 3.3 \end{bmatrix} \begin{matrix} m \\ m\,s^{-1} \end{matrix}; \quad Q_{\hat{x}_{k-1}} = \begin{bmatrix} 4.50 & 0.90 \\ 0.90 & 0.18 \end{bmatrix}$$

15.3 You have been given the following information for a vehicle traveling along a road. The position and velocity at an arbitrary starting time

$t_0 = 0$ second are easting $= 3\,878\,948.9$ m, northing $= 5\,860\,516.3$ m, east velocity $= -4.28$ m s^{-1}, and north velocity $= -5.41$ m s^{-1}. Some of the matrices for the filtering are given as follows:

$$Q_{\tilde{x}_{k-1}} = \begin{bmatrix} 2.25 & 0 & 0 & 0 \\ 0 & 2.25 & 0 & 0 \\ 0 & 0 & 0.0001 & 0 \\ 0 & 0 & 0 & 0.0001 \end{bmatrix}; \; H = \begin{bmatrix} 1 & 0 & \Delta t & 0 \\ 0 & 1 & 0 & \Delta t \\ 0 & 0 & 1 & 0 \\ 0 & 0 & 0 & 1 \end{bmatrix};$$

$$Q_w = \sigma_a^2 \begin{bmatrix} \dfrac{\Delta t^2}{2} \\[2mm] \dfrac{\Delta t^2}{2} \\[2mm] \Delta t \\[2mm] \Delta t \end{bmatrix} \begin{bmatrix} \dfrac{\Delta t^2}{2} & \dfrac{\Delta t^2}{2} & \Delta t & \Delta t \end{bmatrix}$$

Using a constant velocity model, the standard deviation of the vehicle's acceleration is 0.015 m s^{-2}. The position and velocity from GNSS at $t_1 = 0.6$ seconds are easting $= 3\,878\,947.1$ m, northing $= 5\,860\,513.7$ m, east velocity $= -4.22$ m s^{-1}, and north velocity $= -5.39$m s^{-2}. The standard deviations of the GNSS measurements are position 1.2 m (easting and northing) and velocity 0.03m s^{-2} (east and north). Compute the vehicle's filtered state vector and the covariance matrix at t_1 and the predicted state vector at $t_2 = 1.2$ seconds.

16

Introduction to Least Squares Collocation and the Kriging Methods

---

CHAPTER MENU

OBJECTIVES

After studying this chapter, you should be able to:

1) Explain the theories of least squares collocation and kriging.
2) Discuss the steps involved in least squares collocation and kriging techniques.
3) Explain differences and similarities between least squares collocation and kriging.

16.1   Introduction

Least squares collocation and kriging are unbiased minimum variance interpolation methods. The least squares collocation was originally used in geodesy by Krarup (1969) and Moritz (1972) as a statistical tool for interpolating gravity anomalies. Today, some of the applications of the tool include digitizing terrain models, modeling camera distortions and film deformations in photogrammetry, studying land subsidence by photographic approach, coordinate transformations in two or three dimensions, modeling of measurements to artificial

Understanding Least Squares Estimation and Geomatics Data Analysis, First Edition.
John Olusegun Ogundare.
© 2019 John Wiley & Sons, Inc. Published 2019 by John Wiley & Sons, Inc.
Companion website: www.wiley.com/go/ogundare/Understanding-lse-and-gda

satellites, modeling of graduation errors of theodolite circles, etc. In general, the tool is widely applied in many fields to solve for some unknown parameters, observation residuals, observation noise, and signal (pattern) in observations. Kriging is a group of geospatial interpolation methods named (Matheron 1963) after Dr. Daniel G. Krige of South Africa, who is widely credited with developing the original theory of kriging, for mining purposes. The various methods of kriging are now widely applied in many disciplines where measurements are assumed to be spatially correlated with distance or direction, such as in geology, hydrology, remote sensing, geophysics, soil science, landslide monitoring, geographic information system (GIS), etc. In general, kriging is applied in a surface where limited measured points do not allow for an exact description (or continuous surface modeling) of the terrain surface.

Some of the commonly used terms in relation to collocation and kriging techniques are *signal, white noise, random process, data series, least squares adjustment, prediction, interpolation* (or *extrapolation*), *filtering, and stationarity.* A *signal* is a physical quantity, which represents some information that one is interested in; it may vary with time, space, or any other independent variable by which the information is transmitted. *White noise,* in comparison with white light, is an unwanted meaningless random quantity that has normal statistical distribution with zero mean and statistically uncorrelated with each other or other quantities; it is a variation in the value of a random process that cannot be described mathematically. A *random process* is a set of random observables, e.g. the position of image points on an aerial photograph. A *random observable* is a physical quantity, which varies and may assume different values with specified probabilities when measured repeatedly under the similar conditions; a random (or stochastic) process considers all values of the specified random observables. Random processes are specified by their *moments*, especially the first and second moments of their probability distributions. The *first moment* is the vector of their mean values or of the expectation of the random process, and the second is their covariance matrix or the expectation of the random process squared (assuming the vector of their mean value is zero). An *expectation* of a random process X, represented by $E(X)$, is the sum of all possible values of the random process, weighted by their probability values. In probabilistic sense, $E(X)$ is precisely a value that an average of a number of observations of the random process tends to as the number of the observations increases.

Under an assumption of constant measurement conditions, a random process will be called a *stationary random process* to indicate that the statistical properties (mean of the process and its covariance matrix) are not dependent on time or time shift. A situation where the values of a random process can vary without affecting the mean values and the associated covariance matrix is referred to as *stationarity.* In this case, the mean of the random process is constant, and the covariance matrix is only dependent on the increments. According to Mikhail (1976), *homogeneity* and *isotropy* are used together to mean the same thing as stationarity when relating to a *random field* or a *random vector field*. A *random field* consists of one random process, which is dependent on a vector of random observables, while a *random vector field* consists of several random processes, each of which is dependent on a vector of random observables. A random vector field or a random field is homogeneous if the vector of its mean values is the same everywhere in the field and its covariance matrix is not dependent on absolute point in space or time. If, at every point in the field, these properties are the same in all directions, the field is also considered to be isotropic; in this case, the covariance matrix will be dependent only on the vector of separations between points. Note that homogeneity and isotropy are independent from each other; homogeneity is about the material property (such as composition) of the field at a point, while isotropy is about the physical property (such as thermal, density, stress–strain relationship, etc.) along any direction from a given point. For example, an electric field is homogeneous because it is uniform throughout the field, but it is not isotropic since the field is based only on one particular direction; and a radiation field having the same intensity in all directions is isotropic based on its intensity property.

A set of repeated observations of the same observable over time or a set of observations distributed in space (associated with coordinates) is referred to as a *data series*. An investigation of a data series, which may include representing the trend surface by linear or low-order polynomial terms, is known as *trend analysis*. A statistical estimation of parameters and measurement residuals in which the mathematical model for the measurements is known a priori and all the measurements are used at once is known as the *least squares adjustment*; when estimation is made at computation points where no measurements are made, *least squares prediction, least squares interpolation,* or *least squares extrapolation* is used. In time-dependent prediction (e.g. Kalman filtering), estimation is made for the future using all measured data up to and including the present. In interpolation and extrapolation, measurements at the measured points are used to determine values for points where measurements are not provided; if, in determining these interpolated values, the measuring errors at the measured points are minimized, the procedure is referred to as *least squares filtering*, where the measurements of the present are allowed to affect the solution of the past thereby allowing for the screening (filtering) of the measurements of the present as they are added. In *prediction*, measurements relate to the past, while estimated (or predicted) points relate to the future. Trend removal from measurements is usually considered as prerequisite to applying interpolation and filtering to measurements.

It should also be mentioned that there are other techniques of studying time or space series that are not discussed in this book, such as autoregression (AR), moving average (MA), mixed autoregressive moving average (ARMA), radial basis function (RBF), etc. Interested readers can refer Gelb (1974) and Micchelli (1986).

16.2 Elements of Least Squares Collocation

Least squares collocation is a procedure that applies the concepts of classical least squares adjustment (for trend surface determination) with interpolation and filtering techniques. Its main objective is to predict some signal for where measurements have not been made, the secondary objective is to filter out noise (if it is present) from observations, and the third objective is to estimate unknown parameters if present. In this case, the collocation equation can be given in form of linear parametric model involving signal components, as (cf. Moritz 1972)

$$\ell = A\delta + s + n \tag{16.1}$$

where

ℓ is a vector of observations

n is the vector of noise components in the observations (or negative residual vector)

s is a vector of signal components (or defects or additional corrections in the mathematical model) at the observation or data points or the signal components corresponding to the given observations

δ is a vector of unknown parameters

A is the first design matrix, which is the Jacobian of the parametric model with respect to the unknown parameters

$A\delta$ is a vector of the systematic component of the mathematical model, which originally will often be nonlinear but linearized by Taylor series expansion.

From Equation (16.1), a vector of predicted signals ($\bar{s}$) in space, which is different from the data points, can be determined. The signal components (s) represent the systematic effects in the observations, while the noise components (n) are the observation errors, which are free of any systematic biases or systematic effects. If the systematic effects have not been completely removed from measurements at the preprocessing stage, it can be removed at the adjustment stage as signal s. It is also assumed that only the signal s can be predicted as $\bar{s}$ elsewhere, unlike the parameters (δ), which are usually attached to observation points. The signal components are associated with external factors, while the noise is internal to the measuring instrument. For example, in geomatics, geoid undulation, gravity anomalies, and the deflections of the vertical can each be considered as signals. It is also assumed that the signals have known covariance matrix and their values are unknown to be estimated. According to Krakiwsky (1975), the standard deviations of noise and signal components can be illustrated using the standard deviation formula for electromagnetic distance measurement (EDM) of a distance D, expressed as

$$\sigma_s = a + bD \tag{16.2}$$

where constant a can be likened to the standard deviation of the noise and bS as the standard deviation of the signal component. From Equation (16.1), the observations (ℓ) with their uncorrelated covariance matrix (C_{nn}) provide the estimated values of the unknown parameters (δ) in the functional part, while the covariance function provides information for the signal s in the prediction part with the noise component as white noise.

Generally, the mathematical expectations (or the universal averages) of both the signal and noise components are considered as zero. Since the signal components are small, they can be likened to a vector of observation errors in classical least squares adjustment; if these signal components, however, are insignificant, the noise components n can be considered as the vector of observation errors. The classical least squares method uses the variances of the noise and signal components and ignores the covariances in the signal components. The collocation method tends to account for the correlation among the signal components through a fully populated covariance matrix for the signal; it is able to extract the parameters δ and both the signal s and the noise n components of the observations unlike the classical least squares method that deals only with the parameters δ and the noise (n) components. If parameter (δ) estimation is considered as adjustment, noise (n) removal as *filtering*, and signal (s) computation at points other than the observation points as *prediction*, then the least squares collocation can be considered as a combined adjustment, filtering, and prediction method.

16.3 Collocation Procedure

The collocation procedure uses standard errors of measurements, makes assumptions about the magnitude and correlation of signal quantities, and assumes noise is correlation-free so that noise covariance between measurements is zero. Assuming the mean value of observations is zero (i.e. $A\delta$ component is a zero vector) and the expected values of the signal and noise components are zero vectors with the signal and noise components uncorrelated, the observations can be modeled from Equation (16.1) as

$$\ell = s + n \tag{16.3}$$

From Equation (16.3) the following collocation solution vectors can be derived from the classical least squares solution:

$$\bar{s}_\mathrm{p} = C_{s_\mathrm{p}s} C_\ell^{-1} \ell \tag{16.4}$$

$$\hat{s} = C_{ss} C_\ell^{-1} \ell \tag{16.5}$$

$$\hat{n} = C_{nn} C_\ell^{-1} \ell \tag{16.6}$$

$$C_\ell = C_{ss} + C_{nn} \tag{16.7}$$

where

$\hat{s}$ is the vector of estimated signals at the observation points
$C_{s_p s}$ is the covariance matrix for the correlation between the points where the predicted signals $(\bar{s}_p)$ are made and the observation points
C_ℓ is the covariance matrix of the observations
C_{ss} is the covariance matrix between the signal components (s) at the observation points
C_{nn} is the covariance matrix between the noise components (n) at the observation points.

From Equation (16.7), it can be seen that C_ℓ is essentially the sum of the covariance matrix of the signal components and that of the noise components (assuming no correlation between noise and signal components). The collocation estimate $(\bar{s}_p)$ of the signal at the prediction point can be given from Equation (16.4) as

$$\bar{s}_p = W^T \ell \tag{16.8}$$

or

$$\bar{s}_p = C_{s_p s}(C_{ss} + C_{nn})^{-1}\ell \tag{16.9}$$

where W is a matrix of combination weights (or a vector of combination weights if only one signal is involved). If two observations have the same noise variance and the same distance to the prediction point, the weights associated with these observations will be the same. The elements of the covariance matrices $C_{s_p s}$ and C_{ss} are computed using some covariance functions, which are the backbone of collocation. Since signals $\bar{s}_p$ at the prediction points and signals s at the observation points have identical statistical properties, their covariances are determined by the same covariance functions. Least squares collocation is fundamentally dependent on the type of covariance function used.

16.4 Covariance Function

A random process is homogeneous if its mean is constant (or zero) and its covariance function is not dependent on absolute locations or absolute time, but rather on only the length and direction between different locations or time intervals. Covariance functions are assumed in least squares collocation to be *homogeneous* and *isotropic*. A covariance function will be considered both homogeneous and isotropic if its covariance matrix is dependent only on the length of the separation between locations or the time interval.

Covariance function in collocation can be determined empirically from observations or using some analytical techniques. If sufficiently large set of signal values that are representative enough are available, their mean square value can be taken as the signal variance with all the signals assigned to this variance. A type of covariance function for a homogeneous and isotropic random process given some space distance (or time interval) r between points is commonly used in modeling signal covariance. A typical covariance function ($C(r_{ij})$) as a function of distance (or time interval) r_{ij} between any two data points (x_i, y_i) and (x_j, y_j) (in two-dimensional space, for example) is formulated as

$$C(r_{ij}) = C_0 e^{-a^2 r_{ij}^2} \qquad (16.10)$$

where C_0 and a are constants that are known or determined. Equation (16.10) is based on the assumption of stationarity of the random process, which dictates that the covariance function $C(r_{ij})$ only depends on the separation r_{ij} between the two points. The principal properties of the covariance function ($C(r_{ij})$) are (Blais 1984):

1) It has values for all nonnegative values of r_{ij}.
2) It tends to zero value as r_{ij} value tends to infinity.
3) The value of $C(0)$ is the mean square value of the random process and is maximum in magnitude.
4) Least squares collocation uses ordinary covariance function as its covariance model, while kriging methods use generalized covariance functions as the one given in Equation (16.10).

The covariance function is also referred to as *covariogram* in kriging methods (Cressie 1993). Assuming zero mean process and a homogeneous and isotropic conditions, an experimental covariance function with an exponential model as in Equation (16.10) can be estimated, for example, for a case involving two-dimensional (x, y) measured points (with assumed distances, r_{ij} = 0, 0.25, 0.75, 1.0) as follows (cf. Molteni et al. 2009):

1) Represent the observation points graphically by plotting their (x, y) coordinates; if the mean of the process is unknown, it must first be estimated, and the observations reduced appropriately before using them to determine the covariance function.
2) Compute the $N \times N$ matrix (D) of the distances (r_{ij}) between each pair of measured points (x_i, y_i) and (x_j, y_j) with all the diagonal elements of the matrix as zero, where

$$r_{ij}^2 = (x_i - x_j)^2 + (y_i - y_j)^2 \qquad (16.11)$$

and N is the number of measured points. A typical D matrix can be given as

$$D = \begin{bmatrix} r_{11} & r_{12} & \cdots & r_{1N} \\ r_{21} & r_{22} & \cdots & r_{2N} \\ \vdots & & \ddots & \vdots \\ r_{N1} & \cdots & & r_{NN} \end{bmatrix} \qquad (16.12)$$

3) Compute the experimental covariance $C_{ex}(\tau)$ as

$$C_{ex}(0) = \frac{1}{N} \sum_{i}^{N} Z_{0i}^2 \quad \text{for all } N \text{ data points (having } r_{ii} = 0) \qquad (16.13)$$

$$C_{ex}(0.25) = \frac{1}{N_1} \sum_{i} \sum_{j} Z_{0i} Z_{0j} \quad \text{for } 0 < r_{ij} \le 0.5 \qquad (16.14)$$

$$C_{ex}(0.75) = \frac{1}{N_2} \sum_{i} \sum_{j} Z_{0i} Z_{0j} \quad \text{for } 0.5 < r_{ij} \le 1.0 \qquad (16.15)$$

where τ is the midpoint of the given distance intervals, Z_{0i} is the measured value at point (x_i, y_i), and N_k ($k = 0, 1, 2$) is the number of distances (r_{ij}) falling between the given distance intervals with (i, j) corresponding with the (row, column) location of the distance r_{ij} in the distance matrix D.

4) Use the experimental covariances computed in step 3 (except the one at the origin) to determine the covariance function parameters $\hat{C}_0$ and $\hat{a}$ in Equation (16.10) using the least squares adjustment method (a number of iterations may be required before a convergence to final solution is obtained).

5) The variance ($\hat{\sigma}_n^2$) of the white noise for the process is computed as

$$\hat{\sigma}_n^2 = C_{ex}(0) - \hat{C}_0 \qquad (16.16)$$

After determining the covariance function of the signal and the variance of the white noise, they can be used to compute the covariance for any two points separated by a distance r, the signal covariance matrix (C_{ss}) in the observation points and the noise covariance matrix (C_{nn}) in the observation points. The C_{nn} matrix is diagonal since there are no correlations between the white noise components. For example, the estimated covariance function from Equation (16.10) can be used to determine the covariance matrix of the signal components (C_{ss}) between the measured points as follows:

$$C_{ss} = \begin{bmatrix} \hat{C}(r_{11}) & \hat{C}(r_{12}) & \cdots & \hat{C}(r_{1N}) \\ \hat{C}(r_{21}) & \hat{C}(r_{22}) & \cdots & \hat{C}(r_{2N}) \\ \vdots & & \ddots & \vdots \\ \hat{C}(r_{N1}) & & \cdots & \hat{C}(r_{NN}) \end{bmatrix} \tag{16.17}$$

The covariance matrix ($C_{s_p s}$) between the prediction point (x_p, y_p) and the observation points (x_j, y_j), for N observation points, can be given as

$$C_{s_p s} = \begin{bmatrix} \hat{C}(r_{p1}) & \hat{C}(r_{p2}) & \cdots & \hat{C}(r_{pN}) \end{bmatrix} \tag{16.18}$$

For q number of computation points, the size of matrix $C_{s_p s}$ will be $q \times N$. The covariance matrix of noise component (C_{nn}) is a diagonal matrix, which can be expressed as

$$C_{nn} = \begin{bmatrix} \hat{\sigma}_n^2 & & & \\ & \hat{\sigma}_n^2 & & \\ & & \ddots & \\ & & & \hat{\sigma}_n^2 \end{bmatrix} = \hat{\sigma}_n^2 I \tag{16.19}$$

The variance of the signal estimation error at a prediction point (x_p, y_p) (assuming there is no systematic effect due to $A\delta$ in Equation (16.1)) can be given as

$$\sigma_{\bar{s}_p}^2 = \hat{C}(r_{pp}) - C_{s_p s}(C_{ss} + C_{nn})^{-1} C_{s_p s}^T \tag{16.20}$$

In two-dimensional coordinate transformations, for example, there should be separate covariance functions for x- and y-axes and another for cross-covariance between the axes. For simplicity, the same covariance function can be used for x- and y-axes with the correlation between the two axes ignored.

16.5 Collocation and Classical Least Squares Adjustment

Considering the collocation Equation (16.1), if $A\delta$ component is not a zero vector, indicating that the mean $m(x)$ of the process as a function of locations (x) is not zero, the following trend equation can be used to determine the mean:

$$m(x) = G(x)^T \delta \tag{16.21}$$

where $G(x)$ is a vector of some selected base functions and δ is a vector of unknown coefficients. By fitting the equation to the given measurements (ℓ), the unknown vector δ can be solved for by the method of least squares with the linearized version of Equation (16.21) given as

$$\ell = A\delta + n \tag{16.22}$$

where A is the first design matrix with respect to the unknown vector of parameters δ and n is a vector of measurement errors. The least squares solution of Equation (16.22) can be given as

$$\hat{\delta} = \left(A^T C_\ell^{-1} A\right)^{-1} A^T C_\ell^{-1} \ell \tag{16.23}$$

The estimated vector of parameters $\hat{\delta}$ is used in Equation (16.21) to determine the unknown mean value ($\hat{m}$) of the process, and the trend is removed from the measurement vector ℓ using $A\hat{\delta}$. On this basis, when the mean of the process is not zero, the theory of collocation gives the predicted signal $\bar{s}_p$ as

$$\bar{s}_p = C_{s_p s} C_\ell^{-1} \left(\ell - A\hat{\delta}\right) + \hat{m} \tag{16.24}$$

where the collocation procedure is applied to the residual data (after subtracting the estimated mean vector $A\hat{\delta}$ from the observation vector ℓ) and the removed mean value ($\hat{m}$) restored later to the signal estimate. The vector $\bar{s}_p$ may consist of one component or it may consist of many components. If it consists of many components, then $\hat{m}$ will become a vector of mean values of the processes involved, which will correspond to the number of signal quantities being computed. If only one signal s_p is being predicted, then the vector $\bar{s}_p$ will have only one component, and the matrix $C_{s_p s}$ will reduce to a row matrix. Equation (16.24) should only be applied when the process is assumed to have nonzero mean value. Equation (16.24) can be rearranged to obtain the zero mean signal estimate $\bar{s}'_p = \bar{s}_p - \hat{m}$ assuming the observation vector ℓ includes a nonzero mean signal.

According to Moritz (1972), the variance–covariance propagation of least squares collocation results from Equation (16.1) can be given as follows:

$$C_{\hat{\delta}} = \left(A^T C_\ell^{-1} A\right)^{-1} \tag{16.25}$$

$$C_{\bar{s}_p} = C_{s_p s_p} - C_{s_p s} C_\ell^{-1} C_{s_p s}^T + C_{s_p s} C_\ell^{-1} M C_\ell^{-1} C_{s_p s}^T \tag{16.26}$$

$$C_{\hat{\delta}\bar{s}_p} = -C_{\hat{\delta}} A^T C_\ell^{-1} C_{s_p s}^T \tag{16.27}$$

$$M = A C_{\hat{\delta}} A^T \tag{16.28}$$

where

$C_{\hat{\delta}}$ is the covariance matrix for the adjusted parameters $\hat{\delta}$, which is similar to the one used in the classical least squares adjustment except that weight matrix P is used instead of C_ℓ^{-1} based on Equation (16.7)

$C_{\bar{s}_p}$ is the covariance matrix for the predicted signal

$C_{\hat{\delta}\bar{s}_p}$ is the covariance matrix between the parameter $\hat{\delta}$ and the predicted signal $\bar{s}_p$

$C_{s_p s_p}$ and $C_{s_p s}$ are determined from the given covariance function.

Equations (16.10) and (16.23)–(16.27) constitute the basic computational tools for the least squares collocation estimates and their propagated variance–covariance matrices. The last term in Equation (16.26) describes the effect of inaccurately determining the parameters δ; this term will be zero if parameters are not involved in the process. One drawback to least squares collocation is the size of the covariance matrix of the observations C_ℓ since all the filtering, prediction, and parameter estimation formulae are based on it; this drawback, however, is often minimal considering the power of modern computing technologies. From Equation (16.24), the predicted signals $(\bar{s}_p)$ from collocation can be given as

$$\bar{s}_p = W^T \bar{\ell} + b \tag{16.29}$$

where

$$W = C_\ell^{-1} C_{ss_p} \tag{16.30}$$
$$\bar{\ell} = \ell - A\hat{\delta} \tag{16.31}$$
$$b = \hat{m} \tag{16.32}$$

$C_{ss_p} = C_{s_p s}^T$, W is a matrix of combination weights for all the given observations, $\bar{\ell}$ is a vector of measurements reduced for the effect of nonzero mean value of the process, and b is a vector of the additive constants. Since the correlation between observations is usually unknown in a least squares solution, the covariance matrix of observations (C_ℓ) in a least squares solution (Equation (16.23) is a diagonal matrix. In this case, the collocation Equation (16.1) is reduced to a form of Equation (16.22) with the least squares solution as Equation (16.23). In collocation technique, however, correlated covariance matrix expressed by Equation (16.7) is used in forming the weight matrix in Equation (16.23). If, however, the C_{ss} in Equation (16.7) is null matrix, the collocation solution in Equation (16.23) will reduce to that of least squares, which makes C_{ss} the main difference between the two methods. The use of proper covariance function for correlated observations has been shown (Moritz 1972) to have improved the results of adjusted parameters with their standard deviations having the least possible value. The covariance function allows the signal-type systematic errors (s) to be estimated instead of being absorbed in the residuals as it is usually the case in the classical least squares adjustments. The solutions based on collocation procedure are considered optimal in the sense that they are more accurate than those by the classical least squares method.

Generally, least squares collocation is a combination of classical least squares adjustment, smoothing, and signal prediction. Compared with dynamic mode filtering (e.g. Kalman filtering), collocation is static with predictions made at the computation points where signals are being determined, while in Kalman filtering, predictions are made at points where future measurements are to be made. Predictions in Kalman filtering only make sense if measurements for the predicted points are to be made available later.

16.6 Elements of Kriging

Kriging technique is a group of statistical tools for modeling statistical relationships among measured points (using kriging weights) in predicting values and uncertainties for unmeasured locations. In kriging, predicted values at any unmeasured location are obtained as a linear combination of kriging weighted values of observations made at the measured locations. The kriging weights, which are to make the variance of the predicted value as small as possible, are dependent on fitting some models to measured points, distances between measured and prediction locations, and on the overall spatial arrangement (direction) of measured points around the prediction locations. On average, the difference between any kriging prediction and its true value should be zero, making the prediction unbiased. This is enforced by making the kriging weights sum up to one.

Some of the important terms commonly used with kriging are *variography*, which means spatial modeling, and *semivariogram*, which is a measure of spatial dependence between spatial locations. Kriging equations are usually formulated in terms of semivariograms instead of covariance functions or covariograms.

16.7 Semivariogram Model and Modeling

Semivariogram (meaning half a *variogram*) is used in kriging to provide information on spatial autocorrelation of datasets, and its modeling is usually considered the most difficult step in kriging estimation technique. In kriging methods, semivariogram is commonly used instead of covariance function since the mean of the signal process might be unknown and a direct signal covariance function estimation from data can be very complex. The theoretical semivariogram $\gamma(x_i, x_j)$ can be given (Matheron 1963; Cressie 1993) as

$$\gamma(x_i, x_j) = \frac{1}{2} \text{Var}(Z(x_i) - Z(x_j)) \tag{16.33}$$

where Var($\cdot$) represents the variance of the random signal, x_i and x_j are the two spatial locations (or time points) where the spatial stationary process is measured, and $Z(x_i)$ and $Z(x_j)$ are the measured values at the two locations (or time points). The stationarity condition (or homogeneity and isotropy conditions) of the process dictates that

$$\gamma(x_i, x_j) = \gamma(D_{ij}) \tag{16.34}$$

where $D_{ij} = |x_i - x_j|$ is the separation (Euclidean) distance between the spatial locations or time interval. In this case, the theoretical semivariogram $\gamma(D_{ij})$ does not depend on the absolute locations or time points (x_i, x_j), but only on their separation D_{ij}. Kriging requires the knowledge of semivariogram or covariance function in signal prediction. Unfortunately, the true semivariogram is usually unknown, but estimated from the available data as experimental semivariogram $\gamma_{ex}(x_i, x_j)$. An experimental semivariogram is a measure of dissimilarity of a pair of points x_i and x_j, which can be expressed for a case involving one dimension as (cf. Wackernagel 2003)

$$\gamma_{ex}(x_i, x_j) = \frac{1}{2}[Z(x_i) - Z(x_j)]^2 \tag{16.35}$$

In determining the semivariogram for a process, Equation (16.35) is to be formulated for every pair of measured values and the $\gamma_{ex}(x_i, x_j)$ values grouped into classes according to the corresponding separation (Euclidean) distances or time intervals $D_{ij} = |x_i - x_j|$. The averaged distance (or time interval) and the averaged semivariogram for each class are determined; all the averaged semivariogram values are plotted (on the y-axis) against their equivalent averaged distances (or time intervals) (on the x-axis) in order to determine the empirical semivariogram model. Models such as linear, spherical, or exponential curves are usually fitted to the semivariogram plot by the method of least squares or the method of maximum likelihood. The semivariogram modeling procedure can be summarized (cf. Cressie 1993; Wackernagel 2003) as follows for a case where the empirical semivariogram is to be estimated for $D_{ij} < 6$ and the empirical semivariogram is to be averaged over a range or step of $\Delta = 2$ for all the D_{ij} values up to the specified limit, e.g. 6:

1) Compute the matrix of the distances (or time intervals) D_{ij} between possible pairs of points starting with the first point as reference, then the second, etc. In this case, a matrix D of size $N \times N$ will be created, where N is the number of measured points. A typical D matrix is given by Equation (16.12).
2) Compute the corresponding semivariogram value for each distance (or time interval) value in matrix D created in step 1 by using Equation (16.35), producing a matrix Γ_{ex} of semivariogram values with all diagonal elements equal to zero. A typical Γ_{ex} matrix will be as follows:

$$\Gamma_{ex} = \begin{bmatrix} \gamma_{ex}(x_1, x_1) & \gamma_{ex}(x_1, x_2) & \cdots & \gamma_{ex}(x_1, x_N) \\ \gamma_{ex}(x_2, x_1) & \gamma_{ex}(x_2, x_2) & \cdots & \gamma_{ex}(x_2, x_N) \\ \vdots & & \ddots & \vdots \\ \gamma_{ex}(x_N, x_1) & & \cdots & \gamma_{ex}(x_N, x_N) \end{bmatrix} \tag{16.36}$$

3) Create a scatter plot (*semivariogram cloud*) of the empirical semivariogram values by plotting the semivariogram values in matrix Γ_{ex} (along the *y-axis*) against the corresponding Euclidean distances in the matrix D (along the *x-axis*) up to the specified limit of $D_{ij} < 6$.

4) Since the empirical semivariogram is to be interpolated with a proper model with manageable data points, the semivariogram should be averaged in order to identify the shape (linear on nonlinear) of the semivariogram model. For example, if the averaging is done for a case where the range of the averaging (Δ) is equal to 2, this will mean that the averaging should be done between the following intervals (cf. Matheron 1963; Wackernagel 2003): $0 < D_{ij} \leq 2$; $2 < D_{ij} \leq 4$; $4 < D_{ij} \leq 6$ (for a given limit of $D_{ij} < 6$). Every data falling within the specified intervals are grouped together. The sample averaging steps for each group can be given as follows (for the given distance separation in steps of $\Delta = 2$):

$$\bar{\gamma}_{ex}(0) = 0 \quad \text{for all } D_{ij} \text{ with } i = j \tag{16.37}$$

$$\bar{\gamma}_{ex}(H_1) = \frac{1}{N_1}\sum_i\sum_j \gamma_{ex}(x_i, x_j) \quad \text{for } 0 < D_{ij} \leq 2 \tag{16.38}$$

$$\bar{\gamma}_{ex}(H_2) = \frac{1}{N_2}\sum_i\sum_j \gamma_{ex}(x_i, x_j) \quad \text{for } 2 < D_{ij} \leq 4 \tag{16.39}$$

$$\bar{\gamma}_{ex}(H_3) = \frac{1}{N_3}\sum_i\sum_j \gamma_{ex}(x_i, x_j) \quad \text{for } 4 < D_{ij} \leq 6 \tag{16.40}$$

where N_k (for $k = 1, 2, 3$) is the number of D_{ij} in matrix D whose values are within the given intervals, $\bar{\gamma}_{ex}(H_k)$ (for $k = 1, 2, 3$) is the average of all the N_k semivariogram values in the Γ_{ex} corresponding to D_{ij} in matrix D, and H_k (for $k = 1, 2, 3$) is the average of all the distances in the k group. In Equations (16.37)–(16.40), all observations that are within the given intervals for D_{ij} are pooled in order to estimate $\bar{\gamma}_{ex}(H_k)$ based on isotropy assumption on the random signal.

5) Plot the average semivariogram $\bar{\gamma}_{ex}(H_k)$ (along the *y-axis*) against the average distance H_k (along the *x-axis*) to identify the shape (trend) of the semivariogram.

6) After identifying the shape of the semivariogram model in step 5, fit a suitable model to the scatter plot in step 3, and determine the parameters of the semivariogram model by applying the least squares adjustment (with equal weights to the semivariogram values), neglecting the value in the origin of the plot, which is always equal to zero.

7) From the semivariogram plot, the white noise variance ($\hat{\sigma}_n^2$) (if present) can be obtained as the value of the intercept on the *y-axis*, and the semivariogram values for various separations between pairs of points are determined using

the semivariogram model. The semivariogram values (Γ_{ss}) between all pairs of observation points and the semivariogram values (Γ_{ss_p}) between each measured point and the prediction point are based on the model.

Similarly, two-dimensional (x-y) plot of semivariogram values against the corresponding distances between pairs of locations (on the x-axis) will provide empirical semivariogram; an appropriate model fitted to this empirical semivariogram is also used as a prediction model.

Although a semivariogram does not necessarily derive from a covariance function, it is sometimes derived from it. In this case, a typical relationship between a covariance function ($C_s(D_{ij})$) and the semivariogram ($\gamma_s(x_i, x_j)$) of a signal (under stationarity condition or homogeneity and isotropy conditions) can be given (Cressie 1989) as

$$\gamma_s\left(D_{ij}\right) = C_s(0) - C_s\left(D_{ij}\right) \tag{16.41}$$

or

$$\gamma_s\left(x_i, x_j\right) = C_s\left(x_i, x_i\right) - C_s\left(x_i, x_j\right) \tag{16.42}$$

where

$$C_s(0) = C_s\left(x_i, x_i\right) \tag{16.43}$$

$$C_s\left(x_i, x_i\right) = C_s\left(x_i - x_i\right) \tag{16.44}$$

$$\gamma_s\left(x_i, x_j\right) = \gamma_s\left(x_i - x_j\right) \tag{16.45}$$

$$\gamma_s\left(D_{ij}\right) = \gamma_s\left(-D_{ij}\right) \tag{16.46}$$

$\gamma_s(x_i, x_j)$ is the semivariogram value for the signal between the measured points, $C_s(0)$ is the variance of the signal for $D_{ii} = 0$, and D_{ij} is the distance between a pair of measured points x_i and x_j (or time interval) and $\gamma_s(0) = 0$. If the covariance function for the signal is given, Equation (16.41) can be used to locally compute the semivariogram matrix between the measured points and the semivariogram vector between the observation points and the prediction point. It should be mentioned that the relationships in Equations (16.41) and (16.42) might or might not exist in some cases; the semivariogram may exist without the equivalent covariance function according to Cressie (1989).

16.8 Kriging Procedure

For simplicity, only one prediction point or one spatial location where the value of a process will be predicted is considered. In this case, the observations can be modeled in kriging as

$$\ell = s + n \tag{16.47}$$

where s is the vector of signal quantities at the observation points with unknown mean (μ) and n is the vector of noise components with the expected values of zero. The noise and the signal components are assumed uncorrelated. The objective of kriging is to estimate signal values at unmeasured points (the predictions) using a linear combination of the measured values. The predicted signal quantity ($\bar{s}_p$) at a location where no measurement has been made can be given as

$$\bar{s}_p = \sum_i^N w_i \ell_i \tag{16.48}$$

or

$$\bar{s}_p = W^T \ell \tag{16.49}$$

where w_i is the combination weight at point i, ℓ_i is the measurement at point i, W is a vector of weights, and ℓ is a vector of N observations. Equation (16.48) should be in such a way that the expectation of the predicted value at the measured points is the same as the unknown mean value, μ; Equation (16.48) is developed so that it does not depend on the mean value. In order to enforce this condition, the following constraint is imposed on Equation (16.48):

$$\sum_i^N w_i = 1 \tag{16.50}$$

However, according to Blais (personal communication), "The constraint that the statistical weights add up to one at the known observation points is simply requiring that any (signal) prediction should agree with the known values at the observation points. Otherwise, one cannot speak of interpolation in the mathematical sense." Three common types of kriging methods, which are discussed further in the following sections, are *simple kriging*, *ordinary kriging*, and *universal kriging*.

16.8.1 Simple Kriging

Simple kriging assumes that the random process has a constant mean that is known or can be estimated and the covariance function ($C_s(D_{ij})$) for the process exists or can be estimated; like in least squares collocation, it works with the residual part of the prediction model after removing the constant component (estimated by the method of least squares) from the measurements. In this case, like in the least squares collocation, it determines the signal at the prediction point under the condition of minimum error variance. If the mean is unknown, it can be estimated as the weighted arithmetic mean; this, however, will ignore the spatial correlations among the points. A geostatistical technique for estimating the global mean with a consideration for the spatial correlations is known as *kriging the mean* (Wackernagel 2003). This technique assumes that the mean, which is constant, exists at all points of the space and relies on the existence of a known covariance function $(C_s(D_{ij}))$ for the process. It uses the estimated covariance matrix of the signal or the semivariogram function to determine the weights for the linear combination of the measurements, which are used in estimating the mean; the estimated combination weights, however, must sum up to one. After estimating the constant mean $(\hat{m})$, signal value at any arbitrary point can be predicted using the simple kriging technique. The simple kriging predictor is given (Cressie 1993) as

$$\bar{s}_p = \hat{m} + \sum_{i}^{N} w_i(\ell_i - \hat{m}) \tag{16.51}$$

or

$$\bar{s}_p = \hat{m} + W^T(\ell - \hat{m}\kappa) \tag{16.52}$$

where w_i is the simple kriging weight of the corresponding residual $(\ell_i - \hat{m})$, W is a vector of kriging weights, and κ is a vector of ones with the same size as the vector of measurements ℓ. The vector of weights W is determined from simple kriging equation system (Wackernagel 2003) as

$$W = C_\ell^{-1} C_{ss_p} \tag{16.53}$$

which is the same as the W in Equation (16.30) for the least squares collocation. If W in Equation (16.53) is substituted into Equation (16.52), the solution will be identical to the prediction by least squares collocation given in Equation (16.24). The prediction variance by simple kriging is given (Wackernagel 2003) as

$$C_{\bar{s}_p} = C_{s_p s_p} - C_{ss_p}^T C_\ell^{-1} C_{ss_p} \tag{16.54}$$

Equation (16.54) represents the first two terms in the least squares collocation in Equation (16.26) with the last term that describes the effect of inaccurate determination of the mean missing.

16.8.2 Ordinary Kriging

Ordinary kriging assumes that the random process is stationary with known semivariogram function and an unknown constant global mean that should be estimated or treated as nuisance parameter. In this case, if one does not know the mean of a random process and that mean is of no interest, ordinary kriging can be considered as the correct estimation procedure. It is the most frequently used kriging method in practice today. Ordinary kriging predictor can be given as

$$\bar{s}_p = \sum_i^N w_i \ell_i \tag{16.55}$$

with the unbiasedness condition enforced by the kriging or combination weights w_i ($i = 1, ..., N$) summing to one. These weights must be determined using semivariogram values (Γ_{ss_p}) between measured and prediction points and semivariogram values (Γ_{ss}) between pairs of measured points by using some empirical semivariogram models. With regard to ordinary kriging method (under the assumption that the mean of the stochastic process is constant and unknown), the system of equations to be solved in order to determine the kriging weights can be given (cf. Matheron 1971) as

$$\Gamma_{ss} W + \alpha \kappa = \Gamma_{ss_p} \tag{16.56}$$

$$W^T \kappa = 1 \tag{16.57}$$

where Γ_{ss} is the semivariogram matrix between the measured points, expressed as

$$\Gamma_{ss} = \begin{bmatrix} \gamma_s(x_1, x_1) & \gamma_s(x_1, x_2) & \cdots & \gamma_s(x_1, x_N) \\ \gamma_s(x_2, x_1) & \gamma_s(x_2, x_2) & \cdots & \gamma_s(x_2, x_N) \\ \vdots & & \ddots & \vdots \\ \gamma_s(x_N, x_1) & \cdots & & \gamma_s(x_N, x_N) \end{bmatrix} \tag{16.58}$$

Γ_{ss_p} is the semivariogram vector between the observation points (x_i) and one prediction point (x_p), which can be expressed as

$$\Gamma_{ss_p} = \begin{bmatrix} \gamma_s(x_1, x_p) \\ \gamma_s(x_2, x_p) \\ \vdots \\ \gamma_s(x_N, x_p) \end{bmatrix} \tag{16.59}$$

W is a vector of kriging or combination weights (w_i), expressed as

$$W = \begin{bmatrix} w_1 & w_2 & \cdots & w_N \end{bmatrix}^T \tag{16.60}$$

α is a constant (Lagrangian multiplier) with κ as a vector of ones of size $N \times 1$, given as

$$\kappa = \begin{bmatrix} 1 & 1 & \cdots & 1 \end{bmatrix}^T \tag{16.61}$$

If q number of prediction points are of interest, the matrix Γ_{ss_p} in Equation (16.59) will have a matrix size of $N \times q$. The solution of Equations (16.56)–(16.57) will give the values for the vector of the unknown kriging weights (W) and the only unknown constant α. The number of unknowns in kriging solution is always equal to the number of observations plus one unknown constant. The estimated kriging weights (w_i) are then used in Equation (16.55) to determine the predicted signal at the unmeasured point. The variance of the estimation error is given (Cressie 1993) by

$$\sigma_{\bar{s}_p}^2 = 2W^T \Gamma_{ss_p} - W^T \Gamma_{ss} W \tag{16.62}$$

The kriging Equations (16.56)–(16.57) can be rewritten by simply replacing the variogram terms in the computation according to the relation given in Equations (16.41)–(16.46) as follows:

$$\Gamma_{ss} = \begin{bmatrix} \sigma^2 & \sigma^2 & \cdots & \sigma^2 \\ \sigma^2 & \sigma^2 & \cdots & \sigma^2 \\ \vdots & \vdots & \ddots & \vdots \\ \sigma^2 & \cdots & \cdots & \sigma^2 \end{bmatrix} - \begin{bmatrix} C_s(0) & C_s(D_{12}) & \cdots & C_s(D_{1N}) \\ C_s(D_{21}) & C_s(0) & \cdots & C_s(D_{2N}) \\ \vdots & \vdots & \ddots & \vdots \\ C_s(D_{N1}) & \cdots & \cdots & C_s(0) \end{bmatrix} \tag{16.63}$$

$$\Gamma_{sp} = \begin{bmatrix} \gamma_s(D_{1p}) \\ \gamma_s(D_{2p}) \\ \vdots \\ \gamma_s(D_{Np}) \end{bmatrix} = \begin{bmatrix} \sigma^2 \\ \sigma^2 \\ \vdots \\ \sigma^2 \end{bmatrix} - \begin{bmatrix} C_s(D_{1p}) \\ C_s(D_{2p}) \\ \vdots \\ C_s(D_{Np}) \end{bmatrix} \tag{16.64}$$

where

$$\Gamma_{ss} = \begin{bmatrix} \gamma_s(0) & \gamma_s(D_{12}) & \cdots & \gamma_s(D_{1N}) \\ \gamma_s(D_{21}) & \gamma_s(0) & \cdots & \gamma_s(D_{2N}) \\ \vdots & \vdots & \ddots & \vdots \\ \gamma_s(D_{N1}) & \cdots & \cdots & \gamma_s(0) \end{bmatrix}$$

and $\sigma^2 = C_s(0)$. However, since correlation in kriging is modeled in terms of semivariograms, the kriging equation system is usually formulated in terms of semivariograms as given in Equations (16.56)–(16.57) and Equation (16.62).

16.8.3 Universal Kriging

Simple and ordinary kriging techniques assume constant mean of the random process. The mean values of some spatial data sets, however, cannot be assumed constant in general since they vary depending on the absolute locations of the data. *Universal kriging* simultaneously performs trend estimation and prediction and assumes that the mean of a process may vary instead of just being a constant value; in this case, it uses general polynomial trend model and assumes that the mean of the process can be expressed by some equations with unknown coefficients. This technique, which is a generalization of ordinary kriging, is otherwise known as *kriging with a trend*. The universal kriging procedure is more complicated compared with those of the simple and ordinary kriging techniques. Like the other techniques, it also makes use of some form of covariance functions or semivariograms.

16.9   Comparing Least Squares Collocation and Kriging

Least squares collocation and simple kriging are simply interpolation methods that are equivalent in many ways (Dermanis 1984; Ligas and Kulczycki 2010). Collocation theory assumes zero mean and uses covariance function as the main statistic; kriging theory assumes an arbitrary constant mean and uses semivariogram as the main statistic. Under a condition of stationarity of processes involved, the semivariogram function used in kriging is equivalent to the covariance function used in least squares collocation. The semivariogram filters off the mean value of a process and does not require it to be known as in the case of least squares collocation. As opposed to covariance function, which describes similarity, semivariogram determines the differences between pairs of observations. Kriging weights in general sum up to unity and may be negative. The other main differences between kriging and collocation can be summarized as (cf. Dermanis 1984):

- Both methods follow the principle of unbiased minimal error variance prediction when the underlying function is a second-order stochastic process with known or estimated covariance function. As optimal unbiased minimum variance estimation methods, both of them should provide identical results under identical conditions. According to Blais (personal communication), "… in practice, different conditions are assumed, often implicitly, and the results are different. The covariance model assumed in least-squares collocation is an ordinary covariance function while the covariogram model (assumed in kriging) does not always correspond to such an ordinary covariance function, it is a generalized covariance function that can imply real differences in practical cases."
- From Equations (16.29) and (16.48), it can be seen that collocation allows additive constant to the linear combination of the measured values, while kriging does not. If the additive constant is zero, however, the collocation model will be equivalent to that of kriging. The same variogram or covariogram model, however, must be used for the two methods to provide equivalent

results. However, according to Blais (personal communication), "Whenever a trend or functional model is included to achieve a zero expectation for the signal plus noise formulation, it should be the same so that the prediction part can be compared appropriately."

- The kriging techniques are likely to give identical results with least squares collocation with estimated means, where the mean function is first estimated from the available measurements through a trend removal technique.
- Whenever nonzero means of data points are known, the kriging results are usually different from those from least squares collocation.
- The kriging variance of the kriging solution is always larger than (or at best equal to) the corresponding variance of the collocation solution. This is due to the effects of unknown mean of the process that is unestimated and of neglecting measurement error in the kriging solution.

According to Blais (1982), Kriging approach will produce different optimal solutions depending on the conditions of the random process evaluated. Such conditions include assuming a process mean is constant when it is not, which may produce biased estimates; treating the mean as an unknown when, in fact, it is a known constant, which may produce deteriorated estimates; and if the random process is not covariance stationary, various solutions are possible for a number of different situations. In this case, the optimality properties of the estimates are only ensured when the underlying assumptions are reasonable. Further reading on kriging techniques can be found in Webster and Oliver (2007), Wackernagel (2003), Cressie (1993), Blais (1982), and Matheron (1963).

Appendix A

Extracts from Baarda's Nomogram

See Tables A.1–A.4.

Table A.1 For the values $\lambda_0 = \lambda(\alpha_0, \beta_0 = 0.20, 1) = \lambda(\alpha, \beta_0 = 0.20, df)$.

$100\alpha_0 =$	Degrees of freedom (df)	λ_0
5	1	7.8
2.4	2	9.6
1.3	3	11.0
0.9	4	12.0
0.6	5	13.0
0.1	12	17
0.2	10	16
0.3	9	15.5
0.35	8	15
0.40	7	14
0.50	6	13.5
0.07	14	18

(*Continued*)

Understanding Least Squares Estimation and Geomatics Data Analysis, First Edition.
John Olusegun Ogundare.
© 2019 John Wiley & Sons, Inc. Published 2019 by John Wiley & Sons, Inc.
Companion website: www.wiley.com/go/ogundare/Understanding-lse-and-gda

Table A.1 (Continued)

$100\alpha_0 =$	Degrees of freedom (df)	λ_0
0.04	16	• 19
0.03	18	20
0.02	20	21
0.02	22	22
0.01	24	22

Table A.2 For the values of $100\alpha_0 = 0.1$, $\beta_0 = 0.20$, $\lambda_0 = 17.0$.

Alpha (α)	Degrees of freedom (df)	Alpha (α)	Degrees of freedom
0.006	3	0.056	12
0.009	4	0.094	18
0.013	5	0.107	20
0.018	6	0.119	22
0.025	7	0.132	24
0.030	8	0.150	26
0.038	9	0.158	28
0.043	10	0.167	30

Table A.3 For the values of $100\alpha_0 = 0.9$, $\beta_0 = 0.20$, $\lambda_0 = 12.0$.

Alpha (α)	Degrees of freedom (df)	Alpha (α)	Degrees of freedom (df)
0.01	1	0.085	6
0.022	2	0.100	7
0.038	3	0.114	8
0.050	4	0.129	9
0.070	5	0.140	10

Table A.4 $100\alpha_0 = 1.0$, $\beta_0 = 0.20$, $\lambda_0 = 11.7$.

Alpha (α)	Degrees of freedom (df)	Alpha (α)	Degrees of freedom (df)
0.01	1	0.090	6
0.025	2	0.110	7
0.041	3	0.121	8
0.058	4	0.136	9
0.075	5	0.150	10

Appendix B

Standard Statistical Distribution Tables

See Tables B.1–B.4.

Understanding Least Squares Estimation and Geomatics Data Analysis, First Edition.
John Olusegun Ogundare.
© 2019 John Wiley & Sons, Inc. Published 2019 by John Wiley & Sons, Inc.
Companion website: www.wiley.com/go/ogundare/Understanding-lse-and-gda

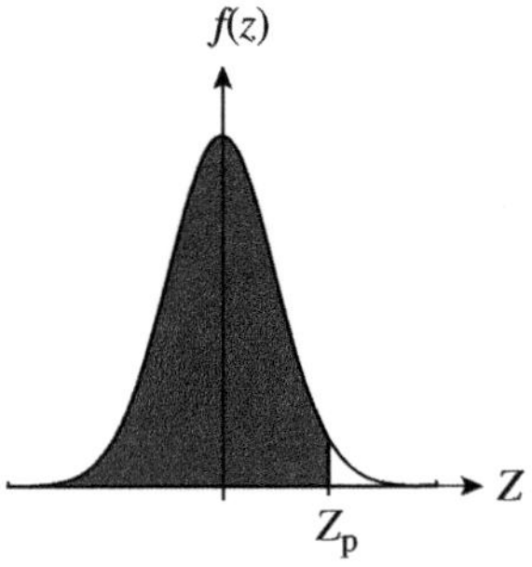

$f(z)$

Figure B.1 Normal distribution curve.

Table B.1 Standard normal distribution.

p	Z_p	p	Z_p	p	Z_p	p	Z_p
0.999	3.090	0.985	2.170	0.935	1.514	0.870	1.126
0.998	2.878	0.980	2.054	0.930	1.476	0.860	1.080
0.997	2.748	0.975	1.960	0.925	1.440	0.850	1.036
0.996	2.652	0.970	1.881	0.920	1.405	0.840	0.994
0.995	2.576	0.965	1.812	0.915	1.372	0.830	0.954
0.994	2.512	0.960	1.751	0.910	1.341	0.820	0.915
0.993	2.457	0.955	1.695	0.905	1.311	0.810	0.878
0.992	2.409	0.950	1.645	0.900	1.282	0.800	0.842
0.991	2.366	0.945	1.598	0.890	1.227	0.790	0.806
0.990	2.326	0.940	1.555	0.880	1.175	0.780	0.772

p is probability or area under the curve, which can be given as

$$p = \int_{-\infty}^{z_p} \frac{1}{\sqrt{2\pi}} e^{-\frac{z^2}{2}} dz$$

where the area (probability p) is between negative infinity ($-\infty$) and the value z_p of the random observable with its value increasing to the right. The Microsoft Excel 2013 program NORM.S.INV was used in forming the standard cumulative normal distribution table.

Figure B.2 *t*-Distribution.

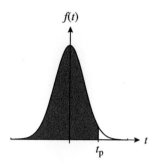

Table B.2 Table for Student's *t*-distribution.

df	$p = 0.80$	0.85	0.90	0.95	0.975	0.99	0.995
1	1.376	1.963	3.078	6.314	12.706	31.821	63.657
2	1.061	1.386	1.886	2.920	4.303	6.965	9.925
3	0.978	1.250	1.638	2.353	3.182	4.541	5.841
4	0.941	1.190	1.533	2.132	2.776	3.747	4.604
5	0.920	1.156	1.476	2.015	2.571	3.365	4.032
6	0.906	1.134	1.440	1.943	2.447	3.143	3.707
7	0.896	1.119	1.415	1.895	2.365	2.998	3.499
8	0.889	1.108	1.397	1.860	2.306	2.896	3.355
9	0.883	1.100	1.383	1.833	2.262	2.821	3.250
10	0.879	1.093	1.372	1.812	2.228	2.764	3.169
11	0.876	1.088	1.363	1.796	2.201	2.718	3.106
12	0.873	1.083	1.356	1.782	2.179	2.681	3.055
13	0.870	1.079	1.350	1.771	2.160	2.650	3.012
14	0.868	1.076	1.345	1.761	2.145	2.624	2.977
15	0.866	1.074	1.341	1.753	2.131	2.602	2.947
16	0.865	1.071	1.337	1.746	2.120	2.583	2.921
17	0.863	1.069	1.333	1.740	2.110	2.567	2.898
18	0.862	1.067	1.330	1.734	2.101	2.552	2.878
19	0.861	1.066	1.328	1.729	2.093	2.539	2.861
20	0.860	1.064	1.325	1.725	2.086	2.528	2.845
25	0.856	1.058	1.316	1.708	2.060	2.485	2.787
30	0.854	1.055	1.310	1.697	2.042	2.457	2.750
35	0.852	1.052	1.306	1.690	2.030	2.438	2.724
40	0.851	1.050	1.303	1.684	2.021	2.423	2.704
45	0.850	1.049	1.301	1.679	2.014	2.412	2.690

p is the probability or area under the curve, and df is the number of degrees of freedom.

The Microsoft Excel 2013 program T.INV was used in forming the Student's *t*-distribution table.

Figure B.3 χ^2-Distribution.

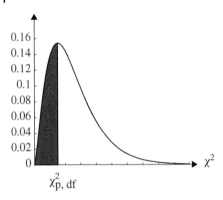

$\chi^2_{p,\,df}$

Table B.3 Distribution table for chi-square.

p	0.005	0.01	0.025	0.05	0.95	0.975	0.99	0.995
df					$\chi^2_{p,df}$			
1	0.000	0.000	0.001	0.004	3.841	5.024	6.635	7.879
2	0.010	0.020	0.051	0.103	5.991	7.378	9.210	10.597
3	0.071	0.115	0.216	0.352	7.815	9.348	11.345	12.838
4	0.207	0.297	0.484	0.711	9.488	11.143	13.277	14.860
5	0.412	0.554	0.831	1.145	11.070	12.833	15.086	16.750
6	0.676	0.872	1.237	1.635	12.592	14.449	16.812	18.548
7	0.989	1.239	1.690	2.167	14.067	16.013	18.475	20.278
8	1.344	1.646	2.180	2.733	15.507	17.535	20.090	21.955
9	1.735	2.088	2.700	3.325	16.919	19.023	21.666	23.589
10	2.156	2.558	3.247	3.940	18.307	20.483	23.209	25.188
11	2.603	3.053	3.816	4.575	19.675	21.920	24.725	26.757
12	3.074	3.571	4.404	5.226	21.026	23.337	26.217	28.300
13	3.565	4.107	5.009	5.892	22.362	24.736	27.688	29.819
14	4.075	4.660	5.629	6.571	23.685	26.119	29.141	31.319
15	4.601	5.229	6.262	7.261	24.996	27.488	30.578	32.801
16	5.142	5.812	6.908	7.962	26.296	28.845	32.000	34.267
17	5.697	6.408	7.564	8.672	27.587	30.191	33.409	35.718
18	6.265	7.015	8.231	9.390	28.869	31.526	34.805	37.156
19	6.844	7.633	8.907	10.117	30.144	32.852	36.191	38.582
20	7.434	8.260	9.591	10.851	31.410	34.170	37.566	39.997
30	13.787	14.953	16.791	18.493	43.773	46.979	50.892	53.672
31	14.458	17.074	17.539	19.281	44.985	48.232	52.191	55.003
32	15.134	17.789	18.291	20.072	46.194	49.480	53.486	56.328
33	15.815	18.509	19.047	20.867	47.400	50.725	54.776	57.648
34	16.501	5.229	19.806	21.664	48.602	51.966	56.061	58.964
35	17.192	5.812	20.569	22.465	49.802	53.203	57.342	60.275

p is the probability or area under the curve from the negative infinity ($-\infty$) to the given point $\chi^2_{p,df}$, whose value increases toward the right. This can be expressed as

$$p = \int_{-\infty}^{\chi^2_{p,df}} \chi^2(x)\,dx$$

where df is the number of degrees of freedom. The Microsoft Excel 2013 CHISQ.INV was used in forming the chi-square distribution table.

Figure B.4 *F*-Distribution.

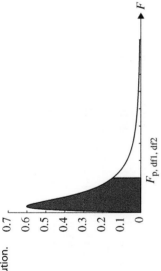

$F_{p, df1, df2}$

Table B.4 Table for *F*-distribution.

df₁	p	1	2	3	4	5	6	7	8	9	10
						df₂					
1	0.995	16 211	198.5	55.552	31.333	22.785	18.635	16.236	14.688	13.614	12.826
	0.99	4 052	98.503	34.116	21.198	16.258	13.745	12.246	11.259	10.561	10.044
	0.95	161.4	18.513	10.128	7.709	6.608	5.987	5.591	5.318	5.117	4.965
	0.05	0.006	0.005	0.005	0.004	0.004	0.004	0.004	0.004	0.004	0.004
	0.01	0.000	0.000	0.000	0.000	0.000	0.000	0.000	0.000	0.000	0.000
	0.005	0.000	0.000	0.000	0.000	0.000	0.000	0.000	0.000	0.000	0.000

(Continued)

Table B.4 (Continued)

df$_1$	p	1	2	3	4	5	6	7	8	9	10
2	0.995	20 000	199	49.799	26.284	18.314	14.544	12.404	11.042	10.107	9.427
	0.99	5 000	99	30.817	18	13.274	10.925	9.547	8.649	8.022	7.559
	0.95	199.5	19	9.552	6.944	5.786	5.143	4.737	4.459	4.256	4.103
	0.05	0.054	0.053	0.052	0.052	0.052	0.052	0.052	0.052	0.052	0.052
	0.01	0.010	0.010	0.010	0.010	0.010	0.010	0.010	0.010	0.010	0.010
	0.005	0.005	0.005	0.005	0.005	0.005	0.005	0.005	0.005	0.005	0.005
3	0.995	21 615	199.2	47.467	24.259	16.530	12.917	10.882	9.596	8.717	8.081
	0.99	5 403	99.166	29.457	16.694	12.060	9.780	8.451	7.591	6.992	6.552
	0.95	215.7	19.164	9.277	6.591	5.409	4.757	4.347	4.066	3.863	3.708
	0.05	0.099	0.105	0.108	0.110	0.111	0.112	0.113	0.113	0.113	0.114
	0.01	0.029	0.032	0.034	0.035	0.035	0.036	0.036	0.036	0.037	0.037
	0.005	0.018	0.020	0.021	0.022	0.022	0.022	0.023	0.023	0.023	0.023
4	0.995	22 500	199.2	46.195	23.155	15.556	12.028	10.050	8.805	7.956	7.343
	0.99	5 625	99.249	28.710	15.977	11.392	9.148	7.847	7.006	6.422	5.994
	0.95	224.6	19.247	9.117	6.388	5.192	4.534	4.120	3.838	3.633	3.478
	0.05	0.130	0.144	0.152	0.157	0.160	0.162	0.164	0.166	0.167	0.168
	0.01	0.047	0.056	0.060	0.063	0.064	0.066	0.067	0.068	0.068	0.069
	0.005	0.032	0.038	0.041	0.043	0.045	0.046	0.046	0.047	0.047	0.048

df$_2$

5	0.995	23 056	199.3	45.392	22.456	14.940	11.464	9.522	8.302	7.471	6.872
	0.99	5 764	99.299	28.237	15.522	10.967	8.746	7.460	6.632	6.057	5.636
	0.95	230.2	19.296	9.013	6.256	5.050	4.387	3.972	3.687	3.482	3.326
	0.05	0.151	0.173	0.185	0.193	0.198	0.202	0.205	0.208	0.210	0.211
	0.01	0.062	0.075	0.083	0.088	0.091	0.094	0.096	0.097	0.098	0.099
	0.005	0.044	0.055	0.060	0.064	0.067	0.069	0.070	0.072	0.073	0.073
6	0.995	23 437	199.3	44.838	21.974	14.513	11.073	9.155	7.952	7.134	6.545
	0.99	5 859	99.333	27.911	15.207	10.672	8.466	7.191	6.371	5.802	5.386
	0.95	234.0	19.330	8.941	6.163	4.950	4.284	3.866	3.581	3.374	3.217
	0.05	0.167	0.194	0.210	0.221	0.228	0.233	0.238	0.241	0.244	0.246
	0.01	0.073	0.092	0.102	0.109	0.114	0.118	0.121	0.123	0.125	0.127
	0.005	0.054	0.069	0.077	0.083	0.087	0.090	0.093	0.095	0.096	0.098
7	0.995	23 715	199.4	44.434	21.622	14.200	10.786	8.885	7.694	6.885	6.302
	0.99	5 928	99.356	27.672	14.976	10.456	8.2600	6.993	6.178	5.613	5.200
	0.95	236.8	19.353	8.887	6.094	4.876	4.207	3.787	3.500	3.293	3.135
	0.05	0.179	0.211	0.230	0.243	0.252	0.259	0.264	0.268	0.272	0.275
	0.01	0.082	0.105	0.118	0.127	0.134	0.139	0.143	0.146	0.149	0.151
	0.005	0.062	0.081	0.092	0.100	0.105	0.109	0.113	0.115	0.117	0.119
8	0.995	23 925	199.4	44.125	21.352	13.961	10.566	8.678	7.496	6.693	6.116
	0.99	5 981	99.374	27.489	14.799	10.290	8.102	6.840	6.029	5.467	5.057
	0.95	238.9	19.371	8.845	6.041	4.818	4.147	3.726	3.438	3.230	3.072

(Continued)

Table B.4 (Continued)

df$_1$	p	1	2	3	4	5	6	7	8	9	10
	0.05	0.188	0.224	0.246	0.261	0.271	0.279	0.286	0.291	0.295	0.299
	0.01	0.089	0.116	0.132	0.143	0.151	0.157	0.162	0.166	0.169	0.172
	0.005	0.068	0.091	0.104	0.114	0.120	0.126	0.130	0.133	0.136	0.139
9	0.995	24 091	199.388	43.882	21.139	13.772	10.391	8.514	7.339	6.541	5.968
	0.99	6 022	99.388	27.345	14.659	10.158	7.976	6.719	5.911	5.351	4.942
	0.95	240.5	19.385	8.812	5.999	4.772	4.099	3.677	3.388	3.179	3.020
	0.05	0.195	0.235	0.259	0.275	0.287	0.296	0.304	0.310	0.315	0.319
	0.01	0.095	0.125	0.143	0.156	0.165	0.172	0.178	0.183	0.187	0.190
	0.005	0.073	0.099	0.115	0.126	0.134	0.140	0.145	0.149	0.153	0.156
10	0.995	24 224	199.400	43.686	20.967	13.618	10.250	8.380	7.211	6.417	5.847
	0.99	6 055.8	99.399	27.229	14.546	10.051	7.874	6.620	5.814	5.257	4.849
	0.95	241.88	19.396	8.786	5.964	4.735	4.060	3.637	3.347	3.137	2.978
	0.05	0.201	0.244	0.270	0.288	0.301	0.311	0.319	0.326	0.331	0.336
	0.01	0.100	0.132	0.153	0.167	0.177	0.186	0.192	0.198	0.202	0.206
	0.005	0.078	0.106	0.124	0.136	0.146	0.153	0.159	0.164	0.168	0.171
11	0.995	24 334	199.409	43.524	20.824	13.491	10.133	8.270	7.104	6.314	5.746
	0.99	6 083.3	99.408	27.133	14.452	9.963	7.790	6.538	5.734	5.178	4.772
	0.95	242.98	19.405	8.763	5.936	4.704	4.027	3.603	3.313	3.102	2.943

df$_2$

	0.350	0.345	0.339	0.332	0.323	0.312	0.298	0.279	0.251	0.206
0.05	0.350	0.345	0.339	0.332	0.323	0.312	0.298	0.279	0.251	0.206
0.01	0.220	0.216	0.211	0.205	0.197	0.188	0.176	0.161	0.139	0.104
0.005	0.185	0.181	0.176	0.171	0.164	0.156	0.145	0.132	0.112	0.082
12 0.995	5.661	6.227	7.015	8.176	10.034	13.384	20.705	43.387	199.416	24 426
0.99	4.706	5.111	5.667	6.469	7.718	9.888	14.374	27.052	99.416	6 106.3
0.95	2.913	3.073	3.284	3.575	4.000	4.678	5.912	8.745	19.413	243.91
0.05	0.363	0.358	0.351	0.343	0.334	0.322	0.307	0.287	0.257	0.211
0.01	0.233	0.228	0.222	0.216	0.207	0.197	0.185	0.168	0.144	0.107
0.005	0.197	0.192	0.187	0.181	0.174	0.165	0.153	0.138	0.118	0.085
13 0.995	5.589	6.153	6.938	8.097	9.950	13.293	20.603	43.271	199.423	24 505
0.99	4.650	5.055	5.609	6.410	7.657	9.825	14.307	26.983	99.422	6 125.9
0.95	2.887	3.048	3.259	3.550	3.976	4.655	5.891	8.729	19.419	244.69
0.05	0.374	0.368	0.361	0.353	0.343	0.331	0.315	0.293	0.263	0.214
0.01	0.244	0.239	0.232	0.225	0.216	0.206	0.192	0.174	0.149	0.110
0.005	0.207	0.203	0.197	0.190	0.182	0.173	0.160	0.144	0.122	0.088
14 0.995	5.526	6.089	6.872	8.028	9.877	13.215	20.515	43.172	199.428	24 572
0.99	4.601	5.005	5.559	6.359	7.605	9.770	14.249	26.924	99.428	6 142.7
0.95	2.865	3.025	3.237	3.529	3.956	4.636	5.873	8.715	19.424	245.36
0.05	0.384	0.378	0.371	0.362	0.351	0.338	0.321	0.299	0.267	0.217
0.01	0.254	0.248	0.242	0.234	0.224	0.213	0.199	0.180	0.153	0.113
0.005	0.217	0.212	0.206	0.199	0.190	0.180	0.167	0.150	0.126	0.090

p is the probability or area under the curve; df_1 and df_2 are the first and second numbers of degrees of freedom, respectively.

The Microsoft Excel 2013 program F.INV was used in forming the sample F-distribution table.

Appendix C

Tau Critical Values Table for Significance Level α

No. obs.	α	Degrees of freedom									
		1	2	3	4	5	6	7	8	9	10
3	0.1	1.000	1.412								
	0.05	1.000	1.414								
	0.025	1.000	1.414								
	0.02	1.000	1.414								
	0.01	1.000	1.414								
	0.009	1.000	1.414								
	0.007	1.000	1.414								
	0.002	1.000	1.414								
	0.001	1.000	1.414								
4	0.1	1.000	1.413	1.687							
	0.05	1.000	1.414	1.710							
	0.025	1.000	1.414	1.721							
	0.02	1.000	1.414	1.723							
	0.01	1.000	1.414	1.728							
	0.009	1.000	1.414	1.728							
	0.007	1.000	1.414	1.729							
	0.002	1.000	1.414	1.731							
	0.001	1.000	1.414	1.732							

(Continued)

Understanding Least Squares Estimation and Geomatics Data Analysis, First Edition.
John Olusegun Ogundare.
© 2019 John Wiley & Sons, Inc. Published 2019 by John Wiley & Sons, Inc.
Companion website: www.wiley.com/go/ogundare/Understanding-lse-and-gda

| No. obs. | α | \multicolumn{10}{c}{Degrees of freedom} |
		1	2	3	4	5	6	7	8	9	10
5	0.1	1.000	1.413	1.696	1.865						
	0.05	1.000	1.414	1.714	1.916						
	0.025	1.000	1.414	1.723	1.948						
	0.02	1.000	1.414	1.725	1.955						
	0.01	1.000	1.414	1.729	1.972						
	0.009	1.000	1.414	1.729	1.974						
	0.007	1.000	1.414	1.730	1.978						
	0.002	1.000	1.414	1.731	1.990						
	0.001	1.000	1.414	1.732	1.994						
6	0.1	1.000	1.414	1.702	1.880	1.991					
	0.05	1.000	1.414	1.717	1.926	2.065					
	0.025	1.000	1.414	1.725	1.954	2.117					
	0.02	1.000	1.414	1.726	1.960	2.129					
	0.01	1.000	1.414	1.729	1.975	2.161					
	0.009	1.000	1.414	1.729	1.977	2.165					
	0.007	1.000	1.414	1.730	1.980	2.173					
	0.002	1.000	1.414	1.731	1.991	2.203					
	0.001	1.000	1.414	1.732	1.995	2.212					
7	0.1	1.000	1.414	1.706	1.892	2.009	2.087				
	0.05	1.000	1.414	1.719	1.933	2.078	2.179				
	0.025	1.000	1.414	1.726	1.958	2.125	2.247				
	0.02	1.000	1.414	1.727	1.964	2.137	2.265				
	0.01	1.000	1.414	1.730	1.977	2.167	2.310				
	0.009	1.000	1.414	1.730	1.979	2.170	2.316				
	0.007	1.000	1.414	1.730	1.982	2.178	2.329				
	0.002	1.000	1.414	1.732	1.992	2.205	2.377				
	0.001	1.000	1.414	1.732	1.995	2.214	2.395				
8	0.1	1.000	1.414	1.709	1.901	2.024	2.106	2.164			
	0.05	1.000	1.414	1.721	1.939	2.088	2.194	2.271			
	0.025	1.000	1.414	1.727	1.962	2.133	2.258	2.351			
	0.02	1.000	1.414	1.728	1.967	2.144	2.274	2.373			
	0.01	1.000	1.414	1.730	1.979	2.171	2.318	2.431			
	0.009	1.000	1.414	1.730	1.981	2.174	2.323	2.439			

No. obs.	α	Degrees of freedom									
		1	**2**	**3**	**4**	**5**	**6**	**7**	**8**	**9**	**10**
	0.007	1.000	1.414	1.731	1.984	2.182	2.335	2.456			
	0.002	1.000	1.414	1.732	1.993	2.207	2.381	2.521			
	0.001	1.000	1.414	1.732	1.996	2.216	2.397	2.547			
9	0.1	1.000	1.414	1.712	1.909	2.036	2.122	2.184	2.229		
	0.05	1.000	1.414	1.722	1.943	2.097	2.206	2.286	2.346		
	0.025	1.000	1.414	1.727	1.965	2.139	2.267	2.363	2.438		
	0.02	1.000	1.414	1.728	1.970	2.149	2.283	2.384	2.463		
	0.01	1.000	1.414	1.730	1.981	2.175	2.324	2.440	2.531		
	0.009	1.000	1.414	1.730	1.982	2.178	2.329	2.447	2.540		
	0.007	1.000	1.414	1.731	1.985	2.185	2.341	2.463	2.561		
	0.002	1.000	1.414	1.732	1.994	2.209	2.384	2.526	2.643		
	0.001	1.000	1.414	1.732	1.996	2.217	2.400	2.551	2.677		
10	0.1	1.000	1.414	1.714	1.915	2.046	2.136	2.200	2.248	2.285	
	0.05	1.000	1.414	1.723	1.947	2.104	2.216	2.298	2.361	2.410	
	0.025	1.000	1.414	1.728	1.967	2.144	2.274	2.373	2.450	2.511	
	0.02	1.000	1.414	1.729	1.972	2.154	2.290	2.393	2.474	2.539	
	0.01	1.000	1.414	1.730	1.982	2.178	2.329	2.447	2.540	2.616	
	0.009	1.000	1.414	1.730	1.983	2.181	2.334	2.454	2.549	2.626	
	0.007	1.000	1.414	1.731	1.986	2.188	2.345	2.470	2.569	2.650	
	0.002	1.000	1.414	1.732	1.994	2.210	2.387	2.530	2.649	2.747	
	0.001	1.000	1.414	1.732	1.996	2.218	2.402	2.554	2.681	2.788	
11	0.1	1.000	1.414	1.716	1.920	2.055	2.148	2.215	2.264	2.303	2.333
	0.05	1.000	1.414	1.724	1.951	2.110	2.225	2.310	2.374	2.425	2.466
	0.025	1.000	1.414	1.728	1.969	2.148	2.281	2.382	2.460	2.523	2.574
	0.02	1.000	1.414	1.729	1.973	2.157	2.296	2.401	2.484	2.550	2.604
	0.01	1.000	1.414	1.730	1.983	2.181	2.334	2.453	2.548	2.625	2.689
	0.009	1.000	1.414	1.731	1.984	2.184	2.338	2.460	2.557	2.635	2.700
	0.007	1.000	1.414	1.731	1.987	2.190	2.349	2.475	2.576	2.659	2.727
	0.002	1.000	1.414	1.732	1.994	2.211	2.389	2.534	2.654	2.753	2.837
	0.001	1.000	1.414	1.732	1.996	2.219	2.404	2.557	2.685	2.793	2.884

Appendix D

General Partial Differentials of Typical Survey Observables

As can be seen in Chapters 2 and 8, the key element of error propagation, variance–covariance propagation and pre-analysis of survey project, is finding partial differentials. In least squares adjustment methods, the partial differentials of models are also required. This appendix presents general partial differentials of typical survey observables, such as azimuth, total station direction, angle, and distance. Refer to Figure D.1 for the illustration of models for the observables discussed in this appendix. In the figure, the coordinates of points A (x_A, y_A), B (x_B, y_B), and O (x_O, y_O) are indicated. Models of survey observables and their partial differentials are given as follows.

D.1   Azimuth Observable

The azimuth observable (ℓ_{OA}) from point O to A can be given from Figure D.1 as follows:

$$\ell_{OA} = \operatorname{atan}\left(\frac{x_A - x_O}{y_A - y_O}\right) \tag{D.1}$$

Understanding Least Squares Estimation and Geomatics Data Analysis, First Edition.
John Olusegun Ogundare.
© 2019 John Wiley & Sons, Inc. Published 2019 by John Wiley & Sons, Inc.
Companion website: www.wiley.com/go/ogundare/Understanding-lse-and-gda

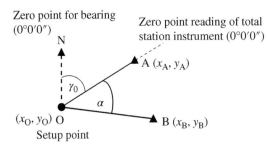

Zero point for bearing (0°0'0")

Zero point reading of total station instrument (0°0'0")

Figure D.1 Relationship between typical survey observables and coordinate parameters.

If the coordinates of points A (x_A, y_A) and O (x_O, y_O) are unknown variables, the partial differentials of the observable (ℓ_{OA}) with respect to the variables can be given based on the rules for finding the partial differentials given as follows:

$$y = \mathrm{atan}(u) \tag{D.2}$$

The partial differential of y with respect to a variable p can be given according to the following general rule:

$$\frac{\partial y}{\partial p} = \left(\frac{1}{1+u^2}\right)\frac{\partial u}{\partial p} \tag{D.3}$$

Applying this general rule to Equation (D.1),

$u = \dfrac{x_A - x_O}{y_A - y_O}$ and $p = x_A, y_A, x_O, y_O$, in turn, and $y = \ell_{OA}$, the following partial differentials are obtained:

$$\frac{\partial \ell_{OA}}{\partial x_A} = \left(\frac{1}{1 + \left(\dfrac{x_A - x_O}{y_A - y_O}\right)^2}\right)\frac{\partial\left(\dfrac{x_A - x_O}{y_A - y_O}\right)}{\partial x_A} \tag{D.4}$$

or

$$\frac{\partial \ell_{OA}}{\partial x_A} = \left(\frac{1}{1 + \left(\dfrac{x_A - x_O}{y_A - y_O}\right)^2}\right)\left(\frac{1}{y_A - y_O}\right) = \frac{(y_A - y_O)}{(x_A - x_O)^2 + (y_A - y_O)^2} \tag{D.5}$$

Similarly,

$$\frac{\partial \ell_{OA}}{\partial y_A} = \left(\frac{1}{1 + \left(\dfrac{x_A - x_O}{y_A - y_O}\right)^2}\right)\left(\frac{-(x_A - x_O)}{(y_A - y_O)^2}\right) = \frac{-(x_A - x_O)}{(x_A - x_O)^2 + (y_A - y_O)^2} \tag{D.6}$$

$$\frac{\partial \ell_{OA}}{\partial x_O} = \left(\frac{1}{1 + \left(\dfrac{x_A - x_O}{y_A - y_O} \right)^2} \right) \left(\frac{-1}{y_A - y_O} \right) = \frac{-(y_A - y_O)}{(x_A - x_O)^2 + (y_A - y_O)^2} \quad \text{(D.7)}$$

$$\frac{\partial \ell_{OA}}{\partial y_O} = \left(\frac{1}{1 + \left(\dfrac{x_A - x_O}{y_A - y_O} \right)^2} \right) \left(\frac{(x_A - x_O)}{(y_A - y_O)^2} \right) = \frac{(x_A - x_O)}{(x_A - x_O)^2 + (y_A - y_O)^2} \quad \text{(D.8)}$$

D.2 Total Station Direction Observable

The total station direction observable (ℓ_{OA}) from point O to A can be given from Figure D.1 as follows:

$$\ell_{OA} = \operatorname{atan}\left(\frac{x_A - x_O}{y_A - y_O} \right) - \gamma_0 \quad \text{(D.9)}$$

where γ_0 is the orientation of the zero direction of the instrument with respect to the north direction. If the orientation (γ_0) and the coordinates of points A (x_A, y_A) and O (x_O, y_O) are unknown variables, the partial differentials of the observable (ℓ_{OA}) with respect to the variables can be given based on the rules for finding the partial differentials given as follows:

$$\frac{\partial \ell_{OA}}{\partial \gamma_0} = -1 \quad \text{(D.10)}$$

$$\frac{\partial \ell_{OA}}{\partial x_A} = \left(\frac{1}{1 + \left(\dfrac{x_A - x_O}{y_A - y_O} \right)^2} \right) \left(\frac{1}{y_A - y_O} \right) = \frac{(y_A - y_O)}{(x_A - x_O)^2 + (y_A - y_O)^2} \quad \text{(D.11)}$$

$$\frac{\partial \ell_{OA}}{\partial y_A} = \left(\frac{1}{1 + \left(\dfrac{x_A - x_O}{y_A - y_O} \right)^2} \right) \left(\frac{-(x_A - x_O)}{(y_A - y_O)^2} \right) = \frac{-(x_A - x_O)}{(x_A - x_O)^2 + (y_A - y_O)^2}$$

$$\text{(D.12)}$$

$$\frac{\partial \ell_{OA}}{\partial x_O} = \left(\frac{1}{1 + \left(\dfrac{x_A - x_O}{y_A - y_O} \right)^2} \right) \left(\frac{-1}{y_A - y_O} \right) = \frac{-(y_A - y_O)}{(x_A - x_O)^2 + (y_A - y_O)^2} \quad \text{(D.13)}$$

$$\frac{\partial \ell_{OA}}{\partial y_O} = \left(\frac{1}{1 + \left(\dfrac{x_A - x_O}{y_A - y_O} \right)^2} \right) \left(\frac{(x_A - x_O)}{(y_A - y_O)^2} \right) = \frac{(x_A - x_O)}{(x_A - x_O)^2 + (y_A - y_O)^2}$$

$$\text{(D.14)}$$

D.3 Horizontal Angle Observable

The angle observable (α) measured at point O between directions ℓ_{OA} and ℓ_{OB} can be given as follows:

$$\alpha = \text{atan}\left(\frac{x_B - x_O}{y_B - y_O} \right) - \text{atan}\left(\frac{x_A - x_O}{y_A - y_O} \right) \quad \text{(D.15)}$$

If the coordinates of points A (x_A, y_A), O (x_O, y_O), and B (x_B, y_B) are unknown variables, the partial differentials of the observable (α) with respect to the variables can be given based on the rules for finding the partial differentials given as follows:

$$\frac{\partial \alpha}{\partial x_A} = \frac{-(y_A - y_O)}{(x_A - x_O)^2 + (y_A - y_O)^2} \quad \text{(D.16)}$$

$$\frac{\partial \alpha}{\partial y_A} = \frac{(x_A - x_O)}{(x_A - x_O)^2 + (y_A - y_O)^2} \quad \text{(D.17)}$$

$$\frac{\partial \alpha}{\partial x_O} = \frac{-(y_B - y_O)}{(x_B - x_O)^2 + (y_B - y_O)^2} - \frac{(y_A - y_O)}{(x_A - x_O)^2 + (y_A - y_O)^2} \quad \text{(D.18)}$$

$$\frac{\partial \alpha}{\partial y_O} = \frac{(x_B - x_O)}{(x_B - x_O)^2 + (y_B - y_O)^2} + \frac{(x_A - x_O)}{(x_A - x_O)^2 + (y_A - y_O)^2} \quad \text{(D.19)}$$

$$\frac{\partial \alpha}{\partial x_B} = \frac{(y_B - y_O)}{(x_B - x_O)^2 + (y_B - y_O)^2} \quad \text{(D.20)}$$

$$\frac{\partial \alpha}{\partial y_B} = \frac{-(x_B - x_O)}{(x_B - x_O)^2 + (y_B - y_O)^2} \quad \text{(D.21)}$$

D.4 Distance Observable

The distance observable (ℓ_{OA}) from point O to A can be given from Figure D.1 as follows:

$$\ell_{OA} = \sqrt{(x_A - x_O)^2 + (y_A - y_O)^2} \tag{D.22}$$

If the coordinates of points A (x_A, y_A) and O (x_O, y_O) are unknown variables, the partial differentials of the observable (ℓ_{OA}) with respect to the variables can be given based on the rules for finding the partial differentials given as follows:

$$\frac{\partial \ell_{OA}}{\partial x_A} = \frac{(x_A - x_O)}{\sqrt{(x_A - x_O)^2 + (y_A - y_O)^2}} \tag{D.23}$$

$$\frac{\partial \ell_{OA}}{\partial y_A} = \frac{(y_A - x_O)}{\sqrt{(x_A - x_O)^2 + (y_A - y_O)^2}} \tag{D.24}$$

$$\frac{\partial \ell_{OA}}{\partial x_O} = \frac{-(x_A - x_O)}{\sqrt{(x_A - x_O)^2 + (y_A - y_O)^2}} \tag{D.25}$$

$$\frac{\partial \ell_{OA}}{\partial y_O} = \frac{-(y_A - y_O)}{\sqrt{(x_A - x_O)^2 + (y_A - y_O)^2}} \tag{D.26}$$

D.5 Zenith Angle Observable

For the measured raw zenith angle (Z_{ij}) reduced to the local plane for line i to j in Figure 5.2, the parametric equation (in three-dimensional case) can be given from Equation (5.13) as

$$Z_{ij} = \text{atan} \left[\frac{\sqrt{(x_j - x_i)^2 + (y_j - y_i)^2}}{(z_j - z_i)} \right] \tag{D.27}$$

Using the rules for finding the partial derivatives of arctangent expressed by Equations (D.2) and (D.3), the following partial derivatives for Equation (D.27) with the variables $p = x_i, y_i, x_j, y_j, z_i, z_j$ and

$$u = \frac{\sqrt{(x_j - x_i)^2 + (y_j - y_i)^2}}{(z_j - z_i)} \tag{D.28}$$

can be obtained as follows. For simplicity sake, let

$$H = \sqrt{\left(x_j - x_i\right)^2 + \left(y_j - y_i\right)^2} \tag{D.29}$$

$$dx = \left(x_j - x_i\right); \quad dy = \left(y_j - y_i\right); \quad dz = \left(z_j - z_i\right) \tag{D.30}$$

$$\frac{\partial z_{ij}}{\partial x_i} = \frac{-dx}{\left(\dfrac{H^2 + dz^2}{dz}\right) \times H} \tag{D.31}$$

$$\frac{\partial z_{ij}}{\partial y_i} = \frac{-dy}{\left(\dfrac{H^2 + dz^2}{dz}\right) \times H} \tag{D.32}$$

$$\frac{\partial z_{ij}}{\partial x_j} = \frac{dx}{\left(\dfrac{H^2 + dz^2}{dz}\right) \times H} \tag{D.33}$$

$$\frac{\partial z_{ij}}{\partial y_j} = \frac{dy}{\left(\dfrac{H^2 + dz^2}{dz}\right) \times H} \tag{D.34}$$

$$\frac{\partial z_{ij}}{\partial z_i} = \frac{H}{\left(H^2 + dz^2\right)} \tag{D.35}$$

$$\frac{\partial z_{ij}}{\partial z_j} = \frac{-H}{\left(H^2 + dz^2\right)} \tag{D.36}$$

D.6 Other General Rules for Partial Differentials

The following rules can be used as demonstrated in the case of *arcsine* of a quantity *u*. Given

$$y = \mathrm{asin}(u) \tag{D.37}$$

the partial differential of *y* with respect to *p* can be given as

$$\frac{\partial y}{\partial p} = \left(\frac{1}{\sqrt{1 - u^2}}\right) \frac{\partial u}{\partial p} \tag{D.38}$$

Given the following in the case of *arccosine*

$$y = \mathrm{acos}(u) \tag{D.39}$$

the partial differential of y with respect to p can be given as

$$\frac{\partial y}{\partial p} = \left(\frac{-1}{\sqrt{1-u^2}} \right) \frac{\partial u}{\partial p} \tag{D.40}$$

Given the following in the case of *arc cotangent*

$$y = \mathrm{acot}(u) \tag{D.41}$$

the partial differential of y with respect to a variable p can be given according to the following general identity:

$$\frac{\partial y}{\partial p} = \left(\frac{-1}{1+u^2} \right) \frac{\partial u}{\partial p} \tag{D.42}$$

Appendix E

Some Important Matrix Operations and Identities

CHAPTER MENU

E.1 Matrix Lemmas

The following matrix inversion lemmas are useful in the partitioning that may be required in formulating the sequential least squares adjustment procedure.

1) Given the matrices A, U, C, V,

$$(A + UCV)^{-1} = A^{-1} - A^{-1}U(C^{-1} + VA^{-1}U)^{-1}VA^{-1} \tag{E.1}$$

2) Given the matrices T, U, V, W, and Q, the matrix inverse of partitioned matrix block can be done as follows (Faddeev and Faddeeva 1963):

$$\begin{bmatrix} T & U \\ V & W \end{bmatrix}^{-1} = \begin{bmatrix} a_{11} & a_{12} \\ a_{21} & a_{22} \end{bmatrix} \tag{E.2}$$

where (assuming T^{-1} and Q^{-1} exist)

$$
\begin{aligned}
a_{11} &= T^{-1} + T^{-1}UQ^{-1}VT^{-1} & a_{12} &= -T^{-1}UQ^{-1} \\
a_{21} &= -Q^{-1}VT^{-1} & a_{22} &= Q^{-1} \\
Q &= W - VT^{-1}U
\end{aligned}
\tag{E.3}
$$

Understanding Least Squares Estimation and Geomatics Data Analysis, First Edition.
John Olusegun Ogundare.
© 2019 John Wiley & Sons, Inc. Published 2019 by John Wiley & Sons, Inc.
Companion website: www.wiley.com/go/ogundare/Understanding-lse-and-gda

If W^{-1} and $(T - UW^{-1}V)^{-1}$ exist, the following will exist:

$$a_{11} = (T - UW^{-1}V)^{-1} \qquad a_{12} = -a_{11}UW^{-1}$$

$$a_{21} = -W^{-1}Va_{11} \qquad a_{22} = W^{-1} - W^{-1}Va_{12} \tag{E.4}$$

3) Given the matrices A, B, C, and D,

$$DC(A + BDC)^{-1} = (CA^{-1}B + D^{-1})^{-1}CA^{-1} \tag{E.5}$$

4) Given the matrices A, B, C, and D,

$$(A + BDC)^{-1}BC = A^{-1}(I + BCDA^{-1})^{-1}BC \tag{E.6}$$

or

$$(A + BDC)^{-1}BC = A^{-1}B(I + CDA^{-1}B)^{-1}C \tag{E.7}$$

If C is invertible,

$$(A + BDC)^{-1}BC = A^{-1}B(C^{-1} + DA^{-1}B)^{-1} \tag{E.8}$$

5) Given matrices A and B,

$$(A^{-1} + B^{-1})^{-1} = A(A + B)^{-1}B = B(A + B)^{-1}A \tag{E.9}$$

$$A - A(A + B)^{-1}A = B - B(A + B)^{-1}B \tag{E.10}$$

6) For an arbitrary matrix A and positive definite matrices B and C, the following matrix theorem are given by Morrison (1969):

$$(C^{-1} + A^T B^{-1}A)^{-1} = C - CA^T(B + ACA^T)^{-1}AC \tag{E.11}$$

$$(C^{-1} + A^T B^{-1}A)^{-1}A^T B^{-1} = CA^T(B + ACA^T)^{-1} \tag{E.12}$$

7) For two matrices A and B that are nonsingular and of the same dimensions, the following will be true:

$$(A + B)^{-1} = A^{-1}(A^{-1} + B^{-1})^{-1}B^{-1} \tag{E.13}$$

8) Given four matrices A, B, C, and D with B and C invertible, the following matrix inversion lemma can be given:

$$(A^T BA + C)^{-1}A^T BD = C^{-1}A^T(AC^{-1}A^T + B^{-1})^{-1}D \tag{E.14}$$

E.2   Generalized Inverses and Pseudo-inverses

Given a linear model,

$$w = v - A\delta \tag{E.15}$$

where A is a matrix of size $n \times u$ of rank $r = u - s$ with $r \leq u \leq n$, v has zero mean and variance matrix $\sigma^2 I$, w is a $n \times 1$ vector of known quantities, and δ is a $u \times 1$ vector of parameters to be estimated. A unique generalized inverse of the matrix A has been defined (Penrose 1955) as a matrix A^- satisfying:

(i) $AA^- A = A$

(ii) $A^- AA^- = A^-$

(iii) $(AA^-)^T = AA^-$

(iv) $(A^- A)^T = A^- A$
$$\tag{E.16}$$

According to Rao and Mitra (1971), the inverse of a singular or rectangular matrix is called pseudo-inverse; however, the Rao's inverse does not necessarily satisfy all of the Penrose conditions. The inverse is called generalized inverse by Rao when only Penrose condition (i) in Equation (E.16) is satisfied. If it is desired to estimate the parameters δ in Equation (E.15) subject to a set of s linearly independent constraints,

$$G^T \delta = 0 \tag{E.17}$$

where G is a $q \times u$ matrix of rank s complementary to matrix A. By applying the Lagrange's method of undetermined multipliers (K) to Equations (E.15) and (E.17) to obtain the constrained minimum of the variation function

$$\varphi = (A\delta + w)^T P(A\delta + w) + 2K^T (G^T \delta) = \text{minimum} \tag{E.18}$$

the following block matrix is obtained:

$$\begin{bmatrix} N & G \\ G^T & 0 \end{bmatrix} \begin{bmatrix} \delta \\ K \end{bmatrix} = \begin{bmatrix} -A^T Pw \\ 0 \end{bmatrix} \tag{E.19}$$

or

$$\begin{bmatrix} \delta \\ K \end{bmatrix} = M^- \begin{bmatrix} -A^T Pw \\ 0 \end{bmatrix} \tag{E.20}$$

where

$$N = A^T PA, \tag{E.21}$$

N is a $u \times u$ symmetric matrix of rank r ($r < u$), and G is $u \times s$ matrix of rank $s = r - u$.

$$M = \begin{bmatrix} N & G \\ G^T & 0 \end{bmatrix} \tag{E.22}$$

and M^- is the generalized inverse of M matrix, given (Pringle and Rayner 1970) as

$$M^- = \begin{bmatrix} m_{11}^- & m_{12}^- \\ m_{21}^- & m_{22}^- \end{bmatrix} = \begin{bmatrix} Q^{-1}NQ^{-1} & Q^{-1}G \\ G^TQ^{-1} & 0 \end{bmatrix} \tag{E.23}$$

with

$$Q = N + GG^T \tag{E.24}$$

$$m_{11}^- = Q^{-1}NQ^{-1} \tag{E.25}$$

$$m_{12}^- = Q^{-1}G \tag{E.26}$$

Equations (E.25) and (E.26) are the generalized inverses of matrices N and G, respectively. According to Pringle and Rayner (1970), the following equation is true:

$$Q^{-1} - Q^{-1}GG^TQ^{-1} = Q^{-1}NQ^{-1} \tag{E.27}$$

According to Searle (1999), there exists a matrix D satisfying the following equations:

$$m_{11}^- = Q^{-1} - D\left(D^T GG^T D\right)^{-1} D^T \tag{E.28}$$

$$m_{12}^- = D\left(G^T D\right)^{-1} \tag{E.29}$$

From Rayner and Pringle (1967), matrix D can be taken as

$$D = Q^{-1}G \tag{E.30}$$

so that

$$G^T D = G^T Q^{-1} G = I_s \tag{E.31}$$

where I_s is the identity matrix of size $s \times s$. Knowing that $Q^{-1}Q = I$ and from Equations (E.20) and (E.24),

$$G^T Q^{-1}\left(A^T PA + GG^T\right) = G^T \tag{E.32}$$

From Equation (E.32), the following must be true (Rayner and Pringle 1967):

$$G^T Q^{-1} A^T PA = 0 \tag{E.33}$$

$$G^T Q^{-1} A^T = 0 \tag{E.34}$$

Substituting Equations (E.30) and (E.31) into Equations (E.28) and (E.29) and simplifying gives the following:

$$m_{11}^- = Q^{-1} - Q^{-1} G G^T Q^{-1} \tag{E.35}$$

$$m_{12}^- = Q^{-1} G \tag{E.36}$$

Based on the Equation (E.27), it can be seen that Equation (E.35) is the same as Equation (E.25). Hence, the generalized inverse of N can be taken as Equation (E.25) or Equation (E.35). These generalized inverses satisfy Penrose conditions (i)–(iii) (Rayner and Pringle 1967). From Equations (E.20), (E.25), and (E.35), the least squares solution for the unknown parameters δ can be given as follows:

$$\delta = -Q^{-1} N Q^{-1} \left(A^T P w \right) \tag{E.37}$$

or

$$\delta = -\left(Q^{-1} - Q^{-1} G G^T Q^{-1} \right) A^T P w \tag{E.38}$$

From Equation (E.37), the generalized inverse of A is $Q^{-1} N Q^{-1} A^T P$. From Equation (E.38), the following can be deduced:

$$\delta = -Q^{-1} A^T P w + Q^{-1} G \left(G^T Q^{-1} A^T \right) P w \tag{E.39}$$

Substituting Equation (E.34) into Equation (E.39) gives the final solution of parameters as

$$\delta = -Q^{-1} A^T P w \tag{E.40}$$

It has been shown by Searle (1999) that $Q^{-1} A^T P$ in Equation (E.40) is the generalized inverse of matrix A, satisfying the first three Penrose conditions for pseudo-inverses. Equation (E.40) can be derived directly by setting $AD = 0$ when Equation (E.28) is used in the solution Equation (E.20). Equation (E.37) or Equation (E.40) can be used to estimate the unknown parameters. The generalized inverses, however, are not unique but dependent on the choice of G and A matrices (Rayner and Pringle 1967).

John (1964) derived the solution to the matrix inversion problem of M in Equation (E.22) by assuming that matrix M is a square nonsingular symmetric matrix with the product $MM^- = I$. In this case, the following equations are obtained:

$$\left(A^T P A \right) m_{11}^- + G m_{21}^- = I_u \tag{E.41}$$

$$\left(A^T P A \right) m_{12}^- + G m_{22}^- = 0 \tag{E.42}$$

$$G^T m_{11}^- = 0 \tag{E.43}$$

$$G^T m_{12}^- = I_s \tag{E.44}$$

Multiply Equations (E.41) and (E.42) on the left side by D^T subject to the following conditions:

$$AD = 0 \tag{E.45}$$

$$G^T D = I \tag{E.46}$$

From Equations (E.41) and (E.42),

$$m_{21}^- = \left(D^T G\right)^{-1} D^T \tag{E.47}$$

$$m_{12}^- = D\left(G^T D\right)^{-1} \tag{E.48}$$

$$m_{22}^- = 0 \tag{E.49}$$

Substituting Equation (E.47) back into Equation (E.41) and rearranging gives

$$Nm_{11}^- = I - G\left(D^T G\right)^{-1} D^T \tag{E.50}$$

Note that

$$\delta = -m^-_{11} A^T P w \tag{E.51}$$

For $N = A^T P A$, there exists $u \times s$ matrix D with properties (Goldman and Zelen 1964):

(a) $D^T N = 0$ (or $ND = 0$ since N is symmetric) $\tag{E.52}$

(b) $\det\left(D^T G\right) \neq 0$ $\tag{E.53}$

(c) $G^T m_{11}^- = 0$ $\left(\text{or } m_{11}^- G = 0 \text{ since } m_{11}^- \text{ is symmetric}\right)$ $\tag{E.54}$

For a symmetric matrix inverse m_{11}^- obeying (c) and Equation (E.50) for every D satisfying (a) and (b), the following additional properties are satisfied:

(d) $m_{11}^- = m_{11}^- N m_{11}^-$ and $N = N m_{11}^- N$

(e) m_{11}^- is of rank q $\tag{E.55}$

Since m_{11}^- and N are symmetric, Equation (E.50) can be written as

$$m_{11}^- N = I - D\left(G^T D\right)^{-1} G^T \tag{E.56}$$

Post-multiplying Equation (E.56) by m_{11}^- and using the relation in (d) gives

$$m_{11}^- = \left[I - D\left(G^T D\right)^{-1} G^T\right] m_{11}^- \tag{E.57}$$

Post-multiplying Equation (E.50) by N and using the relation in (d) gives

$$N = \left[I - G\left(D^T G\right)^{-1} D^T\right] N \tag{E.58}$$

Goldman and Zelen (1964) have shown that the following Equation (E.59) is true:

$$\left(N + GG^T\right)^{-1}G = D\left(G^T D\right)^{-1} \tag{E.59}$$

Transposing Equation (E.57) and multiplying it by $-A^T Pw$ gives the following:

$$\delta = \left[I - G\left(D^T G\right)^{-1} D^T\right]\delta \tag{E.60}$$

Equation (E.60) is a similarity transformation equation; the value of δ will depend on the choice of D and G. Other important identities are

$$\left(I + A\right)^{-1} = I - \left(I + A\right)^{-1}A \tag{E.61}$$

$$\left(I + AB\right)^{-1}A = A\left(I + BA\right)^{-1} \tag{E.62}$$

Appendix F

Commonly Used Abbreviations

The following abbreviations are used in this book:

1) $acos()$ for arc cosine of the number in the bracket. Sometimes, $arccos()$ or $cos^{-1}()$ might have been used to mean the same thing.
2) $acot()$ for arc cotangent of the number in the bracket. Alternative abbreviations for the same thing: $arccotan()$ and $tan^{-1}()$.
3) $asin()$ for arc sine of the number in the bracket. Alternative abbreviations for the same thing: $arcsin()$ and $sin^{-1}()$.
4) $atan()$ for arc tangent of the number in the bracket. Alternative abbreviations for the same thing: $arctan()$ and $tan^{-1}()$.
5) $sin()$ for sine of the angle, $cos()$ for cosine of an angle, and $tan()$ for tangent of an angle.
6) $diag()$ to represent a diagonal matrix. Sometimes the full word "diagonal" is used to mean the same thing.
7) $det()$ for the determinant of a matrix.
8) $E()$ for mathematical expectation.
9) $trace()$ for the trace of matrix, which is the sum of the diagonal elements of the matrix.
10) cdf for cumulative distribution function.
11) χ^2 for chi-square statistic.
12) z for standard normal distribution statistic or score.
13) t for Student's t-distribution score.
14) $|\cdot|$ absolute value of the quantity in the two vertical lines.
15) I_u for identity matrix of the size $u \times u$, which has off-diagonal elements as zeros and the diagonal elements as ones.
16) ISO 17123 – International Standards Organization 17123.

Understanding Least Squares Estimation and Geomatics Data Analysis, First Edition.
John Olusegun Ogundare.
© 2019 John Wiley & Sons, Inc. Published 2019 by John Wiley & Sons, Inc.
Companion website: www.wiley.com/go/ogundare/Understanding-lse-and-gda

References

Adrain, R. (1808). *Research concerning the probabilities of the errors which happen in making observations. The Analyst or Mathematical Museum* 1 (IV): 93–109.

Ashkenazi, V. (1980). *Least squares adjustment: signal or just noise? Chartered Land and Minerals Surveyor* 3 (1): 42–49.

Baarda, W. (1968). *A Testing Procedure for Use in Geodetic Networks*, Publications on Geodesy, New Series, Vol. 2, No. 5. Delft: Netherlands Geodetic Commission.

Barrodale, I. and Roberts, F.D.K. (1974). Algorithm 478: solution of an overdetermined system of equations in the L_1 norm. *Communications of the ACM* 17: 319.

Bektas, S. and Sisman, Y. (2010). The comparison of L_1 and L_2-norm minimization methods. *International Journal of the Physical Sciences* 5 (11): 1721–1727. 18 September.

Bjerhammar, A. (1951). *Application of Calculus of Matrices to Method of Least Squares: With Special Reference to Geodetic Calculations*, 49. Stockholm, Sweden: Elanders Boktr., Transactions/Royal Institute of Technology.

Blais, J.A.R. (1982). Synthesis of Kriging estimation methods. *Manuscripta Geodaetica* 7 (4): 325–352.

Blais, J.A.R. (1984). Generalized covariance functions and their applications in estimation. *Manuscripta Geodaetica* 9 (4): 307–322.

Blais, J.A.R. (2010). Least squares for practitioners. *Mathematical Problems in Engineering* 2010: 508092. 19 pages, Hindawi Publishing Corporation.

Cooper, M.A.R. (1987). *Control Surveys in Civil Engineering.* New York: W.G. Nichols, Inc.

Crandall, C.L. (1901). *The adjustment of a Transit survey as compared with that of a compass survey. Transactions, ASCE* XLV (893): 453–464.

Cressie, N. (1989). Geostatistics. *The American Statistician* 43 (4): 197–202.

Cressie, N.A.C. (1993). *Statistics for Spatial Data*, Wiley Series in Probability and Mathematical Statistics: Applied Probability and Statistics. New York: Wiley.

Understanding Least Squares Estimation and Geomatics Data Analysis, First Edition.
John Olusegun Ogundare.
© 2019 John Wiley & Sons, Inc. Published 2019 by John Wiley & Sons, Inc.
Companion website: www.wiley.com/go/ogundare/Understanding-lse-and-gda

Cross, P.A. (1985). Numerical methods in network design. In: *Optimization and Design of Geodetic Networks* (ed. E.W. Grafarend and F. Sanso), 429–435. Berlin, New York: Springer.

Dermanis, A. (1984). Kriging and collocation: a comparison. *Manuscripta Geodaetica* 9: 159–167.

Dragomir, V., Ghiţău, D., Mihăilescu, M., and Rotaru, M. (1982). *Theory of the Earth's Shape*. Amsterdam: Elsevier Scientific Publishing Company.

Even-Tzur, G. (2006). Datum definition and its influence on the sensitivity of geodetic monitoring networks. Proceedings of the 3rd IAG/12th FIG Symposium, Baden (22–24 May, 2006).

Faddeev, D.K. and Faddeeva, V.N. (1963). *Computational Methods of Linear Algebra* (trans. R.C. Williams). San Francisco, London: Freeman.

Faragher, R. (2012). Understanding the basis of the Kalman Filter via a simple and intuitive derivation. *IEEE Signal Processing Magazine* 29 (5): 128–132.

Förstner, W. (1979). On the internal and external reliability of photogrammetric coordinates. Paper presented at the ASP-ASCM Convention, Washington, DC (18–23 March 1979).

Frankich, K (2006). Least Squares in Geomatics. BCIT Lecture Notes developed for GEOM 8353 Course, March.

Gauss, C.F. (1809). *Theoria Motus Corporum Coelestium*. Lib. 2, sec. 3. Hamburg: Sumtibus F. Perthes et I.H. Besser.

Gelb, A. (ed.) (1974). *Applied Optimal Estimation* (Written by the Technical Staff, The Analytic Sciences Corporation; Principal authors: A. Gelb, J.F. Kaper, Jr., R.A. Nash, Jr., C.F. Price, A.A. Sutherland, Jr.). Cambridge, London: MIT Press.

Goldman, A.J. and Zelen, M. (1964). Weak generalized inverses and minimum variance linear unbiased estimation. *Journal of research of the National Bureau of Standards – B. Mathematics and Mathematical Physics* 68B (4): 151.

Grafarend, E.W. (1974). *Optimization of geodetic networks. Bolletino di Geodesia a Science Affini* 33 (4): 351–406.

Grafarend, E.W., Heister, H., Kelm, R. et al. (1979). *Optimierung Geodetischer Messoperationen*. Karlsruhe: Herbert Wichmann Verlag.

Groves, P.D. (2013). *Principles of GNSS, Inertial, and Multisensor Integrated Navigation Systems*, 2e. Boston: Artech House.

Harvey, B.R. (1993). Survey network adjustments by the L1 method. *Australian Journal of Geodesy, Photogrammetry and Surveying* 59: 39–52.

Heiskanen, W.A. and Moritz, H. (1967). *Physical Geodesy*. San Francisco: W.H. Freeman and Company.

Huber, P.J. (1964). Robust estimation of a location parameter. *The Annals of Mathematical Statistics* 35: 73–101.

John, P.W.M. (1964). Pseudo-inverses in the analysis of variance. *The Annals of Mathematical Statistics* 35: 895–896.

Kavouras, M. (1982). *On the Detection of Outliers and the Determination of Reliability in Geodetic Networks*, Technical Report, vol. 87. Fredericton: Department of Geodesy and Geomatics Engineering, UNB November.

Krakiwsky, E.J. (1975). *A Synthesis of Recent Advances in the Method of Least Squares*, Lecture Notes, vol. 42. Fredericton: University of New Brunswick.

Krarup, T. (1969). *A Contribution to the Mathematical Foundation of Physical Geodesy*, Publication, vol. 44. Copenhagen: Danish Geodetic Institute 80p.

Kuang, S.L. (1991). Optimization and design of deformation monitoring schemes. Ph.D. dissertation. Technical Report No. 157, University of New Brunswick, Fredericton, Canada, July, 179p.

de Laplace, P.S. (1811). *Théoris Analytique des Probabilitiés*. Paris: Courcier.

Legendre, A.-M. (1806). *Nouvelles Méthodes pour la Détermination des Orbites des Comètes*. Paris: Courcier.

Leick, A. (1982). Minimal constraints in two dimensional networks. *Journal of the Surveying and Mapping Division, American Society of Civil Engineers* 108 (SU2): 53–68.

Leick, A. (2004). *GPS Satellite Surveying*, 3e. Hoboken, NJ: Wiley.

Ligas, M. and Kulczycki, M. (2010). Simple spatial prediction: least squares prediction, simple kriging, and conditional expectation of normal vector. *Geodesy and Cartography* 59 (2): 69–81.

Matheron, G. (1963). Principles of geostatistics. *Economic Geology* 58: 1246–1266.

Matheron, G. (1971). *The Theory of Regionalized Variables and Its Applications*, Les Cahiers du Centre de Morphologie Mathématique de Fontainebleau, vol. 5. Fontainebleau: Ecole Nationale Supérieure des Mines de Paris.

Merriman, M. (1877). *On the history of the method of least squares. The Analyst* IV (2): 33–36.

Micchelli, C.A. (1986). *Interpolation of scattered data: distance matrices and conditionally positive functions. Constructive Approximation* 2 (1): 11–22.

Mikhail, E.M. (1976). With contributions by F.Ackerman in:). *Observations and Least Squares*. Lanham: Thomas Y. Crowell Company, Inc., University Press of America, Inc.

Molteni, A., L. Pertusini and M. Reguzzoni (2009). Exercises – Part 5: Collocation and Kriging. Course notes by Prof. Fernando Sanso' on "Statistical Analysis of Environmental Data" for Academic Year 2008–2009. 28p. http://geomatica. como.polimi.it/corsi/statistical-analysis/TrattamentoII_Es.5_FILTERING.pdf (accessed 9 December 2017).

Moritz, H. (1972). *Advanced Least Squares Methods*, Report of the Department of Geodetic Science, vol. 175. Columbus, OH: Ohio State University June, 132p.

Morrison, N. (1969). *Introduction to Sequential Smoothing and Prediction*. Boston: McGraw-Hill.

Nickerson, B.G. (1979). Horizontal network design using computer graphics. M.Sc. E. thesis, Department of Geodesy and Geomatics Engineering, UNB, Fredericton, Canada.

Ogundare, J.O. (2016). *Precision Surveying: The Principles and Geomatics Practice*. Hoboken, NJ: Wiley.

Penrose, P. (1955). A generalized inverse for matrices. *Proceedings of the Cambridge Philosophical Society: Mathematical and Physical Sciences* 51: 406–413.

Pope, A.J. (1976). *The Statistics of Residuals and the Detection of Outliers*, NOAA Technical Report, vol. NOS 65 NGS 1. Rockville, MD: National Geodetic Information Center.

Pringle, P.M. and Rayner, A.A. (1970). Expressions for generalized inverses of bordered matrix with applications to the theory of constrained linear models. *SIAM Review* 12 (1): 107–115. January.

Rao, C.R. and Mitra, S.K. (1971). *Generalized Inverse of Matrices and Its Applications*. New York: Wiley.

Rayner, A.A. and Pringle, R.M. (1967). A note on generalized inverses in the linear hypothesis not of full rank. *The Annals of Mathematical Statistics* 38 (1): 271–273.

Searle, S.R. (1999). On linear models with restrictions on parameters. Departments of Biometrics and Statistics, Cornell University, Ithaca, NY, September. https://dspace.library.cornell.edu/bitstream/1813/32048/1/BU-1450-M.pdf (accessed December 2013).

Sprinsky, W.H. (1987). Traverse adjustment for observations from modern survey instrumentation. *Journal of Surveying Engineering* 113 (2): 70–81.

Stefanovic, P. (1980). Pitfalls in blunder detection techniques. *International Archives of Photogrammetry* 23 (Part B3, Commission III): 687–700.

Tienstra, J.M. (1966). *Theory of Adjustment of Normally Distributed Observations*. Argus: Amsterdam.

Van Mierlo, J. (1981). A testing procedure for analyzing geodetic deformation measurements. In: *Beiträge zum II, Internationalen Symposium über Deformationsmessungen mit geodätischen methoden* (ed. L. Hallerman). Stuttgart: K. Wittwer.

Vanicek, P. and Krakiwsky, E.J. (1986). *Geodesy: The Concepts*, 2e. the Netherlands: Elsevier Science Publishers B.V.

Vincenty, T. (1979). *The Havago Three-Dimensional Adjustment Program*, NOAA Technical Memorandum NOS NGS, vol. 17. Rockville, MD: Department of Commerce.

Vincenty, T. and Bowring, B.R. (1978). *Application of Three-Dimensional Geodesy to Adjustments of Horizontal Networks*, NOAA Technical Memorandum NOS NGS, vol. 13. Rockville, MD: Department of Commerce.

Wackernagel, H. (2003). *Multivariate Geostatistics: An Introduction with Applications*, 3e. Berlin, Heidelberg: Springer.

Webster, R. and Oliver, M.A. (2007). *Geostatistics for Environmental Scientists*, Statistics in Practice, 2e. Chichester: Wiley.

Wolf, H. (1963). Die Grundgleichungen der dreidimensionalen Geodäsie in elementarer Darstellung. *Zeitschrift für Vermessunggswesen* 88 (6): 225–233.

Wolf, H. (1975). *Ausgleichungsrechnung-Formeln zur pracktischen Anwendung*. Bonn, Hannover, München: Ferd. Dümmler Verlag.

Index

Understanding Least Squares Estimation and Geomatics Data Analysis, First Edition.
John Olusegun Ogundare.
© 2019 John Wiley & Sons, Inc. Published 2019 by John Wiley & Sons, Inc.
Companion website: www.wiley.com/go/ogundare/Understanding-lse-and-gda

Printed and bound by CPI Group (UK) Ltd, Croydon, CR0 4YY

17/10/2024

14576049-0001